GEOMETRIC MORPHOMETRICS FOR BIOLOGISTS: A PRIMER

SECOND EDITION

science & technology books

Companion Web Site:

http://www.elsevierdirect.com/companions/9780123869036

Geometric Morphometrics for Biologists: A Primer

Miriam L. Zelditch, Donald L. Swiderski and H. David Sheets

Software Workbook:

The companion site contains an online workbook that covers the practical details of a morphometric analysis. The 14 chapter workbook includes detailed instructions for conducting analyses with freely available, easy-to-use software.

GEOMETRIC MORPHOMETRICS FOR BIOLOGISTS: A PRIMER

SECOND EDITION

MIRIAM LEAH ZELDITCH
Museum of Paleontology, University of Michigan, Ann Arbor, MI, USA

DONALD L. SWIDERSKI
University of Michigan Medical Center, Ann Arbor, MI, USA

H. DAVID SHEETS
Department of Physics, Canisus College, Buffalo, NY, USA

AMSTERDAM • BOSTON • HEIDELBERG • LONDON
NEW YORK • OXFORD • PARIS • SAN DIEGO
SAN FRANCISCO • SINGAPORE • SYDNEY • TOKYO
Academic Press is an imprint of Elsevier

Academic Press is an imprint of Elsevier
32 Jamestown Road, London NW1 7BY, UK
225 Wyman Street, Waltham, MA 02451, USA
525 B Street, Suite 1800, San Diego, CA 92101-4495, USA

Second edition

British Library Cataloguing-in-Publication Data
A catalogue record for this book is available from the British Library

Library of Congress Cataloging-in-Publication Data
A catalog record for this book is available from the Library of Congress

ISBN: 978-0-12-386903-6

For information on all Academic Press publications
visit our website at store.elsevier.com

Typeset by MPS Limited, Chennai, India
www.adi-mps.com

Printed and bound in United States of America

12 13 14 15 10 9 8 7 6 5 4 3 2 1

Contents

Contributors

Miriam Leah Zelditch is an Associate Research Scientist at the Museum of Paleontology, University of Michigan, MI. She studies the developmental and evolution origins of morphological variation and its impact on evolution over short and long time scales.

Donald L. Swiderski is a Research Laboratory Specialist in the Kresge Hearing Research Institute and an Adjunct Research Investigator in the Mammal Division of the Museum of Zoology, University of Michigan, MI. He is an evolutionary morphologist interested in relationships between the morphological, functional and ecological diversity of mammals.

H. David Sheets is a Professor of Physics at Canisius College, Buffalo NY, Adjunct Professor of Geology, SUNY at Buffalo, Buffalo NY and a member of the Graduate Program in Ecology, Evolution and Behavior, SUNY at Buffalo. A physicist by training, his interest in dynamical processes in a wide range of contexts led to work on the processes and patterns of evolutionary changes, and thus to the analysis of shape.

Preface

The second edition of *Geometric Morphometrics for Biologists*, like the first edition, is a textbook on shape analysis for biologists, covering both its basic theory and its practice. We revised the first edition for the same selfish reasons that we wrote it: we teach morphometrics to advanced undergraduate and graduate students and need a textbook. Our students continue to ask sophisticated biological questions and require methods that can answer those questions, but they have little (if any) experience with matrix algebra, non-Euclidean geometry or multivariate statistics. And they continue to want to apply the new methods that they learn immediately on learning them. Our students want practical tools that they can apply to their data, not just theoretical rationales or justifications for methods. Accordingly, we emphasize the biological questions answered by various methods and provide examples of applications to both simple and complex biological questions. The second edition also emphasizes applications to biological questions and illustration of results, and presumes that the reader's background consists of only a basic course in statistics and some familiarity with elementary geometry and algebra. We provide instructions on how to use available software to conduct analyses explained in the book and we still strongly encourage students (and faculty) to begin collecting their own data as soon as possible. Even though we provide data that you can use for learning to run programs, that cannot compare to analyzing your own data. Your own data will be more familiar and far more interesting than any that we can provide.

The second edition does differ substantially from the first edition in several important respects. The most obvious difference between the two editions is that we include a few pages at the end of each chapter that explain how to run the software and we provide an online workbook that covers the practical details of a morphometric analysis. We decided on an online workbook so that we could more regularly update it. Rather than freezing software so that it will match the manuals contained in the book, we are unfreezing the manuals so that they can match the updated software. We also revised the order of some of the chapters, include a chapter on complex statistical designs (Chapter 9, General Linear Models), and cover a wider range of biological applications, reflecting the increasing use of geometric morphometrics in diverse biological disciplines. Among the applications added to the second edition are methods for incorporating phylogenies into morphometric analyses (Chapter 10, Evolutionary and Ecological Morphology), methods for analyzing developmental stability, morphological integration and modularity (Chapter 12, Variational Properties), and the most socially significant additional chapter is the application of geometric morphometrics to forensics (Chapter 14, Forensic Applications of Geometric Morphometrics). Of course, one of the major changes is the complete rewriting of the manuals for conducting morphometric analyses.

Other changes in the book are more structural. The first five chapters are restructured because of a change in the student population – to many students now entering the biological disciplines, geometric morphometrics *is* traditional morphometrics. They do not need to be eased into the subject by appealing to intuitions grounded in older methods. We have therefore streamlined the first three chapters and incorporated superimposition methods into the third chapter rather than beginning with the one that seemed simplest to grasp, then deriving another from the theory of shape, then reprising the discussion of superimposition methods given that theory. There are many other more subtle changes, including the addition of some newer methods (e.g., between-group principal components analysis [Chapter 6], methods for analyzing fluctuating asymmetry [Chapter 12] and methods for preserving information about size within shape analysis [Chapter 14]).

We are very grateful to our students, both those in our regular courses at the University of Michigan, and the State University of New York, Buffalo and Canisius College and to participants at (and organizers of) our regular workshop at the University of California, Berkeley. We are indebted to readers of the first edition who pointed out errors and ambiguities, especially, Jim Rohlf, Ian Dworkin, and Mark Webster, and to those who provided (or rewrote) code for conducting morphometrics in R, including Dean Adams, Ian Dworkin, Annat Haber, Adam Rountrey, Aaron Wood and Nathan Young.

M.L.Z., D.L.S., H.D.S.
May 2012

CHAPTER 1

Introduction

Shape analysis plays an important role in many kinds of biological studies. A variety of biological processes produce differences in shape between individuals or their parts, such as disease or injury, mutation, ontogenetic development, adaptation to local geographic factors, or long-term evolutionary diversification. Differences in shape may signal differences in processes of growth and morphogenesis, different functional roles played by the same parts, different responses to the same selective pressures, or differences in the selective pressures themselves. Shape analysis is an approach to understanding those diverse causes of morphological variation and transformation.

Sometimes, differences in shape are adequately summarized by comparing the observed shapes to more familiar objects such as circles, kidneys or letters of the alphabet (or even, in the case of the Lower Peninsula of Michigan, a mitten). Organisms, or their parts, are then characterized as being more or less circular, reniform, C-shaped or mitten-like. Such comparisons can be extremely valuable because they help us to visualize unfamiliar organisms or to focus attention on biologically meaningful components of shape. However, they can also be vague, inaccurate or even misleading, especially when the shapes are complex and do not closely resemble familiar icons. Even under the best of circumstances, we still cannot say precisely how much more circular, reniform, or C-shaped or mitten-like one shape is than another. When we need that precision, we turn to measurement.

Morphometrics is a quantitative way of addressing the shape comparisons that have always interested biologists. This may not seem to be the case because the morphological approaches once typical of the quantitative literature appeared very different from the qualitative descriptions of morphology; whereas the qualitative studies produce pictures or detailed descriptions (in which analogies figure prominently), morphometric studies usually produced tables with disembodied lists of numbers. Those numbers seemed so highly abstract that we could not readily visualize them as descriptors of shape differences, and the language of morphometrics also seemed highly abstract and mathematical. As a result, morphometrics seemed closer to statistics or algebra than to morphology. In one sense that perception is entirely accurate: morphometrics *is* a branch of mathematical shape analysis. The way that we extract information from morphometric data involves mathematical operations rather than concepts rooted in biological intuition or classical morphology.

Geometric Morphometrics for Biologists
DOI: http://dx.doi.org/10.1016/B978-0-12-386903-6.00001-0

Indeed, the pioneering work in modern geometric morphometrics (the focus of this book) had nothing at all to do with organismal morphology; the goal was to answer a question about the alignment of megalithic "standing stones" like Stonehenge (Kendall and Kendall, 1980). Nevertheless, morphometrics *can* be as much a branch of morphology as it is a branch of statistics. It is that when the tools of shape analysis are turned to organismal shapes, illustrating and even explaining shape differences that have been mathematically analyzed.

The tools of geometric shape analysis have a tremendous advantage when it comes to these purposes: not only because it offers precise and accurate description, but also because it enables rigorous statistical analyses and serves the important purposes of visualization, interpretation and communication of results. Geometric morphometrics allows us to visualize differences among complex shapes with nearly the same facility as we can visualize differences among circles, kidneys and letters of the alphabet (and mittens).

In emphasizing the biological component of morphometrics, we do not discount the importance of its mathematical component. Mathematics provides the models used to analyze data, both the general linear models exploited in statistical analyses and the algebraic models underlying exploratory methods such as principal components analysis. Additionally, mathematics provides a theory of measurement that we use to obtain the data in the first place. It may not be obvious that any theory governs measurement because very little theory (if any) underlays traditional measurement approaches. Asked the question "What are you measuring?", we could give many answers based on our biological motivation for measurement – such as (1) "functionally important characters"; (2) "systematically important characters"; (3) "developmentally important characters"; or (4) "size and shape". However, when asked "what do you mean by "character" and how is that related either mathematically or conceptually to what you are measuring?" or if asked "what do you mean by "size and shape"?", it was difficult to provide coherent answers. A great deal of experience and tacit knowledge went into devising measurement schemes, but that knowledge and experience had very little to do with any general theory of measurement. Rather than being grounded in a general theory of measurement, each study appeared to devise its approach to measurement according to the biological questions at hand, as guided by the particular tradition within which that question arose. There was no general theory of shape nor were there any analytic methods adapted to the characteristics of shape data.

Owing largely to developments in measurement theory over the past two decades, there has been remarkable progress in morphometrics. That progress resulted from first precisely defining "shape" and then pursuing the mathematical implications of that definition. We therefore now have a theory of measurement. Below we offer a critical overview of the recent history of measurement theory, presenting it first in terms of exemplary data sets and then in more general terms, emphasizing the core of the theory underlying geometric morphometrics – the definition of shape. We conclude the conceptual part of this Introduction with a brief discussion of methods of data analysis. The rest of the Introduction is concerned with the organization of this book and where you can find more information about available software and other resources for carrying out morphometric analyses.

A CRITICAL OVERVIEW OF MEASUREMENT THEORY

Traditionally, morphometric data were measurements of length, depth and width, such as those shown in Figure 1.1, based on a scheme presented in a classic ichthyology text (Lagler et al., 1962). Such a data set contains relatively little information about shape and some of it is fairly ambiguous. These kinds of data sets contain less information than they appear to hold because many of the measurements overlap or run in similar directions. What may be most obvious is that several measurements radiate from a single point so that their values cannot be completely independent; any error in locating that point affects all of these measurements. Such a data set contains less information than could have been collected with no greater effort because some directions are measured redundantly and many measurements overlap. For example, there are many measurements of length along the anteroposterior body axis and most of them cross some part of the head, whereas there are only two measurements along the dorsoventral axis and both are of post-cranial dimensions. In addition, because most of the measurements are long, it is difficult to localize shape differences to any region, such as any change in the proportions of the pre- and postorbital head or the position of the dorsal fin relative to the back of the head. Also, some of the information that is missing from this type of measurement scheme, but which is necessary for morphological analysis, concerns the spatial relationships among measurements. That information might be in the descriptions of the measurements, i.e. the line segments, but it is not captured by the data. The data consist solely of a list of observed values of those lengths. Finally, the measurements may not sample homologous features of the organism, making it difficult to interpret the results. For example, body depth can be measured by a line extending between two well-defined points (e.g. the anterior base of the dorsal fin to the anterior base of the anal fin), but it can also be measured wherever the body is deepest, yielding a measurement of "greatest body depth" wherever that occurs. That second measurement of depth might not be comparable anatomically from species to species, or even from specimen to specimen, so it provides almost no useful information (except for maximal depth). Considering these many limitations of traditional measurements, it is clear that the number of measurements greatly overestimates the amount of shape information that is actually collected.

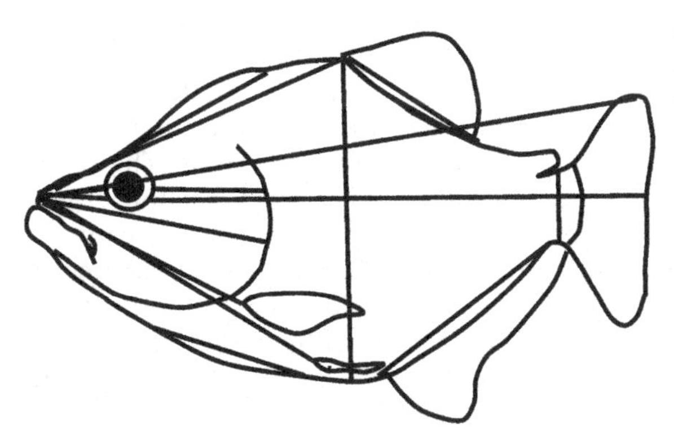

FIGURE 1.1 Traditional morphometric measurements of external body form of a teleost, adapted from the scheme in Lagler et al., 1962.

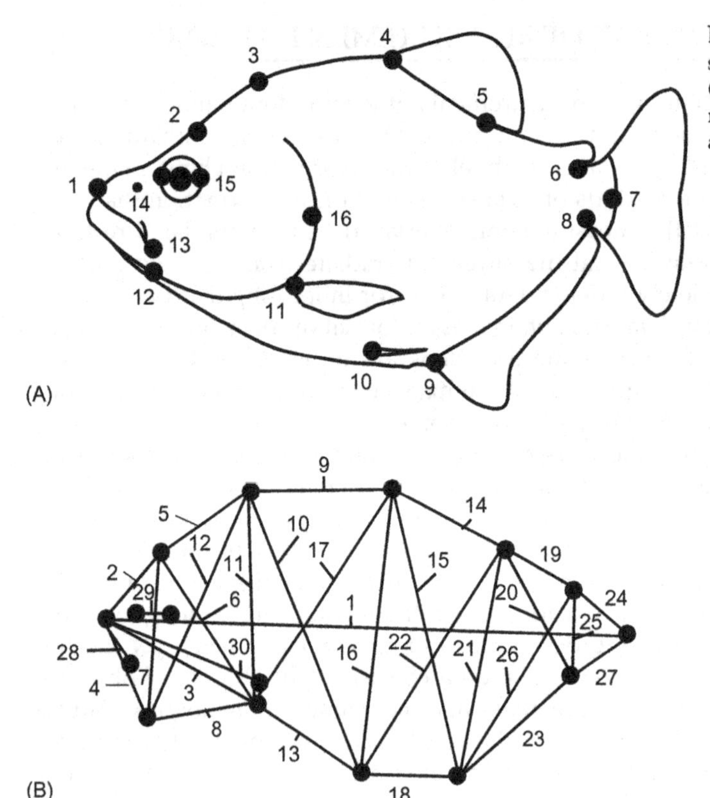

FIGURE 1.2 Truss measurement scheme of external body form of a teleost: (A) well-defined endpoints of measurements; (B) a selection of 30 lengths, arranged in a truss.

The classical measurement scheme can be greatly improved without altering its basic mathematical framework, by the box truss (Figure 1.2) – a scheme developed by Strauss, Bookstein and colleagues (Strauss and Bookstein, 1982; Bookstein et al., 1985). This set of measurements samples more directions of the organism, the measurements are more evenly spaced, and there are also many short measurements. Moreover, all the endpoints of the measurements are biologically homologous anatomical loci – landmarks. But even though the truss is a clear improvement over classical measurement schemes when it comes to describing shape differences, the result is still just a list of numbers (i.e. the lengths of the truss elements), with all the attendant problems of visualization and communication.

One general problem shared by both those measurement schemes is that they fail to collect all the information available from the endpoints of the measurements. The truss scheme shown in Figure 1.2 contains 30 measurements, but 30 is only a fraction of the 120 that could be made among the same 16 landmarks (Figure 1.3). Of course, many of the 120 are redundant, and several of them span large regions of the organism, making it difficult to localize where changes occur. Additionally, we would need extraordinarily large samples in order to test hypotheses about shape and the results would be incredibly difficult to interpret because there would be 120 pieces of information (e.g. regression

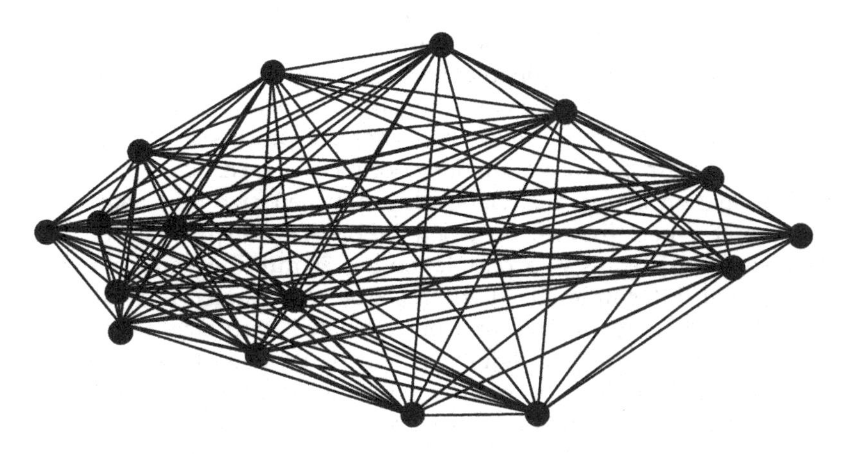

FIGURE 1.3 All 120 measurements between endpoints defined by the 16 landmarks of Figure 1.2.

coefficients, principal component loadings) for each trend or difference. Analyzing all 120 requires specialized methods beyond the scope of this book (see Lele and Richtsmeier, 1991, 2001; Richtsmeier and Lele, 1993). We might be tempted to cull the 120 measurements, retaining only those that seem most likely to be informative but, until we have done the analysis, we cannot know which can be safely culled. Clearly, we need another way to get the same shape information as the 120 measurements, but without the excessive redundancy.

Another problem common to the truss and more traditional schemes is that all the measurements are of size – each measurement is the magnitude of some dimension, such as length, width, area, all of which are measures of size. That does not mean that the data include no information about shape – they do. But that information is contained in the ratios among the lengths, and it can be surprisingly difficult to extract it because that requires separating information about shape from that about size. Some studies have analyzed ratios directly but ratios pose serious statistical problems (debated by Atchley et al., 1976; Corruccini, 1977; Albrecht, 1978; Atchley and Anderson, 1978; Dodson, 1978; Hills, 1978). The more usual approach is to construct shape variables from linear combinations of length measurements, such as Principal Component (PC) loadings. Here, one component, usually the first (PC1), is interpreted as a measure of size and all the others are interpreted as measures of shape. However, PC1 includes information about both shape and size, as do all the other PCs. The raw measurements include information about both shape and size and so do their linear combinations.

Not only are the methods of separating size from shape problematic; the whole *idea* of "size and shape" has been one of the most controversial subjects in traditional morphometrics. One reason for this controversy is the multiplicity of definitions of size (and therefore also of shape), several of which are articulated by Bookstein (1989). Virtually any approach to effecting this separation can be disputed on the grounds that the notion of "size" being separated from "shape" is not really "size". Also, some workers argue that no such separation is biologically reasonable; see, for example, the discussion of studies of heterochrony based on growth models by Klingenberg (1998). However, even if we accept that size and shape are intimately linked by biological processes, we still want to know more about their relationship than the mere fact of its existence.

Extracting information about the relationship between size and shape from a set of measurements can be especially difficult when the organisms span a broad size range. For example, when some organisms in a population are 20 mm long and others are 250 mm, *all* measurements will differ in length. Even if shape is not much influenced by a tenfold change in size, all measurements will still be correlated with size. Quantifying that fact is merely restating the obvious. Consequently, we should expect size to be the dominant source of variance in traditional morphometric data because these measurements *are* measurements of size. We have redundantly measured the same factor and it will therefore be the dominant factor in the data. But we should be concerned about the possibility that the variance in shape is not so much explained by the variance in size, as it is simply overwhelmed by it. For instance, in analyses of ontogenetic series of two species of piranha (one being the running example throughout this chapter), we find that 99.4% of the variance within each species is explained by PC1. That suggests that there is nothing else to explain because it is hard to imagine that the remaining 0.6% is anything but noise. And yet, we do not actually know what proportion of shape variation is actually explained by size. What that quantity, 99.4%, tells us is the proportion of variation in measurements of *size* that is explained by size.

Finally, one serious limitation of traditional morphometrics is that the measurements convey no information about their geometric structure. If we strip off the line segments connecting the landmarks in Figure 1.3 and just look at the position of the landmarks on the page (Figure 1.4), we can see that some are close to each other (e.g. 12 and 13) and others are far apart (e.g. 1 and 7); some are ventral (9 and 10) and some are dorsal (3 and 5), others are anterior (e.g. 1 and 2) and still others are posterior (6–8). That information about position, which is so important to morphologists, is contained in the coordinates of the landmarks but not in the list of distances among them – not even in the more comprehensive list of 120 measurements. But the *x*- *y*-coordinates of the 16 landmarks contain *all* that positional information in addition to all the information contained in the 120 distances between all pairs of landmarks. Those distances can be reconstructed from the coordinates if the units of the coordinate system are known. More importantly, simple algebraic manipulations allow us to partition the information into size and shape and to strip off irrelevant information like the position of the organism in the photograph and its orientation in the photographic plane. After we have removed that irrelevant information, and separated shape from size, we have fewer than 32 shape variables but we have all the information about the geometric structure of our landmarks that was captured when we digitized the specimens. We also have all the information present in the full list of 120 measurements without its redundancy. Consequently, we do not need to cull the data in advance of the analysis, and we do not thereby lose any information we might have in the data. In addition, partitioning morphological variation into size and shape means that variance in size does not overwhelm variance in shape even when variance in size is relatively large. In the two species mentioned above (in which PC1 accounts for 99.4% of

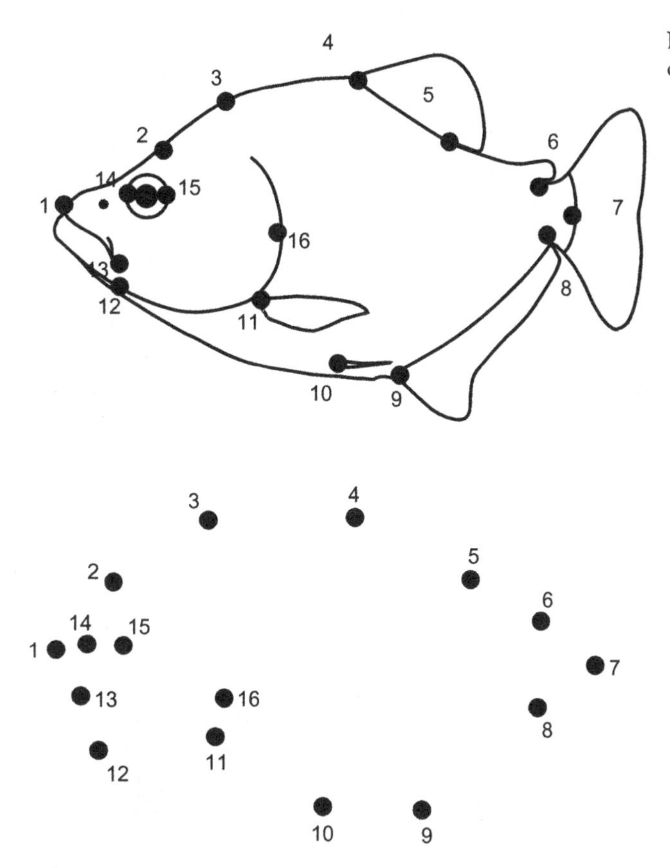

FIGURE 1.4 The 16 landmarks, stripped of the line segments connecting them.

the variance), size explains 72.3% of the variance in shape in one species but only 21.7% in the other.

An important advantage of analyzing landmark coordinates is that it is relatively easy to draw informative pictures to illustrate the results. In Figure 1.5, the shape changes that occur during the ontogeny of one species of piranha are shown as vectors of relative landmark displacement and as a deformed grid that shows the changes between those vectors. In both representations, it is quite clear that the middle of the body becomes relatively deeper while the postanal region becomes relatively short, especially the caudal peduncle (between landmarks 6 and 7). Both pictures also show that the posterodorsal region of the head (above and behind the eye) becomes relatively longer and deeper while other regions of the head become relatively shorter. (We emphasize that these are *relative* changes, because the piranha becomes *absolutely* larger in every dimension and region mentioned.)

It is possible to present traditional morphometric results in graphic form by placing the values of the allometric coefficients on the organisms, as in Figure 1.6. This, like

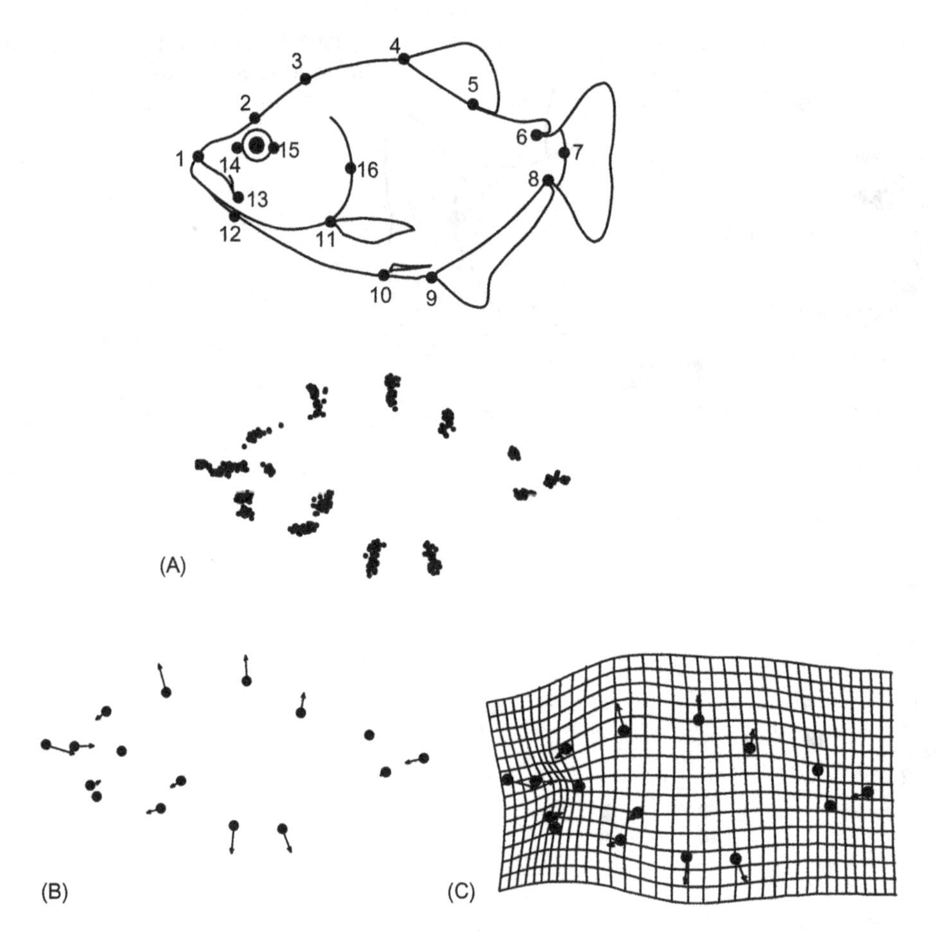

FIGURE 1.5 Ontogenetic shape change depicted in two visual styles. (A) Landmarks of all specimens; (B) vectors of relative landmark displacement; (C) deformed grid.

Figure 1.5, shows that the middle of the body grows faster and becomes deeper than the rest of the animal. The limitation of this representation (and of the analysis) is exemplified by the difficulty of interpreting the large coefficient (1.23) of the posterior segment of dorsal head length – it is not clear whether the head is just elongating rapidly in this area or if it is mainly deepening or if it is both elongating and deepening. We also cannot tell if the pre- and postorbital head increase at the same rate because the measurement scheme does not include distances from the eye to other landmarks. None of these ambiguities arose from the geometric analysis of the landmark coordinates; the figure illustrating that result showed the ontogenetic changes in all those specific regions. This ability to extract and communicate information about the spatial localization of morphological variation, including its magnitude, position and spatial extent, is among the more important benefits of geometric morphometrics. Following the statistical analysis, which allows us to determine which factors have an effect on shape, we can diagram the effects of the factors.

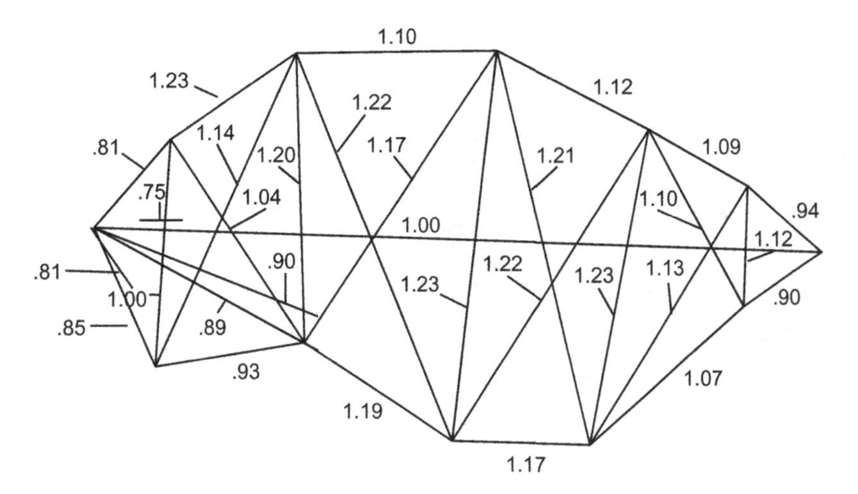

FIGURE 1.6 Allometric coefficients of traditional morphometric measurements, plotted on the organism.

Using landmarks as data does not solve all of the problems confronting traditional methods, and one remaining problem becomes evident as soon as we try to examine the changes in head profile over the piranha's ontogeny (Figure 1.7). We can see that the average slope on either side of landmark 2 must get steeper, but we cannot tell whether the profile becomes more S-shaped, C-shaped or any other shape. This uncertainty arises because the three landmarks provide too little information about the curve between them; they are no better a sample of the curve's shape than the line segments connecting them. For this reason, we need to incorporate information about points on the curve between landmarks (Figure 1.8). These points (called semilandmarks) are not landmarks because they are not individually homologous anatomical loci even though they sample points along homologous curves. The utility of semilandmarks is even more apparent if we look at structures whose curvature is both complex and critical to its function, like those on the margins of rodent mandibles (Figure 1.9). The regions on which the muscles insert lack landmarks but the information about the depth of the jaw, the positioning and orientations of the muscles provided by semilandmarks is necessary for any explanation of jaw shape. As we predicted in the Introduction to the first edition of this book, the limitation of geometric morphometrics to landmarks was transitory.

Geometric morphometrics may also have appeared to have a limitation not confronting traditional methods: the restriction to two-dimensional data. But that limitation was purely a matter of the technology for obtaining the coordinate data. The mathematical theory poses no obstacle to the analysis of three-dimensional shapes. The obstacles lie in cost of the equipment for obtaining three-dimensional coordinates and the details of using it, such as the need to transport either the equipment or the specimens so that they are in the same place at the same time. The other problem results from the difficulty of depicting results of the analysis on a static, two-dimensional medium such as the pages of a journal. Traditional morphometric studies did not face these obstacles because specimens could always be measured with calipers if the equipment needed for three-dimensional

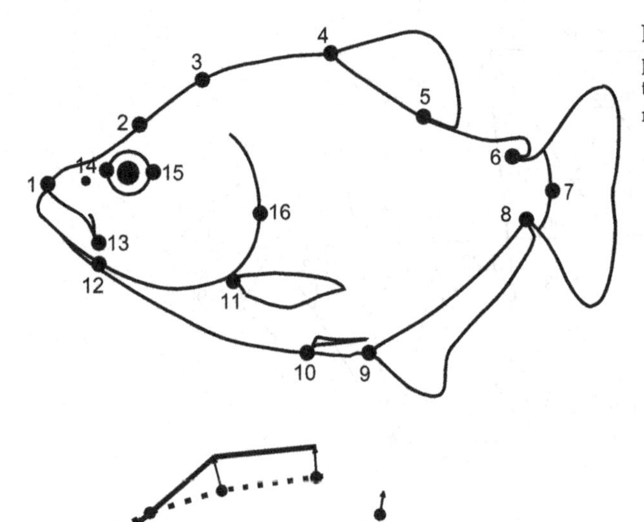

FIGURE 1.7 Ontogenetic change in head profile as implied by changes in the orientation of straight lines drawn between landmarks of the head.

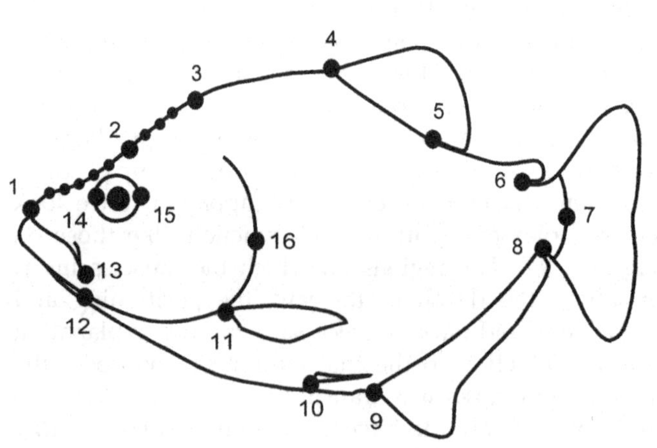

FIGURE 1.8 Additional points on the head profile that capture information about its curvature. These points, which are not individually homologous, are "semilandmarks".

digitizing is exorbitant (in either time or money) and the results were not depicted at all so the problems posed by projecting three-dimensional data onto a two-dimensional page did not arise. However, when using calipers we do not collect three-dimensional coordinates so the technology sidesteps rather than solves the problem. And tabulating numbers rather than depicting results also sidesteps rather than solves the problem. Although the equipment is costly and not always portable, and the results can be difficult to show, three-dimensional morphometrics is just as feasible mathematically as two-dimensional morphometrics.

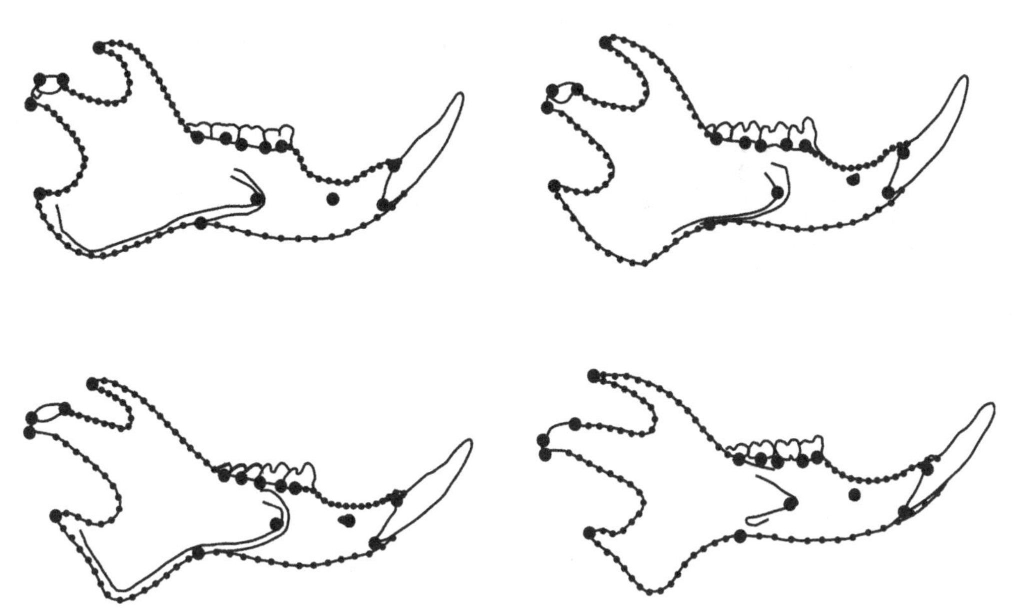

FIGURE 1.9 Semilandmarks on four squirrel mandibles to capture information about the curvature of the jaw, which reflects both the curvature of the incisor and the length and orientation of muscle insertions in regions where there are no landmarks.

SHAPE AND SIZE

The rapid progress in geometric morphometrics resulted from a coherent mathematical theory of shape which, in turn, resulted from a precise definition of the concept. Like the definition of any word, that of "shape" is entirely a matter of semantics. However, semantics is not trivial. We cannot have a coherent theory about an ambiguous concept and we cannot have a coherent mathematical theory of shape until shape is unambiguously defined. The definition of shape is thus the foundation for the mathematical theory of shape. Whether that theory applies to our biological questions depends on whether it captures what we mean by shape. Thus, it is important to understand the definition of shape underlying geometric morphometrics; also, because the meaning of "size" depends on the meaning of "shape" (and vice versa), we cannot understand one without understanding the other.

Shape

In geometric morphometrics, shape is defined as "all the geometric information that remains when location, scale and rotational effects are filtered out from an object" (Kendall, 1977). The earliest work that depends on this definition of shape began the analysis with the coordinates of points; consequently, the "objects" are sets of those coordinates – i.e. configurations of landmarks, such as that shown in Figure 1.4. An important implication of Kendall's definition is that removing the differences between configurations

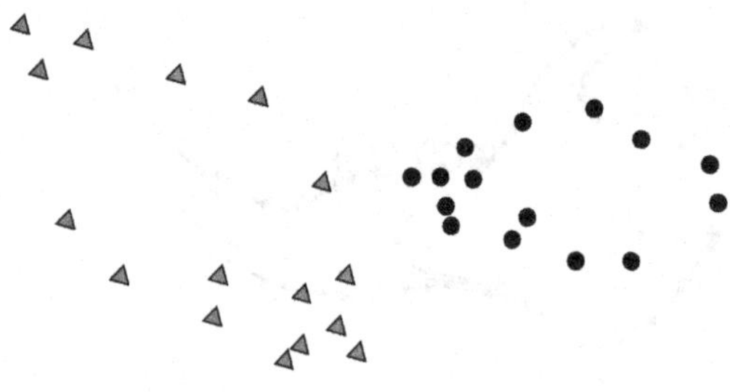

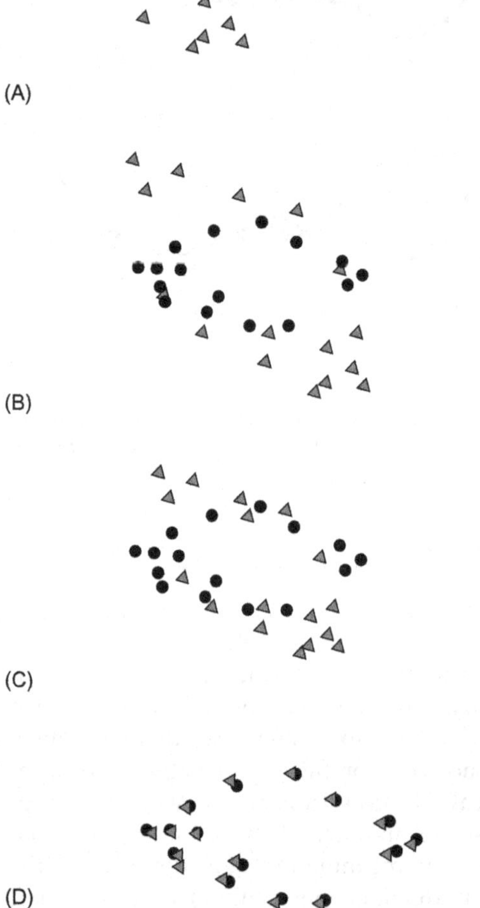

FIGURE 1.10 Removing variation due to differences in position, scale and orientation. (A) Two original configurations; (B) after removing differences in location; (C) after removing differences in scale; (D) after removing differences in orientation, leaving only differences in shape.

that are attributable to differences in location, scale and orientation leaves only differences in shape. These operations and their consequences are illustrated in Figure 1.10. In Figure 1.10A, there are two configurations, side by side. This difference in location has no bearing on their shape difference so, in Figure 1.10B, both have been translated to the same location. The two configurations still differ in scale, which also has no bearing on their shape difference so, in Figure 1.10C, they are converted to the same scale. The two configurations still differ in orientation (their long axes are about 45° apart). That, too, has no bearing on their shape differences so, in Figure 1.10D, they are rotated to an alignment that leaves shape as the only difference between them. After removing all the differences that are *not* in shape, we are left with only the differences in shape. We can now use the coordinates of the final configurations (Figure 1.10D) to analyze these shape differences. When we have sampled the curves between landmarks, we also have to remove any non-shape variation resulting from our (arbitrary) choice of where to sample those points along a curve.

Size

Kendall's definition of shape mentions scale as one of the effects to be filtered out. The implication is that "scale" is the definition of size that is complementary to shape under some models of error. The two are ideally geometrically independent (i.e. orthogonal). The concept underlying geometric scale is quite simple, and may be intuitively obvious by visual inspection. As you can see in Figure 1.10A, the landmarks are generally further apart in one configuration than in the other. That is obviously what we expect when a configuration is larger, whether because the organism is larger or the photograph of it is larger. To calculate geometric scale, we compute the distances of all the landmarks to the center of the form (its "centroid"); Figure 1.11 shows the location of the centroid and the segments connecting the landmarks to it. Now we can compute geometric scale by calculating the square of each distance from landmark to the centroid, summing those squares and taking the square root of their sum. This quantity is called "centroid size".

Centroid size is the one measure of size that is *mathematically* independent of shape. It is, more precisely, orthogonal to shape. That is a matter of definition, not biology. In biological data, centroid size may often be empirically correlated with shape because larger organisms often are shaped differently than smaller ones. The fact that we have defined and measured shape and size separately does not mean that we are assuming them to be biologically separate. Nor does their separation cause us to lose information about the relationship *between* size and shape. We can easily analyze that relationship by conventional statistical methods.

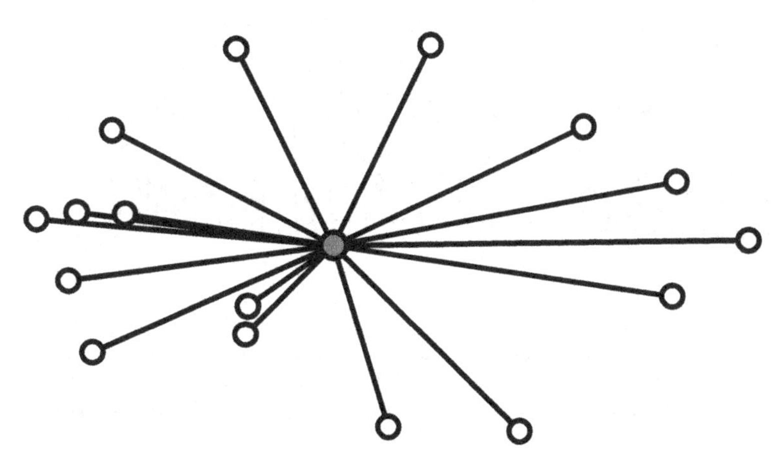

FIGURE 1.11 A visual representation of centroid size as computed for 16 landmarks on a piranha. The open circle is the centroid; the segments connecting the centroid to the landmarks represent the distances used to compute centroid size.

METHODS OF DATA ANALYSIS

Replacing traditional morphometric variables with landmark coordinates does not deprive us of the methods we have long used in statistical analyses of morphological data. We can ask all the questions that we ever asked about morphology. Such questions often comprise two parts, the first of which Bookstein (1991) termed the "existential question": *is* there an effect on shape? We answer that by determining the probability that the association between variables is no greater than could have arisen by chance. The second question, "*what* is the effect?" calls for description. In the ontogenetic series of piranhas discussed earlier, we can analyze the relationship between shape and size by computing the centroid

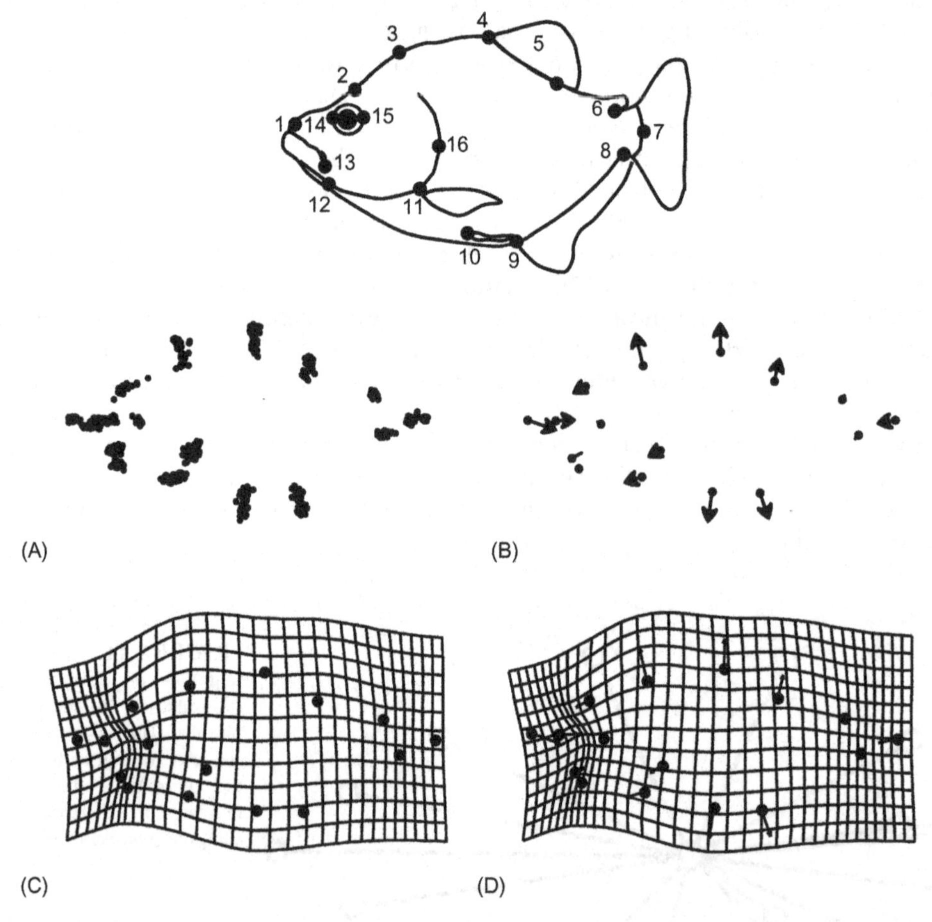

FIGURE 1.12 Analyzing the impact of size on shape by multivariate regression. (A) Configurations of landmarks from which differences in position, scale and orientation have been removed; (B) the covariance between size and shape depicted by vectors of relative landmark displacements; (C) the covariance between size and shape depicted by a deformed grid; (D) the covariance between size and shape depicted by a deformed grid plus vectors of relative landmark displacement.

size of each configuration and the coordinates of the landmarks from which differences in position, scale and rotational effects have all been removed. These new configurations, shown in Figure 1.12A, represent the shapes of all the specimens. To answer the first question about the existence of an effect, we regress shape on centroid size using a multivariate regression in which "shape" is the dependent variable and "centroid size" (or its logarithm) is the independent variable. For this example, we can conclusively reject the null hypothesis of no effect; we obtain an F-ratio of 94.02 with 28 and 1008 degrees of freedom; $P < 1 \times 10^{-5}$. We can also determine that 72.3% of the shape variation is explained by size. To answer the second question, we depict the changes either by relative landmark displacement (Figure 1.12B), a deformed grid (Figure 1.12C) or both (Figure 1.12D).

Replacing distances with coordinates also does not require us to abandon familiar ordination methods, such as principal components analysis and canonical variates analysis. These methods are often used to explore patterns in the data in the hope that their results will suggest the factors responsible for variation among individuals or differences among groups. At the very least, these analyses can extract the dimensions along which individuals vary most and groups differ most. The results include scatter-plots of specimens that depict patterns of variation or differences. The interpretation of these scatter-plots is by the accompanying graphics of the dimensions along which specimens most vary (Figure 1.13) or groups most differ (Figure 1.14).

The important distinction between analyses of geometric shape data and conventional morphometric data is that analyses of landmark configurations are necessarily multivariate. By definition, shape is a feature of the whole configuration of landmarks. Even the simplest shape, a triangle, cannot be analyzed univariately. Shape data are multidimensional in that each individual datum, i.e. each configuration, is described by multiple coordinates. Because we have defined shape in terms of a whole configuration of landmarks, our analyses must be of that

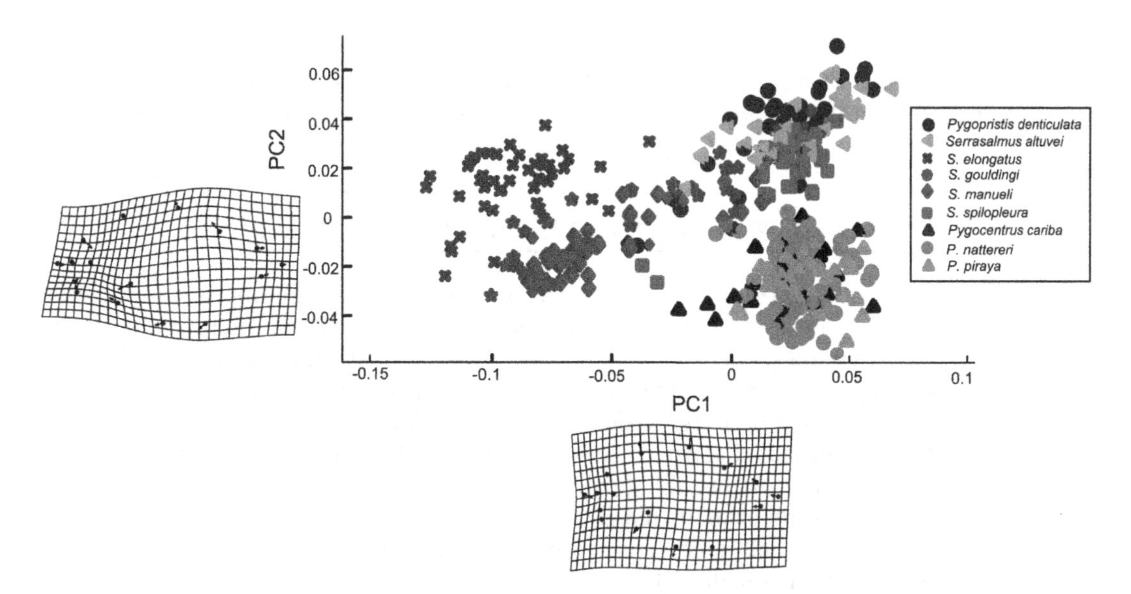

FIGURE 1.13 Principal components analysis of piranha body shape.

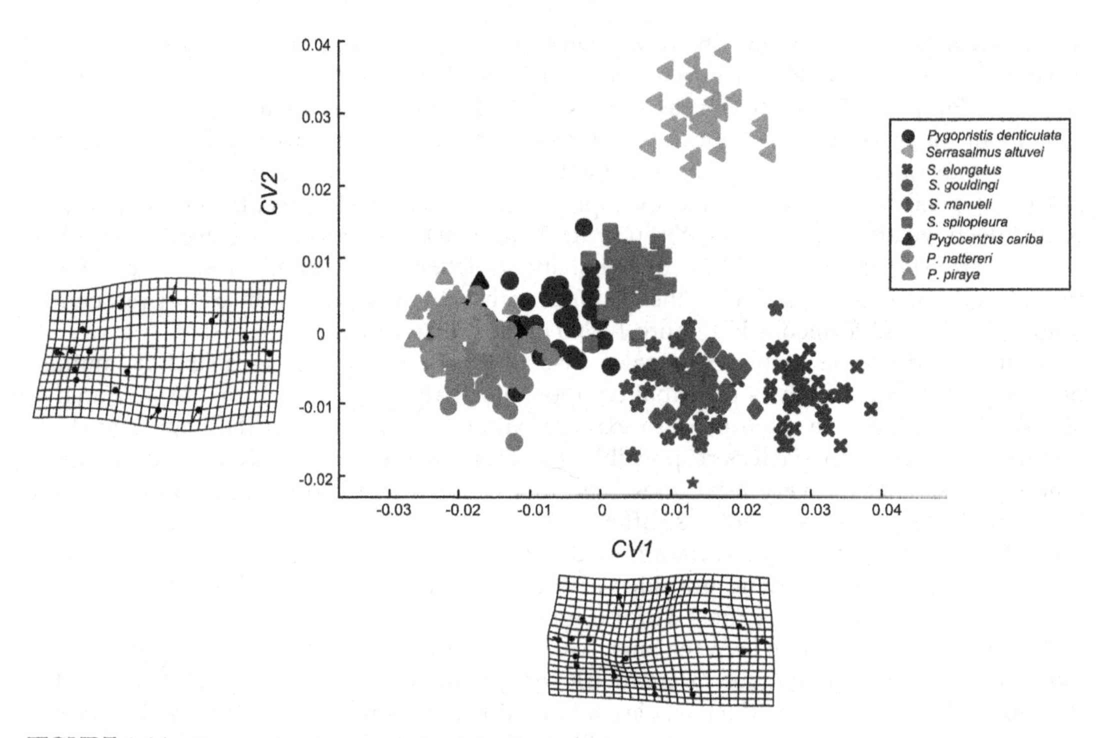

FIGURE 1.14 Canonical variates analysis of piranha body shape.

whole. However, this does not prevent us from subdividing an *organism* to analyze relationships between parts. For example, we could divide the piranha into the cranial and post-cranial regions and analyze landmarks from each region as a separate configuration to ask whether two regions covary. The requirement that *configurations* be analyzed multivariately does not force us to treat *organisms* as unitary wholes (although we may find out that they are).

BIOLOGICAL AND STATISTICAL HYPOTHESES

Few hypotheses of interest to biologists are as simple as the allometric hypothesis examined earlier. Only rarely can the more complex hypotheses be wrestled into the form of a statistical null hypothesis and its alternatives. The first difficulty is that the statistical null merely states that the factor of interest has no effect; this is the hypothesis we hope to reject in favor of the alternative hypothesis that the factor does have an effect. In this situation, we have two hypotheses that are diametrically opposed, meaning that they are mutually exclusive. In contrast, many biological hypotheses are more complex because they state multiple alternative theories of *causation*. These alternative causal theories may not be mutually exclusive and all predict that the factor has an effect. Thus, the real goal of many studies is to discriminate between expected effects, not to reject a hypothesis of no effect. For example, perhaps we are interested in the evolution of claw shape in crabs.

We probably already know that claw shape has evolved, so we are not aiming to test the null hypothesis that they have not. The more interesting (and difficult) question is whether the derived claw shape arose to enhance the ability to burrow into a muddy substrate or was intrinsically constrained by development (or both). We may also have multiple hypotheses in addition to these, including others regarding the function of claws (e.g. that the derived claw shape enhances the ability to block a burrow entrance or even to attract mates). We might also have several alternative theories about how development could constrain the evolution of claw shape.

Yet another obstacle to translating a biological hypothesis into a statistical one is that the complexity of the biological hypotheses rarely allows for adequate testing by any single method. To test whether the evolution of crab claw shape was intrinsically constrained by development, we must first determine whether there is any evidence of a developmental constraint on variation, and we would also need to show that the various adaptive hypotheses predict different evolutionary transformations than those specified by the developmental constraint hypothesis. Otherwise we cannot rule out any of the competing biological alternative hypotheses.

In emphasizing the complexity of biological hypotheses we do not mean to say that they cannot be tested rigorously. The point is that doing so requires far more effort and creativity than testing the simple hypothesis that size affects shape or that species differ in shape. It also requires understanding what various analytic methods do, what their limits are, and how they are mathematically related. Far too often biologists use a limited array of techniques to analyze multivariate data, regardless of their questions. Throughout this book we emphasize the biological questions that prompt the morphometric analysis, and underscore the applications of each method as we discuss them in turn. However, only after a variety of methods has been introduced (and mastered) can we begin to address questions of realistic biological complexity.

ORGANIZATION OF THE BOOK

This book is divided into three main sections. The first is a series of chapters covering the basics of shape data – what landmarks and semilandmarks are, how to select landmarks as well as how to incorporate semilandmarks into the measurement scheme (Chapter 2), how the coordinates of both landmarks and semilandmarks are transformed into shape variables that will be used in subsequent analyses (Chapter 3), the mathematical theory of shape (Chapter 4) and the thin-plate spline interpolation function (Chapter 5). The second section covers analytic methods, including exploratory methods (Chapters 6 and 7) and formal methods of hypothesis testing (Chapters 8 and 9). The final section discusses the application of these methods to complex biological questions, ones that will require using multiple methods, both exploratory and hypothesis testing (Chapters 10–14).

The first section begins with what will be your own first step – deciding what to measure (Chapter 2), and then turning the coordinates that you digitize into data (Chapter 3). Only after you have some experience with these two steps will the abstractions of the theory (Chapter 4) make sense (or be interesting). In our discussion of shape variables, we present three methods for superimposing the data: first the two-point registration that yields Bookstein's shape coordinates, which are the easiest to understand; next Procrustes superimposition, the most widely used method; and third, resistant fit Procrustes superimposition. The main reason for presenting all three is so that you can see how the same three operations that remove non-shape information from the data are applied. We discuss these first because doing so allows us to discuss a number of general issues (including the interpretation of results) before presenting the more abstract theory of shape analysis in Chapter 4. That theory provides the framework for generating (as well as analyzing) shape variables. After reviewing the basic theory, we introduce the thin-plate spline (Chapter 5), an interpolation function useful for depicting results by means of a deformed grid (as in Figures 1.11–1.13). All the chapters in this first section are about the data.

The second section of the book is about the methods for analyzing the data. These methods produce the biologically interesting variables – the ones that covary with the biological factors of interest. Unlike the variables produced by the methods of the first section, the variables produced by these analytic methods have a biological meaning. They answer such fundamental questions as "What impact does size have on shape?", or "By how much, and in what way, do these species differ in their ontogenies?", or "Do these populations vary along a single latitudinal gradient?", or even "What shape has the highest fitness in this population?" Each of these questions is answered in terms of a shape variable – the vector of coordinates that covaries with size or age, or that covaries with latitude or fitness, or that expresses the difference between the means of two groups. When we do not know what factors might be present in the data, a common problem in studies that analyze the variation within and between natural populations, we can explore the data algebraically, using methods of matrix algebra to determine if any interesting patterns emerge. Principal components analysis (PCA) is one example of this kind of exploratory technique; the main purpose is to determine what varies and to look for biological explanations for that variation. Canonical variates analysis (CVA) and between-group principal components analysis (BGPCA) are also exploratory methods, but they presume that you are asking questions about differences between groups, so there is a factor of interest: your grouping variable.

Because many biologists begin a study by exploring patterns in the data, the section on analytic methods begins with an overview of the exploratory methods (Chapter 6). These are useful for extracting simple patterns from complex multidimensional data because they provide a space of relatively low dimensionality that captures most of the variation among specimens (PCA), or most of the differences among groups (canonical variates analysis, CVA and between-groups PCA). We explain the algebra underlying these methods, compare them, and discuss when each is appropriate in light of particular biological questions. We also discuss a method for analyzing the covariance between two multivariate blocks of data, partial least squares, PLS (Chapter 7).

The two following chapters cover methods of hypothesis testing. We begin with simple statistical models (Chapter 8), regression of shape onto a single continuous independent variable (multivariate regression) or a single independent categorical variable (one-way multivariate analysis of variance, MANOVA). We discuss how the hypotheses are formulated and tested using both analytic tests and resampling-based methods (bootstrapping and permutations). The next chapter (Chapter 9) introduces the General Linear Model and discusses more complex hypotheses, those that include multiple factors, both continuous and categorical (and both fixed and random, including nested terms). These models are introduced first in the context of univariate data, then in the context of multivariate data. Testing complex hypotheses requires understanding how to construct design matrices and also, in some cases, how to determine the appropriate numerator and denominator sums of squares for the *F*-ratio (because some programs require that you input design matrices, and you may even need to input the design matrix for the numerator and denominator sums of squares for every term of interest).

The third section covers applications of morphometric methods to realistically complex biological hypotheses. Among these complex questions are those posed by the relationship between ecology and morphology, and the evolution of morphology, a broad subject that encompasses studies of phylogenetic patterns, morphological diversity (disparity) and the covariance between evolutionary changes in traits such as size and shape (Chapter 10). The next two chapters focus on applications of morphometrics to evolutionary developmental biology. The first concerns the evolution of ontogeny, focusing on a series of hypotheses, such as ontogenetic scaling, heterochrony and diversification of ontogenetic allometries and how these can be tested as well as how their impact on disparity can be assessed (Chapter 11). The second (Chapter 12) concerns variational properties, i.e. canalization, phenotypic plasticity, developmental stability and morphological integration and modularity. The next chapter discusses methods useful in systematic studies (Chapter 13). The final chapter covers methods useful in forensics (Chapter 14). Forensic applications of geometric morphometrics have included the determination of sex and age of human remains and forensic odontology, especially bitemark analysis, as well as in other areas that use impressions as evidence, including tool and footwear marks. Forensic applications require adapting geometric morphometric methods to include information about size.

The terminology of statistical shape analysis can be daunting because there are many unfamiliar words and many terms that differ by only a single letter or subscript. Thus we conclude this book with a glossary of terms, including general statistical terms (e.g. population, sample) and more specialized terms of shape analysis (e.g. Procrustes distance, partial warps).

SOFTWARE AND OTHER RESOURCES

Geometric morphometrics studies require fairly specialized software, not so much to analyze the data as to depict the results. Fortunately, the software necessary for most analyses is readily and freely available. Information about finding, downloading and installing software, as well as about running the programs, is available in the workbook on the companion site for this book. We recommend that you get the workbook even before you have your own data, using the datasets that we provide to become familiar with programs before you need them for your own analyses.

All the details about conducting analyses that we describe in this book are in the workbook rather than at the end of the chapters in this edition. We keep them separate because the software develops more rapidly than the mathematical theory of shape and keeping the workbook separate allows us to update information about software more regularly than we revise the book.

References

Albrecht, G. (1978). Some comments on the use of ratios. *Systematic Zoology, 27*, 67–71.

Atchley, W. R., & Anderson, D. (1978). Ratios and the statistical analysis of biological data. *Systematic Zoology, 27*, 71–78.

Atchley, W. R., Gaskins, C. T., & Anderson, D. (1976). Statistical properties of ratios. I. Empirical results. *Systematic Zoology, 25*, 137–148.

Bookstein, F. L. (1991). *Morphometric tools for landmark data: Geometry and biology*. Cambridge: Cambridge University Press.

Bookstein, F. L. (1989). "Size and shape": A comment on semantics. *Systematic Zoology, 38*, 173–190.

Bookstein, F. L., Chernoff, B. L., Elder, R. L., Humphries, J. M., Jr., Smith, G. R., & Strauss, R. E. (1985). *Morphometrics in evolutionary biology*. Philadelphia: The Academy of Natural Sciences Philadelphia.

Corruccini, R. S. (1977). Correlation properties of morphometric ratios. *Systematic Zoology, 26*, 211–214.

Dodson, P. (1978). On the use of ratios in growth studies. *Systematic Zoology, 27*, 62–67.

Hills, M. (1978). On ratios – a response to Atchley, Gaskins and Anderson. *Systematic Zoology, 27*, 61–62.

Kendall, D. G. (1977). The diffusion of shape. *Advances in Applied Probability, 9*, 428–430.

Kendall, D. G., & Kendall, W. S. (1980). Alignments in two-dimensional random sets of points. *Advances in Applied Probability, 12*, 380–424.

Klingenberg, C. P. (1998). Heterochrony and allometry: The analysis of evolutionary change in ontogeny. *Biological Reviews, 73*, 99–123.

Lagler, K. F., Bardach, J. E., & Miller, R. R. . (1962). *Ichthyology*. New York: John Wiley & Sons.

Lele, S., & Richtsmeier, J. T. (1991). Euclidean distance matrix analysis – A coordinate-free approach for comparing biological shapes using landmark data. *American Journal of Physical Anthropology, 86*, 415–427.

Lele, S., & Richtsmeier, J. T. (2001). *An invariant approach to statistical analysis of shapes*. London: Chapman & Hall-CRC Press.

Richtsmeier, J. T., & Lele, S. (1993). A coordinate-free approach to the analysis of growth-patterns: Models and theoretical considerations. *Biological Reviews of the Cambridge Philosophical Society, 68*, 381–411.

Strauss, R. E., & Bookstein, F. L. (1982). The truss – body form reconstructions in morphometrics. *Systematic Zoology, 31*, 113–135.

PART 1

BASICS OF SHAPE DATA

CHAPTER 2

Landmarks and Semilandmarks

Landmarks are discrete anatomical loci that can be recognized as the same point in all specimens in the study. They are often termed "homologous points" because these points can be matched up, one by one, as "the same point" in all individuals in the study. For example, the mental foramen of the squirrel lower jaw (Figure 2.1A) is a discrete point which corresponds to the mental foramen on any other mammalian jaw. "Homology" may seem like the crucial term in the definition of a landmark, but the idea that these are discrete points is equally crucial and these are notably sparse on some structures, including the jaw. We could restrict the analysis to these few points (Figure 2.1B.), but we would miss a great deal of the morphology if we did. How much we would miss becomes more obvious by removing the photograph from the picture and looking only at landmarks (Figure 2.1C). Landmarks are scarce, particularly in the regions of greatest interest, such as where the muscles insert on the jaw. The complex curves that contain critical information about the morphology motivates including additional points that can capture information about curvature (Fig. 2.1D). These additional points are not discrete anatomical loci (much less homologous); rather than being landmarks these points are "semilandmarks".

There are some obvious differences between landmarks and semilandmarks. As noted above, one difference is their discreteness, another is their homology. Related to these is another: the position of semilandmarks along the curve is arbitrary – we could just as easily sample the curve according to the scheme shown in Figure 2.2A or 2.2B. The question every researcher must face is whether to include semilandmarks at all and, should they decide "yes", then the next question is how to sample them. How these questions will be answered depends on the weight given to each of the several criteria for selecting landmarks. That is the primary topic of the present chapter. Throughout much of our discussion of these criteria, we will talk about selecting landmarks one by one, treating each as an individual point, but landmarks (and semilandmarks) are not analyzed or interpreted one by one. The information about shape is contained in the entire constellation of landmarks and semilandmarks, i.e. the configuration of points. That configuration is a single datum. This may seem intuitively obvious to many readers, but readers experienced in traditional morphometrics may find this view of a configuration as a single datum counterintuitive because, in traditional morphometrics, each individual measurement was often viewed as a trait in its own right. Even when analyzed multivariately, the individual

DOI: http://dx.doi.org/10.1016/B978-0-12-386903-6.00002-2

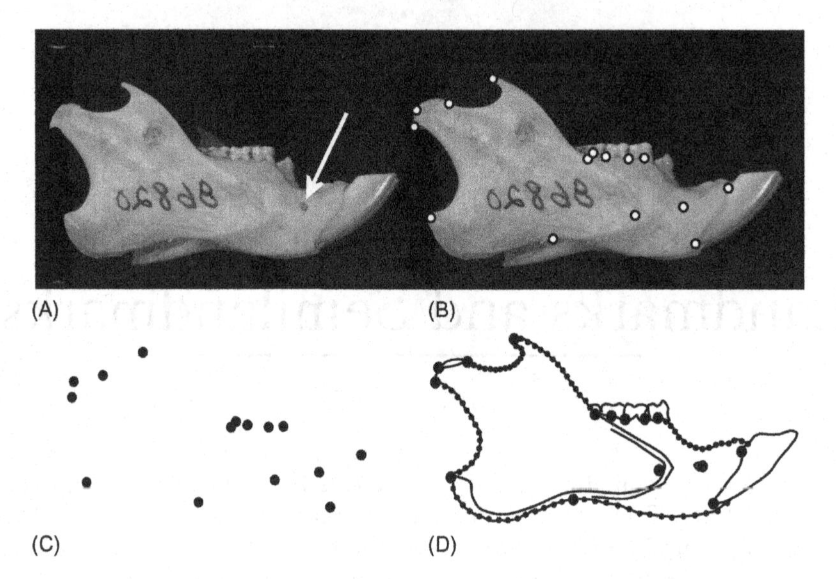

FIGURE 2.1 Measurement schemes for the lower jaw of a squirrel. (A) Mental foramen; (B) landmarks on photograph; (C) landmarks without the photograph; (D) landmarks plus semilandmarks.

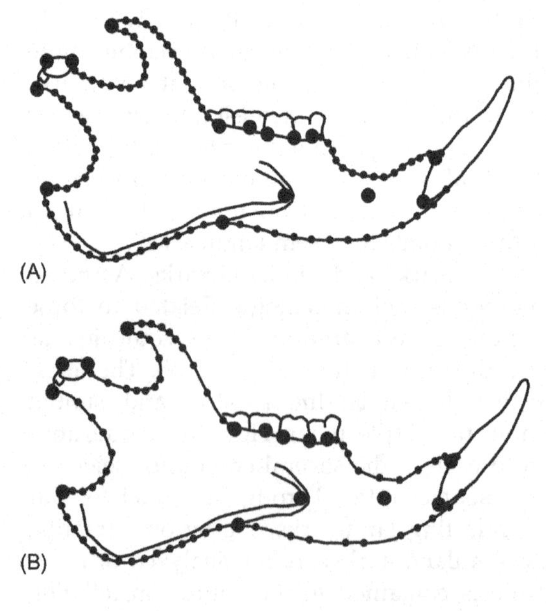

FIGURE 2.2 Two semilandmark measurement schemes. The choice between them is essentially arbitrary because no anatomical features specify the location of points along the curve.

traits retained their identity and biological meaning. Interpreting the results of a multivariate analysis often involved relating the multivariate results to the correlations between traits. As a result, one of the most important considerations that directed the section of measurements was the biological meaning of that trait in the context of a given study. But landmarks do not correspond to "traits" and landmarks are rarely meaningful as individual points.

In geometric morphometrics, a configuration, not a landmark, is a datum. What we measure on individuals is their configuration of landmarks. The "trait" might be the impact of a mutation on the average configuration, or the impact of some experimentally manipulated factor on the average configuration. Analyses of configurations are necessarily multivariate, but it may be more useful to think of a configuration as a multidimensional datum rather than as a collection of multiple variables. One reason for stressing the point is that individual landmarks are not expected to be individually meaningful biologically; their function is to delimit where changes occur. It is the response of the configuration to some factor that conveys biological meaning. Thus, when designing a measurement scheme, what we are after is a configuration that allows us to delimit where those responses occur. Landmarks should therefore provide a sufficiently comprehensive sampling of morphology that the features of biological significance can be discovered. If you are interested in the biomechanics of lever arms, then you should locate landmarks at the endpoints of those lever arms else you will not have the data required to analyze them. However, you will not lose or dilute biomechanical information by including other landmarks of unknown relevance – if they are not functionally relevant, they will not covary with measures of performance. However, if your *only* question is "What is the mechanical advantage of this jaw compared to that one?", then there is no reason to do a shape analysis – the question you are asking is about mechanical advantage, not shape. As Bookstein (1996) pointed out, geometric methods might be "overkill" in such purely biomechanical studies. When you want to place those lever arms in a broader morphological context, geometric morphometrics helps to provide one.

CRITERIA FOR CHOOSING LANDMARKS

Ideally, landmarks are (1) homologous anatomical loci that (2) provide adequate coverage of the morphology, and (3) can be found repeatedly and reliably. Two other criteria may also be important under some conditions, that landmarks (4) do not switch positions relative to each other and (5), in the case of two-dimensional (2D) landmarks, lie within the same plane. That last criterion is the only one specific to two-dimensional data, and information that is either lost by restricting landmarks to one plane or made difficult to interpret when landmarks are not coplanar, is what makes three-dimensional data so useful.

Homology

In the context of landmarks, the criterion of homology means that the points on one specimen correspond (as the "same" point) to that point on all individuals. For example, a landmark located in the middle of the mental foramen of the mandible of one individual is homologous to a landmark located in the middle of the mental foramen of another individual. Similarly, a landmark located in the middle of the mandibular foramen of the mandible is homologous to a landmark located in the middle of the mandibular foramen of another individual. Conversely, a landmark located in the middle of the mental foramen of the mandible of one individual is not homologous with a landmark located in the middle of the mandibular foramen of another individual. The mental and mandibular foramina do not correspond to each other, as may be obvious from the fact that they have different names (it is even more obvious anatomically because one is located on the external/lateral side of the mandible, the other on the internal/medial side; one is located anteriorly and the other posteriorly. That these two points do not correspond is obvious, perhaps so obvious as to go without saying. But there are many examples of traditional morphometric measurements that disregard homology, and it is in contrast to them that the criterion of homology is so important for ensuring comparability of shapes. For example, measurements such as "greatest skull breadth" or "least interorbital width" may be taken between different pairs of endpoints in different specimens because they are measured where the skull is widest, wherever that is, or where the interorbital region is narrowest, wherever that is. The endpoints of those measurements need not correspond one-to-one from one specimen to another; it is only that the length is greatest or least that makes the measurements comparable at all. These measurements are not made between homologous points and, in this context, the converse of "homology" is "not the same point".

Calling landmarks "homologous" may seem to be a curious usage of the term for two reasons. First, biologists often use homology for similarities due to common ancestry, implying more than that the points correspond. Second, the term is usually applied to structures rather than to points. If using "homology" for corresponding points bothers you, you can always call them "corresponding" rather than "homologous" points.

In geometric morphometrics, homology has been stressed above all other criteria for selecting landmarks for both mathematical and biological reasons. The mathematical reasons are important to understand because semilandmarks, which are not usually argued to be homologous biologically, are nonetheless treated as if they are homologous mathematically. The mathematical issues are discussed in more depth in the next two chapters, but you will likely select your landmarks before you read them so you need an intuitive feel for the mathematical issues before choosing them. The primary mathematical issue is that the coordinates of points (whether landmarks or semilandmarks) are treated as if they correspond one-to-one when computing the difference between shapes. That is, the coordinates are averaged, and deviations between individuals are quantified by summing the squares of the differences between the coordinates of the points. If the points in one specimen do not correspond to the points in another, averaging them is like averaging apples and oranges. The calculations make sense when each point on one specimen corresponds to the same point on another; more specifically, when landmark 1 in one organism corresponds to landmark 1 in another, as do landmarks 2 and so forth. That assumption of correspondence is built into the methods whether they are applied to landmarks or semilandmarks.

Semilandmarks are usually not regarded as homologous points. It is the curve that they sample, not the semilandmarks individually, that are viewed as if corresponding one-to-one. The positions of the semilandmarks along a curve are not viewed as informative about shape and so information about their position along the curve will be removed when removing other kinds of non-shape information from the coordinates. This is the major distinction between landmarks and semilandmarks – for landmarks, their position (in all directions) contains information about shape. In contrast, for semilandmarks, their position in the direction of the curve contains no information about shape. Once that non-shape information is removed, semilandmarks are treated just like landmarks, meaning that they are also matched up one-to-one when computing the difference between shapes. For that reason, semilandmarks are treated as if they are homologous, and one criticism of semilandmarks is that decisions about their homology are not based on comparative anatomy, as in the case of landmarks, but rather are delegated to algorithms (Klingenberg, 2008). However, that contrast between landmarks and semilandmarks may be framed too starkly because comparative anatomy is often not enough to establish the homology of landmarks. For example, two points often used in studies of mandibular shape are the superior and inferior "angular notches" (Figure 2.3A). These notches are inflection points along curves whose curvature can vary considerably, even within species (Figure 2.3B,C) and more so across disparate species (Figure 2.3B,D,E). The notches correspond as inflection points on the curves but it is not clear where the anatomically corresponding points are along those curves. Comparative anatomy may suffice to establish the homology of the notches, but it does not specify where, precisely, a homologous point can be found within the notch.

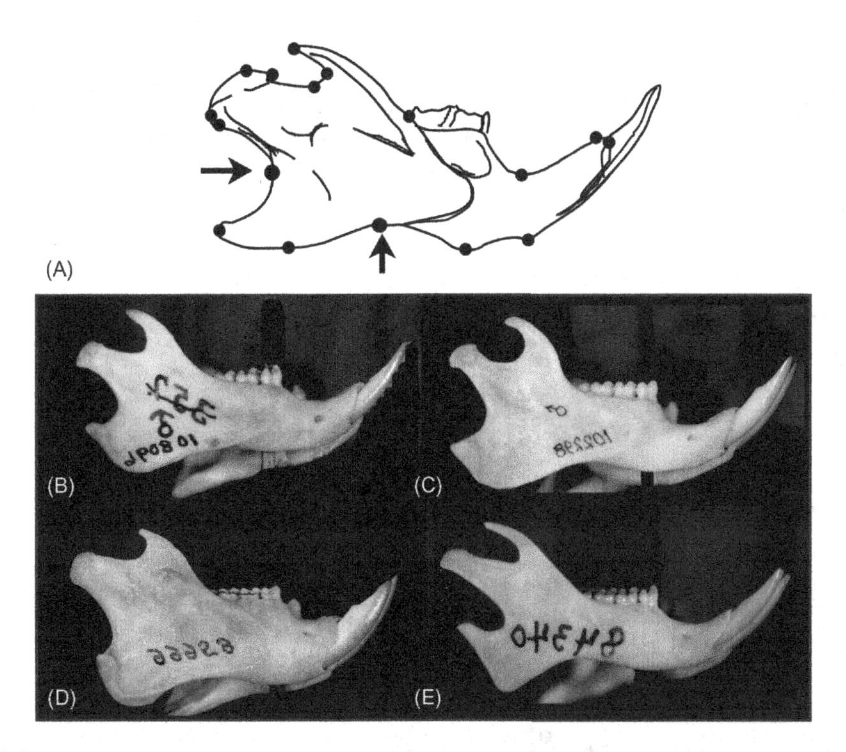

FIGURE 2.3 Angular notches. (A) Angular notches shown on the mouse mandible; (B and C) mandibles of two conspecific ground squirrels, *Spermophilus beecheyi*; (D) mandible of another ground squirrel, *Ammospermophilus leucurus*; (E) mandible of a tree squirrel, *Sciurus nayaritensis*.

Adequate Coverage of the Form

A second important criterion, arguably as important as homology, is adequate coverage of the form or, as Roth (1993) put it, comprehensive coverage. The importance of this criterion should be self-evident because we cannot detect changes in shape without data, and the landmarks are the data. Additionally, we cannot find changes within particular regions unless we have landmarks within them. One way to decide if you have met this criterion is to draw a picture of the landmarks without tracing the rest of the organism or, as we did above, removing the photograph and looking only at the configuration of the landmarks. Another example is a squirrel scapula, one of the examples discussed later in this chapter. Whereas in Figure 2.4A the form of the scapula is present, even if the outline of the structure is erased, in Figure 2.4B, it is virtually impossible to tell that the structure is a scapula. Given the landmarks shown in Figure 2.4B, we cannot tell what is happening between the peripheral points (meaning those on the outline). Therefore, if there are any localized changes in scapula shape, we will not find them.

Sometimes we cannot find landmarks where we need them. For example, there are no discrete points just anterior and posterior to the eye of a piranha (another of the examples discussed below). We need that information to analyze changes in relative size of the eye,

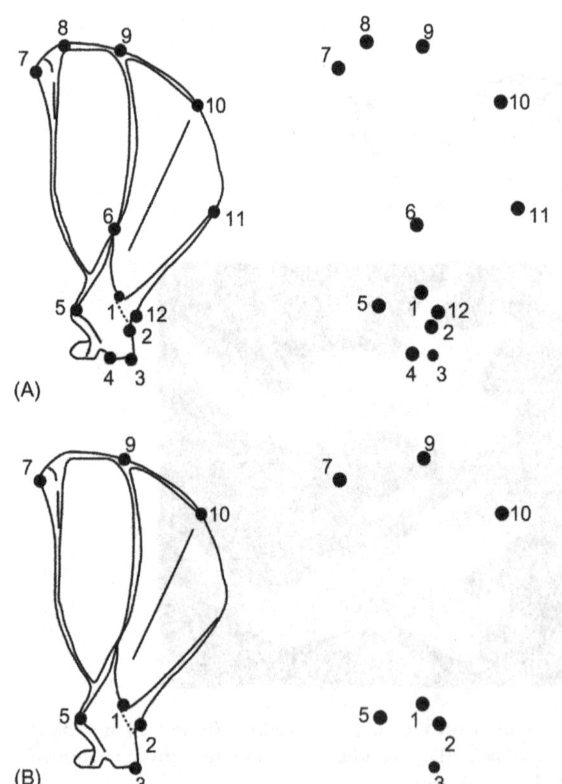

FIGURE 2.4 Landmark schemes for a scapula. (A) The form of the scapula is evident in the landmarks; (B) present even if the outline of the structure is erased; in (B) the form of the scapula is barely detectable in this configuration of landmarks.

which is one of the most visually obvious ontogenetic changes in shape. Because we do not want to sacrifice that information, we place points marking the anterior and posterior boundaries of the structure. In effect, to obtain that information, we sample the same geometric points in all specimens. In other cases, we cannot sample complex curves that lack landmarks, as exemplified by the jaw (see Figure 2.1C). Thus, to obtain adequate coverage of the form, we sample those curves with semilandmarks (see Figure 2.1D).

Repeatability

The third criterion for selecting landmarks is that they can be found reliably. If they are difficult to locate even on the same specimen measured multiple times, they can induce measurement error. Sometimes, these difficult points are easy to recognize even before digitizing them; for example, the landmarks on the mandibular notches mentioned above (see Figure 2.3). These landmarks are difficult to locate reliably for two reasons: first, the landmarks are defined in terms of a change in curvature along curves that change curvature more than once, so finding the landmarks requires assessing curvature by eye, and recognizing which particular changes in curvature are the ones to digitize. Second, the curvature (and where it changes) varies among individuals of the same species and sometimes even between the two sides of the same individual. In the case of landmarks such as these, the definition of the landmark may migrate over the course of digitizing many specimens – the precise point that is recognized as the right change in curvature may differ between the first and hundredth specimen digitized. Sometimes, the reliability of a landmark is not so easily anticipated in advance of measurement. For example, some points seem as though they ought to be difficult to find repeatedly, such as the anterior and posterior points on the piranha eye, but they may actually be less prone to error than points that are more discrete and well defined. Also, points that seem very fuzzy (such as blurs on x-rays) can sometimes be more reliable than you might imagine. In the case of landmarks that are obviously difficult to locate, such as those points on the mandibular notches, it is important to check and recheck your digitizing, going back to the beginning of the file and scrolling through to see if the definition of the landmark migrated. But for others, it may be best to avoid prejudging them, and check their repeatability empirically.

Some landmarks are prone to measurement error in only one dimension because the landmark is easy to locate along an axis, e.g. the anteroposterior axis, but difficult to locate along another. This ambiguity along one direction can be a real problem for points that might otherwise be well defined, such as those on a suture. Sutures that generally follow a body axis sometimes wander, taking a complex path. It may be easy to pin down the anteroposterior location of a point along the suture, but more difficult to decide its mediolateral position. When a landmark is difficult to find in only one direction, the error will be concentrated in that direction, inducing biased rather than random error. Biased error is a more serious problem than a large random error because biased error will look like something that merits an explanation. However, the difficulty that you perceive in the course of digitizing may not be reflected in the actual variability of the point. At the outset of the analysis, before deciding that a point is unrepeatable in one or both directions, digitize it and then check its error. You can always delete it if you find that the error is biased.

In some cases, it is not that the landmarks themselves are difficult to measure but rather that they are subject to other sources of error, some of which are biased. One potentially important source of measurement error is preservational artifact. One study of preservational artifacts affecting the variation of landmarks of the trilobite cephalon found that compaction of shale-preserved specimens tripled the scatter of the landmarks around their mean positions, but not randomly (Webster and Hughes, 1999). The more lateral landmarks were most affected, as were the most convex forms. The technology used to capture images and also to record the data can be important sources of measurement error. Even very subtle variation in the orientation of a three-dimensional specimen can have visible effects on the shape when it is projected onto the photographic plane. The distance between the specimen and the camera can also be a source of measurement error; the projection of three-dimensional objects onto a plane may distort points at different perpendicular (normal) distances from the plane, especially when the camera is too close to the specimen (Mullin and Taylor, 2002). Parallax is always a concern when three-dimensional specimens are projected onto a plane but three-dimensional measurement systems (including computer tomography, laser scanning systems and articulating arm digitizers) can also introduce error. One study of repeated measurements of a single specimen (mounted in fixed position) found that the Polhemus 3Space digitizer produced the same errors along all three axes, but the authors note that other approaches, such as computer tomography, might differ in their resolution along the Z axis because that is the axis on which the slices of the scan are stacked (Corner et al., 1992).

At the outset of any study, as part of the process of selecting landmarks, it is a good idea to capture multiple images of the same specimen, repositioning it between each session, and to measure each image multiple times so that you can assess the error introduced by the image capture and measurement process. Because the magnitude and distribution of measurement error may be related to the size and/or shape of the specimen, it is useful to examine several specimens that span the variety of sizes and shapes that you will be analyzing. If you have more than one instrument available to use, you can also check whether they differ in the magnitude or distribution of measurement errors. We will return to the analysis of measurement error, more specifically, of how its various sources can be decomposed by a multivariate analysis of variance (MANOVA) in Chapter 9. If your preliminary assessments of measurement error suggest that variation due to measurement error is large relative to the variation of biological interest, we recommend imaging and measuring each individual several times so that you can use the average value as your data.

Consistency of Relative Position

The issue, in this case, is landmarks that switch their position relative to each other. This can occur even when changes in shape appear to be very modest, such as when a foramen near a suture is sometimes anterior and sometimes posterior to it. These two landmarks may thus move past each other. From the anatomy, it may not look like there is much variation in shape because the bones do not differ in their proportions, what

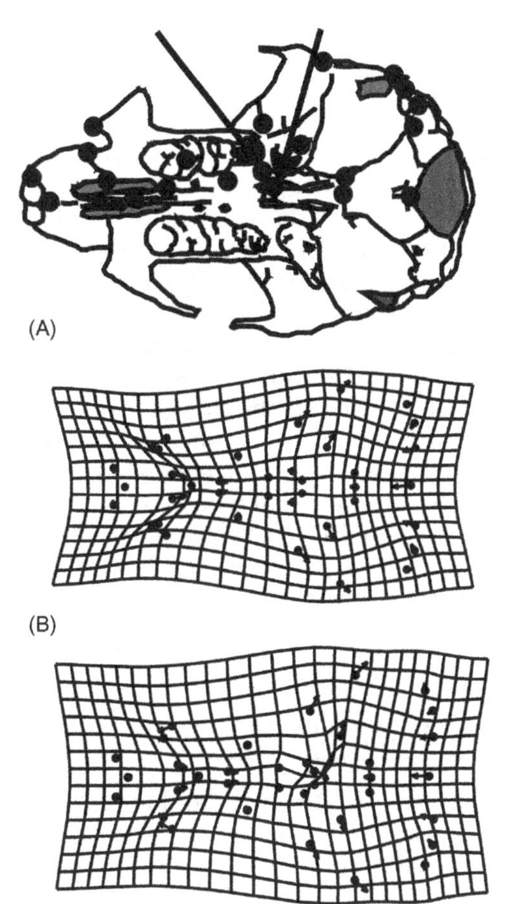

FIGURE 2.5 Landmarks that switch positions. (A) The landmarks that switch position; (B) the third principal component, in the absence of any switch in position; (C) the third principal component when the landmarks are switched.

differs is only those two points that switch their positions. Figure 2.5 shows a manufactured example, highlighting two landmarks that switch their position (Figure 2.5A). In this case, the coordinates of the landmarks for four of the 28 specimens were simply exchanged in the data matrix. The effect is fairly subtle; instead of the component that accounts for approximately 13% of the variation in the actual sample (Figure 2.5B), there is another that also accounts for approximately 13% of the variance (Figure 2.5C). The effects, however, can be far larger, just like digitizing points out of order can produce dramatic outliers.

Landmarks that switch position may seem to be a rather trivial problem, one that arises rarely. Yet, when densely sampling a landmark-rich structure, it can become a real issue, especially when many of the landmarks are located on sutures and foramina. You may find this kind of variation interesting, but if not, you have two choices. The first is to exclude one of the two landmarks from the analysis, and the other is to exclude some of the individuals from the analysis.

Coplanarity of Landmarks

This is the sole criterion that is specific to two-dimensional data, and it arises from the distortion caused by projecting a three-dimensional organism into a two-dimensional plane. To minimize this distortion, specimens must be consistently oriented under the camera, and one particular plane must be chosen for that orientation. Points not in that plane may be inconsistently oriented or difficult to interpret. The two-dimensional analysis will suggest that the points have moved within the plane of photography, but it is possible that they actually have moved toward or away from that plane. What you will see is the projection of a change that is actually in the third dimension onto the plane of the photograph. Foreshortening can look like shortening. The distortions resulting from projection can be a serious problem, as it turned out to be in the analysis of cotton rat (*Sigmodon fulviventer*) skull ontogeny (Zelditch et al., 1992). One characteristic feature of mammalian skull ontogeny is the change in orientation of the skull base: points that initially are on the posterior end of the ventral surface move dorsally, out of the picture plane. Some points could not be included in the data set because they were not visible at all ages (in consistently oriented photographs) but worse, other points apparently on the lateral boundary of the skull (in the photograph) are actually on the lateral *surface* of the skull. It was not possible to tell if they moved in the anteroposterior and mediolateral directions (the plane of photography) or if they instead moved dorsoventrally. In hindsight, those lateral points should have been excluded as too ambiguous.

BOOKSTEIN'S TYPOLOGY OF LANDMARKS

Bookstein (1991) introduced an influential classification of landmarks into three categories: Type 1, Type 2 and Type 3 (see Roth, 1993, for another discussion of these types). According to this scheme, Type 1 landmarks are optimal, Type 2 are more problematic and Type 3 might not even be considered landmarks at all. The classification is based on two interrelated considerations: one is that landmarks ought to be locally defined, the other is the type of explanation into which they can enter. The first consideration is relatively easy to summarize because it is a matter of the degree to which landmarks are locally defined. The second consideration, however, is more difficult to summarize because it depends on a classification of explanations.

Bookstein categorizes Type 1 landmarks as points at discrete juxtapositions of tissues, which need not be juxtapositions of different tissue types – by his usage, the juxtaposition of three bones is a juxtaposition of tissues, and a foramen is also a Type 1 landmark, although it is not a juxtaposition of observed tissues so much as the consequence of the passage of neural or vasculature tissue through the bone. Type 1 landmarks may be more clearly distinguished as points whose definition refers solely to structures close to the point. For example, the intersection between three bony sutures is locally defined, as is a foramen. For Type 1 landmarks you do not need to mention any structures far away from that point. They are surrounded by tissue on all sides. At the other extreme are Type 3 landmarks. The definition of these points depends on structures far removed from the landmark, and meaningful variation is usually limited to a single direction, the one stipulated by the definition. Type 3 landmarks are often constructed geometrically. For example, Figure 2.6 shows a classic measurement scheme for

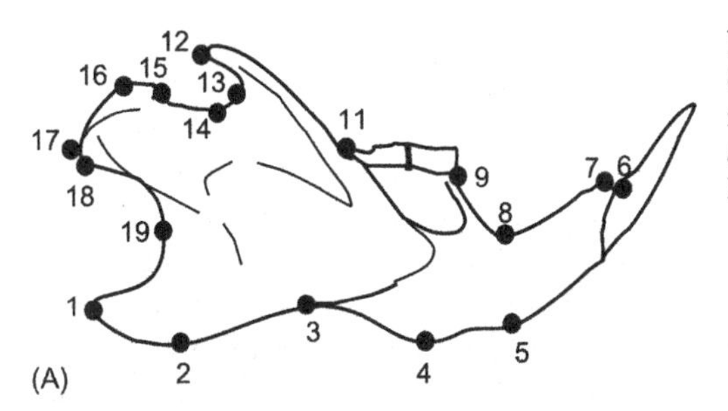

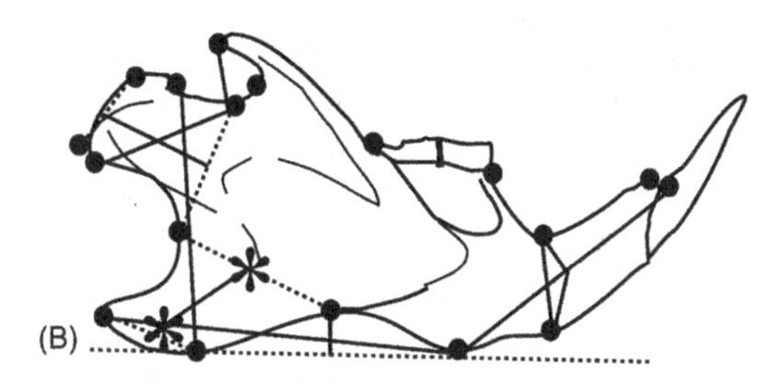

FIGURE 2.6 Landmarks on a mouse mandible. (A) Schematic showing the locations of landmarks; (B) two measurements defined by Type 3 landmarks (represented as asterisks). (After Atchley et al., 1985. Genetics of mandibular form in the mouse.)

the mouse mandible (Atchley et al., 1985). One of the measurements is the length of the angular process, which is the Euclidean distance of a line segment that extends from the midpoint of a line extending from landmarks 1 and 2, to the midpoint of a line extending from landmarks 3 to 19. These midpoints, shown as asterisks in Figure 2.6B are Type 3 landmarks. Another measurement is of the incisive process, which is defined as the shortest distance from landmark 8 to a line extending from landmarks 4 to 6. The point at which those two lines intersect is also a Type 3 landmark. Type 2 landmarks are intermediate between these extremes. They are locally defined, but not as locally as Type 1 landmarks, and they often refer to geometric constructs, but these constructs are the tip of a structure, a bulge, or local maxima or minima of curvature. All but two of the landmarks shown on the mandible are Type 2 (the exceptions are the landmark at the emergence of the incisor from the alveolar bone, and the point on the intersection between the posterior of the articulating surface of the condyle and the condylar process, although both of these could be regarded as Type 2 landmarks because their position depends, at least in part, on the orientation of the specimen.

As well as ranking these types according to the localization of their definining criteria, Bookstein (1991, pp. 64–66) also distinguishes them according to their explanatory role. In his view, Type 1 landmarks can enter into familiar valid functional explanations, more specifically, the accounts of a deformation, such as conservation or optimization of biomechanical strength or stiffness under loading, or conservation of enclosing structures under changes in their content. Type 1 landmarks allow you to identify directions of forces that impinge on a structure, or to recognize the effects of processes moving the landmarks, being surrounded by tissue on all sides. Type 2 landmarks lack information from surrounding tissues in at least one direction. These are often the points at which forces are applied but, in the absence of information from at least one direction, it is not possible to distinguish displacements lateral to the boundary direction from a combination of inward and outward displacements. Type 3 may not be meaningful as landmarks and their displacements are meaningful in only one direction (along the length of the defining segment). For example, a Type 3 landmark could be defined as the point furthest away from some structure (e.g. the hypoglossal foramen) in the direction parallel to the skull midline axis. That point has only one real coordinate (the other is fixed by the definition of the landmark) so the only meaningful displacements of the point are in the direction of the one free coordinate. Semilandmarks are similarly deficient, having only one meaningful direction of change because any variation in their position *along* the curve is arbitrary.

This classification discourages the use of both Type 2 landmarks and semilandmarks but, in our experience, Type 2 landmarks are at least as valuable as Type 1 for developmental and biomechanical studies. Landmarks along the boundary of a growing structure, which are often Type 2, may even provide most of the information available about developmental processes when growth occurs by deposition at the margins (i.e. by processes such as periosteal growth). Type 1 landmarks that are surrounded by tissue get trapped in the growing extracellular matrix and may therefore be nearly invariant in their position – their displacements are duc to growth happening elsewhere. Landmarks along the periphery are the ones that record the changes in proportion caused by bone deposition. Biomechanical studies similarly benefit from information about the locations where the forces are applied. The explanations supported by Type 2 landmarks may not be as elegant as accounts of deformations, but to serve a useful explanatory role, landmarks should provide a rich description of the effects of a process. They should covary with the process and inform us about its effects on shape. That Type 2 landmarks and semilandmarks serve that purpose should be evident in Figure 2.7, which

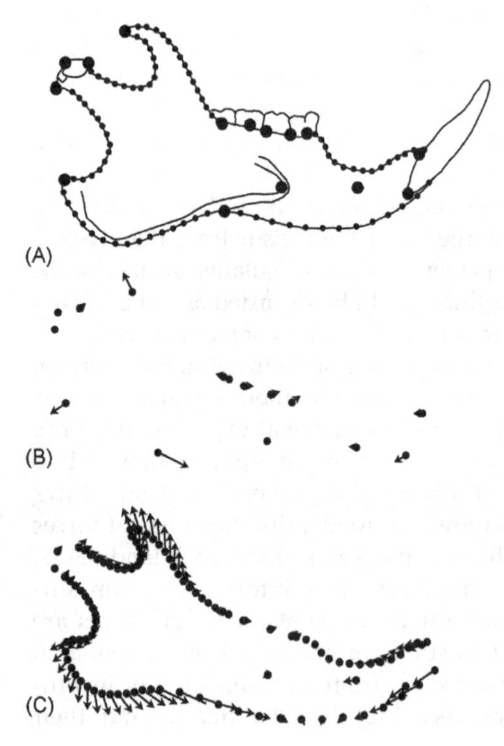

FIGURE 2.7 Ontogenetic shape change of the mandible of a ground squirrel, *Spermophilus beecheyi*. (A) Landmark and semilandmark measurement scheme (the larger dots are the landmarks); (B) ontogenetic change described by landmarks alone; (C) ontogenetic change described by landmarks, supplemented by semilandmarks.

shows the latter part of postnatal ontogeny of a ground squirrel mandible. The few Type 1 landmarks are the mental foramen and juxtapositions between teeth and bone (Figure 2.7A). These are nearly invariant over ontogeny (relative to each other and to the centroid of the form). The changes in proportion are evident in the regions sampled by Type 2 landmarks (Figure 2.7B) and a far richer description is obtained when that sparse set of landmarks is supplemented by semilandmarks; there are dramatic changes in the proportions and orientations of the angular and coronoid process (Figure 2.7C).

EXAMPLES: APPLYING IDEALS TO ACTUAL CASES

Having discussed some general principles and theory, we now turn to specific and concrete examples of data. They display an obvious vertebrate bias, especially a bias towards mammal skeletons. That is because these examples are taken from our own work rather than from a review of the literature. We focus on our own examples for two reasons. First, we can explain our own reasoning. Second, because we have the data for these examples, we can use that data to demonstrate methods throughout this book.

Landmarks on the Lateral Surface of the Squirrel Scapula

Figure 2.8 shows the major anatomical features of a tree squirrel scapula. Also shown are 12 landmarks that were digitized in a study of changes in scapula shape associated with the evolution of burrowing in chipmunks and ground squirrels (Swiderski, 1993). Studies of scapulae of other mammals have found important changes in the blade, acromion and metacromion associated with functional shifts (Oxnard, 1968; Taylor, 1974; Stein, 1981). These same studies found little or no change in the coracoid process and the bell-shaped structure that articulates with the humerus (hidden behind the metacromion

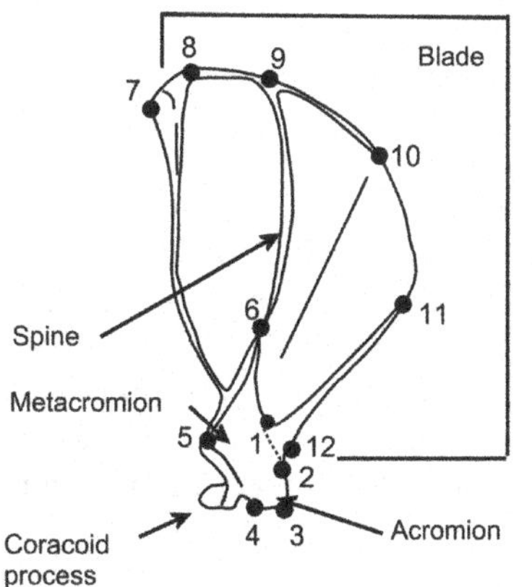

FIGURE 2.8 The major anatomical features of a tree squirrel scapula, shown are the 12 landmarks analyzed in a study of the evolution of burrowing in chipmunks and ground squirrels (Swiderski, 1993).

in Figure 2.8). A preliminary survey of squirrel scapulae indicated that they may have a similar anatomical distribution of changes. This pattern dictated that the squirrel scapulae should be digitized from the lateral view, because this is the only view in which the blade, acromion and metacromion could be seen in all taxa. Fortunately, the one feature of the bell that was considered potentially relevant to a functional analysis was also visible in the lateral view. That feature, the "neck" between the blade and the bell, is expected to change in thickness to reflect the magnitude of the forces transmitted to the scapula from the humerus. Thus, before any decisions were made about inclusion of specific landmarks, functional considerations were used to decide which general aspects of scapula shape would be analyzed.

The anticipated importance of changes in the acromion and metacromion meant that concerns about the distortion of three-dimensional aspects of shape could not be ignored, and also that landmarks could not be deleted if the distortion was expected to be large. Instead, concerns about distortion were addressed by standardizing the protocol used to capture the images that were digitized. As is usual for morphometric analyses based on photographs or video images, the scapula was placed in a standard orientation so that differences in orientation would not be interpreted as differences in shape. In addition, the distance of the camera lens from the scapula was adjusted for each specimen so that the blade always occupied the same proportion of the field. Then, if the height of the spine and sizes of the acromion and metacromion were proportional to the size of the blade, the acromion and metacromion would also occupy a constant proportion of the field. More importantly, the pattern of landmark displacement that would occur if these proportions changed could be predicted and tests for these patterns could be performed. No evidence of such patterns was found in the data.

After deciding which view to digitize, a major concern was coverage: finding enough landmarks to represent adequately the shape of the scapula. Structurally, the scapula is rather simple, which means there are few points that can be uniquely defined. This is especially true of the main portion of the scapula, the semicircular or triangular "blade"; the blade is nearly flat and has only two ridges crossing it – the large scapular spine on the lateral surface, and the smaller subscapular ridge on the medial surface. The margin of the blade is also rather featureless, having few corners and no spines, only more ridges or thickenings.

Despite the shortage of potential landmarks, it was still considered important to define them so that they could reasonably be considered homologous. For example, the ends of ridges may seem to be good landmarks, but quite often these are gently tapered, making it difficult to define precisely where they end. Usually, when a ridge ends abruptly, it ends at an intersection with some other structure. On the scapula blade, landmarks 8, 9 and 10 are points where two ridges intersect. Landmark 6, on the metacromion, is another intersection, marking the attachment of the metacromion to the spine. Landmarks 7 and 11 are points on the margin of the blade where the end of a marginal ridge is associated with a corner. Landmark 5, on the metacromion, is another corner associated with the end of a marginal ridge. Landmark 1 is one of the few places on the blade where a ridge (the scapular spine) ends abruptly without intersecting another structure.

Concern for homology extended to the corners as well as the ends of the ridges. Landmarks 2, 3 and 4 are at the only corners that are not associated with the ends of ridges. Other anatomical information was used to infer their homology. Landmarks 2 and 3 are corners where the acromion terminates in a flat surface that articulates with the clavicle. The corner labeled as landmark 4 appears to mark the boundary between the acromion and metacromion. This interpretation is reinforced by the point's proximity to the line of the scapular spine, which separates anterior and posterior components of both the scapula and the attached muscles.

The grounds for inferring homology are weakest for landmark 12. This is the only point on the articulating structure, the "bell", that could be seen in lateral view in all taxa. If more points on this structure were visible, landmark 12 might not have been used. This point is identified only as the cranial edge of the neck, which is the narrowest region between the blade and the bell of the articulating structure. This criterion for recognizing a landmark is harder to apply than the criteria for recognizing the other 11 landmarks because the boundary between bell and blade is not marked by a corner or other distinctive feature. In this regard, the neck of the scapula may seem similar to the least interorbital width of the skull, as being poorly defined and of doubtful homology. However, unlike least interorbital width, the neck of the scapula marks the boundary of two functionally distinct components of the scapula. In addition, analysis of digitizing error indicated that this point was not substantially harder to locate than other landmarks. Therefore, doubts about the homology of this point were set aside in favor of having at least one landmark on this structure.

Landmarks on the External Body of Piranhas

Figure 2.9 illustrates the landmarks used in several studies of shape change in piranhas. These points were originally intended for analyses of shape by trusses (see Strauss and Bookstein, 1982), so they were chosen to allow for constructing a series of boxes and diagonals over the form. In addition, because the truss analysis was to be compared

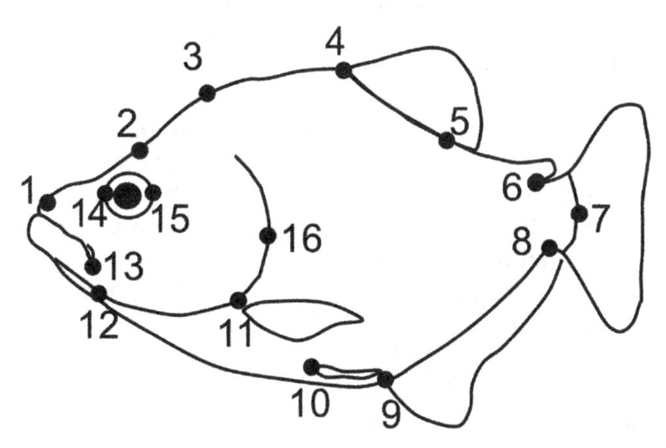

FIGURE 2.9 Landmarks on the external body of a piranha, used in several studies of ontogeny and disparity of piranhas (e.g. Fink and Zelditch, 1995, 1996; Zelditch et al., 2000, 2003a).

to more traditional measurement schemes in ichthyology, some landmarks were chosen to allow duplication of those measures. Traditional measurements between these landmarks were used in a systematic study of *Pygocentrus* (Fink, 1993), and in several geometric morphometric studies of the evolution of piranha ontogeny and the diversification of their body forms (e.g. Zelditch et al., 2000, 2003a).

Selecting landmarks on the lateral body of piranhas is relatively straightforward because specimens are essentially two-dimensional. Most of the shape variation can be seen in that view, and little distortion is caused by viewing the animal in a plane. Specimen bending can occur at fixation or during preservation, and such bent specimens were not included in analyses unless they could be manually straightened with no resulting distortion in the lateral body shape. Data acquisition consisted of placing the specimen in a standard view, using a specially designed container that kept the animal's midline in the plane defined by the top edges of the container. A piece of metric graph paper was placed on the container's edge in the same plane for calculating size. In some cases, insect pins of various sizes were used to make landmarks more visible. The camera was placed so that each specimen occupied approximately the same area in the viewing field, in order to minimize distortion.

There are few landmarks on the post-cranial lateral body of piranhas, and almost all landmarks chosen are from around the perimeter of the body. Had the data been taken from radiographs, some internal osteological landmarks could have been used. However, it was decided that data would be taken from entire specimens, partly to facilitate application to identification keys. Most of the landmarks chosen are at boundaries or extremes of structures, or are skeletal features accessible without x-rays.

Landmark 1 represents the anterior point of the head, and is taken where the two premaxillary bones articulate at the midline. Because this point is directly on a vertical from the plane of the specimen, no special marking is required. The landmark involves soft tissues, and thus could be affected by desiccation of the specimen.

Landmarks 2, 3, 7 and 12–16 all represent skeletal features, representing extremal points, intersections of structures, or borders of bones. Landmark 2 is the anterior border of the epiphyseal bar – a small extension of bone that spans a large fossa in the dorsal neurocranium – and was chosen to provide information on the shape of the head. The landmark is found by inserting a pin through the skin of the midline dorsal to the orbital region, where the pin just penetrates past the bar into the brain cavity. Although this landmark is constantly available in piranhas, some related fishes show ontogenetic change in the width of the bar, such that the bone grows anteriorly as the fish grows, independent of head shape changes. Landmark 3 lies at the posterior tip of the supra-occipital bone of the neurocranium. It lies just under the skin at the dorsal midline, and is found by moving a fingernail along the midline until the junction between bone and muscle is found. A pin is inserted at that point for purposes of digitizing. Landmark 7 represents the posterior termination of the hypural bones of the caudal skeleton, traditionally a point used in the calculation of standard length (tip of snout to base of caudal fin). In piranhas, there is a concavity in the hypural bones at the lateral midline such that the bone lies anterior to the rest of the posterior border of the caudal skeleton, so the actual point measured is where the bone would be in other teleosts. This is less problematic than it might seem, since the actual measurement is done at the area where the caudal fin base can be bent laterally.

Until some experience in finding this landmark is gained, it may be difficult to be consistent in reproducing this point. An inexperienced person usually has error in the anteroposterior axis. This is a landmark for which some argument regarding homology must be made. This is because the internal skeleton may not be consistent with the point used externally. However, consistently measured as the posterior termination of the body at the lateral midline, the point may be considered homologous.

Landmark 12 represents the ventral side of the articulation between the quadrate bone and the mandible. It thus lies lateral to the midline, although it usually lies on a vertical from the ventral midline. This point is located by placing a fingernail in the joint between the two bones, and then a pin is inserted in the joint. Landmark 13 lies at the intersection of the maxillary bone and the infraorbital bone that defines the "cheek" area of the face. The point lies well lateral to the midline, but marks an important area of the skull, approximating the length of the upper jaw. This point is marked by slipping the pin under the infraorbital bone adjacent to the posterior border of the maxillary. This landmark is composed of an extreme point (the posterior maxillary border) as well as an intersection of two structures. The homology of this landmark may be questioned. Landmarks 14 and 15 capture the width of the bony orbit. Each point lies at the extreme of the orbit along the anteroposterior body axis. Both of these landmarks are of questionable homology, but are taken because the eye has been shown to be highly allometric and has been used in traditional measurement schemes. With practice, these landmarks can be taken with little error.

Landmark 16 is perhaps the most difficult to justify in this analysis. It occupies the most posterior point of the bony opercle, the bone that forms the bulk of the gill cover, and its original purpose was to duplicate the landmark used in traditional ichthyology measurements of head length. This landmark was expected to be an articulation point between the opercle and subopercle bones. However, comparisons among several species showed that the position of the articulation varied excessively, and inaccurately represented the posterior of the head. The landmark now taken is simply the extreme along the bone border as measured from the tip of the snout. No reasonable homology argument can be made for this landmark; it may be that it is partially redundant with landmark 11. However, our analyses have shown that this landmark can be consistently digitized and is often informative about alterations in head shape.

Landmarks 4–6 and 8–11 represent points where the fins insert on the body, at the anterior or posterior of the fin base. In most cases, these points are measured where the bony fin ray intersects the body. Together, these landmarks provide a great deal of information on post-cranial body shape. Landmarks 4 and 5 lie at the anterior and posterior of the dorsal fin base, respectively. Ontogenetic variation in anterior fin ray morphology can reduce the repeatability of landmark 4, as discussed in Fink (1993).

Landmarks 8 and 9 represent the posterior and anterior of the anal fin base. Often the fin is collapsed, and a pin must the inserted to make landmark 8 visible. In some piranhas, there are accessory spines at the anterior of the fin base, and they are not included. Landmarks 10 and 11 represent the insertion onto the underlying skeletal girdles of the pelvic and pectoral fins, respectively. Both lie dorsolateral to the ventral midline. Landmark 10 is easily visible in larger specimens but, in some smaller specimens, the transparency of the fin makes it difficult to find; in this case it can be located by raising the fin laterally and placing a pin at the anterior fin-ray's base.

Landmark 6 lies at the posterior base of the fleshy adipose fin, where the fin meets the skin of the dorsal midline. This point may be difficult to locate unambiguously because it may be obscured by the fin overlapping the skin of the peduncle, so a pin is inserted to mark its location for digitizing. In some of our studies we have attempted to use the anterior insertion of the adipose fin, but its broadly curving profile in many species renders it too difficult to repeat.

Note that landmarks 9–12 represent the ventral area of the body form, but they do not capture the actual convex belly shape of these fishes. A great deal of effort was spent in attempting to find appropriate landmark locations along the ventral profile, but no repeatable and consistent landmarks were found that could be located on all piranha species.

Landmarks on the Skull of *Sigmodon fulviventer* and *Mus musculus domesticus*

The landmarks on the skull of cotton rats, *Sigmodon fulviventer* (Figure 2.10), were selected to cover the skull as evenly as possible for the purpose of determining whether

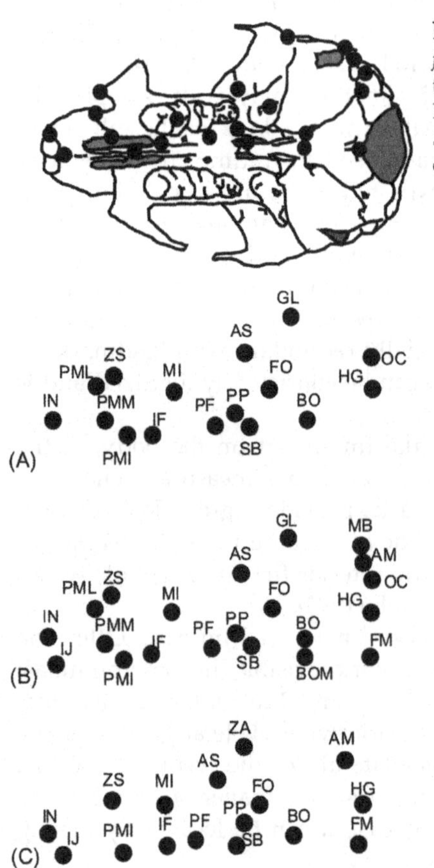

FIGURE 2.10 Landmarks on the skull of cotton rats, *Sigmodon fulviventer* and *Mus musculus domesticus*. (A) Landmarks used in the initial analysis of *S. fulviventer* (Zelditch et al., 1992, 1993; Figure 2.11A); (B) landmarks added for a comparative study of *S. fulviventer* and *M. m. domesticus*; (C) landmarks used in the analysis of *M. m. domesticus*.

ontogenetic changes in skull form are spatially integrated or localized (Zelditch et al., 1992) and to study developmental constraints on variability in that species (Zelditch et al., 1993). Because the studies were designed to analyze the ontogeny of skull shape and its variation, and the data come from photographs, the only landmarks that could be included are the ones that are visible in (approximately) the same plane at all ontogenetic stages. Mammalian skulls are highly three-dimensional structures, and the cranial base rotates during ontogeny, so landmarks that are parallel to the camera at one stage may rotate out of that plane later. This produces what appears to be a change in shape (within the plane). However, omitting *all* the landmarks that might be affected by such a rotation would mean losing vital information about cranial length and width. The landmarks most strongly affected by the extension of the cranial base are the ones marking the juncture between the anterior and posterior cranial base, as well as those located on the posterolateral braincase. Consequently, landmarks were placed on those locations even though that complicates distinguishing between changes in shape caused by differential growth and apparent changes in shape due to rotation in the third dimension.

A subsequent study was undertaken to compare skull shape ontogeny of *S. fulviventer* to that of another rodent, the house mouse *Mus musculus domesticus*. A major objective of that study was to examine the relationship between life-history strategy and timing of skull morphogenesis (Zelditch et al., 2003b). Ideally, we would have sampled both skulls densely, selecting homologous landmarks that provide a richly detailed description of the ontogeny of both species. However, some landmarks could be seen in only one species or another. For example, in *S. fulviventer* we can locate a landmark on the posterior of the glenoid fossa, but the curve of the glenoid is so smooth in *M. m. domesticus* that we cannot find a distinct point anywhere comparable to the glenoid landmark of *S. fulviventer*. To capture information about skull width in the region of the zygomatic arch of *M. m. domesticus*, a different point had to be chosen, complicating the comparative analysis. Several other points that are readily visible in *S. fulviventer* also cannot be found in *M. m. domesticus*. However, the problem posed by the inability to find landmarks in *M. m. domesticus* that are homologous with those already measured in *S. fulviventer* is partly mitigated because there are landmarks in *S. fulviventer* that had not been previously sampled, but which can be recognized in both species. Thus, in the comparative study, additional landmarks were sampled on *S. fulviventer*. Even so, the set of landmarks common to both species comprises a rather sparse sample of each skull. Therefore, analyses were done separately for each species, using the landmarks providing the densest coverage possible for each species, and the comparative analyses exploited the subset of landmarks common to both.

The original analyses of *S. fulviventer* (Zelditch et al., 1992, 1993; see Figure 2.10A) include 16 landmarks. Landmark 1 is the lateral margin of the incisive alveolus where it intersects the outline of the skull in the photographic plane (IN). Landmark 2 is the anteriormost point on the zygomatic spine (ZS). Landmark 3 is the premaxilla–maxilla suture where it intersects the outline of the skull in the photographic plane (PML). Landmark 4 is the premaxilla–maxilla suture lateral to the incisive foramen (PMM). Landmark 5 is the posteriormost point of the incisive foramen (IF). Landmark 6 is the median mure of the first molar (M1). Landmark 7 is the posterolateral palatine pit (PP). Landmark 8 is the junction between squamosal, alisphenoid and frontal on the squamosal– alisphenoid side of the suture (AS). Landmark 9 is the midpoint along the posterior margin of the glenoid fossa (GL). Landmark 10 is the anteriormost point of the foramen ovale (FO). Landmark 11 is the most lateral point on the presphenoid–basisphenoid suture where it intersects the sphenopalatine vacuity in the photographic plane (SB). Landmark 12 is the most lateral point on the basisphenoid–basioccipital suture (BO). Landmark 13 is the hypoglossal foramen (HG). Landmark 14 is the juncture between the paraoccipital process and mastoid portion of the temporal bone (OC). Landmark 15 is where the premaxilla-maxilla suture intersects the midline (PMI). Landmark 16 is the posterior palatine foramen (PF).

Several landmarks were added to these in the later study, designed to compare *S. fulviventer* to *M. m. domesticus* (Zelditch et al., 2003b). These additional landmarks (see Figure 2.10B) include the juncture between the incisors on the premaxillary bone (IJ), the midpoint of the basisphenoid–basioccipital suture along the sagittal axis (BOM), the midpoint of foramen magnum (FM), the juncture of mastoid, squamosal and bullae (MB) and the juncture between the mastoid and the medial end of the auditory tube (AM). The landmarks of *M. m. domesticus* include a subset of the original *Sigmodon* landmarks, plus the newly added ones, and a point at the interior corner formed by the intersection of the zygomatic arch with the braincase (ZA) (see Figure 2.10C).

Three-Dimensional Landmarks on a Marmot Skull

As stressed in the previous example, the mammalian skull is obviously not a two-dimensional structure. Like many structures of interest to biologists, the marmot skull is not only three-dimensional (Figure 2.11A) but also has relatively few landmarks (Figure 2.11B). The marmot skull is strongly curved anteroposteriorly and mediolaterally, so features on the same bone may be as far apart in the dorsoventral dimension as they are in the mediolateral or anteroposterior dimensions. In addition, the skull is composed of a small number of relatively large bony plates, so points that can be used as landmarks are sparsely distributed, occurring primarily at locations where at least three bones meet.

Landmarks on a Squirrel Mandible

The landmarks on the eastern fox squirrel mandible (*Sciurus niger*, Figure 2.12) are a general scheme for analyzing mandibular development, modularity, and evolution. The rodent mandible has become one of the favorite model systems for studies of complex morphologies, especially for studies of developmental and evolutionary modularity (e.g. Atchley and Hall, 1991; Cheverud et al., 1991; Mezey et al., 2000; Ehrich et al., 2003; Klingenberg et al., 2003; Monteiro et al., 2005; Marquez, 2008; Zelditch et al., 2008, 2009; Monteiro and Nogueira, 2009; Willmore et al., 2009). The attraction of this model system lies partly in the contrast between its developmental complexity and its structural and functional integration (Atchley et al., 1985; Atchley and Hall, 1991). How that integration could be achieved developmentally, and how it facilitates and/or constrains mandibular evolution are questions long motivating studies of rodent mandibles. One of the reasons for selecting the mandible as the model system for these studies is that its development is relatively well understood, although much remains to be explained. The mandible is also interesting from a functional perspective because it is obviously crucial for feeding

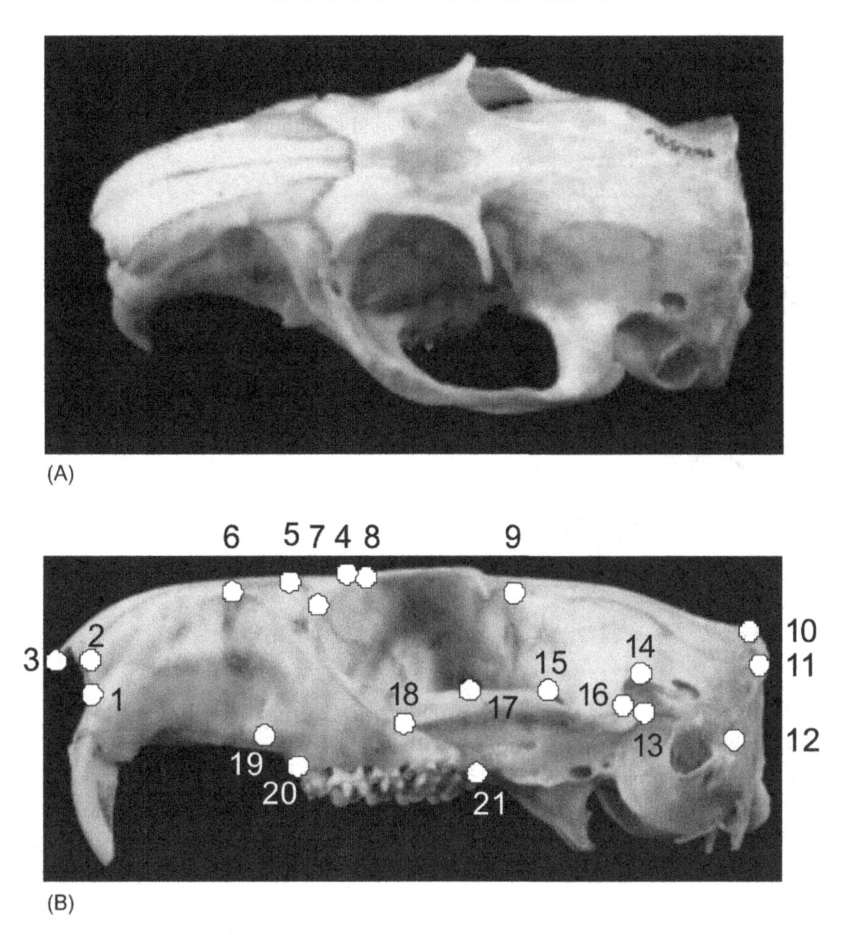

FIGURE 2.11 Landmarks for a three-dimensional analysis of marmot skull shape. (A) A marmot skull; (B) the configuration of three-dimensional landmarks.

function; motivating studies of mandibular shape are questions about the evolutionary relationships between mandibular form, function and ecology and about the discrepancies between morphological and biomechanical disparity (e.g. Velhagen and Roth, 1997; Caumul and Polly, 2005; Barrow and Macleod, 2008; Michaux et al., 2008; Perez et al., 2009; Swiderski and Zelditch, 2010).

The mandible is not only interesting from a variety of biological perspectives, it is also apparently quite simple to analyze, being a single bone that is nearly two-dimensional. However, as may already be evident from comments about mandibular landmarks earlier in this chapter, the mandible lacks landmarks where they are most needed. In the case of studies of mandibular function, the lack is most significant in the regions where muscles insert. For studies of integration and modularity, the lack is more widespread; such studies assess the covariances within hypothesized modules relative to covariance between them. Thus, every module needs to be sampled densely enough to be able to

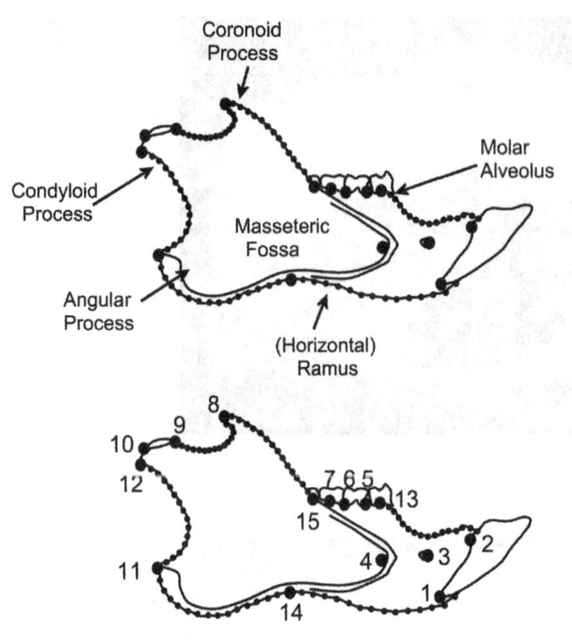

FIGURE 2.12 The major anatomical features of the rodent mandible and the landmarks used in studies of mandibular developmental, modularity and evolution.

assess the within-modular covariances. That obviously requires having more than one or two landmarks per module. A classic hypothesis of mandibular modularity posits six such modules (Atchley and Hall, 1991), but there are only 15 landmarks in the entire data set, and four of them are within one hypothesized module (the molar alveolus) and three are within another (the condyloid process). By simple arithmetic, it is obvious that the remaining eight landmarks cannot be enough to sample the other four hypothesized modules.

The landmark scheme shown in Figure 2.12 is designed to analyze modularity, both developmental and evolutionary, for dissecting the developmental origins of morphological disparity and for exploring the evolution of jaw function in relation to ecology. Because these are interrelated issues, a single scheme is needed that can be applied across disparate morphologies without sacrificing the ability to detect subtleties of intraspecific variation. This scheme has evolved; two prior studies used different landmarking schemes. The initial one was designed for a myomorph, the prairie deer mouse *Peromyscus maniculatus bairdii* (Zelditch et al., 2008) and then the analysis was extended to a sciuromorph, the eastern fox squirrel, *Sciurus niger* (Zelditch et al., 2009). Myormorphs and sciurumorphs differ in the number of cheek teeth so the scheme was modified to include the premolar, but the more consequential difference is in the position of the landmark on the ventral curve of the ramus (see Figure 2.12, landmark 14). In deer mice, there are two points where the landmark could have been located; one is located where the curve of the mandible ceases to follow the curve of the incisor and instead follows the curve of the angular process; this was the boundary between incisor and ramal modules. No landmark was placed more proximally, at the most anterior point on the angular process. The scheme devised for the sciuromorph included two ventral landmarks, one that marked that transition in curvature from incisor to ramus (which is located considerably more anteriorly in the fox squirrel than deer mouse) and the other at the most anterior angular point. Efforts to reconcile these two schemes and to apply it to other squirrels revealed the difficulty of designing one general-purpose scheme for rodent jaws. That difficulty arises from the combination of sparse landmarks and morphological disparity. That disparity makes the mandible interesting but it also makes it difficult to rely on Type 2 landmarks that are defined by inflection points along curves. Those inflection points, when visible at all, are not necessarily in corresponding anatomical regions. Thus, the scheme that we devised has only 15 landmarks and relies heavily on semilandmarks.

The first two landmarks are the points at which the incisor emerges from the alveolus, marking the upper and lower edges of the alveolar opening. Landmark 3 is the mental foramen. Landmark 4 is a point on the masseteric fossa where it changes direction – the most anterior point of the masseter muscle attachment. This point is often difficult to locate reliably because the masseteric fossa is often effaced, the change in direction is not always abrupt, and seeing it clearly depends on the depth of field and lighting, but photographing the mandible to make this point most visible would make other points difficult to visualize. Landmark 5 is on the molar alveolus between the fourth premolar (p4) and the first molar (m1), and landmarks 6 and 7 are also on the molar alveolus, 6 is between m1 and m2, and 7 is at the midpoint of m2. Landmark 8 is the "tip" of the coronoid process although, in many species, this process does not taper to a point but rather ends bluntly. The landmark is then placed at the midpoint of the blunt end. The next two landmarks are on the condyle, the part that articulates with the upper skull to form the jaw joint. Landmark 9 is on the anterior of the condyle and landmark 10 is at the posterior end of the articulating surface. Landmark 11 is at the posterior "corner" of the angular process which is the most posterior point of the superficial masseter muscle attachment. Landmark 12 is another point on the condyle, the tip of the zygomaticomaseteric fossa. This point, like others that are out of anatomical order were added after finding that the original scheme did not suffice. Landmark 13 is on the molar alveolus at the midpoint of p4. The reason for using the midpoint of the tooth rather than its most anterior point is that the alveolar bone is sometimes so heavily eroded that there is no bone at the anterior point of the tooth. Using that point would mean having to find a point on the tooth where the alveolar bone would be had it not eroded. Landmark 14, the most difficult one to locate reliably, is on the ventral margin of the angular process, where the ramus curves towards the angular process (this marks the boundary between ramus and angular process). The landmark is the most anterior point that is definitively on the angular process. Landmark 15 is the most anterior point on the base of the coronoid process, where it departs from the plane of the molar tooth row (it forms a "V" with alveolus).

Given such a sparse and uneven coverage of the jaw, and the lack of landmarks where we need them, we sampled the six complex curves as semilandmarks. In Figure 2.12, one curve may seem *too* densely sampled – the one between landmarks 8 and 9. It hardly takes 14 semilandmarks to describe the shape of that curve. But that is because the landmark scheme is shown on a tree squirrel's mandible. In other squirrels, such as the ground squirrels and chipmunk, shown in Figure 2.13, that coronoid process is far longer and the curve is much deeper. The measurement scheme is designed to capture that kind of information about jaw shape regardless of where it occurs.

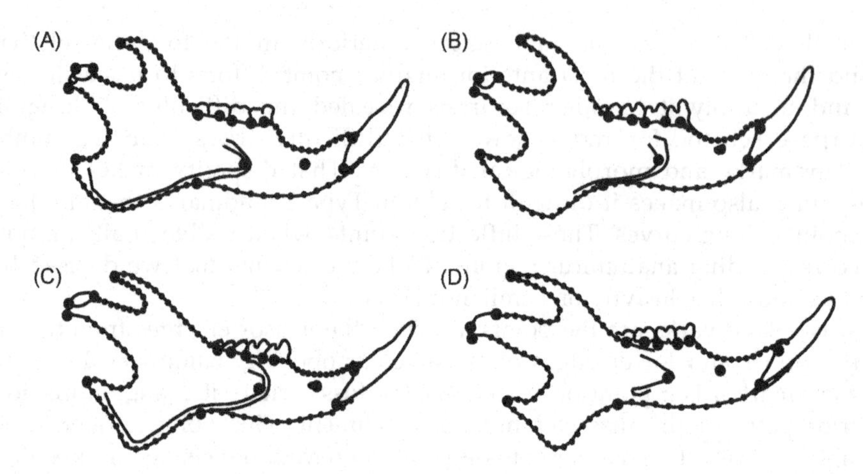

FIGURE 2.13 Landmarks and semilandmarks shown on the jaws of four squirrels. (A) *Spermophilus beecheyi*; (B) *S. lateralis*; (C) *Ammospermophilus leucurus*; (D) *Tamias alpinus*.

DESIGNING YOUR OWN MEASUREMENT SCHEME

None of our examples will provide much guidance if you are trying to find landmarks on a tadpole, a raptorial appendage of a shrimp, a fiddler crab carapace, trilobite cephalon, a flower, a tooth or an insect wing, or any of the great variety of other systems. Fortunately, there are numerous excellent examples of landmark-based studies of these and other systems in the recent literature. In fact, there are far too many to provide a comprehensive list of good examples, so we mention a few recent studies. Morphometric studies of tadpoles are difficult in light of their sparse landmarks, but thoughtful analyses have been done that look at the relationship between body form and spring speed (Arendt, 2010) and a seven-year investigation into the associations between year-to-year changes in shape and pond environments (Van Buskirk, 2009). A landmark-based analysis of functional modularity in the power-amplification system of mantis shrimp raptorial appendages tested the hypothesis that each component of that functional system constitutes a developmental module (Claverie et al., 2011). The spatial structure of geographic variation of fiddler crab carapace shape was examined to test whether geographically widespread species exhibit more intraspecific variation and morphological divergence (Hopkins and Thurman, 2010). Several studies have used landmark-based morphometrics to examine the trilobite cephalon, including its ontogeny (Kim et al., 2002; Webster, 2007, 2009) and variation along an environmental gradient (Webber and Hunda, 2007). Tooth shape has been analyzed two-dimensionally in several studies (Caumul and Polly, 2005; Wood et al., 2007; Laffont et al., 2009; Piras et al., 2010) and there are now also studies of three-dimensional tooth shape (Skinner et al., 2008; Singleton et al., 2011). Corolla shape has been the focus of recent studies exploring its genetic covariance structure in the monocarpic herb *Erysimum mediohispanicum* (Gomez et al., 2009) and the adaptive significance of its bilateral symmetry in the same species (Gomez et al., 2006). Insect wings have been the focus of several studies, including one that examined two factors that might shape dragonfly wings: long-distance migration and high-maneuverability mate guarding (Johansson et al., 2009), the impact of temperature and insertional mutations of 16 genes involved in the formation of the *Drosophila* wing on wing shape, its variance and developmental stability (Debat et al., 2009) and the modularity of insects wings (Klingenberg and Zaklan, 2000; Klingenberg et al., 2001). An especially challenging problem for landmark-based studies was confronted by Benitez-Vieyra and colleagues (2009) who analyzed sexual mimicry in orchids. The orchid's flowers attract sexually excited male wasps that pollinate them; although chemical signals play a major role in attracting the wasps, the question addressed by this study is whether plants that are shaped more like female wasps are favored. The authors thus had to find landmarks on the distal part of the orchid labellum corresponding to those on the head and most of the thorax of the wasp!

Many more examples of measurement schemes can be found by searching on "geometric morphometrics" and a keyword relevant to your study, such as "geographic variation", "ontogeny", "quantitative genetics" or "phenotypic plasticity" or the name of a study system. Reading several of them before you begin your study can help solve one of the major problems confronting a researcher, that you need to know how to measure your organisms before you can analyze your data but you may not determine what you need to measure until you finish your data collection. Fortunately, this is not as insurmountable a problem as it may appear to be even if the relevant literature is sparse. That is because the process of digitizing helps you discover where that variation is. In general, by the time that you have finished digitizing (and redigitizing) your specimens, you will not be surprised by the results. That is especially true if you use one of the nice features of the most widely-used digitizing program, tpsDig (discussed in more detail in the workbook). That feature is the "template mode" which allows you to copy the landmarks from one specimen to the next. You need only to reposition them. That ability to copy all the landmarks, in approximately the correct positions, is very useful when your landmarks are not in a sensible anatomical order; one of the common measurement errors is to digitize landmarks out of order. But the template mode is not only a convenience, it is also a way to discern what varies. Once you have positioned and scaled the copied landmarks to suit the photograph onto which they are copied, you will find that some landmarks need very little adjustment but others regularly do. As you scroll through the images, copying and adjusting the landmarks, you will soon recognize where the variation lies. You can thus design your measurement scheme by a pilot study, which begins with a preliminary set of landmarks, perhaps based on a published study, which you modify by adding or repositioning landmarks and adding curves, until you can measure the variation that you see. As should be evident from at least two of the examples above, the skull landmarks of the two rodents, cotton rat (*S. fulviventer*) and laboratory mouse (*M. m. domesticus*), as well as the mandibular landmarks of rodents, measurement schemes evolve.

A pilot study will be especially valuable if you plan to obtain three-dimensonal data directly from specimens using devices such as a Reflex Microscope or Microscribe (see the workbook for summaries of methods for capturing three-dimensonal coordinate data). Such devices offer less flexibility than other technologies that produce reconstructions of the three-dimensonal images whose coordinates can be digitized. When all that you have are the coordinates that you chose to measure before beginning the study, you obviously cannot just reload a file of images and either add or move landmarks. Instead, you need to redigitize each specimen. Consequently, you will need to know what to measure before you start. If you have ready access to specimens that span the variation that you hope to study, you can do a three-dimensonal pilot study, experimenting with landmark (and semilandmark) selection.

References

Arendt, J. (2010). Morphological correlates of sprint swimming speed in five species of spadefoot toad tadpoles: Comparison of morphometric methods. *Journal of Morphology, 271*, 1044–1052.

Atchley, W. R., & Hall, B. K. (1991). A model for development and evolution of complex morphological structures. *Biological Reviews of the Cambridge Philosophical Society, 66*(2), 101–157.

Atchley, W. R., Plummer, A. A., & Riska, B. (1985). Genetics of mandible form in the mouse. *Genetics, 111*, 555–577.

Barrow, E., & Macleod, N. (2008). Shape variation in the mole dentary (Talpidae : Mammalia). *Zoological Journal of the Linnean Society, 153*, 187–211.

Benitez-Vieyra, S., Medina, A. M., & Cocucci, A. A. (2009). Variable selection patterns on the labellum shape of *Geoblasta pennicillata*, a sexually deceptive orchid. *Journal of Evolutionary Biology, 22*, 2354–2362.

Bookstein, F. L. (1991). Morphometric tools for landmark data: Geometry and biology. Cambridge: Cambridge University Press.

Bookstein, F. L. (1996). Combining the tools of geometric morphometrics. In L. F. Marcus, M. Corti, A. Loy, G. J. P. Naylor, & D. E. Slice (Eds.), *Advances in morphometrics* (pp. 131–151). New York: Plenum.

Bookstein, F. L., Chernoff, B. L., Elder, R. L., Humphries, J. M., Jr., Smith, G. R., & Strauss, R. E. (1985). *Morphometrics in evolutionary biology*. Philadelphia: The Academy of Natural Sciences Philadelphia.

Caumul, R., & Polly, P. D. (2005). Phylogenetic and environmental components of morphological variation: Skull, mandible, and molar shape in marmots (Marmota, Rodentia). *Evolution, 59*, 2460–2472.

Cheverud, J. M., Hartman, S. E., Richtsmeier, J. T., & Atchley, W. R. (1991). A quantitative genetic analysis of localized morphology in mandibles of inbred mice using finite-element scaling analysis. *Journal of Craniofacial Genetics and Developmental Biology, 11*, 122–137.

Claverie, T., Chan, E., & Patek, S. N. (2011). Modularity and scaling in fast movements: Power amplication in mantis shrimp. *Evolution, 65*, 443–461.

Corner, B. D., Lele, S., & Richtsmeier, J. T. (1992). Measuring precision of three-dimensional landmark data. *Journal of Quantitative Anthropology, 3*, 347–359.

Debat, V., Debelle, A., & Dworkin, I. (2009). Plasticity, canalization, and developmental stability of the *Drosophila* wing: Joint effects of mutations and developmental temperature. *Evolution, 63*, 2864–2876.

Ehrich, T. H., Vaughn, T. T., Koreishi, S., Linsey, R. B., Pletscher, L. S., & Cheverud, J. M. (2003). Pleiotropic effects on mandibular morphology I. Developmental morphological integration and differential dominance. *Journal Of Experimental Zoology Part B – Molecular And Developmental Evolution, 296B*, 58–79.

Fink, W. L. (1993). Revision of the piranha genus *Pygocentrus* (Teleostei, Characiformes). *Copeia*, 665–687.

Fink, W. L., & Zelditch, M. L. (1995). Phylogenetic analysis of ontogenic shape transformations – a reassessment of the piranha genus *Pygocentrus* (Teleostei). *Systematic Biology, 44*(3), 343–360.

Fink, W. L., & Zelditch, M. L. (1996). Historical patterns of developmental integration in piranhas. *American Zoologist, 36*, 61–69.

Gomez, J. M., Abdelaziz, M., Munoz-Pajares, J., & Perfectti, F. (2009). Heritability and genetic correlation of corolla shape and size in *Erysimum mediohispanicum*. *Evolution, 63*(7), 1820–1831.

Gomez, J. M., Perfectti, F., & Camacho, J. P. M. (2006). Natural selection on *Erysimum mediohispanicum* flower shape: Insights into the evolution of zygomorphy. *American Naturalist, 168*, 531–545.

Hopkins, M. J., & Thurman, C. L. (2010). The geographic structure of morphological variation in eight species of fiddler crabs (Ocypodidae: genus Uca) from the eastern United States and Mexico. *Biological Journal of the Linnean Society, 100*, 248–270.

Johansson, F., Soderquist, M., & Bokma, F. (2009). Insect wing shape evolution: independent effects of migratory and mate guarding flight on dragonfly wings. *Biological Journal of the Linnean Society, 97*, 362–372.

Kim, K., Sheets, H. D., Haney, R. A., & Mitchell, C. E. (2002). Morphometric analysis of ontogeny and allometry of the Middle Ordovician trilobite *Triarthrus becki*. *Paleobiology, 28*(3), 364–377.
Klingenberg, C. P. (2008). Novelty and "homology-free" morphometrics: What's in a name? *Evolutionary Biology, 35*(3), 186–190.
Klingenberg, C. P., Badyaev, A. V., Sowry, S. M., & Beckwith, N. J. (2001). Inferring developmental modularity from morphological integration: Analysis of individual variation and asymmetry in bumblebee wings. *American Naturalist, 157*, 11–23.
Klingenberg, C. P., Mebus, K., & Auffray, J. C. (2003). Developmental integration in a complex morphological structure: How distinct are the modules in the mouse mandible? *Evolution & Development, 5*, 522–531.
Klingenberg, C. P., & Zaklan, S. D. (2000). Morphological integration between developmental compartments in the Drosophila wing. *Evolution, 54*, 1273–1285.
Laffont, R., Renvoise, E., Navarro, N., Alibert, P., & Montuire, S. (2009). Morphological modularity and assessment of developmental processes within the vole dental row (Microtus arvalis, Arvicolinae, Rodentia). *Evolution & Development, 11*, 302–311.
Marquez, E. J. (2008). A statistical framework for testing modularity in multidimensional data. *Evolution, 62*, 2688–2708.
Mezey, J. G., Cheverud, J. M., & Wagner, G. P. (2000). Is the genotype-phenotype map modular? A statistical approach using mouse quantitative trait loci data. *Genetics, 156*, 305–311.
Michaux, J., Hautier, L., Simonin, T., & Vianey-Liaud, M. (2008). Phylogeny, adaptation and mandible shape in Sciuridae (Rodentia, Mammalia). *Mammalia, 72*, 286–296.
Monteiro, L. R., Bonato, V., & dos Reis, S. F. (2005). Evolutionary integration and morphological diversification in complex morphological structures: Mandible shape divergence in spiny rats (Rodentia, Echimyidae). *Evolution & Development, 7*, 429–439.
Monteiro, L. R., & Nogueira, M. R. (2009). Adaptive radiations, ecological specialization, and the evolutionary integration of complex morphological structures. *Evolution, 64*, 724–743.
Mullin, S. K., & Taylor, P. J. (2002). The effects of parallax on geometric morphometrics data. *Computers in Biology and Medicine, 32*, 455–464.
Oxnard, C. E. (1968). The architecture of the shoulder in some mammals. *Journal of Morphology, 126*, 249–290.
Perez, S. I., Diniz, J. A. F., Rohlf, F. J., & Dos Reis, S. F. (2009). Ecological and evolutionary factors in the morphological diversification of South American spiny rats. *Biological Journal of the Linnean Society, 98*, 646–660.
Piras, P., Maiorino, L., & Raia, P., et al. (2010). Functional and phylogenetic constraints in Rhinocerotinae craniodental morphology. *Evolutionary Ecology Research, 12*, 897–928.
Roth, V. L. (1993). On three-dimensional morphometrics, and on the identification of landmark points. In L. F. Marcus, M. Corti, A. Loy, G. J. P. Naylor, & D. E. Slice (Eds.), *Advances in morphometrics* (pp. 41–62). New York: Plenum.
Singleton, M., Rosenberger, A. L., Robinson, C., & O'Neill, R. (2011). Allometric and metameric shape variation in *Pan* mandibular molars: A digital morphometric analysis. *Anatomical Record – Advances in Integrative Anatomy and Evolutionary Biology, 294*, 322–334.
Skinner, M. M., Gunz, P., Wood, B. A., & Hublin, J. J. (2008). Enamel–dentine junction (EDJ) morphology distinguishes the lower molars of *Australopithecus africanus* and *Paranthropus robustus*. *Journal of Human Evolution, 55*, 979–988.
Stein, B. R. (1981). Comparative limb myology of two opossums, *Didelphis* and *Chironectes*. *Journal of Morphology, 169*, 113–140.
Strauss, R. E., & Bookstein, F. L. (1982). The truss–body form reconstructions in morphometrics. *Systematic Zoology, 31*, 113–135.
Swiderski, D. L. (1993). Morphological evolution of the scapula in tree squirrels, chipmunks, and ground squirrels (Sciuridae): An analysis using thin-plate splines. *Evolution, 47*, 1854–1873.
Swiderski, D. L., & Zelditch, M. L. (2010). Morphological diversity despite isometric scaling of lever arms. *Evolutionary Biology, 37*, 1–18.
Taylor, M. E. (1974). The functional anatomy of the forelimb of some African Viverridae (Carnivora). *Journal of Morphology, 143*, 307–336.
Van Buskirk, J. (2009). Natural variation in morphology of larval amphibians: Phenotypic plasticity in nature? *Ecological Monographs, 79*, 681–705.
Velhagen, W. A., & Roth, V. L. (1997). Scaling of the mandible in squirrels. *Journal of Morphology, 232*, 107–132.

Webber, A. J., & Hunda, B. R. (2007). Quantitatively comparing morphological trends to environment in the fossil record (cincinnatian series; upper ordovician). *Evolution, 61*, 1455–1465.

Webster, M. (2007). Ontogeny and evolution of the early Cambrian trilobite genus *Nephrolenellus* (Olenelloidea). *Journal of Paleontology, 81*, 1168–1193.

Webster, M. (2009). Ontogeny, systematics, and evolution of the effaced early Cambrian trilobites peachella Walcott, 1910 and Eopeachella new genus (Olenelloidea). *Journal of Paleontology, 83*, 197–218.

Webster, M., & Hughes, N. C. (1999). Compaction-related deformation in Cambrian olenelloid trilobites and its implications for fossil morphometry. *Journal of Paleontology, 73*(2), 355–371.

Willmore, K. E., Roseman, C. C., Rogers, J., Cheverud, J. M., & Richtsmeier, J. T. (2009). Comparison of mandibular phenotypic and genetic integration between baboon and mouse. *Evolutionary Biology, 36*, 19–36.

Wood, A. R., Zelditch, M. L., Rountrey, A. N., Eiting, T. P., Sheets, H. D., & Gingerich, P. D. (2007). Multivariate stasis in the dental morphology of the Paleocene-Eocene condylarth. Ectocion. *Paleobiology, 33*, 248–260.

Zelditch, M. L., Bookstein, F. L., & Lundrigan, B. L. (1992). Ontogeny of integrated skull growth in the cotton rat *Sigmodon fulviventer*. *Evolution, 46*, 1164–1180.

Zelditch, M. L., Bookstein, F. L., & Lundrigan, B. L. (1993). The ontogenetic complexity of developmental constraints. *Journal of Evolutionary Biology, 6*, 121–141.

Zelditch, M. L., Sheets, H. D., & Fink, W. L. (2000). Spatiotemporal reorganization of growth rates in the evolution of ontogeny. *Evolution, 54*, 1363–1371.

Zelditch, M. L., Sheets, H. D., & Fink, W. L. (2003a). The ontogenetic dynamics of shape disparity. *Paleobiology, 29*, 139–156.

Zelditch, M. L., Lundrigan, B. L., Sheets, H. D., & Garland, J. T. (2003b). Do precocial mammals have a fast developmental rate? A comparison between *Sigmodon fulviventer* and *Mus musculus domesticus*. *Journal of Evolutionary Biology, 16*, 708–720.

Zelditch, M. L., Wood, A. R., Bonett, R. M., & Swiderski, D. L. (2008). Modularity of the rodent mandible: Integrating bones, muscles, and teeth. *Evolution & Development, 10*(6), 756–768.

Zelditch, M. L., Wood, A. R., & Swiderski, D. L. (2009). Building developmental integration into functional systems: Function-induced integration of mandibular shape. *Evolutionary Biology, 36*, 71–87.

CHAPTER

3

Simple Size and Shape Variables: Shape Coordinates

This chapter presents methods for obtaining shape variables. One is particularly simple and easily understood, and we present it first because the method is so accessible. This method is sometimes called the "two-point registration" and it produces coordinates that are termed "Bookstein shape coordinates", which can be used both for graphical displays and formal statistical tests. The second method, the Procrustes superimposition is perhaps less intuitive, but the method is the one most widely used, for reasons that will become apparent in the next chapter. It is the one that we will use throughout the rest of this book so we introduce it now rather than laying its theoretical foundations in the next chapter and deriving the superimposition method from theory. We focus on the simplest possible application of the methods, the analysis of shapes with only three two-dimensional landmarks (triangles). We also discuss how information about size can be restored (because it is removed in the course of obtaining shape coordinates). We then extend the analysis to three-dimensional landmarks and, in the case of the Procrustes superimposition, to semilandmarks, points along outlines or curves. As well as presenting the methods for obtaining the coordinates, we also discuss the graphical description of results because, to a large extent, it is the descriptive power of geometric morphometrics – the visualization of shape change – that makes these methods so useful. The graphical results can differ depending on the methods for obtaining the shape variables, so we show how apparent inconsistencies can be reconciled.

SHAPE COORDINATES

In Chapter 1, we discussed the meanings of shape and size as they are defined in geometric morphometrics. We defined shape in terms of operations that do *not* alter shape – specifically, translation, rotation and rescaling. These operations can be applied to a simple form, a triangle, allowing us to obtain a coordinate system. Because there is more than one way to apply these operations, we can obtain different sets of coordinates from

Geometric Morphometrics for Biologists
DOI: http://dx.doi.org/10.1016/B978-0-12-386903-6.00003-4

the same data. One way to apply these operations is shown in Figure 3.1: we can translate the triangle so that one landmark is at the origin (0, 0) (Figure 3.1A). We can then rotate the triangle so that the side AB is along the X-axis (Figure 3.1B) and, finally, we can scale it so that the coordinate of landmark B is at point (1, 0) (Figure 3.1C). We can then calculate the coordinate of the third landmark, C, in the coordinate system that we have just defined. All of these operations can be applied without worrying about the consequences for shape, because we have defined shape such that none of the operations alter it. Notice that we have used a particular set of operations, translations of the *X*- and *Y*-coordinates of point A to (0,0), a rotation about A to place point B on the *X*-axis, and then a scaling to make the distance from A to B equal to one. The different superimpositions use different choices of how to carry out these basic operations. These three are the only operations involved in calculating the coordinates of point C, which is done according to the following formula, in which A_x, A_y, B_x, B_y, C_x, and C_y are the original digitized coordinates, and SC_x, and SC_y are the coordinates of landmark C in the new coordinate system:

$$SC_x = \frac{(B_x - A_x)(C_x - A_x) + (B_y - A_y)(C_y - A_y)}{(B_x - A_x)^2 + (B_y - A_y)^2}$$
$$SC_y = \frac{(B_x - A_x)(C_y - A_y) - (B_y - A_y)(C_x - A_x)}{(B_x - A_x)^2 + (B_y - A_y)^2} \tag{3.1}$$

(The numerators for the two equations really do differ in sign, as well as subscripts; that is not a misprint.)

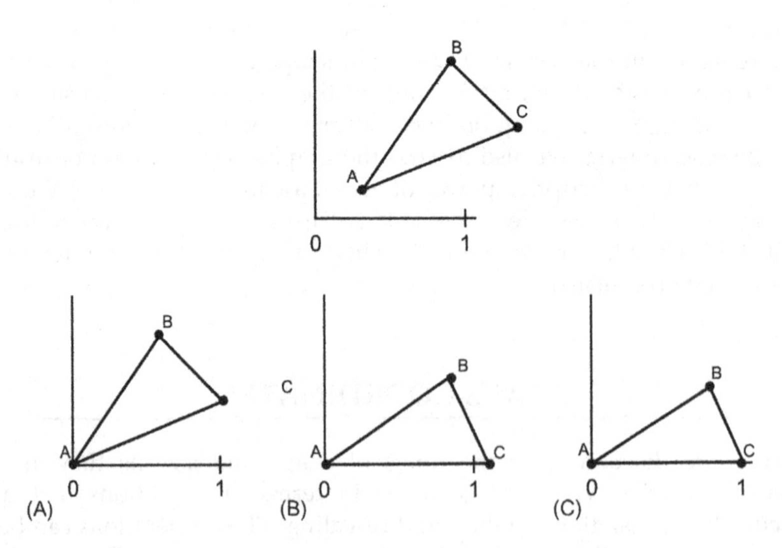

FIGURE 3.1 Three operations that do not alter shape, applied to a triangle. (A) Translation; (B) rotation; (C) rescaling.

SC_x and SC_y are the "shape coordinates" of landmark C (which from now on we will simply call C_{xy}). This relatively simple set of operations will be important when we compare the shapes of two triangles.

Comparing Shapes of Two Triangles

To compare the shape of two triangles, we apply the operations outlined above to both of them and calculate the shape coordinates of landmark C. That is, we assign the coordinates (0, 0) to landmark A in both triangles, and we assign the coordinates (1, 0) to landmark B in both triangles (Figure 3.2B). As a result, the difference between the two triangles is entirely represented by the difference in the location of the third vertex, landmark C. We can now draw both triangles on the same coordinate system (Figure 3.2C).

While there are programs to do these calculations, they are easily done in any spreadsheet or statistical program that manipulates formulae. As an exercise, take the following three pairs of coordinates for points of a triangle (in the format produced by a common digitizing program), compute the shape coordinates, and draw the triangle. For the

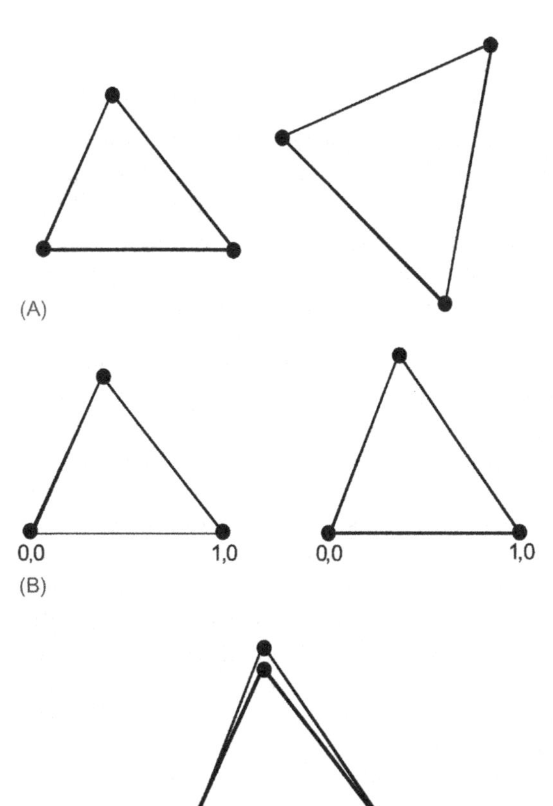

FIGURE 3.2 Two triangles whose shape difference is the subject of investigation. (A) The two triangles as initially recorded; (B) the same two triangles after being translated, rotated and rescaled by the two-point registration; (C) the same two triangles, superimposed.

moment, pick any two points as the endpoints of the baseline (A and B); we will discuss how to choose them later.

1.)	54	306
2.)	223	447
3.)	632	300

Now take the next three coordinate pairs, and draw that triangle:

1.)	11	342
2.)	251	520
3.)	769	318

Now draw both triangles using the same baseline (with point A and B superimposed), and draw the vector extending between the one free landmark (C_{xy}) on both triangles. That vector is the shape variable describing the difference between the triangles.

Comparing Many Individual Triangles

Of course, we rarely (if ever) compare only two specimens (or triangles). We now consider how to compare many individual triangles (below we discuss comparing forms more complex than triangles). The same procedure (and formulae) applies no matter how many triangles or individuals are examined. For example, given a collection of triangles (Figure 3.3A), we assign points A and B the coordinates (0, 0) and (1, 0) and compare all these triangles as whole triangles (Figure 3.3B), or as scatter-plots of the one free point (Figure 3.3C).

The scatter-plot is useful for checking the repeatability of your landmarks, as well as for studying the variability of shape or differences in shape. For all these purposes, it is important that the axes of the scatter-plot be sized so that a square shape is shown as a square – that is, the length of the interval from 0 to 1 on the *X*-axis should be the same as the length of the interval from 0 to 1 on the *Y*-axis. Many programs do not do this scaling of axes automatically, so you may have to scale the axes yourself. Often this can be done by first calculating the maximum and minimum values for the *X*- and *Y*-coordinates; the difference between those values, i.e. the range of values, should be set equal for both coordinates. For example, if the X-coordinate ranges from 0.030 to 0.060 and the *Y*-coordinate ranges from 0.020 to 0.060, both axes should be 0.040 units long (the *Y*-coordinate has the slightly larger range). In this case, the minimum on the *X*-axis could be set to 0.025 and the maximum on the *X*-axis to 0.065. This distributes the extra length equally above and below the observed values, and should enforce a 1:1 aspect ratio for the graph.

When the axes are on the same scale, an approximately circular scatter of points indicates that there is a reasonably equal amount of variation in all directions. Random digitizing error should be circular; systematic errors, in contrast, will look elliptical.

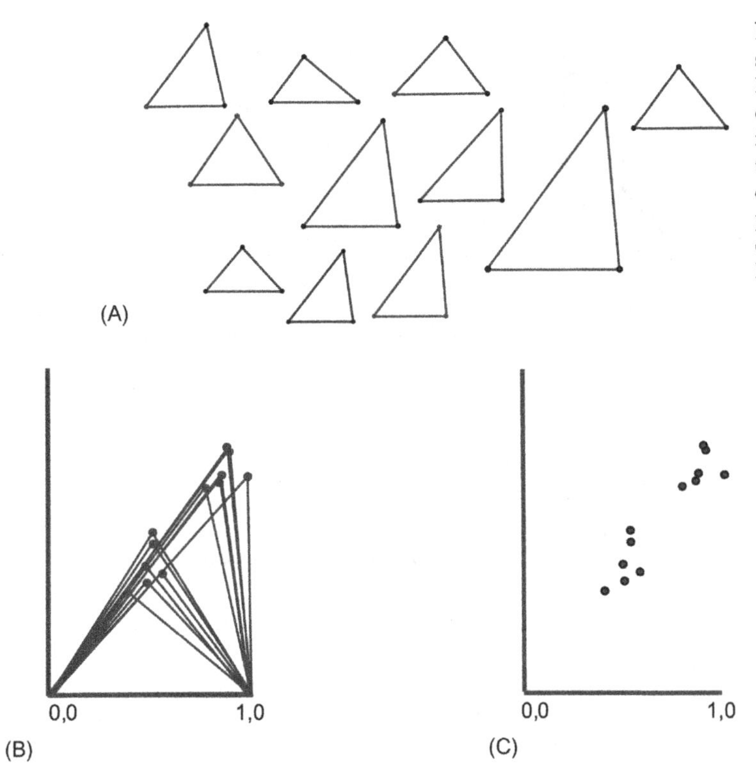

FIGURE 3.3 Comparing shapes of triangles. (A) The collection of triangles whose shape differences are the subject of investigation; (B) the same collection of triangles, put in a common coordinate system by the two-point registration; (C) scatter-plot depicting the location of the free landmark.

If you have already digitized landmarks, now would be a good time to compute shape coordinates, scale the axes appropriately, and check that your digitizing error is circular. Should you find points that depart substantially from circularity, you should either delete that landmark from your analysis, or take its biased error into account in subsequent analyses.

When the scatter is circular, it is said to be isotropic, or uniform with respect to direction. If it is uniform at all landmark locations, the scatter (or variance) at the landmark is said to be homogeneous. Don't worry too much at this point if the scatter in your data isn't homogeneous, as we will discuss later, the baseline registration procedure can produce inhomogeneity.

Multiple Triangles on Each Individual

So far we have concentrated on the simplest possible case: comparisons of a triangle. This is because multiple landmarks can all be transformed into shape coordinates using the formulae introduced for computing the shape coordinates of a single moveable point, C. We just apply that same formula to all the additional points. It is not necessary to use the same baseline for all points, but it does ease the task of reporting the changes. Not only is the same formula applicable to the more complex case, but the same basic

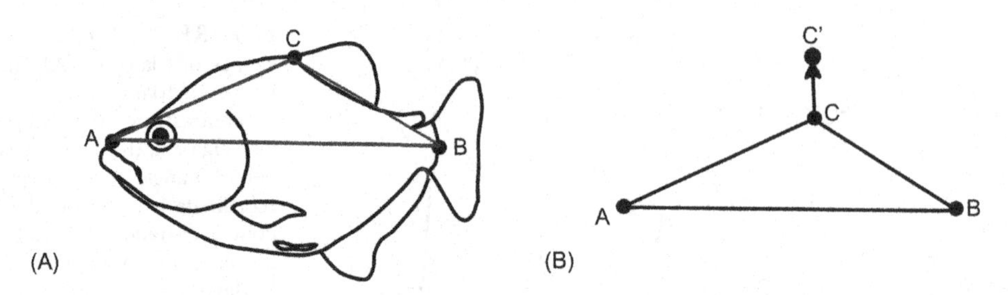

FIGURE 3.4 (A) Triangle with the baseline along the anteroposterior body axis and the free point at the anterior dorsal fin base; (B) the difference between two shapes depicted by a vector extending between C and C′.

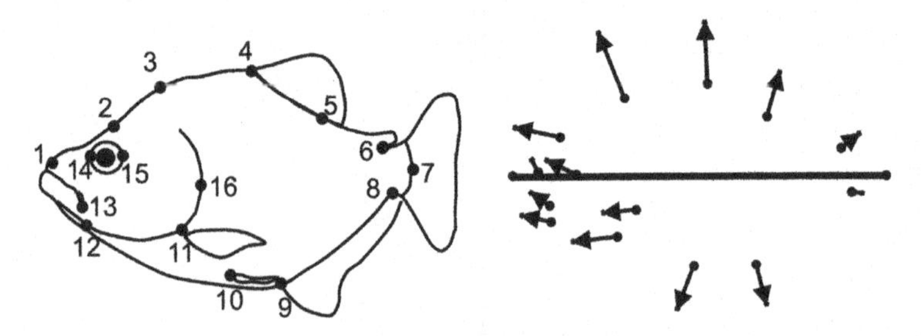

FIGURE 3.5 Ontogenetic changes in the shape of a piranha, *Serrasalmus gouldingi*, represented by vectors depicting the change in location of Bookstein shape coordinates from their position in a young juvenile.

statistical machinery also applies. Another procedure also extends without difficulty from one to many triangles – the depiction of shape differences by vectors at the free landmarks. In the case of a single triangle, we can depict a change in shape by a single vector depicting the displacement of one landmark relative to the baseline (Figure 3.4). In the case of multiple triangles, we can similarly depict the change in shape by vectors that show the displacements of points relative to the baseline (Figure 3.5). If this baseline dependence is not seen as a serious problem, the description can proceed in terms of the displacements of landmarks relative to each other, relative to the baseline. In describing the ontogenetic change in shape depicted in Figure 3.5, we would need to take the relative lengths of all the vectors into account. The most anterior free point on the dorsal margin (landmark 2, at the epiphyseal bar) is displaced anteriorly, indicating that the region between it and the baseline point at the tip of the snout is shortened relative to the length of the baseline. The point immediately posterior to landmark 2 (landmark 3, at the tip of the supra-occipital process) is also displaced anteriorly, although most of its displacement is along the dorsoventral body axis. Because the anteroposterior component of this vector is short relative to that of the more anterior point, the region between the epiphyseal bar and supra-occipital process is relatively elongated (relative both to the length of baseline and to the more anterior region just described). Such descriptions can be useful, even if they depend on the baseline.

There is one special case of multiple triangles that is worth singling out: multiple triangles describing two sides of a bilaterally symmetric organism. If we are interested specifically in asymmetry, both sides contain relevant information because we are then interested in the difference between sides. Otherwise, the two sides are redundant and using both sides in the analysis implicitly treats them as if they were independent and inflates our degrees of freedom. We can avoid this problem using the coordinates we have obtained, reflecting one side across the midline and averaging the coordinates of the two sides. This approach also allows us to use partially fragmentary specimens with landmarks present on only one side or the other because, for these specimens, the "average" for a landmark is obtained from the one side on which it was preserved.

Choosing the Baseline

When we calculated shape coordinates, we chose one side of the triangle to serve as a baseline. An obvious question is whether our results might depend on that choice. One important consideration is that variation will be transferred from the baseline landmarks to the others. There are few, if any, truly invariant landmarks and when two are fixed, their variance must be put *somewhere*. Some landmarks are difficult to digitize and these should not serve as an endpoint of the baseline because their noise will be transferred to all the other landmarks. What makes the transfer of variance really worrisome is that it is not necessarily unbiased, rather it is related to the distances of the free landmarks to the baseline (Dryden and Mardia, 1998). A baseline that runs through the centroid of the form is preferable to one that is far from most other landmarks, as it minimizes this effect. Additionally, when choosing the endpoints of the baseline, we do not want points that are too close to each other because any highly localized variation in shape may be common to both those points and the variance local to those baseline landmarks will be transferred to all the other landmarks. As Bookstein (1991) has argued, the scatters for different sets of shape coordinates of the same triangle to different baselines differ mainly by translation, rotation and rescaling but the inhomogeneity of variance and the correlations among landmarks induced by fixing two points could be problematic for statistical analysis. Another consideration, but one that is primarily a matter of interpretability, is the orientation of the baseline. If the baseline rotates relative to a body axis it does not compromise the statistical analyses, but it can make interpretations based on graphics difficult – it might seem that all the landmarks are moving away from the baseline in the posterodorsal direction, for example, when the baseline rotates in the anteroventral direction. Ideally, therefore, we want endpoints of the baseline to be along the longest diameter of the form that passes through the centroid of the form, so long as those points are not especially unreliable and the longest diameter does not rotate.

To see the consequences of fixing various endpoints, we can consider the ontogenetic series of the piranha *Serrasalmus gouldingi*. We have used landmarks 1 and 7 as the endpoints of the baseline; both the scatter of coordinates and the depiction of ontogenetic change resulting from that choice can be seen in Figure 3.6A,B. We could have used two dorsal landmarks (3 and 5), producing a strikingly different scatter-plot that implies considerably more variation (Figure 3.6C), as well as a strikingly different picture of ontogenetic change (Figure 3.6D). The dorsoventral component of shape change has been removed from all three dorsal landmarks and is expressed as a displacement of every other landmark away from the dorsal edge. In addition, the elongation of the middle of the body relative to the rest of the piranha is now expressed as a relative contraction of the ends towards the middle. If we had used another baseline (landmarks 1 and 3) that rotates relative to most of the other landmarks, the variation seems to be even further increased (Figure 3.6E) and the piranha's body seemingly rotates as it grows (Figure 3.6F). Clearly, the baselines used in Figures 3.6C,D and 3.6E,F are spectacularly bad choices. But they simply exaggerate the general problem that the variance of baseline points is transferred to the other landmarks. It should be intuitively obvious, even if not visibly so,

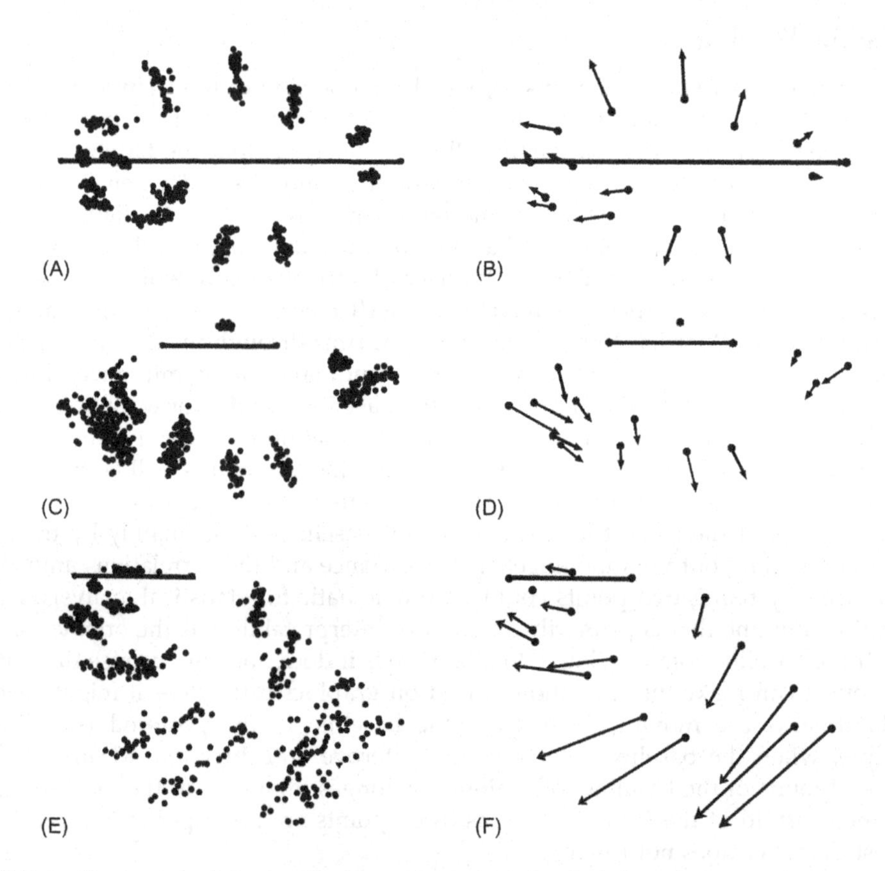

FIGURE 3.6 The impact of the baseline on variation and the depiction of ontogenetic shape change of a piranha, *S. gouldingi*. (A) Coordinates of landmarks relative to a fixed baseline that extends between landmarks 1 and 7; (B) vectors indicating displacements of landmarks relative to the fixed baseline; (C) coordinates relative to a fixed baseline that extends between two dorsal landmarks (3 and 5); (D) vectors indicating displacements of landmarks relative to the fixed baseline; (E) coordinates of landmarks relative to the baseline that extends between landmarks 1 and 3; (F) vectors indicating displacements of landmarks relative to the fixed baseline.

that the actual anatomical landmarks are really no more variable than in Figure 3.6A and B – changing the baseline does nothing but rotate and rescale the data. Although the consequences for our perception of the shape differences can be dramatic, particularly when it makes the data seem inordinately noisy, those consequences can be understood as the result of a change in perspective. In general, it is easiest to interpret results when the baseline lies along an organismal body axis. Even though results can be interpreted in a baseline-invariant way, the interpretations still refer to sides of the triangle. It is most convenient when at least one side is a conventional and familiar reference. Thus, even though we *can* interpret shape changes without reference to organismal body axes, we might still wish to orient our findings with respect to them. This motivates choosing a baseline along one of those axes.

SIZE

To this point we have talked only about shape. In the course of obtaining shape coordinates, we lost no information about shape, but we removed all the information about size. Specifically, we removed it by rescaling the baseline to a length of one. We can restore the information about size by using a measure that captures the notion of scale – the property that changes when an image is enlarged or reduced. There are several other meanings of size, including a simple measure of the length of an organism (e.g. snout–vent length), or area, volume, weight or even a linear combination of all measured quantities that captures the positive correlations among them (as such as the first principal component). In geometric morphometrics, we use a specific concept of size, one related to geometric scale. One reason for choosing such a measure is that it is geometrically independent of shape, at least under some models of error (Bookstein, 1991). To clarify this idea of geometric independence, consider what happens when every dimension is enlarged by the same proportion so that the organism gets larger without altering its shape. In this case, each coordinate is moved away from the center in proportion to its original distance from the center. The size variable that captures this radial notion of scale is centroid size, graphically illustrated in Figure 3.7.

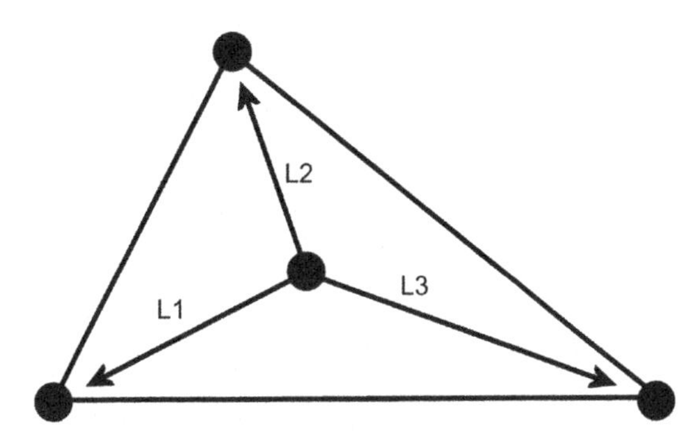

FIGURE 3.7 A geometric depiction of the calculation of centroid size: the square root of the summed squared lengths of distances of landmarks from the centroid: L1, L2, L3.

Centroid size is the square root of the summed squared distances of each landmark from the centroid, giving a linearized measure of size (in a few very early publications, the measure was not linearized). To compute centroid size, first compute the centroid (center) of the form, whose location is the mean position of all coordinates. The mean is found by simply averaging all the *X*-coordinates and all the *Y*-coordinates. For example, the three landmarks of the triangle might be at (0, 0), (1, 0) and (0.3, 0.8) so the average *X*-coordinate is the arithmetic mean of the three *X*-coordinates (0, 1, and 0.3), which is 0.433, and the average *Y*-coordinate is the arithmetic mean of the three *Y*-coordinates (0, 0, 0.8), which is 0.267. Then calculate the squared distance of each landmark from the center, using the standard formula for a squared distance between two points $(X_2 - X_1)^2 + (Y_2 - Y_1)^2$. This sum gives a measure of size related to area; taking the square root gives the linearized measure of size.

Size is thus measured separately from shape, and is geometrically independent of shape. It is also statistically independent of shape (i.e. uncorrelated with shape, when the landmarks vary independently and equally in all directions). This is a useful attribute of a size measure because we do not want size to be intrinsically correlated with shape simply by virtue of its formula. Rather, we want a measure of size that is correlated with shape only when size and shape covary, that is, in the presence of allometry. Allometry is a common phenomenon, so we might expect that size and shape would usually be statistically correlated. But that correlation is an empirical finding, not an outcome of the formula for size. Centroid size, and other measures that are variants on centroid size, is the only size variable that is statistically uncorrelated with shape *in the absence of allometry*. This is one of the main reasons why centroid size is used as a size variable. The other reason is that centroid size plays a crucial role in defining the metric for a distance between two shapes (see Chapter 4).

Bookstein Shape Coordinates in Three Dimensions

Shape coordinates for three-dimensional data can be calculated by an extension of the formula for two-dimensional shape coordinates, but the formula is more complex because three-dimensional objects can rotate around the three orthogonal axes, *X*, *Y* and *Z* (Figure 3.8). The baseline is again translated, scaled to unit length oriented along the *X*-axis and rotated so that the third landmark (C) is in the *X*, *Y* plane. Computing the

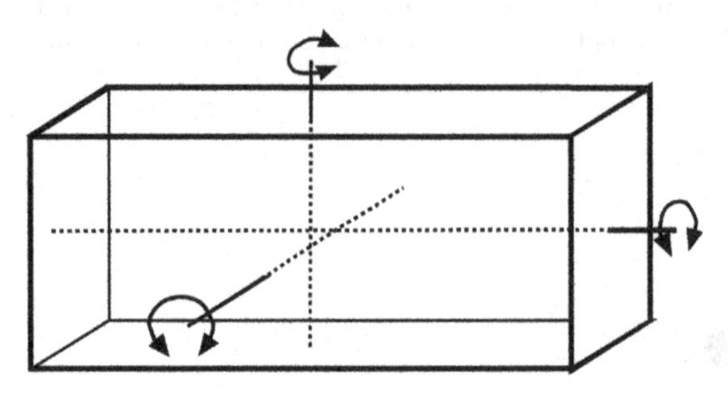

FIGURE 3.8 Rotation in 3D: axes of rotation.

coordinates of the free coordinates involves (clockwise) rotations around the *X*, *Y* and *Z* axes (through angles ϕ, ω, and θ, respectively). The three rotation matrices are:

$$R_x = \begin{bmatrix} 1 & 0 & 0 \\ 0 & \cos\phi & \sin\phi \\ 0 & -\sin\phi & \cos\phi \end{bmatrix}$$
$$R_y = \begin{bmatrix} \cos\omega & 0 & \sin\omega \\ 0 & 1 & 0 \\ -\sin\omega & 0 & \cos\omega \end{bmatrix} \quad (3.2)$$
$$R_z = \begin{bmatrix} \cos\theta & \sin\theta & 0 \\ -\sin\theta & \cos\theta & 0 \\ 0 & 0 & 1 \end{bmatrix}$$

So, designating the translated and scaled coordinates by A_{ts}, B_{ts}, C_{ts}, the three-dimensional shape coordinates are $R_xR_yR_z(A_{ts}, B_{ts}, C_{ts})^T$ (for a more detailed presentation of the calculation of three-dimensional Bookstein shape coordinates see Dryden and Mardia, 1998; Claude, 2008).

STATISTICS OF SHAPE COORDINATES

Now that we have shape coordinates, we can answer the basic "existential" questions as defined in Chapter 1, such as "do these samples differ in shape?" All conventional statistical methods and tests can be applied to shape coordinates and centroid size. For example, an average value for the shape coordinate at point C is computed by averaging the *X*-coordinates for that point across all individuals within a sample, then dividing that sum by the total number of individuals in that sample and applying that same procedure to the *Y*-coordinates. Variances and standard deviations are also calculated by standard formulae. Because the two endpoints of the baseline are fixed, they have no variance and should not be included in statistical analyses. If you use conventional statistical packages to analyze these coordinates, remember to exclude them from the analysis because many programs will not run if the variables do not vary.

Because every landmark has two dimensions (its *X*-, and *Y*-coordinates), statistical analyses are necessarily multivariate. Even if we are asking whether two samples of triangles differ in average shape, we must use a multivariate test. In particular, we would use the multivariate form of the familiar Student's *t*-test, Hotelling's T^2 test (see, for example, Morrison, 1990). When comparing two samples of triangles, the test is applied to the two coordinates of landmark C. When we are comparing more than two samples, we can use Wilks' Λ (Rao, 1973) or one of the related statistics obtained by a multivariate analysis of variance (MANOVA). In studies of allometry, we use multivariate regression. However, an important consideration that needs to be taken into account when applying statistical tests to these shape coordinates is that the variances of landmark locations are not independent of the mean location of that landmark relative to the baseline. Variance thus will not be homogeneous at all landmarks. That is because the variance of the baseline endpoints is transferred to the other coordinates. Additionally, the use of the baseline induces correlations between landmarks so methods like principal components analysis should not be applied to Bookstein shape coordinates (Bookstein, 1996; Dryden and Mardia, 1998).

PROCRUSTES SUPERIMPOSITION

In Greek mythology, Procrustes was a bandit who fit his visitors (victims) to a bed by stretching or truncating them, minimizing the difference between his visitors and the bed. The method that we will now use to obtain shape coordinates is Procrustean in the sense that it minimizes the differences between landmark configurations. Unlike the mythological Procrustes, who altered the shape of his victims, the mathematical Procrustes superimposition method does not alter shape because it uses the three operations that do not alter shape: translation, scaling and rotation. Presumably Procrustes' guests would have preferred that he had done likewise!

A step-wise description of the method was presented by Rohlf and Slice (1990):

1. Center each configuration of landmarks at the origin by subtracting the coordinates of its centroid from the corresponding (X or Y) coordinates of each landmark. This translates each centroid to the origin (and the coordinates of the landmarks now reflect their deviation from the centroid).
2. Scale the landmark configurations to unit centroid size by dividing each coordinate of each landmark by the centroid size of that configuration.
3. Choose one configuration to be the reference, then rotate the second configuration to minimize the summed squared distances between homologous landmarks (over all landmarks) between the forms.

When there are more than two forms, all are rotated to optimal alignment on the first; the average shape is then calculated and all are rotated to optimal alignment on the average shape, which is the new reference. At this point, the average shape is recalculated. If it differs from the previous reference, the rotations are recalculated using this newest reference. When the newest reference is the same as the previous, the iterations stop (usually only a few iterations are necessary). The final reference is the one that minimizes the average distances of shapes from the reference. Note that this result does not depend on the shape of the first specimen used in the alignment; instead, it depends on the distribution of shapes in the sample.

While the terminology and different variations on Procrustes superimposition will be discussed in greater detail later, it is worth mentioning that the procedure just discussed is called partial Procrustes Superimposition, which simply indicates that the centroid size was fixed at one. This is currently the standard Procrustes technique, and a lot of workers will use the term Procrustes to refer to what is really a partial Procrustes superimposition. The key is to look for a centroid size of one in the description of the methods used. When an iterative method is used to estimate the mean form, as discussed above, the entire process is called a Generalized Procrustes Analysis (GPA).

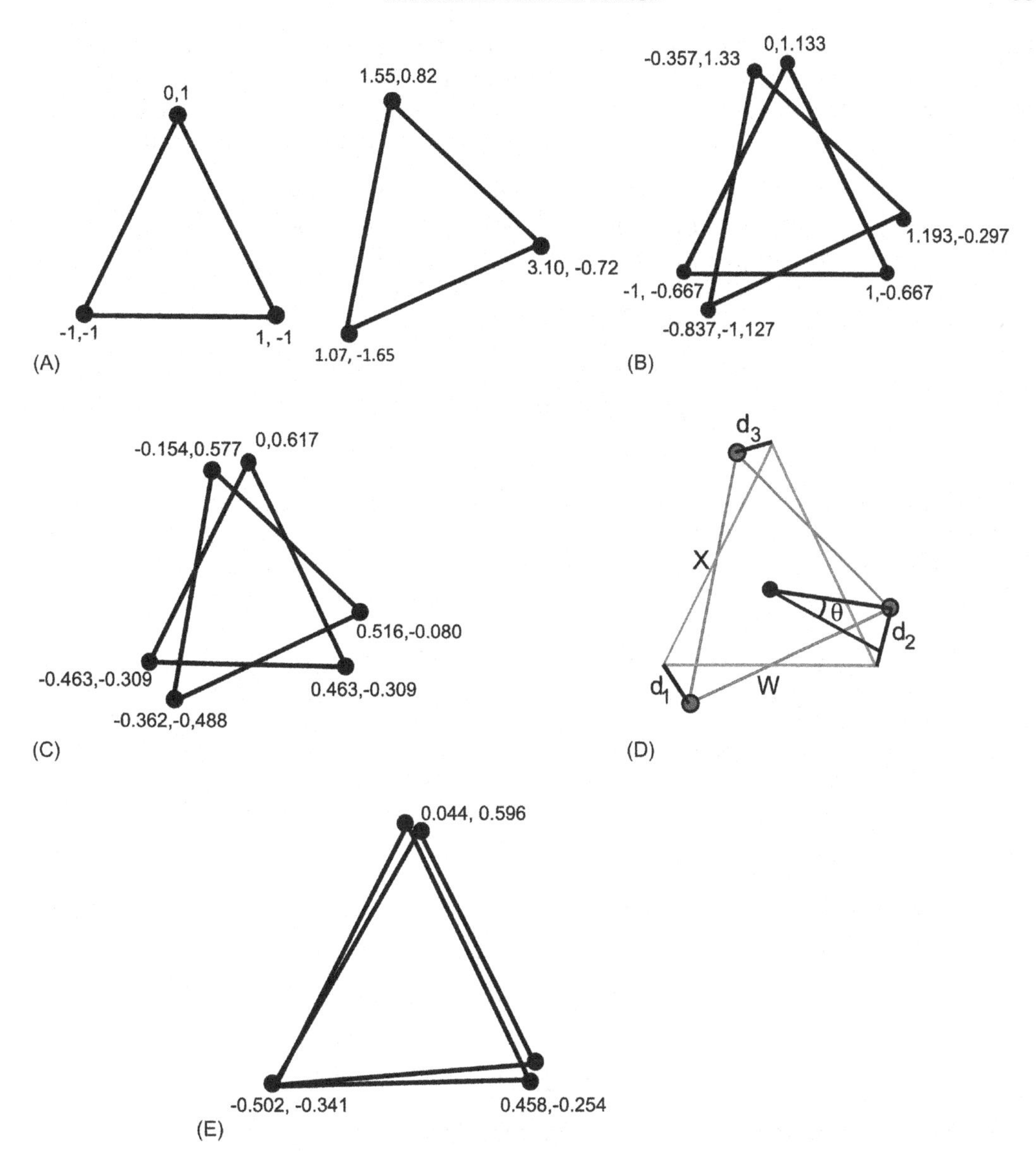

FIGURE 3.9 Procrustes superimposition. (A) Two triangles; (B) two centered triangles; (C) centered and scaled triangles; (D) finding the angle of rotation; (E) centered, scaled and rotated triangles.

We can apply these operations to two triangles shown in Figure 3.9A. One has coordinates (−1, −1), (1, −1) and (0, 1) and the other has coordinates (1.07, −1.64), (3.10, −0.72) and (1.55, 0.82). The first step is to center them by subtracting the coordinates of the centroid, and again we calculate the coordinates of the centroid by computing the arithmetic averages of the *X*- and *Y*-coordinates. For the first triangle, the *X*-coordinate of the centroid is

$(1/3)(-1+1+0)=0$, and the Y-coordinate of the centroid is $(1/3)(-1+-1+1)=-0.333$. For the other triangle, the X-coordinate of the centroid is $(1/3)(1.07+3.10+1.55)=1.907$ and the Y-coordinate of the centroid is $(1/3)(-1.64+-0.72+0.82)=-0.513$. We now subtract these centroid coordinates from the coordinates of the landmarks. For the first triangle, this gives us:

$$=\begin{bmatrix}(-1-0) & (-1-(-0.333))\\ (1-0) & (-1-(-0.333))\\ (0-0) & (1-(-0.333))\end{bmatrix}=\begin{bmatrix}-1 & -0.667\\ 1 & -0.667\\ 0 & 1.333\end{bmatrix} \tag{3.3}$$

And for the other this gives us:

$$\begin{bmatrix}(1.07-1.907) & (-1.64-(-0.513))\\ (3.10-1.907) & (-0.72-(-0.513))\\ (1.55-1.907) & (0.82-(-0.513))\end{bmatrix}=\begin{bmatrix}-0.837 & -1.127\\ 1.193 & -0.207\\ -0.357 & 1.333\end{bmatrix} \tag{3.4}$$

So the two triangles are now both centered on the same coordinates (see Figure 3.9B).

We now need to scale them by dividing each coordinate by centroid size. The calculation of centroid size is simple now that the centroids are at 0,0 – it is the square root of the sum of the squared coordinates. For the first triangle, centroid size is calculated as:

$$\sqrt{(-1.0)^2+(-0.667)^2+(1.0)^2+(-0.667)^2+(0)^2+(1.333)^2}\\ =2.160 \tag{3.5}$$

And for the second triangle it is calculated as:

$$\sqrt{(-0.837)^2+(1.127)^2+(1.193)^2+(-0.207)^2+(0.357)^2+(1.333)^2}\\ =2.311 \tag{3.6}$$

So we now divide each coordinate by centroid size. For the first triangle, dividing each coordinate by 2.16 gives us:

$$\frac{1}{2.160}\begin{bmatrix}-1 & -0.667\\ 1 & -0.667\\ 0 & 1.333\end{bmatrix}=\begin{bmatrix}-0.463 & -0.309\\ 0.463 & -0.309\\ 0.000 & 0.617\end{bmatrix} \tag{3.7}$$

And for the second triangle, dividing each coordinate by 2.311 gives us:

$$=\frac{1}{2.311}\begin{bmatrix}-0.837 & -1.127\\ 1.193 & -0.207\\ -0.357 & 1.333\end{bmatrix}=\begin{bmatrix}-0.362 & -0.488\\ 0.516 & -0.089\\ -0.154 & 0.577\end{bmatrix} \tag{3.8}$$

So now we have centered and scaled the triangles (see Figure 3.9C).

The next step is to rotate the triangles to minimize the sum of the squared differences of the coordinates (summed over all three coordinates). To do this, we will pick one of the triangles to serve as the reference; we will arbitrarily pick the first one. So we now rotate the second, around its centroid, through angle θ to minimize that sum of squared differences (see Figure 3.9D). After the rotation, the X- and Y-coordinates of each landmark will have the coordinates $(X \cos\theta - Y \sin\theta)$, $(X \sin\theta + Y \cos\theta)$. Thus, the rotated form the second triangle will be:

$$\begin{bmatrix} (-0.362\cos\theta) - (-0.488\sin\theta) & (-0.362\sin\theta) + (-0.488\cos\theta) \\ (0.516\cos\theta) - (-0.089\sin\theta) & (0.516\sin\theta) + (-0.089\cos\theta) \\ (-0.154\cos\theta) - (0.577\sin\theta) & (-0.154\sin\theta) + (0.577\cos\theta) \end{bmatrix} \quad (3.9)$$

The value of θ that gives us the minimum sum of squared deviations is $-19.2°$ so, inserting that into Equation 3.9, gives us the coordinates for the second triangle:

$$\begin{bmatrix} -0.502 & -0.341 \\ 0.458 & -0.254 \\ 0.044 & 0.596 \end{bmatrix} \quad (3.10)$$

We now have the superimposed triangles (see Figure 3.9E).

PROCRUSTES SUPERIMPOSITION IN THREE-DIMENSIONS

Differences in location, scale and orientation of three-dimensional configurations are removed by exactly the same operations; the only substantive difference is that we work with larger matrices, making the computations more tedious (especially for the programmer). We will not go through them all again because the only difference is the number of columns to be averaged to compute the centroid, the number of columns from which the centroid coordinates are subtracted, and the number of coordinates that are divided by centroid size. The one step that is complicated by three-dimensional data is rotation, just as this was the one complication encountered when we extended the formula for two-dimensional Bookstein shape coordinates to three-dimensions. We now have to solve for the particular combination of angles that minimizes that distance. Still, the solution remains conceptually simple and it is obtained by a singular value decomposition (SVD) of the matrix $\mathbf{X}_\mathbf{R}^t\mathbf{X}_\mathbf{T}$ in which $\mathbf{X}_\mathbf{R}$ and $\mathbf{X}_\mathbf{T}$ are the centered and scaled configuration matrices of the reference and target, respectively (Rohlf, 1990). As Rohlf points out, this is just one example of the general utility of SVD for finding the angular relationship between two matrices.

SEMILANDMARK SLIDING

As discussed earlier, there may be studies in which we want to incorporate information about outlines or curves as well as landmarks into the analysis. This is done using semilandmark techniques, in which points are placed along the curve or outline. A curve is simply an infinite set of points so, in using semilandmarks, we are approximating this infinite set with a finite number of points placed along the curve using some algorithmic approach to optimize this approximation.

One method for superimposing semilandmarks, suggested by Sampson and colleagues (1996), slides semilandmarks to the position that minimizes the summed squared deviations between each individual and the reference form. The semilandmarks are not free to move in any direction; each is confined to slide along the line tangent to the curve at that semilandmark (Figure 3.10). The shape of the curve is not actually known, so the tangent is estimated as the line parallel to the segment connecting a series of adjacent landmarks or semilandmarks. Each semilandmark of the target slides along its tangent to align with the perpendicular at the corresponding semilandmark of the reference. Using this method,

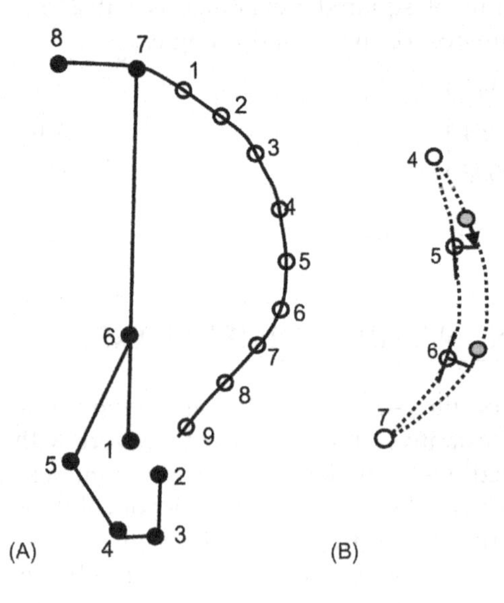

FIGURE 3.10 Semilandmark superimposition. (A) Landmarks (black) and semilandmarks (white) on a squirrel scapula; (B) finding tangents to the curve and sliding semilandmarks.

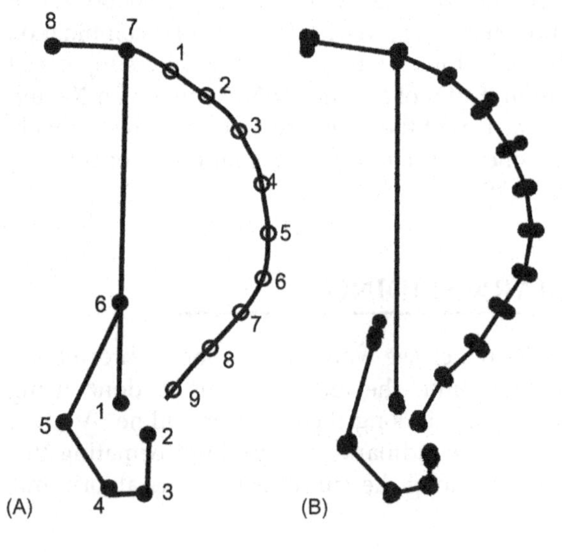

FIGURE 3.11 Semilandmark superimposition result. (A) Landmarks (black) and semilandmarks (hollow) on a squirrel scapula; (B) slid semilandmarks.

the only information extracted from the semilandmarks is about the bowing of the curve (displacement perpendicular to the tangent) because any variation along the curve is treated as non-shape (nuisance) information resulting from the arbitrary choices made when digitizing the semilandmarks. The outcome is shown in Figure 3.11 – all the semilandmarks line up on their respective perpendiculars.

RESISTANT-FIT SUPERIMPOSITION

Resistant-fit superimposition methods (RFTRA) are similar to GPA but there is no goodness-of-fit criterion to be minimized. In recognition of the general similarity, the resistant-fit methods have also been characterized as "Procrustes methods" (see Chapman, 1990). The rationale for such methods is that GPA, like any method that uses least squares, is very sensitive to large displacements at few landmarks. In statistical procedures like regression, a few cases with unusually large deviations from the general pattern (i.e. "outliers" or "influential observations") can have a large effect on the results of a procedure that minimizes the sum of the squared deviations. In shape analysis, a large change limited to one or a few influential landmarks is sometimes called the Pinocchio effect, although those influential landmarks need not be at the tip of a long process. When GPA is used to superimpose landmark configurations, the Pinocchio effect can have a large effect because the least squares criterion distributes the displacements of the few landmarks across all the other landmarks. In graphical displays, the Pinocchio effect appears to be "smeared out" over all landmarks, which can be unsettling for some workers. Figure 3.12 shows a hypothetical example in which the only shape change in a tree squirrel scapula is the ventral displacement of the three most ventral landmarks. The more severe consequence is that

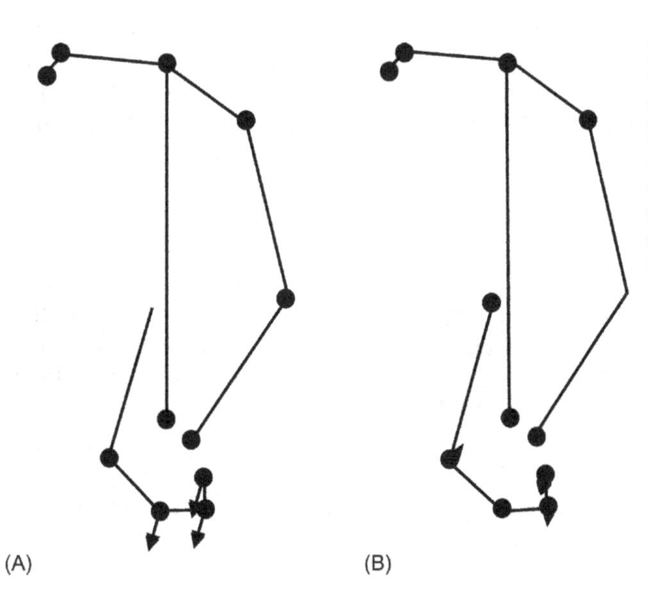

FIGURE 3.12 Hypothetical example of the Pinocchio effect as exemplified by ventral displacements of the three most ventral landmarks of a tree squirrel scapula. (A) Configurations superimposed by a resistant-fit method (RFTRA); (B) configurations superimposed by GLS. The outline of the scapula is approximated by lines connecting the landmarks.

the least squares criterion causes the variances of the influential landmarks to be allocated to other points, inducing covariances (Walker, 2000).

Resistant-fit methods reduce the influence of the Pinocchio effect by taking a "robust" approach to superimposition. In statistics, "robust" means that the method is relatively insensitive to outliers in the data. Similarly, a robust superimposition method is relatively insensitive to a few landmarks with large relative displacements. A wide variety of error functions has been used as criteria for robust fitting procedures, the interested reader is referred to Press et al. (1988) for a discussion of several alternatives. None of them allow analytic solutions for the rotation and scaling parameters needed to carry out a superimposition; instead they use numerical methods (simplex searches) to find the rotation and scaling necessary to minimize the error function.

The robust approach implemented by RFTRA uses the method of "repeated medians" to determine the scaling and rotation necessary to superimpose one shape on another (Chapman, 1990). We describe the steps used to find the scaling factor in some depth; then more briefly describe the steps to find the rotation. For the scaling factor:

1. Compute the pairwise interlandmark distances in both shapes and then compute the ratio of each pair of corresponding distances.
2. For each landmark, find the median of ratios for all segments radiating from that landmark. This will yield one ratio for each landmark.
3. Find the median of the medians generated by step 2. This median of medians is the scaling factor used in the superimposition – in other words, all coordinates of the second shape are scaled by this factor.

After scaling the second form, the rotation angle used by RFTRA can be determined in a similar fashion from the same set of line segments. The first step is to compute the angles between the corresponding segments; the remaining steps find the median angle associated with each landmark and then the median of the medians. Rohlf and Slice (1990) present a generalized resistant-fit method that centers and scales coordinates to a common size (computed as the median squared interlandmark distances) yielding a matrix X_j'. The initial step uses least squares to fit X_j' to the coordinates of the first specimen **Y**, which is the initial reference. A new reference **Y** is then calculated as the median of the rotated specimens, and X_j' is then rotated to fit that **Y**. This procedure is iterated until the change in **Y** is smaller than the chosen stopping criterion. The lack of change in the reference, **Y**, determines when the iterations cease because there is no explicit fitting criterion being minimized.

Resistant fit methods are robust because medians are relatively insensitive to outliers. Consequently, large changes at one or a few landmarks will not appreciably alter the median scaling factor or the median rotation angle. This makes the resistant-fit methods resistant to the Pinocchio effect, which helps to highlight the region where the effect occurs, as in Figure 3.12. However, in the absence of the Pinocchio effect, superimpositions produced by resistant-fit methods usually do not differ greatly from those produced by GPA. Figure 3.13 shows both GPA and resistant-fit superimpositions of the real scapulae that were the basis of the hypothetical example. The real scapulae differ in the relative length of the ventral process, as in the hypothetical case, but they also differ in the shape of the anterior edge of the scapula (producing large relative displacements of the two

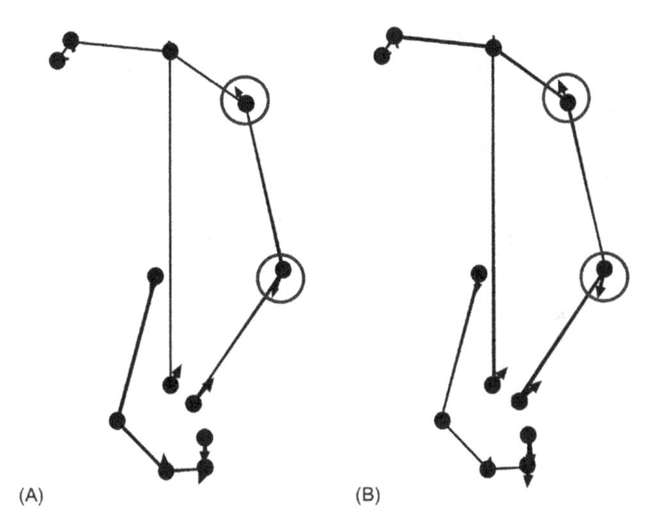

FIGURE 3.13 Comparison of (A) GPA and (B) RFTRA superimpositions for data that lack a Pinocchio effect: real differences in scapula shape between two squirrels. The circles denote the two landmarks that undergo large relative displacements, in addition to the three most ventral landmarks. The outline of the scapula is approximated by lines connecting the landmarks.

circled landmarks). The difference between the superimpositions is subtle; it is most noticeable at the ventral end, where RFTRA attributes somewhat greater anterior displacements to the more ventral landmarks. The principal advantage of RFTRA and other resistant-fit methods lies in their ability to address the Pinocchio effect. The principal disadvantage of these methods will become clearer in the next chapter.

Interpreting the Graphical Results

The graphical representation of results is one of the main reasons why geometric methods are so useful and popular. Different superimposition methods can yield strikingly different visual displays from the same data, which can be both disconcerting and useful. It is useful because seeing different depictions of the same results can help you avoid drawing conclusions that depend on a particular visual display. For example, one method might show large displacements at some landmarks whereas another superimposition method might show large displacements at other landmarks. Seeing two or more pictures helps to avoid a common error, that of interpreting the changes as if they are *at* the landmarks. To illustrate the variety of pictures that can be obtained by different superimposition methods, we use three (BC, GPA and RFTRA) to depict the ontogeny of body shape in the piranha *S. gouldingi* (Figure 3.14).

Perhaps the most obvious difference among the three panels is the degree to which post-cranial landmarks are vertically displaced. It might appear that Bookstein shape coordinates either exaggerate the degree to which the post-cranial body is deepened, or else that the other superimpositions understate it. However, this is not the case; all the other superimpositions show a relative shortening of the body, which is equivalent to a relative deepening. Both mean exactly the same thing. Relative body depth is a ratio between depth and length, so it is just as reasonable to think of it as a decrease in length relative to depth as to think of it as an increase in depth relative to length. Increasing body depth increases the ratio by increasing the numerator; decreasing body length also increases the

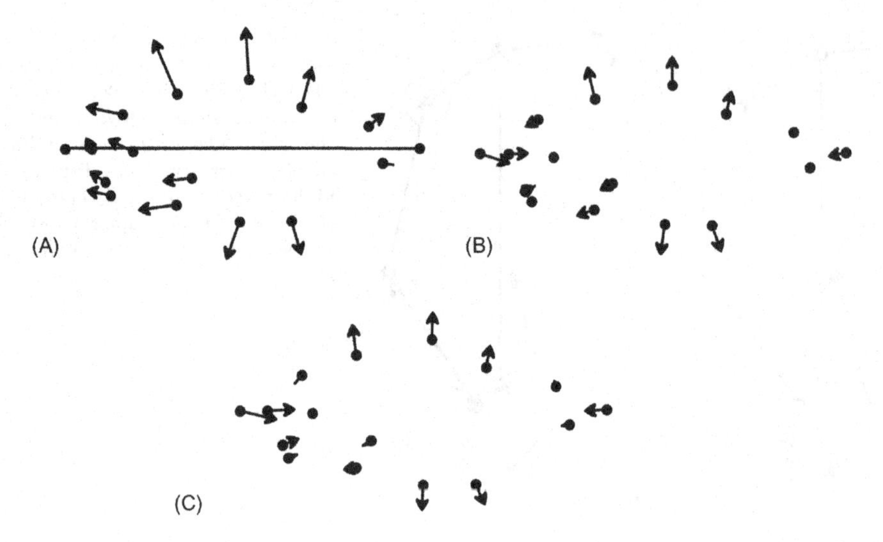

FIGURE 3.14 Ontogenetic change in body shape of *S. gouldingi* depicted by vectors of relative landmark displacement computed from three superimpositions. (A) BC coordinates from the 1–7 baseline; (B) GPA; (C) RFTRA.

ratio, but by decreasing the denominator. Because we come to the pictures informed by our knowledge that body length increases over ontogeny, it may be difficult to grasp that it decreases relative to depth. We would probably avoid saying that body length decreases relative to depth, simply because that phrasing is disconcerting to biological intuition; instead, we would say that depth increases relative to length. When pictures show a *relative* decrease in a feature that is increasing in *absolute* length, readers may need some explanation of the unexpected contrast. In particular, it is important to explain that the decrease is in *relative* (not absolute) length.

Other apparent inconsistencies between pictures can also be reconciled, usually by concentrating on the changes in *relative positions* of landmarks rather than on the vectors at individual landmarks. It may take a lot of practice before this is easy. For example, look at the circled landmark in Figure 3.15. If you look only at this landmark, the results from the different superimpositions appear to be inconsistent. That landmark appears to "move" quite far anterodorsally in the BC superimposition (Figure 3.15A), but much less and in two different directions in the other superimpositions: anteroventrally in GPA (Figure 3.15B), and almost entirely ventrally in RFTRA (Figure 3.15C). However, none of these statements actually reflect what the pictures show. None of the pictures shows the independent movement of any one point in isolation; rather, what they show is the relative displacements of all points.

We get a better indication of the displacement of the circled landmark relative to neighboring landmarks by "connecting the dots" – drawing line segments between landmark locations to approximate the profile of *S. gouldingi*'s head. In Figure 3.16, we show the same superimpositions and ontogenetic displacements of landmarks as in Figure 3.15, just

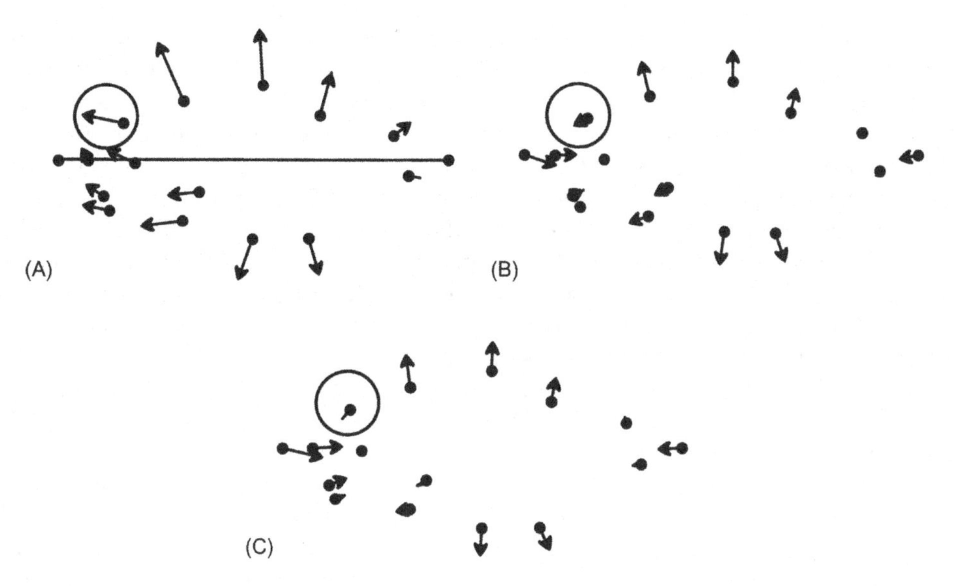

FIGURE 3.15 Ontogenetic change in body shape of *S. gouldingi*, highlighting the circled landmark at the epiphyseal bar. Displacements are shown in three superimpositions: (A) BC coordinates from the 1–7 baseline; (B) GPA; (C) RFTRA.

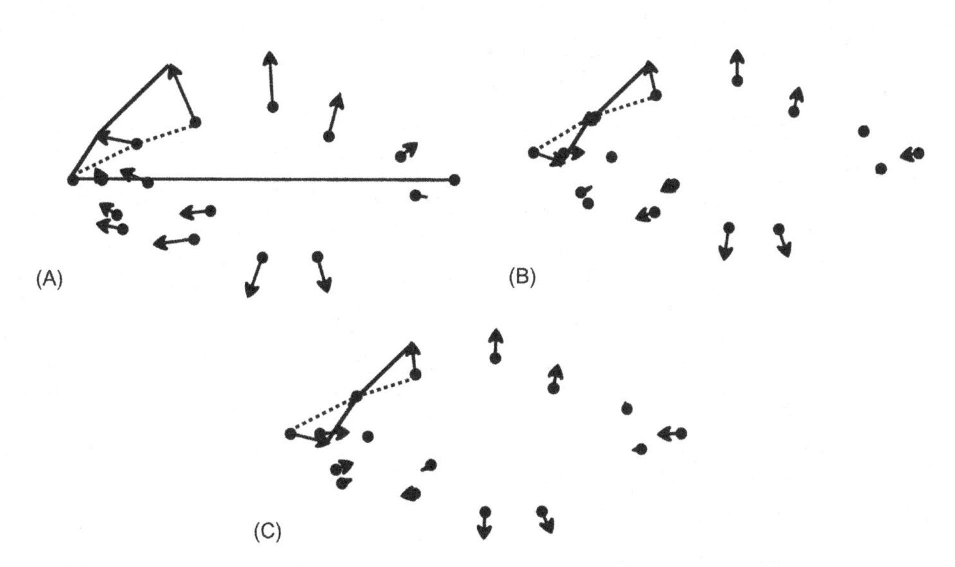

FIGURE 3.16 Ontogenetic changes in dorsal head profile are highlighted by drawing in line segments between the locations of landmarks at two different ages: early in ontogeny (dotted line) and late in ontogeny (solid line). Displacements are shown in three superimpositions: (A) BC coordinates from the 1–7 baseline; (B) GPA; (C) RFTRA.

adding lines to show the relative positions of the landmarks early in ontogeny (dotted lines connecting the bases of the arrows) and late in ontogeny (solid lines connecting the tips of the arrows). Now we can see that the profile of the head is initially fairly shallow (nearly a straight line across all three points), and becomes much steeper (particularly between the tip of the snout and the second landmark – the one that was circled). All three superimpositions show this same change in profile. Despite apparent discrepancies in the displacements of individual landmarks, the relationships among the landmarks are consistently represented. Before we can interpret the results in terms of these vectors of relative landmark displacement, we must become accustomed to what these vectors represent. The individual vectors do not show changes *at* landmarks; rather, the differences between vectors show changes between the landmarks.

GPA can yield some visually unsettling results, such as rotated axes of symmetry. For example, in analyses of rodent skulls (Zelditch et al., 2003), the coordinates of the bilaterally homologous landmarks on the right and left side were averaged to avoid inflating degrees of freedom. When the results are shown by a GPA (Figure 3.17A), the midline of the skull appears to rotate, but that cannot happen; the midline is the midline regardless of variation in shape. Not only is this apparent rotation of the midline visually troubling, it also complicates the interpretation of the results. One superimposition method that is designed to overcome such a problem is the "Sliding Baseline Registration" (Figure 3.17B). The sliding baseline registration (SBR), developed by David Sheets in collaboration with Mark Webster (Webster et al., 2001) and Keonho Kim (Kim et al., 2002) reduces the

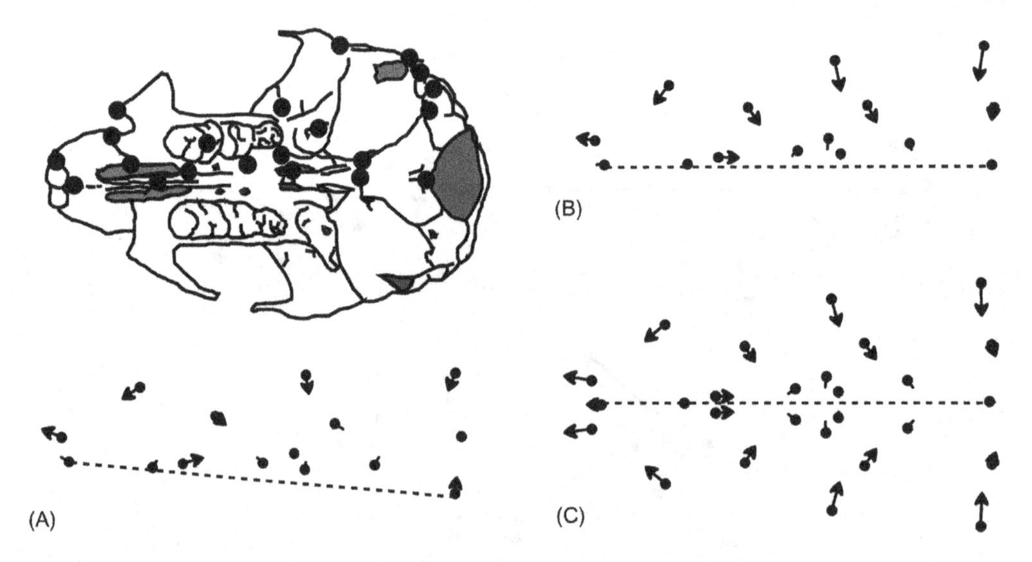

FIGURE 3.17 Superimposition of forms with an axis of symmetry – ontogenetic changes in a cotton rat skull. Dotted lines connect landmarks on the mid-sagittal axis. (A) Landmark displacements inferred from GPA which appears to indicate translation and rotation of the midsagittal axis; (B) landmark displacements inferred from SBR, which does not appear to suggest translation and rotation of the mid-sagittal axis; (C) landmark displacements inferred from GPA on symmetrized and back-reflected configurations, which also does not appear to suggest translation and rotation of the mid-sagittal axis.

disadvantages of aligning landmark configurations along one edge, as in the two-point registration. Configurations are scaled to unit centroid size and, because of that, baselines will usually vary in length. Consequently, the two endpoints cannot be superimposed simultaneously. Instead, their Y-coordinates are fixed at zero and their X-coordinates are allowed to vary as necessary to align the X-coordinates of the centroids at zero, in effect sliding the baseline along the X-axis (the Y-coordinate of the centroid is the average perpendicular distance of the landmarks from the baseline after scaling to unit centroid size). Because SBR prevents rotation of the baseline, it yields a more realistic representation of the data – in this case, of the ontogenetic change in skull shape. Actually, a very similar picture can be obtained by a GPA on the unreflected data, or on the back-reflected data obtained by duplicating the averaged and reflected data back across the midline (Figure 3.17C). In general, reconstructing the whole skull makes a more interpretable picture (one that looks more like the organism), so it might be useful to present results in these terms even if the statistical analyses used the GPA coordinates computed for the reflected and averaged half skull.

Of the various superimposition methods discussed in this chapter, the one that is most widely used is GPA for reasons that will become clearer in the next chapter – this method is grounded in the mathematical theory of shape. Configurations of landmarks are manipulated using the three operations that do not alter shape as defined by Kendall. These operations are used in a manner that removes all differences that are not shape differences. The configurations produced by this procedure are those that map to points in the shape spaces implied by Kendall's definition of shape. The computed distances between these configurations are the distances between points in those spaces, or certain linear approximations of those spaces. The characteristics of these metrics are well known, providing a secure and stable foundation for biological shape analysis, and the pictures embody the results.

References

Bookstein, F. L. (1996). Combining the tools of geometric morphometrics. In L. F. Marcus, M. Corti, A. Loy, G. J. P. Naylor, & D. E. Slice (Eds.), *Advances in morphometrics* (pp. 131–151). New York: Plenum.

Bookstein, F. L. (1991). *Morphometric tools for landmark data: Geometry and biology.* Cambridge: Cambridge University Press.

Chapman, R.E. (1990). Conventional Procrustes methods. In F. J. Rohlf, & F. Bookstein (Eds.), *Proceedings of the Michigan Morphometrics Workshop* (pp. 251–267). Ann Arbor: University of Michigan Museum of Zoology.

Claude, J. (2008). *Morphometrics. R.* New York: Springer.

Dryden, I. L., & Mardia, K. V. (1998). *Statistical shape analysis.* New York: John Wiley & Sons.

Kendall, D. (1977). The diffusion of shape. *Advances in Applied Probability, 9*, 428–430.

Kim, K., Sheets, H. D., Haney, R. A., & Mitchell, C. E. (2002). Morphometric analysis of ontogeny and allometry of the middle ordovician trilobite *Triarthrus becki*. *Paleobiology, 28*, 364–377.

Morrison, D. F. (1990). *Multivariate statistical methods.* New York: McGraw Hill.

Press, W. H., Flannery, B. P., Teukolsky, S. A., & Vetterling, W. T. (1998). *Numerical Recipes in C: The Art of Scientific Computing.* Cambridge University Press.

Rao, C. R. (1973). *Linear statistical inference and its applications.* New York: John Wiley & Sons.

Rohlf, F. J., & Slice, D. E. (1990). Extensions of the Procrustes method for the optimal superimposition of landmarks. *Systematic Zoology, 39*, 40–59.

Sampson, P. D., Bookstein, F. L., Sheehan, H., & Bolson, E. L. (1996). Eigenshape analysis of left ventricular outlines from contrast ventriculograms. In L. F. Marcus, M. Corti, A. Loy, G.J.P. Naylor, & D.E. Slice (Eds.), *Advances in Morphometrics* (pp. 131–152). Nato ASI Series, Series A: Life Science. New York.

Walker, J. A. (2000). Ability of geometric morphometric methods to estimate a known covariance matrix. *Systematic Biology, 49*, 686–696.

Webster, M., Sheets, H. D., & Hughes, N. C. (2001). Allometric patterning in trilobite ontogeny: testing for heterochrony in *Nephrolenellus*. In M. L. Zelditch (Ed.), *Beyond heterochrony: The evolution of development* (pp. 105–144). New York: John Wiley & Sons.

Zelditch, M. L., Lundrigan, B. L., Sheets, H. D., & Garland, T. (2003). Do precocial mammals develop at a faster rate? A comparison of rates of skull development in *Sigmodon fulviventer* and *Mus musculus domesticus*. *Journal of Evolutionary Biology, 16*, 708–720.

CHAPTER

4

Theory of Shape

This chapter covers the basic theory of shape, beginning with the definition of shape and proceeding through the characterization of several theoretical spaces. Some of the mathematics may look a bit difficult, but it is important to grasp the basic ideas, which we present verbally as well as mathematically. These ideas will reappear in the next chapter, because they form the core of geometric morphometrics. Interestingly, many of the techniques used in geometric morphometrics were developed independently of this theory even though they are justified by it. As the theory matured, it became possible to synthesize a large body of techniques that had been developed independently of each other and to explicate their interrelationships. Perhaps most importantly, this theory also allows us to judge whether or not particular methods are valid. The theory provides the underlying justifications for all our techniques, thereby allowing us to make inferences about shape without worrying that those inferences are somehow based on arbitrary or mathematically faulty choices that we happened to make in the course of our analyses. Freed of such concerns, we can concentrate on the biological meaning of the results.

It would be possible to learn techniques without understanding any of this theory – but don't. Without the theory it is impossible to say why some methods are right and others are not. In effect, you would have to memorize a list of "dos" and "don'ts" by rote without understanding why the "dos" are "dos" and the "don'ts" are "don'ts." Learned in that way, it might seem that there are lots of picky rules and dogma, but these rules are not picky and they are not a matter of dogma. Rather, they all logically (and mathematically) follow from the mathematical theory of shape. In fact, they follow from the definition of shape. Because this definition is central to geometric morphometric theory, we begin there, developing it further than in previous chapters.

Geometric Morphometrics for Biologists
DOI: http://dx.doi.org/10.1016/B978-0-12-386903-6.00004-6

THE DEFINITION OF SHAPE

David Kendall's (1977) definition of shape is the basis of all that will follow in this chapter, and indeed of any consideration of the meaning of shape:

> Shape is all the geometrical information that remains when location, scale and rotational effects are filtered out from an object.

This statement is both intuitively reasonable and mathematically useful. It suits our intuitions because we can all agree that moving an object from one place to another does not change its shape; that operation, called translation, obviously does not alter shape. For example, Figure 4.1A shows the translation of a shape along an axis, and this motion has no consequences for shape. Likewise, rotating the object does not change shape (Figure. 4.1B), and neither does enlarging or reducing an image (a manipulation called rescaling; Figure 4.1C). Although it may be obvious that translation, rotation and rescaling do not alter shape, it may not be obvious that this fact provides a mathematically useful definition of shape.

To a non-mathematician this definition may seem a bit odd, because it defines shape by what *does not* alter it rather than in terms of what shape *is* or by the operations that *do* alter it. However, the definition is useful because it means that any operation *not* on

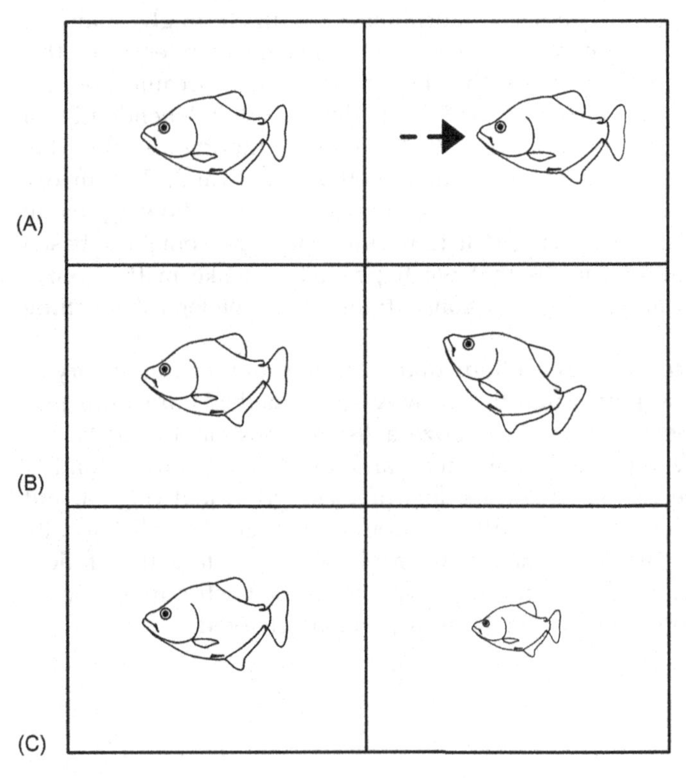

FIGURE 4.1 The operations that do not alter shape: (A) translation; (B) rotation; (C) scaling.

that list *does* affect shape. Also, the list of operations that do not alter shape is useful because we know that we are free to use those operations when we compare shapes mathematically.

The entire theory of geometric morphometrics follows from the definition of shape, so we need to develop it further. First, we need a more precise definition of a landmark. When we discussed the criteria for choosing them in Chapter 2, we emphasized that the criterion of homology has mathematical as well as biological implications. The mathematical implication follows from the formal definition of a landmark (Dryden and Mardia, 1998):

> A landmark is a point of correspondence on each object that matches between and within populations.

The concept of matching encoded in that passage is not necessarily one of biological homology, but the idea of correspondence is essential to the mathematical theory of shape. If the landmarks do not correspond, we cannot compare shapes.

Another crucial idea is that of a *configuration of landmarks*; the full set of landmarks recorded for each specimen. All comparisons of shapes are between matching configurations of landmarks, not between individual landmarks (analyzed separately). An individual landmark is not an object of comparison because it does not satisfy the definition of shape. The objects of comparison are entire configurations comprised of K landmarks (where K refers to the number of landmarks), each of which has M coordinates (i.e. $M = 2$ for planar shapes). For example, in the case of the piranhas introduced in the second chapter, $K = 16$ and $M = 2$. Whatever the number of landmarks and coordinates, our analyses and conclusions are based on the *entire set*. Thus, if we have 16 landmarks with two coordinates apiece, we have one shape – not 32 variables. No one landmark (and no one coordinate) is a shape variable in its own right. Instead, we view each shape as the entire configuration and we analyze samples of entire configurations.

This is a very different view of measurement (and variables) from that commonly encountered in traditional morphometrics, where a single measurement might be viewed as a variable, meriting analysis in its own right. It is common to analyze measurements separately and to draw biological conclusions from them individually. Sometimes, the conclusions based on one measurement conflict with conclusions based on another, and the inference often drawn in such situations is that the processes are trait-specific. In geometric morphometrics, individual measurements are not traits or even variables. Rather, a shape variable is the entire vector of coefficients representing the complete difference in landmark configurations between samples or, alternatively, the entire vector of coefficients measuring the covariance between the landmark configurations and some other variable (e.g. size).

This view of shape as a configuration of landmarks is central to the theory of geometric morphometrics. Recognizing that, and conforming to the requirements it imposes on analytic methods, is crucial. It may seem biologically unreasonable to treat an entire shape as a single entity, but the pay-off for doing so is the guarantee that our results do not depend on arbitrary choices we happened to make in the course of an analysis. The reward for following what might seem like a rigid set of rules is the rigor and power of these methods, as well as the visual appeal of the graphics.

MORPHOMETRIC SPACES

Given the definition of shape, we can now develop the mathematical idea of morphometric spaces. We begin by defining some additional terms.

The Configuration Matrix

A configuration matrix represents an entire configuration of landmarks. It is a $K \times M$ matrix of Cartesian coordinates that describes a particular set of K landmarks in M dimensions (Dryden and Mardia, 1998). When we talk about a $K \times M$ matrix, we mean that the matrix has K rows and M columns; each of the K rows represents a specific landmark on a specimen, with M Cartesian coordinates. For example, the simplest shape we might want to study is a triangle with landmarks located at the three vertices of the triangle. Calling the coordinates of the first vertex X_1 and Y_1, and those of the second vertex X_2 and Y_2, and those of the third vertex X_3 and Y_3, the configuration matrix of triangle $\mathbf{X}$ is:

$$\mathbf{X} = \begin{bmatrix} X_1 & Y_1 \\ X_2 & Y_2 \\ X_3 & Y_3 \end{bmatrix} \tag{4.1}$$

It is often useful to represent this same landmark configuration as a *row vector*, in which the landmark coordinates are listed along a single row in $K \times M$ columns:

$$\mathbf{X} = [X_1 \quad Y_1 \quad X_2 \quad Y_2 \quad X_3 \quad Y_3] \tag{4.2}$$

This contains exactly the same information, represented slightly differently. Given a set of landmark coordinates in row vector form, you can easily convert it to a configuration matrix (the representation you might prefer at any given time depends on the particular task or software at hand).

For example, the configuration matrix of the triangle shown in Figure 4.2 is:

$$\mathbf{X} = \begin{bmatrix} -1 & -1 \\ 1 & -1 \\ 0 & 1 \end{bmatrix} \tag{4.3}$$

The row vector representing the same triangle would be:

$$\mathbf{X} = [-1 \quad -1 \quad 1 \quad -1 \quad 0 \quad 1] \tag{4.4}$$

Configuration Space

The configuration space is a set of all possible $K \times M$ matrices describing all possible sets of landmark configurations for that given K and M. For example, a 16×2 dimensional configuration space is the space of all configurations having 16 two-dimensional landmarks. That space encompasses *all* possible configurations for those 16 landmarks with two coordinates. Should we record the locations of 16 landmarks on a two-dimensional

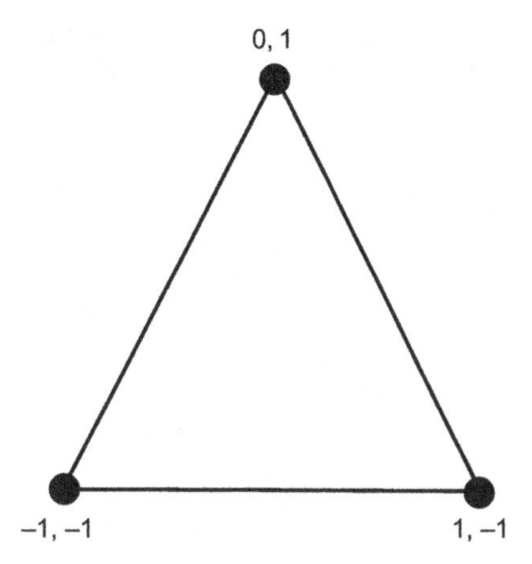

FIGURE 4.2 Example of a triangle.

image of the body of a piranha, and 16 landmarks on a two-dimensional image of a rat skull, both configuration matrices are in the same configuration space. Obviously, the landmarks on these two structures don't have any kind of homology with one another, so comparisons would be meaningless, but the measurements are in the same configuration space. Clearly, any group of biologically similar organisms (with matched landmarks) will occupy a relatively small part of configuration space because the locations of their corresponding landmarks will be fairly similar. For example, in the 16×2 configuration space, piranhas will occupy a very small part of the space – that space also contains the 16×2 two-dimensional coordinates of rat skulls.

The configuration space of K landmarks with M coordinates per landmark has $K \times M$ dimensions. To specify the location of any shape in that space, we must specify $K \times M$ components of a vector (or elements in a matrix).

Position or Location of a Configuration Matrix

The position of a configuration matrix is the location of the centroid of that matrix. This centroid is the M-dimensional vector (two in the case of the two-dimensional landmarks of piranhas) whose components are the averages of the X and Y coordinates of the landmarks (in the two-dimensional case), so the centroid position is given by:

$$X_C = \frac{1}{K}\sum_{j=1}^{K} X_j$$
$$Y_C = \frac{1}{K}\sum_{j=1}^{K} Y_j \tag{4.5}$$

For example, Figure 4.3 shows the centroid position of the triangle seen earlier, which is located at (0, −0.333).

A configuration matrix is said to be *centered* if the average of all the coordinates is zero. Centering is useful because it often simplifies the mathematics; it is done by translating the configuration along the X- and Y-axes. That translation is done by adding a constant

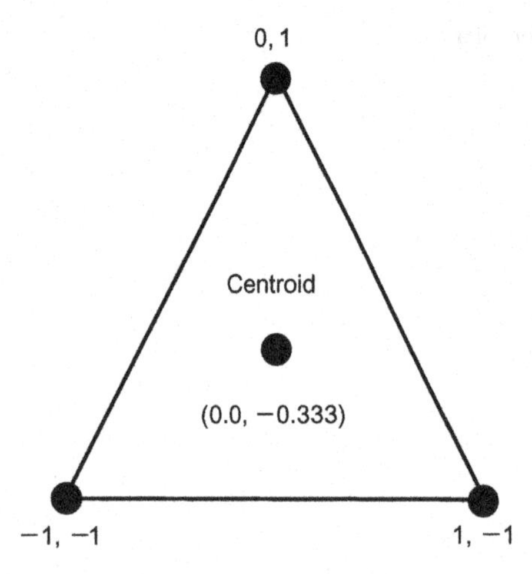

FIGURE 4.3 The centroid of the triangle in Figure 4.2. The coordinates of the centroid are the averaged coordinates of the three vertices.

(positive or negative) to the *X*- and *Y*-coordinates. To do this we first calculate the *X* and *Y* centroid coordinates of the configuration matrix **X** as in Equation 4.5, then subtract the centroid positions from each coordinate to form the centered configuration matrix **XC**:

$$\mathbf{XC} = \begin{bmatrix} (X_1 - X_C) & (Y_1 - Y_C) \\ (X_2 - X_C) & (Y_2 - Y_C) \\ \vdots & \vdots \\ (X_K - X_C) & (Y_K - Y_C) \end{bmatrix} \tag{4.6}$$

Two configuration matrices that differ only in the position of the centroid are not different shapes (they differ only by translation, one of the operations that do not alter shape).

Size of a Configuration Matrix

Before we can coherently talk about *scale*, we need to define what we mean (mathematically) by the term *size*. For configuration matrices, a number of different, non-equivalent size measures have been used. It is not possible to say that one size measure is "correct" or "preferable", but it is important to explain the consequences of making a particular choice. The most commonly used size measure in geometric morphometrics is called *centroid size*, which is favored because it does not induce a correlation between size and shape (at least under some error models, Bookstein, 1991), hence we restrict our discussion of size to that particular measure. The centroid size (*CS*) of a configuration (**X**) is:

$$CS(\mathbf{X}) = \sqrt{\sum_{i=1}^{K}\sum_{j=1}^{M}(\mathbf{X}_{ij} - C_j)^2} \tag{4.7}$$

where the sum is over the rows *i* and columns *j* of the matrix **X**. $\mathbf{X}_{ij}$ is a standard notation from linear algebra specifying the value located on the *i*th row and *j*th column of the

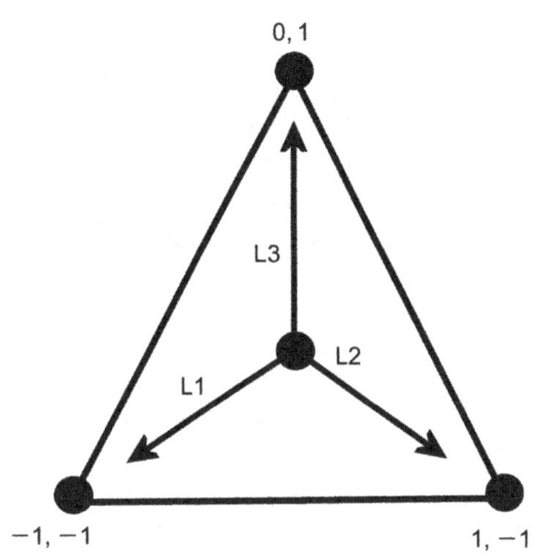

FIGURE 4.4 Centroid size of the triangle in Figure 4.2, calculated as the sum: $(L1^2 + L2^2 + L3^2)^{1/2} = 2.16$.

matrix **X** and, in this case, C_j stands for the location of the jth component of the centroid. C_1 is the X-coordinate of the centroid and C_2 is its Y-coordinate.

Centroid size is thus the square root of the sum of the squared distances of the landmarks from the centroid. The distances from the centroid to each landmark of the triangle are shown in Figure 4.4; the centroid size of this triangle is simply the square root of the sum of the squared lengths of these lines. Centroid size is not altered by changing the position of the configuration, because this leads to all landmarks (and the centroid) changing by a common amount. Similarly, multiplying the configuration matrix **X** by a constant factor increases centroid size by the same factor. Two configurations of landmarks that differ only in centroid size do not differ in shape (they differ only in scale).

PRE-SHAPE SPACE

As we stated above, every configuration of K landmarks having M coordinates can be thought of as a point in a space with $K \times M$ dimensions. (To avoid confusion, we should make it clear that by "point" in this context we mean an individual shape, an entire configuration of landmarks, not one landmark.) Some of the configurations in this space differ only in centroid size; others differ only in location (coordinates of the centroid). We can define a subset of configurations that do not differ in location or size by placing two restrictions on each configuration matrix: (1) that it be centered, and (2) that centroid size be one. These restrictions define a space called pre-shape space (Dryden and Mardia, 1998). In practice, we translate and scale each of the original configurations in our data so that the new configurations meet the restrictions of pre-shape space. In doing this, we are using two of the three operations that do not alter shape. Each of the new configurations is a centered pre-shape.

The Shape of Pre-Shape Space

The two requirements imposed on this space mean that the summed squared landmark positions add up to one. The consequences of that property can be understood by considering the set of points satisfying the restriction in an ordinary two-dimensional space: the set of points is centered on the origin (0,0), and each point in the set has coordinates satisfying the equation $X^2 + Y^2 = 1$. The set of points is a circle of radius one, centered on the origin. This circle is a one-dimensional subspace (a curve) inhabiting a two-dimensional space (a plane). Knowing that all points are equidistant from the center means that we need specify only the direction of a point from the center to define it uniquely; thus, the location of any point on the circle can be described sufficiently by a single dimension (direction). Extending this to a three-dimensional space, we now have the set of all points (X, Y, Z) centered on the origin (0,0,0) such that $X^2 + Y^2 + Z^2 = 1$. This is the surface of a sphere of radius one, centered on the origin, and it is a two-dimensional subspace within a three-dimensional space. Again, the constraint that all points are on the surface allows us to describe the location of a point by giving a direction from the center; the only difference from the circle is that we now need two components to describe that direction (e.g. latitude and longitude). So, in talking about a pre-shape space, we are talking about the surface of a hypersphere centered on the origin, which is the generalization of an ordinary sphere in $K \times M$ dimensions. In that general case, we have:

$$\sum_{i=1}^{K} \sum_{j=1}^{M} (\mathbf{X}_{ij})^2 = 1 \tag{4.8}$$

which states that the sum of all squared landmark coordinates is one. That hypersphere is simply the equivalent of a sphere in more than three dimensions.

We can determine the number of dimensions in pre-shape space by considering the number of dimensions that were lost in the transition from configuration space. One dimension is lost in fixing centroid size to one, eliminating the size dimension of the configuration space. Another M dimensions are lost in centering the configurations; eliminating the M dimensions needed to describe location (the coordinates of the centroid). Thus, in moving from configuration space to pre-shape space, we moved to a space that has $M + 1$ fewer dimensions, which is:

$$KM - (M + 1) = KM - M - 1 \tag{4.9}$$

For two-dimensional configurations of landmarks, pre-shape spaces have $2K - 3$ dimensions; so the pre-shape space for triangles has three dimensions. For three-dimensional configurations of landmarks, pre-shape spaces have $3K - 4$ dimensions.

Returning to the three-dimensional sphere (because most of us have trouble imagining spaces having more than three dimensions), you should be imagining pre-shape space to be a hollow ball of radius one, centered at the origin (0, 0, 0). Arrayed on the two-dimensional surface of this ball are points representing individual configurations of landmarks. The two restrictions we have imposed on our configuration matrices mean that the configurations in this set do not differ in scale or location; we have used the operations of translation and scaling to remove the effects of (differences in) location and scale. We have not yet rotated the shapes to remove the effects of rotation (that comes later, as we move from pre-shape space to shape space). Thus, configurations of landmarks that differ only by a rotation are located at different points in pre-shape space, as are configurations that differ only in shape. This underscores an important point (which some may find counter-intuitive): as we said earlier, configurations that differ only by a rotation (such as those shown in Figure 4.1B) do not differ in shape. Because we have not yet removed all three effects mentioned in Kendall's definition of shape (location, scale and rotation), we have not yet reached shapes. At present we are concerned with pre-shapes, i.e. configurations that may differ by a rotation, by a shape change or by some combination of the two. In pre-shape space, configurations that differ only by rotation are different points, as are configurations that differ only in shape.

Fibers in Pre-Shape Space

To visualize the locations in pre-shape space of configurations that differ only in rotation, we introduce the term *fiber*. A fiber (in the context of our particular discussion of pre-shape space) consists of the set of all the points in pre-shape space that can be obtained by rotating a particular centered pre-shape. The fiber is a circular arc that comprises the set of all points in pre-shape space that can be "reached" by rotating the pre-shape matrix. Figure 4.5 depicts the concept of fibers as an arc on the surface of a sphere (ignoring the higher dimensionality of a pre-shape hypersphere). Two fibers are shown: arcs 1 and 2. Arc 1 is the set of all possible rotations of the pre-shape $\mathbf{Z}_1$, and arc 2 is the set of all possible rotations of the pre-shape $\mathbf{Z}_2$. For a less abstract visualization of the concept of fibers, we have drawn a cartoon (Figure 4.6) representing four fibers (in columns); the triangles within a column differ solely by a rotation, whereas those in different columns also differ in shape. (This visualization is somewhat limited, because a row does not accurately

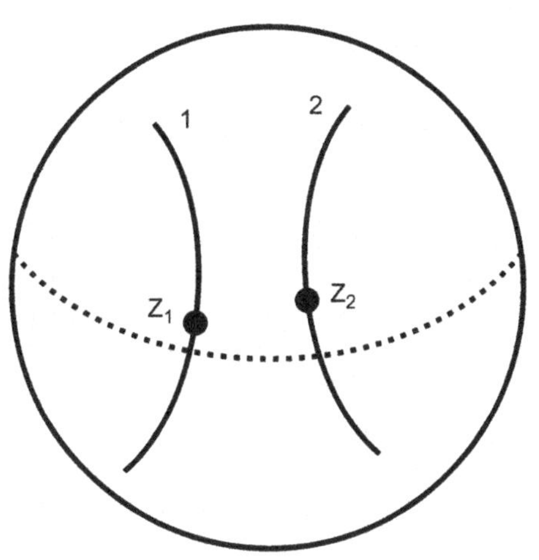

FIGURE 4.5 Fibers in pre-shape space. The points Z_1 and Z_2 are the locations of pre-shapes on the hypersphere (centered and scaled matrices computed from two original matrices X_1 and X_2, which are not shown). Curve 1 passing through Z_1 is a fiber, the set of all centered and scaled pre-shapes differing from Z_1 only by rotation. Curve 2 is a fiber of pre-shapes differing from Z_2 only by rotation. (The dotted curve is the "equator" of the hypersphere, and does not represent a fiber.)

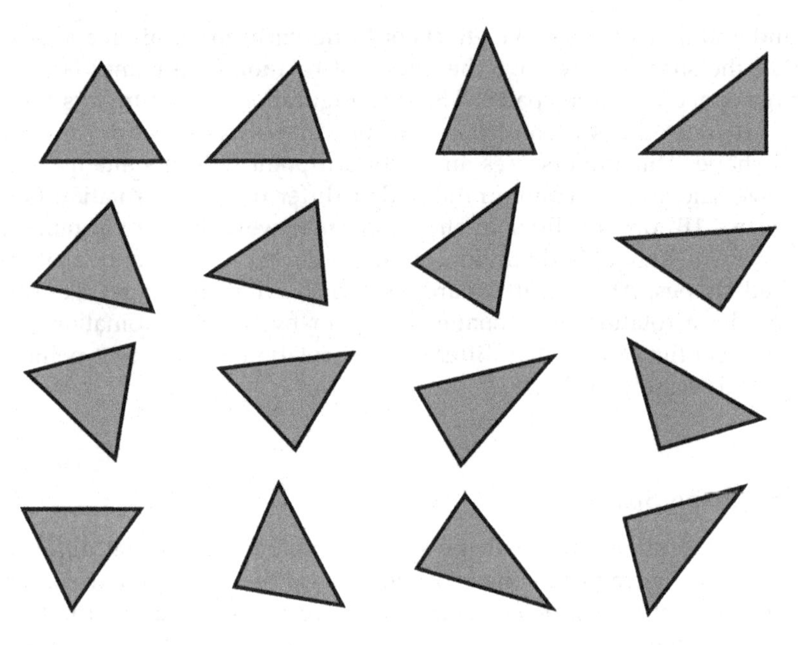

FIGURE 4.6 An alternative visualization of the concept of a fiber. Each column shows rotations of a single shape; triangles in different columns differ in shape. Each column represents a single fiber.

represent the number of dimensions needed to describe shapes of triangles, as explained in the next section.)

With the concept of fiber in hand, it is now possible to talk about the separation of shapes and the distance between them. Figure 4.7 shows the same two fibers on the curved surface of the pre-shape space hypersphere as in Figure 4.5. In addition, Figure 4.7 shows an arc (ρ) crossing the surface from one fiber to the other, and the chord (D_p) that passes through the interior of the hypersphere between the same two surface points. We can draw many such arcs connecting a rotation of the pre-shape $\mathbf{Z}_1$ with a rotation of the pre-shape $\mathbf{Z}_2$. The arc we want is the shortest one – that is, the one connecting fibers at their "point of closest approach". Finding the shortest possible distance between points is a common tactic for defining distances between objects in spaces. When we find that distance, we will find the rotation that is optimal in the sense of being the minimum distance between shapes. The length of this arc is known as the *Procrustes distance*, and it is quantified by determining the angle between the radii that connect the center of the hypersphere to the point at which the fibers most closely approach each other. Figure 4.8 shows the cross-section through the pre-shape space in the plane defined by those two radii. The angle subtended by the arc is ρ; the chord length is D_p. The length of the arc is equal to ρ (in radians) times the length of the radius. Because we have constrained the radius to a length of one, the length of the arc is the value of the angle. This value ranges from zero to $\pi/2$; at $\pi/2$, the hemisphere may always be oriented so that one specimen is at the pole, the farthest location the second may be located at is the equator.

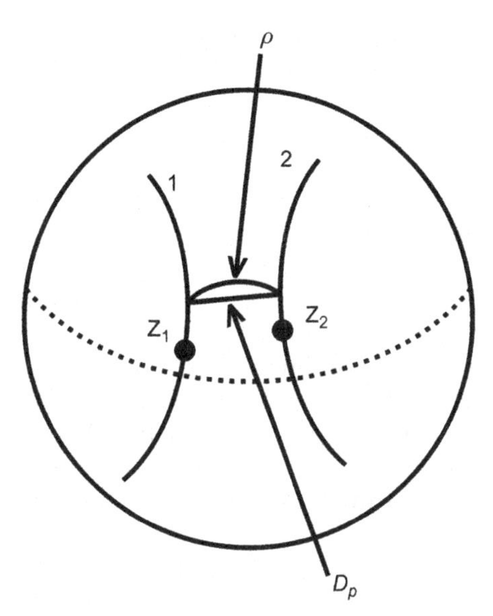

FIGURE 4.7 Determining the distance between the fibers of pre-shapes. The arc ρ is the shortest distance across the surface of the hypersphere from fiber 1 to fiber 2. The length of the arc is the Procrustes distance. The length of the chord (D_p) is the partial Procrustes distance.

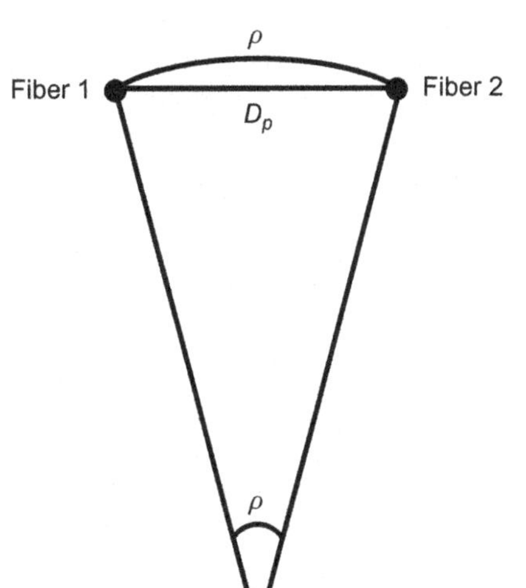

FIGURE 4.8 A slice through pre-shape space showing the Procrustes distance (ρ) and the partial Procrustes distance (D_p).

SHAPE SPACES

In the previous section, we used the points of closest approach on the pre-shape fibers to define the distance between two shapes. Now, we use the same criterion to construct a shape space. This shape space contains one configuration from each fiber, one rotation of a centered pre-shape. Conventionally, we select a convenient orientation of one pre-shape to serve as the *reference* configuration; every other *target* (or subject) configuration is selected as the rotation corresponding to the point of closest approach of its pre-shape fiber to the reference. That is, the orientation is chosen to minimize the Procrustes distance between the target and reference. The points on those fibers that are farther from the reference differ from it in both shape and rotational effects. By selecting the point of closest approach, we reduce each fiber of pre-shapes to a single point (a shape); consequently, configurations in this set differ only in shape.

The shape space we just described has fewer dimensions than the pre-shape space from which it was derived. The number of dimensions lost in the transition is given by:

$$\frac{M(M-1)}{2} \tag{4.10}$$

where M is the number of landmark coordinates. For two-dimensional landmarks, Equation 4.10 simplifies to one, which reflects the fact that a planar shape can only be rotated about its centroid on one axis (the axis perpendicular to the plane of the shape) and still stay in the same plane. Consequently, shape spaces of two-dimensional configurations of K landmarks have $2K-4$ dimensions. The four lost dimensions are those describing differences in size (−1), translation (−2) and rotation (−1). For three-dimensional landmarks, Equation 4.10 simplifies to three, which reflects the fact that a three-dimensional shape can be rotated about its centroid on three distinct orthogonal axes in the three-dimensional coordinate space. Subtracting three from the $3K-4$ dimensions of the pre-shape space (from Equation 4.9) yields $3K-7$ dimensions for shape spaces of three-dimensional shapes, which simplifies to five dimensions for the shape space of tetrahedra. The seven lost dimensions are those describing differences in size (−1), translation (−3) and rotation (−3).

In the special case of triangles, the shape spaces defined above with centroid size still equal to one, are the familiar two-dimensional surfaces of three-dimensional spheres. Because this is a reasonably simple geometry to visualize and illustrate, we will focus on triangles before returning to the general case. In Figure 4.9, we show half of a space determined by using the equilateral triangle as the reference. Because we retain the constraints that each triangle is centered and scaled to centroid size of one, the hemisphere has a radius of one. For convenience, the space is oriented so that the point representing the equilateral triangle configuration is located at the pole. At the equator with maximal difference from the reference are various reflections of the reference (Rohlf, 1999, 2000 [see Figure 1]). Collinear triangles (with all points along a single line) are located at a Procrustes distance of $\pi/4$ from the reference. Although the shape space just described is a useful construction, it does not satisfy the mathematician's urge to find the smallest distances between configurations with those shapes. To illustrate this point, we consider a slice through the polar axis of the hemisphere of triangles just described (Figure 4.10). As in pre-shape space, the distance of a shape (A) from the reference is ρ. The angle and the arc length are unchanged because the dimension eliminated in the transition from pre-shape space to this shape space did not contribute to the measurement of the shape difference. It should be apparent in Figure 4.10 that the arc across the surface is not the shortest possible distance between the two shapes. The chord passing through the interior of the hemisphere would be shorter, but it is still not the shortest possible distance between configurations with those shapes. We obtain that shortest possible distance, and the relevant configurations, by changing the constraint on the centroid sizes of the two

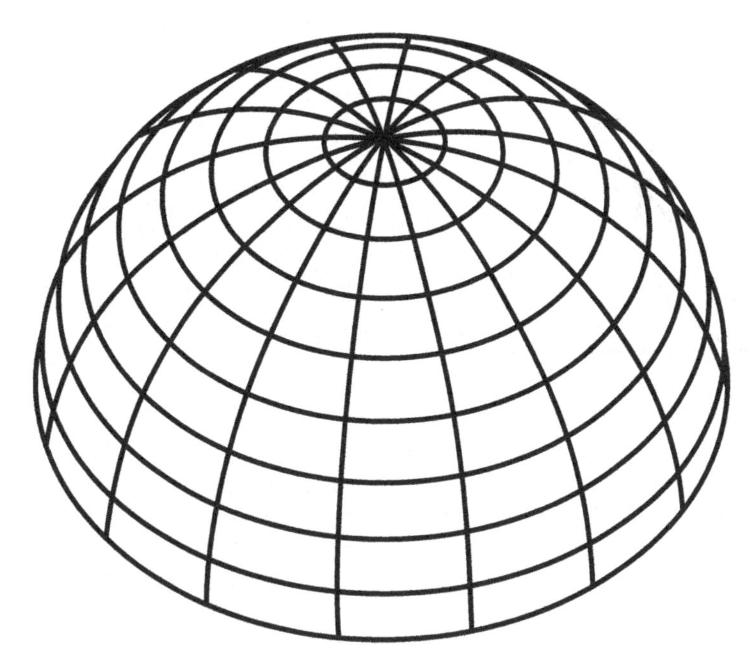

FIGURE 4.9 Half of the space of triangles that have been centered, scaled to unit centroid size and aligned with a centered, scaled equilateral triangle. The equilateral triangle is at the pole. Lines of "latitude" represent shapes equidistant from the equilateral triangle. The "equator" corresponds to the set of triangles with most different from the reference, which for the case of equilateral triangles, is a reflected version of the reference (Rohlf, 1999).

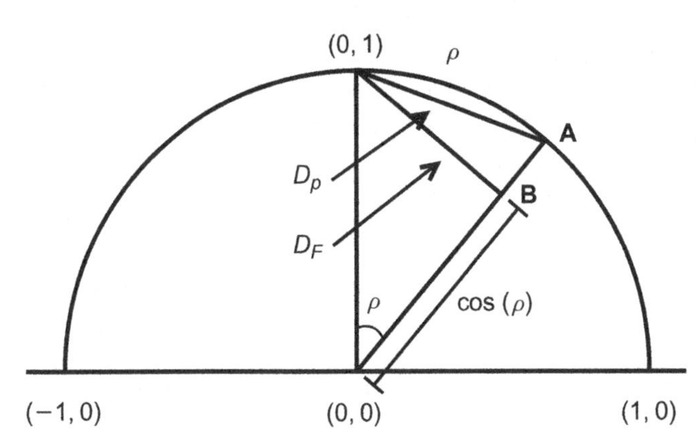

FIGURE 4.10 A slice through part of the space of aligned triangles at unit centroid size, showing the relationships among the distances between the reference shape (at 0, 1) and **A**. The semicircle is a cross-section of the space, which is a hemisphere of radius one. The length of the arc is the Procrustes distance (ρ), the length of the chord is the partial Procrustes distance (D_p), and the shortest possible distance (obtained by relaxing the constraint on centroid size, producing the configuration **B**) is the full Procrustes distance (D_F).

configurations. Conventionally, we keep the centroid size of the reference at one, and allow the centroid size of the target to adopt the value that minimizes its distance from the reference. This is equivalent to allowing the target to travel along its radius while the reference stays on the surface. The point along the radius where the second shape is closest to the target is some distance below the surface of the shape space, reflecting a reduction of the centroid size of the target. This point (B) is defined by the line that is perpendicular to the target's radius and passes through the reference's position on the surface. The corresponding centroid size of the target is $\cos(\rho)$; the distance between configurations is $\sin(\rho)$ and is called the *full Procrustes distance* (D_F).

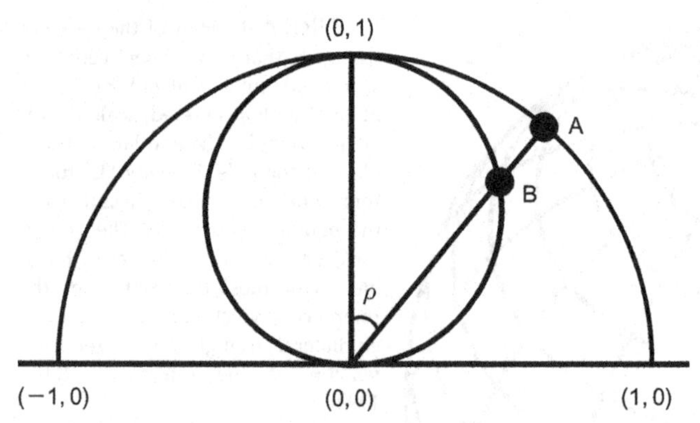

FIGURE 4.11 The relationship of Kendall's shape space to the space of aligned triangles scaled to unit centroid size. The outer semicircle is the cross-section of the space of aligned triangles scaled to unit centroid size, as in Figure 4.10. The inner circle is a cross-section through Kendall's shape space, which is the sphere of aligned triangles scaled to $\cos(\rho)$. Kendall's shape space has a radius of one-half. Points **A** and **B** represent the same shape at $CS = 1$ and $CS = \cos(\rho)$, respectively.

Because $\cos(\rho)$ decreases as ρ increases, scaling each configuration in the shape space to $\cos(\rho)$ (where ρ is its distance from the reference) produces a new shape space sphere with a radius of 1/2, tangent to the previous shape space at the reference shape (Figure 4.11). This new space is *Kendall's shape space* for triangles; it is the set of centered shapes in which each is at the size and orientation that minimizes its distance from the reference. It may appear that Kendall's shape space is dramatically different from the previous shape space, but certain key properties remain the same. One of these properties is the distance of the target shape from the reference shape across the surfaces of the shape spaces. In the first shape space, the distance of the target from the reference was ρ, the angle subtended by the arc. In Kendall's shape space, the angle subtended by the arc is now 2ρ, but the radius is 1/2, so the arc length is $2\rho/2$. Although distances between the reference and the targets are not altered, distances between targets are (Slice, 2001). Another key property that remains the same is the number of dimensions. In the transition between shape spaces, the constraint on centroid size was changed; in Kendall's shape space the constraint is $\cos(\rho)$ instead of one. This still specifies a single value for each shape; configurations that differ only in size are represented by a single point in Kendall's shape space. Thus, Kendall's shape space for triangles is also the two-dimensional surface of a three-dimensional sphere.

For configurations of landmarks that are more complex than triangles, we can apply the same set of operations to move from pre-shape space to the two shape spaces. Regardless of the number of landmarks and the number of coordinates of those landmarks, the transitions involve: (1) selecting the rotations that are at the minimum distance from the reference in pre-shape space, and (2) finding the centroid sizes that fully minimize the distance from the reference. Describing the geometric relationship of these spaces at higher dimensions is rather demanding (Small, 1996), but near their poles (i.e. near the reference configurations) these spaces are expected to have similar properties to the spaces for triangles (Slice, 2001).

Kendall's shape space and all of the spaces described above are curved, non-Euclidean spaces. This is important because the conventional tools of statistical inference assume a linear, Euclidean space. Consequently, we cannot use those tools to analyze shapes in Kendall's shape space. Much of Kendall's own work concerns statistical inference within the curved space that bears his name, but most biologists do not need to work in that space. As discussed in a later section of this chapter, it is possible to map locations in Kendall's shape space to locations in a Euclidean space tangent to Kendall's shape space. Like planar maps of the Earth, the Euclidean "maps" of shape space distort the relative positions of shapes far from the tangent point. This becomes important when comparing extremely dissimilar shapes. In most biological studies, the range of shapes will be small relative to the curvature of the space, so the distortion will be mathematically trivial for any well-considered choice of the tangent point (we discuss criteria for selecting the tangent point in a later section). If you are comparing such highly dissimilar shapes that you need to work in Kendall's shape space, you will need a more detailed understanding of this space than presented here. The excellent texts by Dryden and Mardia (1998) and Small (1996) discuss the variables and procedures for carrying out inference in Kendall's shape space.

Finding the Angle of Rotation That Minimizes the Euclidean Distance Between Two Shapes

To determine the angle of rotation required to place one pre-shape at a minimum Procrustes distance from a second, it is sufficient to rotate the first shape (the target) to minimize the summed squared distance between it and the reference. This distance we are minimizing is the partial Procrustes distance. Because the Procrustes distance is a monotonic function of the partial Procrustes distance, this minimization of the partial Procrustes distance also minimizes the Procrustes distance.

An arbitrary rotation of the target form (of two-dimensional landmarks, $M = 2$) by an angle θ maps the paired landmarks (X_{Tj}, Y_{Tj}) of the target to the coordinates $((X_{Tj} \cos\theta - Y_{Tj} \sin\theta), (X_{Tj} \sin\theta + Y_{Tj} \cos\theta))$. The sum of the squared Euclidean distances between the K landmarks of this rotated target and the reference is:

$$D^2 = \sum_{j=1}^{k} [(X_{Rj} - (X_{Tj}\cos\theta - Y_{Tj}\sin\theta))^2 + (Y_{Rj} - (X_{Tj}\sin\theta + Y_{Tj}\cos\theta))^2] \tag{4.11}$$

where (X_{Rj}, Y_{Rj}) are the coordinates of the landmark in the reference. To minimize this squared distance as a function of θ, we take the derivative with respect to θ and set it equal to zero:

$$-\sum_{j=1}^{K} \begin{bmatrix} 2(X_{Rj} - (X_{Tj}\cos\theta - Y_{Tj}\sin\theta))(-X_{Tj}\sin\theta - Y_{Tj}\cos\theta) \\ +2(Y_{Rj} - (X_{Tj}\sin\theta + Y_{Tj}\cos\theta))(X_{Tj}\cos\theta - Y_{Tj}\sin\theta) \end{bmatrix} = 0 \tag{4.12}$$

and solve for θ:

$$\theta = \text{arctangent}\left(\frac{\sum_{j=1}^{K} Y_{Rj}X_{Tj} - X_{Rj}Y_{Tj}}{\sum_{j=1}^{K} X_{Rj}X_{Tj} + Y_{Rj}Y_{Tj}}\right) \tag{4.13}$$

which gives us the angle by which to rotate the target to minimize its distance from the reference.

THE SPACES OF THREE-DIMENSIONAL CONFIGURATIONS

As discussed above, the set of all possible configurations of K landmarks with M coordinates is called a configuration space, and this space has $K \times M$ dimensions. Centering, scaling and rotating to a specific alignment all select subspaces with fewer dimensions. Because the same operations were used to select these subspaces, the same formulae can be used to determine their dimensions. Centering removes M dimensions because the centroid has M coordinates, so the space of centered coordinates has $KM - M$ dimensions, which is $3K - 3$ when $M = 3$. Scaling removes one dimension because we are still using centroid size, which is a one-dimensional scalar. Consequently, the space of centered and scaled configurations (pre-shapes) has $KM - M - 1$ dimensions (Equation 4.9), which is $3K - 4$ when $M = 3$. Rotation to a standard orientation removes $M(M - 1)/2$ dimensions (Equation 4.10), which are the number of orthogonal axes on which an M-dimensional configuration can be rotated. When $M = 3$, there are three axes, and the space of aligned configurations (a shape space) has $3K - 7$ dimensions.

When we impose on two-dimensional configurations of landmarks ($K \times 2$ matrices) the requirements of centering at the origin and scaling to unit centroid size, we generate a pre-shape space that has the form of the surface of a hypersphere with a radius of one, centered on the origin. When we impose the same requirements on three-dimensional configurations, we again get a pre-shape space that is the surface of a hypersphere with a radius of one, centered on the origin. Pre-shape spaces generated by these operations have the same general shape (differing only in the number of dimensions), regardless of the values of K and M.

The pre-shape spaces described above contain every possible rotation of every possible M-dimensional shape that can be formed of K landmarks. Each shape is represented by the set of all possible rotations of that shape, and the distance between shapes is the minimum distance between these sets. As mentioned above, the set of all possible rotations of a shape is called a *fiber*. This name seems apt when $M = 2$; there is only one axis of rotation, so we can visualize a one-dimensional string lying in the pre-shape space. When $M = 3$, calling the set of rotations a fiber may seem less appropriate because there are now three orthogonal axes of rotation, which does not fit our mental image of a one-dimensional string. However, the actual concept is still the same (the set of all possible rotations), and it is just as useful. Because different fibers represent different shapes, they do not intersect; and if they do not intersect, we can find the shortest distance between them. That distance is the difference between centered and rescaled configurations that is not due to the rotation of one relative to the other. Therefore, regardless of the values of K and M, the distance between two shapes in the same pre-shape space is the distance between two points on the surface of a hypersphere. Now that we are again on (relatively) familiar ground, we can see that we must solve for the rotation of the target that minimizes the partial Procrustes distance (the chord length), which can then be converted to the Procrustes distance (arc length) or the full Procrustes distance (the cosine of the angle subtended by the arc). Having a third set of coordinates makes the computation more tedious, but the procedure is the same.

The shape spaces we generate by the operations described above are hyperspheres tangent to their respective pre-shape spaces at the location of the reference shape. If centroid size is fixed at one, the space is the surface of a hypersphere of radius one. If centroid size is scaled to the cosine of the Procrustes distance, the space is Kendall's shape space, the surface of a hypersphere of radius one-half.

A NUMERICAL EXAMPLE FOR THE SIMPLEST CASE

To make the preceding discussion of theory more concrete and accessible, we apply the ideas to the simplest useful case, the space of triangles (this space has been discussed extensively in Small, 1996; Dryden and Mardia, 1998; Rohlf, 2000; Slice, 2001). We have used this example throughout this chapter, but we now pull all the information together. There are other approaches to constructing the matrices representing shapes in Kendall's shape space, but the sequence of steps we follow here is easily illustrated and requires relatively simple computations.

We begin with two triangles, **X** and **W**, drawn on a flat surface (Figure 4.12). **X** is the triangle from Figure 4.2, with coordinates (−1, −1), (1, −1) and (0, 1); triangle **W** has coordinates (1.07, −1.64), (3.10, −0.72) and (1.55, 0.82). Each triangle has $K = 3$ landmarks with $M = 2$ coordinates; thus the configuration matrix for each has six entries:

$$\mathbf{X} = \begin{bmatrix} -1 & -1 \\ 1 & -1 \\ 0 & 1 \end{bmatrix} \quad \mathbf{W} = \begin{bmatrix} 1.07 & -1.64 \\ 3.10 & -0.72 \\ 1.55 & 0.82 \end{bmatrix} \tag{4.14}$$

The six landmark coordinates of each triangle contain six pieces of information needed to determine all the properties of that triangle: size, shape, location, and rotation. Not only do we need all six coordinates to determine these properties; we cannot infer the value of any one coordinate from the other five. Because we need all six coordinates to determine the triangle, we can say there are six *degrees of freedom*. This also helps to explain why the configuration space of triangles has six dimensions.

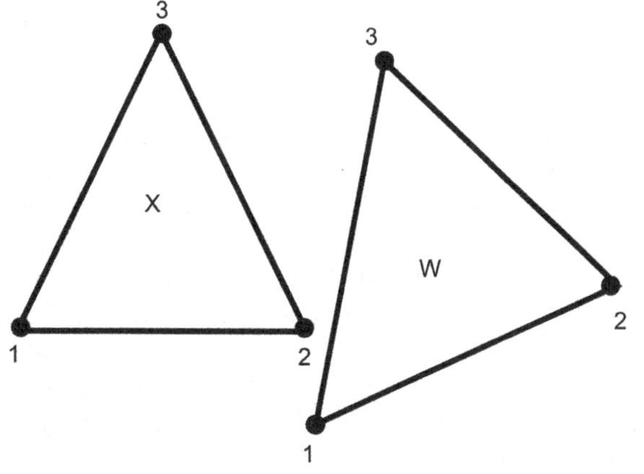

FIGURE 4.12 Two triangles, **X** (from Figure 4.2) and **W**. The vertices are numbered to indicate their homologies.

We can infer from the coordinates that the two triangles have different locations, as suggested in the figure. We confirm this by calculating the coordinates of the centroid using Equation 4.5, reproduced here:

$$X_C = \frac{1}{K}\sum_{j=1}^{K} X_j$$
$$Y_C = \frac{1}{K}\sum_{j=1}^{K} Y_j \tag{4.15}$$

For triangle **X**, the coordinates of the centroid are $X_C = (1/3)(-1 + 1 + 0) = 0$, and $Y_C = (1/3)(-1 + -1 + 1) = -0.333$. For triangle **W**, the coordinates of the centroid are $X_C = (1/3)(1.07 + 3.10 + 1.55) = 1.907$ and $Y_C = (1/3)(-1.64 + -0.72 + 0.82) = -0.513$. We use the coordinates of the centroid to form the centered configuration matrix **XC** by subtracting the centroid coordinate from the corresponding coordinate of each landmark:

$$\mathbf{XC} = \begin{bmatrix} (X_1 - X_C) & (Y_1 - Y_C) \\ (X_2 - X_C) & (Y_2 - Y_C) \\ \vdots & \vdots \\ (X_K - X_C) & (Y_K - Y_C) \end{bmatrix} \tag{4.16}$$

This produces the centered configuration matrices:

$$\mathbf{X}_{\text{centered}} = \begin{bmatrix} (-1-0) & (-1-(-0.333)) \\ (1-0) & (-1-(-0.333)) \\ (0-0) & (1-(-0.333)) \end{bmatrix} = \begin{bmatrix} -1 & -0.667 \\ 1 & -0.667 \\ 0 & 1.333 \end{bmatrix} \tag{4.17}$$

and

$$\mathbf{W}_{\text{centered}} = \begin{bmatrix} (1.07-1.907) & (-1.64-(-0.513)) \\ (3.10-1.907) & (-0.72-(-0.513)) \\ (1.55-1.907) & (0.82-(-0.513)) \end{bmatrix} = \begin{bmatrix} -0.837 & -1.127 \\ 1.193 & -0.207 \\ -0.357 & 1.333 \end{bmatrix} \tag{4.18}$$

The centered triangles are shown in Figure 4.13. One consequence of centering is that the two triangles are now *superimposed*; another is the loss of two degrees of freedom. Knowing that the centroid has coordinates (0, 0), which are the means of the landmark coordinates, we can use the coordinates of any two landmarks to determine the coordinates of the third landmark. Accordingly, the space of centered triangles (which we have not discussed previously) is a four-dimensional space. Another way to think of this is that the two coordinates of the centroid, specifying the location of the triangle, account for two of the six dimensions of the configuration space. Also, now that all individuals have the same value for their centroid coordinates, the variation due to position disappears, collapsing that dimension of variation to a point at the origin.

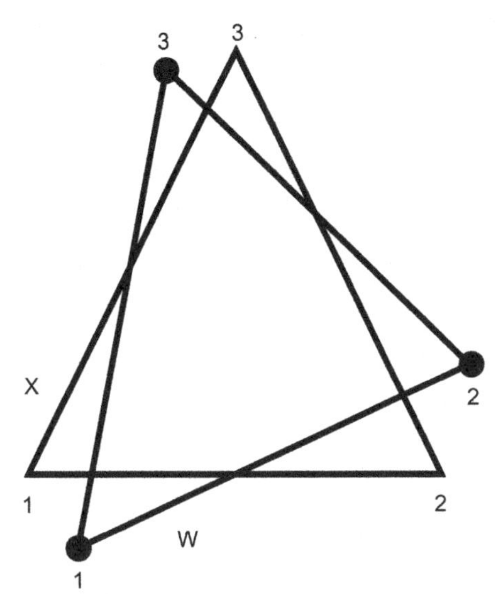

FIGURE 4.13 Centered triangles computed from **X** and **W**. Computation of the centroids of **X** and **W** is given by Equation 4.15; computation of the landmark coordinates after centering is given by Equations 4.16–4.18. Vertices are numbered to indicate their homology.

The centered triangles are not in pre-shape space. To put them there, we need to rescale each so that its centroid size is one. The formula for centroid size is:

$$CS(\mathbf{X}) = \sqrt{\sum_{i=1}^{K}\sum_{j=1}^{M}(X_{ij}-C_j)^2} \tag{4.19}$$

which is the square root of the sum of the squared distances of the landmarks from the centroid. Given that the centroids of $\mathbf{X}_{\text{centered}}$ and $\mathbf{W}_{\text{centered}}$ are both at (0, 0), we can simply sum the squared coordinates:

$$\begin{aligned} CS(\mathbf{X}_{\text{centered}}) &= \sqrt{(-1.0)^2+(-0.667)^2+(1.0)^2+(-0.667)^2+(0)^2+(1.333)^2} \\ &= 2.160 \end{aligned} \tag{4.20}$$

$$\begin{aligned} CS(\mathbf{W}_{\text{centered}}) &= \sqrt{(-0.837)^2+(1.127)^2+(1.193)^2+(-0.207)^2+(0.357)^2+(1.333)^2} \\ &= 2.311 \end{aligned} \tag{4.21}$$

Dividing each coordinate of the centered triangle by its centroid size produces the pre-shape matrices:

$$\mathbf{X}_{\text{pre-shape}} = \frac{1}{2.160}\begin{bmatrix} -1 & -0.667 \\ 1 & -0.667 \\ 0 & 1.333 \end{bmatrix} = \begin{bmatrix} -0.463 & -0.309 \\ 0.463 & -0.309 \\ 0.000 & 0.617 \end{bmatrix} \tag{4.22}$$

$$\mathbf{W}_{\text{pre-shape}} = \frac{1}{2.311}\begin{bmatrix} -0.837 & -1.127 \\ 1.193 & -0.207 \\ -0.357 & 1.333 \end{bmatrix} = \begin{bmatrix} -0.362 & -0.488 \\ 0.516 & -0.089 \\ -0.154 & 0.577 \end{bmatrix} \tag{4.23}$$

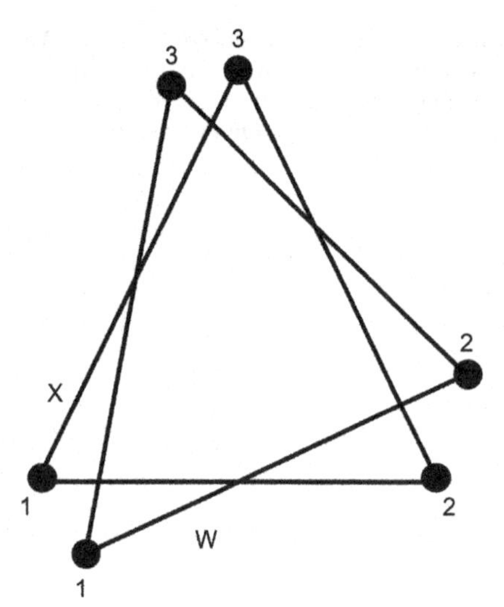

FIGURE 4.14 Centered triangles from Figure 4.13, scaled to unit centroid size. Computation of centroid size is given in Equations 4.19–4.21. Computation of landmark coordinates after scaling is given by Equations 4.22 and 4.23.

These centered and scaled triangles are shown in Figure 4.14.

Because size differences do not contribute to the differences between $\mathbf{X}_{\text{pre-shape}}$ and $\mathbf{W}_{\text{pre-shape}}$, another degree of freedom has been lost (this is the third degree of freedom lost). In other words, size is no longer a dimension of possible variation; configurations that differ only in size are considered equivalent. After subtracting the three degrees of freedom representing differences in location and centroid size, we are left with three degrees of freedom to describe differences among triangle pre-shapes – triangles that are centered and scaled to unit centroid size. Accordingly, the pre-shape space of triangles is a three-dimensional space. As explained above, it is the three-dimensional surface of a four-dimensional hypersphere, so it is not an easy space to visualize or illustrate.

To make the transition from pre-shape space to shape space, we begin by choosing one shape and placing it in a convenient orientation; this configuration will be the reference. For this demonstration it is convenient to use **X** in the orientation shown in the last few figures. Choosing **X** as the reference means that **W** will be the target, so the next step is to rotate **W**, in the plane of the page around its centroid through some angle (θ). The rotation places it in the orientation that minimizes the difference between the two sets of landmark coordinates (Figure 4.15). After the rotation, the X- and Y-coordinates of each landmark will be mapped to the new coordinates ($X \cos\theta - Y \sin\theta$), ($X \sin\theta + Y \cos\theta$). Thus, the rotated form of $\mathbf{W}_{\text{pre-shape}}$ will be:

$$\mathbf{W}_{\text{pre-shape,rotated}} = \begin{bmatrix} (-0.362 \cos\theta) - (-0.488 \sin\theta) & (-0.362 \sin\theta) + (-0.488 \cos\theta) \\ (0.516 \cos\theta) - (-0.089 \sin\theta) & (0.516 \sin\theta) + (-0.089 \cos\theta) \\ (-0.154 \cos\theta) - (0.577 \sin\theta) & (-0.154 \sin\theta) + (0.577 \cos\theta) \end{bmatrix} \tag{4.24}$$

Before we can pick the value of θ that will minimize the difference between the reference ($\mathbf{X}_{\text{pre-shape}}$) and the rotated target ($\mathbf{W}_{\text{pre-shape,rotated}}$), we need a criterion to define

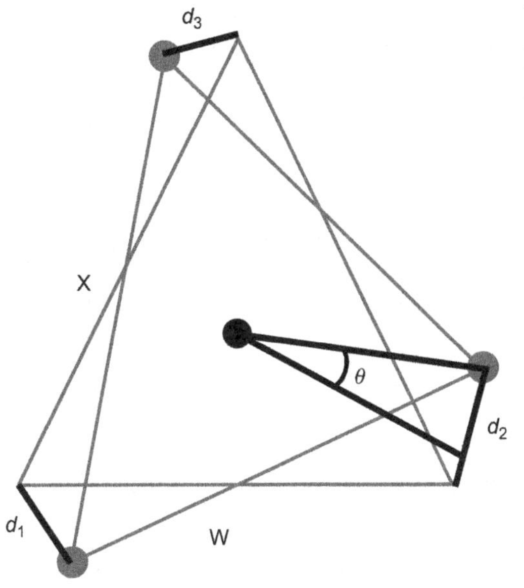

FIGURE 4.15 Optimal alignment of **W** to **X** will be achieved by rotating **W** around its centroid through an unknown angle θ to minimize the square root of the sum of the squares of distances d_1, d_2, and d_3.

what is being minimized. The criterion that leads to the shape space discussed earlier is minimization of the square root of the sum of the squared distances between the corresponding landmarks (the distances d_1, d_2, and d_3 shown in Figure 4.15). This quantity can be computed directly from the squared differences between the corresponding coordinates of the landmarks:

$$D = \sqrt{(X_{11}-X_{21})^2 + (Y_{11}-Y_{21})^2 + \cdots + (X_{13}-X_{23})^2 + (Y_{13}-Y_{23})^2} \tag{4.25}$$

(There are other criteria that lead to other superimpositions of the two triangles; one is discussed below, others in Chapter 5.)

With this criterion in hand, we can solve for the unique value of θ at which D is minimized. In our example, that value is $\theta = -19.2°$. When we insert this value into the matrix for $\mathbf{W}_{\text{pre-shape, rotated}}$ (Equation 4.24), we get:

$$\mathbf{W}_{\text{pre-shape, rotated}} = \begin{bmatrix} -0.502 & -0.341 \\ 0.458 & -0.254 \\ 0.044 & 0.596 \end{bmatrix} \tag{4.26}$$

Under the conditions set out above, this is the optimal alignment to the reference form:

$$\mathbf{X}_{\text{pre-shape}} = \begin{bmatrix} -0.463 & -0.309 \\ 0.463 & -0.309 \\ 0.000 & 0.617 \end{bmatrix} \tag{4.27}$$

Figure 4.16 shows the two triangles under these conditions.

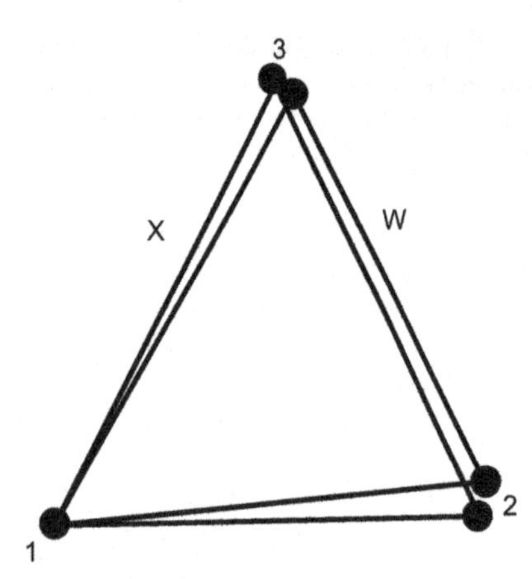

FIGURE 4.16 Triangles **X** and **W** after rotation of **W** to minimize the Procrustes distance. Computation of the landmark coordinates of **W** after rotation is given in Equation 4.24; the result is given in Equation 4.26. Vertices are numbered to indicate their homology.

The distance minimized above is the partial Procrustes distance, so we will label it D_p from this point forward. The value of D_p in this particular case is:

$$\begin{aligned} D_p &= [(-0.502-(-0.463))^2 + (-0.341-(-0.309))^2 + (0.458-0.463)^2 + (-0.254-(-0.309))^2 \\ &\quad + (0.044-0) + (0.596-0.617)]^{1/2} \\ &= 0.089 \end{aligned} \tag{4.28}$$

This is the minimum length of the chord connecting the pre-shape fibers of **X** and **W** in the pre-shape space of triangles. Because **W** is superimposed to meet the criterion of minimizing the partial Procrustes distance, $\mathbf{W}_{\text{pre-shape,rotated}}$ is said to be in *partial Procrustes superimposition* on the reference form $\mathbf{X}_{\text{pre-shape}}$. We can solve for the Procrustes distance, the arc length across the surface between $\mathbf{X}_{\text{pre-shape}}$ and $\mathbf{W}_{\text{pre-shape,rotated}}$, because the radius of the hypersphere is constrained to be one. The perpendicular from the chord to the center of the hypersphere bisects the angle ρ (Figure 4.17), which has the same value (in radians) as the arc length. Thus, there is a very simple relationship between D_p and ρ; specifically, $\rho = 2 \arcsin (D_p/2)$. In our example, D_p and ρ are so small they cannot be distinguished with fewer than 4 decimal places (0.08941 and 0.08943, respectively), which is not surprising given that ρ represents a very small angle of just 5.1°.

Because rotational effects do not contribute to the differences between $\mathbf{X}_{\text{pre-shape}}$ and $\mathbf{W}_{\text{pre-shape, rotated}}$, another degree of freedom has been lost (the fourth). Rotation, or orientation, is no longer a dimension of possible variation; configurations that differ only by rotation are considered equivalent. After subtracting the four degrees of freedom representing differences in location and centroid size and rotation, we are left with two degrees of freedom to describe differences among triangles. Accordingly, the shape space of triangles is a two-dimensional space. As explained above, it is the two-dimensional surface of a three-dimensional sphere, and is a relatively easy space to visualize or illustrate.

$\mathbf{X}_{\text{pre-shape}}$ and $\mathbf{W}_{\text{pre-shape, rotated}}$ are configurations in a shape space, but they are not yet in Kendall's shape space. To make this final transition, we need to solve for the centroid size

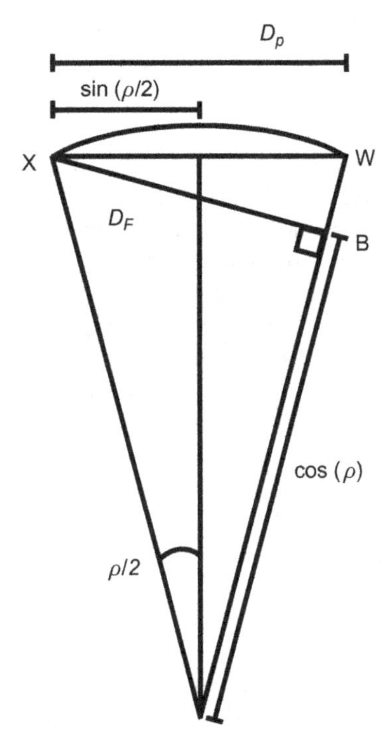

FIGURE 4.17 The relationships among the Procrustes distance, ρ, full Procrustes distance $D_F = \sin(\rho)$, and partial Procrustes distance $D_p = 2\ \sin(\rho/2)$. The configuration at point B represents a triangle in Kendall's shape space.

that would further reduce the distance between the shapes **X** and **W**; we are taking **W** to **B** (see Figure 4.17). As indicated in Figure 4.17, that distance (D_F, the full Procrustes distance) is measured along a line segment orthogonal to the radius of $\mathbf{W}_{\text{pre-shape,rotated}}$, passing through $\mathbf{X}_{\text{pre-shape}}$. In our example, ρ is small (0.0894 radians); its cosine is near one (0.996) so we need make only a very slight adjustment to convert the coordinates of $\mathbf{W}_{\text{pre-shape, rotated}}$ to $\mathbf{W}_{\text{shape}}$:

$$\mathbf{W}_{\text{shape}} = \cos(0.089)\begin{bmatrix} -0.5021 & -0.3414 \\ 0.4583 & -0.2542 \\ 0.439 & 0.5956 \end{bmatrix} = \begin{bmatrix} -0.5001 & -0.3401 \\ 0.4564 & -0.2532 \\ 0.0437 & 0.5932 \end{bmatrix} \tag{4.29}$$

This is the triangle with the same shape as **W**, but it is now in Kendall's shape space with the reference at triangle $\mathbf{X}_{\text{pre-shape}}$. Because the full Procrustes distance was used to determine the coordinates of the landmarks in $\mathbf{W}_{\text{shape}}$, we can say that $\mathbf{W}_{\text{shape}}$ is in full Procrustes superimposition on the reference form $\mathbf{X}_{\text{pre-shape}}$.

TANGENT SPACES

The curvature of shape space makes statistical inference more difficult in this space than it is in Euclidean spaces and most of the familiar methods of multivariate statistical analysis assume a Euclidean space. As mentioned earlier in this chapter, the mathematics of statistical inference in Kendall's shape space has been developed by Kendall and others. However, in this section we discuss the replacement of Kendall's shape space with a Euclidean approximation.

The problem of replacing a curved space with a Euclidean approximation is illustrated for the special case of triangles in Figure 4.18. As before (see Figure 4.11), the outer hemisphere is the space constructed by aligning pre-shapes (with centroid size fixed at one) to minimize the partial Procrustes distance (the square root of the summed squared distances between corresponding landmarks). The inner sphere is Kendall's shape space, constructed by scaling the aligned target shapes to centroid size = $\cos(\rho)$. These two spaces share a common point, the reference shape, because the distance of the reference from itself is zero, so $\cos(\rho)$ is one. Tangent to both of these spaces, at the reference shape, is a Euclidean plane. We also need to decide how we will construct the projection of shapes onto the tangent plane, which includes deciding (1) which space will be the source of the configurations projected onto the tangent plane, and (2) what rule we will use to determine the direction of the projection. (We also need to decide how to choose an appropriate reference configuration to serve as the tangent point, which is discussed in the next section.)

Figure 4.18 illustrates two common approaches to projecting from one space onto another. One approach is to project to the new space from the centroid of some reference space. In this case, the reference space is the hemisphere of aligned pre-shapes, so the projections are along the radii of this hemisphere to the tangent space. In this stereographic projection, the shape represented by points **B** and **A** (at centroid sizes $\cos(\rho)$ and one, respectively) map to the same location (**C**) in the tangent space. The distance in the plane from the reference to **C** is *greater* than the arc length from the reference to **B** (the

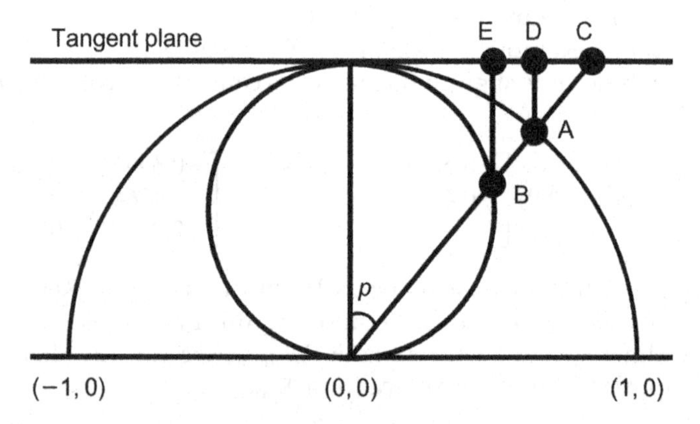

FIGURE 4.18 Tangent space to shape spaces of triangles and projections onto the tangent space (Rohlf, 1999). As in Figure 4.11, the outer hemisphere is a section through the space of centered and aligned shapes scaled to unit centroid size, and the inner circle is a section through Kendall's shape space of centered and aligned shapes scaled to $\cos(\rho)$. The plane is tangent to the sphere and the hemisphere at the point of the reference shape. The configuration at point **B** represents a triangle in Kendall's shape space; **A** is the same shape scaled to unit centroid size. **C** is a stereographic projection of **B** onto the tangent plane. **D** is the orthogonal projection of **A** onto the tangent plane, and **E** is the orthogonal projection of **B** onto the tangent plane.

Procrustes distance); and the discrepancy between these distances increases as ρ increases and the distance in the tangent plane approaches infinity. The other approach to projecting from one space onto another is to project along lines that are orthogonal to the new space. Point **E** represents the orthogonal projection of **B** onto the tangent plane, and this projection produces distances from the reference in the tangent plane that are *less* than the Procrustes distance. As in the stereographic projection, the magnitude of the discrepancy between the distances increases as ρ increases, but in the orthogonal projections, distances in the tangent plane asymptotically approach the maximum equal to the radius of the shape space.

The different projection methods do have different statistical properties (Rohlf 1999, 2000; Slice 2001). Most of the applications discussed in this text will be using partial Procrustes superimposition, which is an orthogonal projection from the oriented pre-shape space hemisphere onto the linear tangent space (see Figure 4, Rohlf, 1999 or Figure 4.18 of the text).

Selecting the Reference Configuration

Many of the steps involved in placing target configurations in shape space, or in the Euclidean space tangent to it, are functions of the reference shape (although the strict definition of Kendall's shape space does not require a reference, we use this approach to describe one way to construct such a space). For example, in the construction of a shape space, each target configuration is rotated to the orientation that minimizes its distance from the reference. Also, in the construction of Kendall's shape space, the scaling of each target configuration is a function of its distance from the reference. Moreover, the tangent space is tangent to shape space at the reference. Perhaps most important, the discrepancies between distances in the tangent space and those in shape space increase as a function of distance between target and reference. Thus, the choice of reference can have important consequences.

Most interesting biological questions will be concerned with differences among more than two specimens. The inferences based on analyses of multiple specimens will be based on all of the distances among specimens, not just their distances from the reference. Accordingly, the choice of a reference must consider the effects of that choice on approximating distances among target specimens, not just distances of target specimens from the reference. Not only will distances from the reference be distorted, so too will the distances among target specimens, and this distortion will also be a function of their distances from the reference. If these distortions are large, inferences based on distances in the Euclidean tangent space will be unreliable.

One possible reference is the average shape of the entire sample (computed using methods discussed in Chapter 5). This approach has the advantage that it minimizes the average distance from the reference, which minimizes the average distortions of interspecimen distances projected to the tangent plane (Bookstein, 1996; Rohlf, 1998). However, Marcus et al. (2000) analyzed differences in skull shape among representatives of several mammalian orders and found that most Procrustes distances are closely approximated by the Euclidean distance in the tangent space. The principal exceptions were the distances from a muskrat to a dolphin (which is not surprising, given the extraordinary reorganization of the cetacean head). This result suggests that most biologists are unlikely to encounter any cases in which the differences among specimens are large enough to worry about the adequacy of the linear approximations. It is unlikely that distances in the tangent space will poorly approximate distances in shape space. Even so, using the average shape of all specimens in the data minimizes the risk that such a problem will occur. The use of any other reference carries the responsibility to ensure that Euclidean distances in the tangent space are accurate approximations of the distances in shape space.

Dimensions and Degrees of Freedom

The issue of degrees of freedom (or the number of independent measurements in a system) is important for statistical analyses, but it can be confusing, especially when talking about shape. To clarify it, we can consider a simple example. Suppose we wish to describe the location of a notebook in a room. We could give its location in terms of three distances from a reference point (such as the corner of the door of the room), and this is equivalent to defining its position by three Cartesian coordinates relative to that reference point. In this example, there are three degrees of freedom for the location of the notebook because three variables are required to describe it. Knowing those variables and the reference suffices to find the notebook. However, if the notebook is on a chair, and all chairs are known to be the same height, specifying the height conveys no more information than saying that the notebook is on a chair. Knowing what we do about the chairs, we only need two additional pieces of information, the X- and Y-coordinates, to specify the location of the notebook in the room. Thus, by specifying the constraint that the notebook is on a chair of fixed height, we have removed one of the three degrees of freedom.

We can take this example a step further by specifying that all the chairs are located along walls of the room, with every chair touching the wall. Now, the X- and Y-coordinates can be replaced by the distance (L) around the perimeter of the room from the door to the notebook, and the direction of the measurement (clockwise or counter-clockwise). If we agree that distances around a perimeter are always measured in the same direction, then the value of L is sufficient to describe the location of the notebook. The additional constraints (chairs against the wall, perimeter measured in clockwise direction) have reduced the degrees of freedom from two (X and Y) to one (L). We have not actually eliminated either X or Y; rather, we have merely replaced that pair by L. Nor have we lost any information; given L, and the direction in which L is measured, as well as the height of the chairs, we can reconstruct the original three Cartesian coordinates (X, Y, and Z) of the notebook.

In the case of two-dimensional shapes, we start out with K landmarks in two dimensions, so we have $2K$ coordinates, which constitute $2K$ independent measurements (because each coordinate is independent of the others, in principle). In the course of superimposing the shapes on the reference form, we perform three operations: (1) we center the matrix on the centroid, thereby losing two degrees of freedom; (2) we set centroid size to one, thereby losing another; and (3) we compute the angle through which to rotate the specimen, thereby losing one more. By the end, we have lost four degrees of freedom as a consequence of applying these constraints to the data. However, unlike the notebook example, we still have $2K$ variable coordinates in our data matrix; none of them have been removed or constrained. We have not lost degrees of freedom by removing coordinates, because the loss of degrees of freedom is shared by *all* coordinates – each coordinate has lost some fraction of a degree of freedom because each is partially constrained by the operations of centering, scaling and rotation. Consequently, we have too many variable coordinates for the degrees of freedom. The primary advantage of the thin-plate spline methods (discussed in Chapter 5) is that we can work with $2K-4$ variables, so that the number of variables and the number of degrees of freedom are the same. The situation is even worse in the case of three-dimensional data, when we have $3K$ variable coordinates but only $3K-7$ degrees of freedom.

SUMMARY

Because there are several different morphometric spaces and distances, some with only slightly different names, we summarize them below.

The *configuration space* is the set of all matrices representing landmark configurations that have the same number of landmarks and coordinates. This space has $K \times M$ dimensions, where K is the number of landmarks and M is the number of coordinates.

The *pre-shape space* is the set of all $K \times M$ configurations with a centroid size of one, centered at the origin. This space is the surface of a hypersphere of radius one. Because of the centering, configurations that differ only in position are represented as the same point in pre-shape space. Similarly, because of the scaling, configurations that differ only in centroid size are represented by the same point in pre-shape space. Consequently, this space has $KM-(M+1)$ dimensions; M dimensions are lost due to centering, and one dimension is lost due to scaling. In pre-shape space, the set of all configurations that may be converted into one another by rotation lies along a circular arc called a *fiber*, which lies on the surface of the pre-shape hypersphere. The distance between shapes in pre-shape space is the length of the shortest arc across the surface connecting the fibers representing those shapes, and is called the *Procrustes distance*. Because the radius of the pre-shape hypersphere is one, the length of the arc is also the value (in radians) of the angle subtended (ρ).

To construct a *shape space*, we select one point on each fiber, removing differences in rotation. The number of axes on which a configuration can be rotated is a function of the number of landmark coordinates: $M(M-1)/2$. This also specifies the number of dimensions that are lost in the transition from pre-shape space to shape space (1 if $M=2$, 3 if $M=3$). The construction of a shape space begins with the selection of one shape in a convenient orientation to serve as the *reference* configuration. Every other shape (called a *target* configuration) is placed in the orientation that corresponds to the location on its fiber that is closest to the reference. This orientation is the position that minimizes the square root of the sum of the squared differences between the coordinates of corresponding landmarks. When minimized simply by rotation, this quantity is called the *partial Procrustes distance.* Configurations that satisfy this condition are said to be in *partial Procrustes superimposition* on the reference. The partial Procrustes distance is the length of the chord of the arc between the fibers in pre-shape space.

After rotation to partial Procrustes superimposition, the square root of the sum of the squared differences between the coordinates of corresponding landmarks can be further reduced by rescaling the target to centroid size of $\cos(\rho)$. Configurations that satisfy this condition are said to be in *full Procrustes superimposition* on the reference; and the resulting distance between shapes (square root of the sum of the squared differences between the coordinates of corresponding landmarks) is the *full Procrustes distance*. The set of shapes in full Procrustes superimposition comprises a hypersphere of radius one-half, inside the hypersphere of shapes in partial Procrustes superimposition, and tangent to the larger hypersphere at the reference. This smaller, inner hypersphere is *Kendall's shape space*.

References

Bookstein, F. L. (1996). Combining the tools of geometric morphometrics. In L. F. Marcus, M. Corti, A. Loy, G. J. P. Naylor, & D. E. Slice (Eds.), *Advances in morphometrics* (pp. 131–151). New York: Plenum.

Dryden, I. L., & Mardia, K. V. (1998). *Statistical shape analysis*. New York: John Wiley & Sons.

Kendall, D. G. (1977). The diffusion of shape. *Advances in Applied Probability, 9*, 428–430.

Marcus, L. F., Hingst-Zaher, E., & Zaher, H. (2000). Applications of landmarks morphometrics to skulls representing the orders of living mammals. *Hystrix (n.s.), 11*, 24–48.

Rohlf, F. J. (1998). On applications of geometric morphometrics to studies of ontogeny and phylogeny. *Systematic Biology, 47*, 147–158.

Rohlf, F. J. (1999). Shape statistics: Procrustes superimpositions and shape spaces. *Journal of Classifiation, 16*, 197–225.

Rohlf, F. J. (2000). Statistical power comparisons among alternative morphometric methods. *American Journal of Physical Anthropology, 111*, 463–478.

Slice, D. E. (2001). Landmark coordinates aligned by Procrustes analysis do not lie in Kendall's shape space. *Systematic Biology, 50*, 141–149.

Small, C. G. (1996). *The statistical theory of shape*. New York: Springer-Verlag.

CHAPTER

5

The Thin-plate Spline: Visualizing Shape Change as a Deformation

In geometric morphometrics, the thin-plate spline serves three functions. First, it is used as a tool for visualizing changes in shape. Second, it provides a coordinate system for the tangent space, a particularly convenient one for landmark data because the number of variables is $2K-4$ (for two-dimensional data) or $3K-7$ (for three-dimensional data). Third, the metric underlying the thin-plate spline is also used for superimposing semilandmarks. In this chapter, we briefly discuss each of these three purposes, then present a basic overview of the mathematical idea of a deformation and the mathematical metaphor underlying the thin-plate spline, which is a particular model of a deformation. We then discuss how it can be decomposed to yield a useful set of variables. We then summarize a method for sliding semilandmarks based on the thin-plate spline. In general, we present a largely intuitive overview before delving more deeply into the mathematics, reserving more technical details to the appendix of this chapter.

The graphical depiction of shape coordinates is fundamentally limited because they cannot tell us what happens *between* landmarks because we have no measurements at those locations. Sometimes it is obvious what happens between landmarks, as in Figure 5.1, where we can see that the snout elongates relative to the eye. That is obvious because the posterior eye landmark is displaced towards the anterior eye landmark, and that anterior eye landmark is *not* displaced towards the snout – so the snout must be lengthening relative to the eye. However, it is not so obvious whether the postorbital region is elongating (relative either to the head or body). Similarly, it is difficult to judge whether the head (as a whole) elongates relative to the post-cranial body. The problem is not that we lack landmarks in the relevant regions; rather, it is that so many landmarks are displaced relative to the others that it is mentally exhausting to track what happens between them all. That tracking requires examining the lengths of *all* the vectors to determine whether several landmarks are displaced to a similar degree in concert or, instead, if some are displaced relatively more than others (thereby increasing or decreasing the

Geometric Morphometrics for Biologists
DOI: http://dx.doi.org/10.1016/B978-0-12-386903-6.00005-8

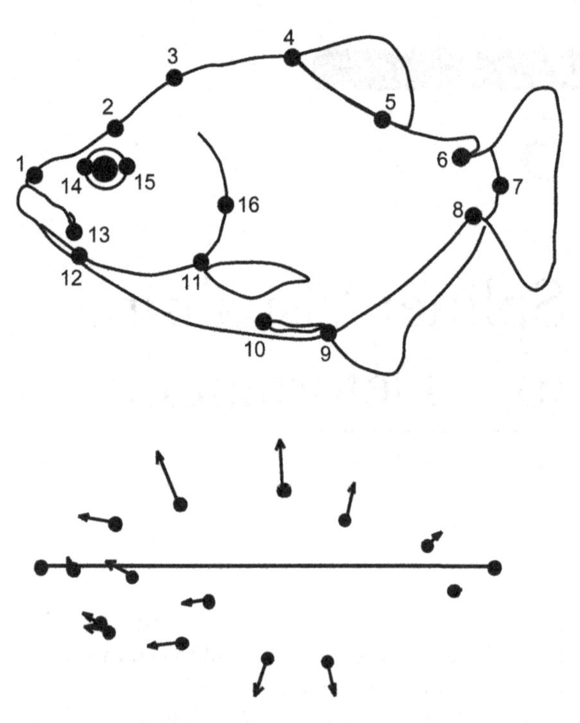

FIGURE 5.1 Ontogenetic change in body shape of *Serrasalmus gouldingi*, depicted by relative displacements of Bookstein shape coordinates.

distance between them). Even the landmarks that are not displaced relative to others must be considered. What we need is a method for visualizing changes between landmarks over the entire form.

That visualization is the primary purpose of the thin-plate spline. Using it, we can interpolate between landmarks, taking all displacements of all landmarks relative to all others into account (Figure 5.2). There are many different types of interpolation functions available (see Dryden and Mardia, 1998), all of which attempt to estimate unknown displacements *between* landmarks from known displacements *at* the landmarks. The thin-plate spline method uses a mathematical approach to this interpolation that is optimally smooth, producing interpolations with the fewest possible abrupt changes or differences. Other spline methods are available, but have seen little, if any, use in geometric morphometrics. Deformation grid plots based on thin-plate splines are very effective for two-dimensional data. Effective presentation of three-dimensional data using the spline is challenging and other methods, such as wireframes, tend to be used to represent changes in three-dimensional data.

The other major purpose of the spline has been mentioned previously in this text. As mentioned several times in this text, we need a set of shape variables to use in conventional statistical tests, by which we mean that the number of variables matches the degrees of freedom in the data. Specifically, when working with two-dimensional landmark data, we need a set of variables that spans the entire space of our data but numbering only $2K-4$ for two-dimensional data (more generally, numbering $(KM-1-M-(M(M-1)/2))$ where K is the number of landmarks in M dimensions). The thin-plate spline method can

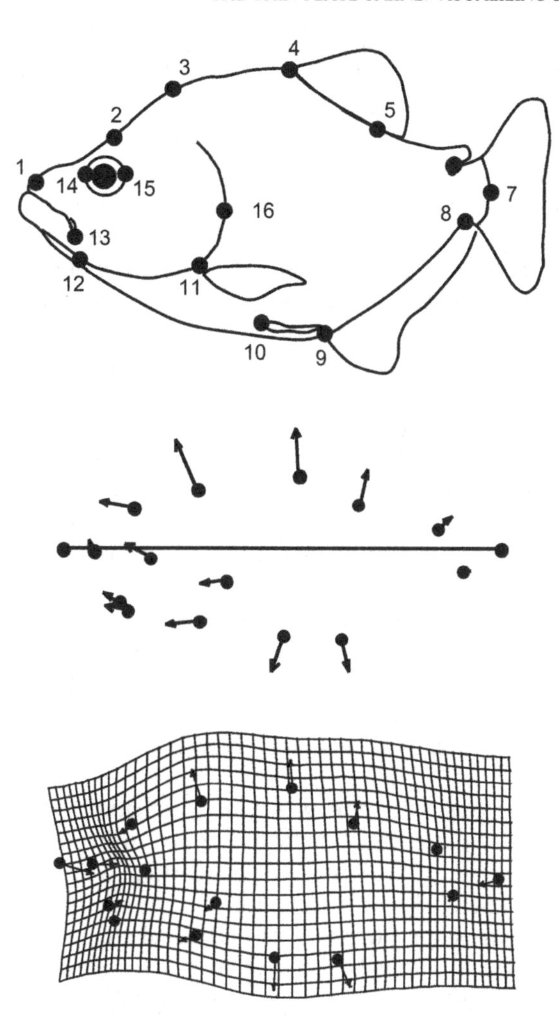

FIGURE 5.2 Ontogenetic change in body shape of *S. gouldingi*, depicted both by relative displacements of Bookstein shape coordinates and by the thin-plate spline.

be used to produce $2k-4$ (or $3k-7$ in three dimensions) orthogonal basis vectors, which allow us to express any shape change in the k landmarks using only $2k-4$ or $3k-7$ coordinates, called partial warp scores, as described in detail later. There is no loss of information in this process, and no changes in distances (in the linear tangent space) between specimens caused by this change from landmark locations relative to the reference to locations along the axes produced by the thin-plate spline method. This is a dimensionality reduction, from $2k$ to $2k-4$, or $3k$ to $3k-7$, with no loss of information, leaving us with variables well suited to conventional, analytic multivariate statistical techniques.

The situation becomes more complex when we start to use semilandmarks, as semilandmarks have lost 1 degree of freedom per semilandmark after alignment. If we have a two-dimensional system with k landmarks, and l semilandmarks, there will be $2k+l-4$ degrees of freedom in the data, but $2k+2l$ coordinates. These data can be expressed as

$2k-4$ partial warp scores, but this no longer matches the degrees of freedom in the data. Therefore, other approaches to dimensionality reduction must be used, such as Principal Components Analysis.

The spline allows a decomposition of a bending energy matrix describing the differences into *partial warps*, which provides such a set of orthogonal vectors; the partial warps supply a basis for the tangent space of all possible shape differences relative to the reference. Unlike the coordinates obtained by the Procrustes-based superimposition methods, the thin-plate spline coefficients (called *partial warp scores*) can be used in conventional statistical tests without adjusting the degrees of freedom, so long as the data contains no semilandmarks, as noted above. Also unlike the coordinates produced by the two-point registration, which also have the appropriate number for statistical tests, the partial warp scores enable using the correct tangent space measure of distance – the partial Procrustes distance. Changing from one orthonormal basis to another does not alter the distances in this linear tangent space to shape space. Using partial warp scores you will get the same distances between specimens as you get using the coordinates obtained by the Procrustes (GPA) superimposition, in a linear tangent space approximation. To get the same statistical results, you do need to adjust correctly the degrees of freedom present in the landmark data.

Additionally, the thin-plate spline provides an approach for superimposing (sliding) semilandmarks. One approach to semilandmark alignment is to minimize the distance between corresponding semilandmarks, alternatively, the approach based on the thin-plate spline slides the semilandmarks to produce the smoothest (least localized) deformation of one curve into another. The thin-plate spline method thus provides a smooth deformation criterion for semilandmarks.

In summary, the thin-plate spline provides a visually interpretable description of a deformation, with the same number of variables as there are statistical degrees of freedom so long as the data consist solely of landmarks. Even if we were not concerned with the advantages of the spline for graphical analysis, nor wished to use it for sliding semilandmarks, we might still want to use the partial warps for purposes of statistical inference. Many of the popular programs for statistical shape analysis use partial warp scores in their internal calculations, although this may change in the future as semilandmarks are increasingly used. Even if we were not concerned with the advantages of the partial warps for statistical analysis, we might still wish to use the thin-plate spline for its graphical capabilities or to slide semilandmarks. You can use the spline to depict your results, and you can use partial warps in your statistical analyses without worrying that the mathematical details (and complexities) will have any impact on your results, although you will have to be aware of the difficulties posed by the dimensionality of semilandmarks. The spline is a convenient tool for visual display and for obtaining variables with the correct degrees of freedom – it is nothing more (or less) than that.

In this chapter, we begin with a basic overview of the mathematical idea of a deformation. We then discuss the mathematical metaphor underlying one particular model of a deformation, the thin-plate spline, and how we can decompose it to yield variables. In general, we present a largely intuitive overview before delving more deeply into the mathematics. At the end of this chapter, we summarize a method for sliding semilandmarks based on the thin-plate spline.

MODELING SHAPE CHANGE AS A DEFORMATION

A deformation is a smooth function that maps points in one form to corresponding points in another form. Intuitively, smoothness means that the function goes on without interruptions or abrupt changes. More precisely, it means that the function is continuously differentiable (it can be differentiated, its first derivative can be differentiated, and so can its second, and so forth). To be differentiable, a function must be continuous. For example, the function $Y = X^3$ is continuous, but the absolute value function $Y = |X|$ is not because it has a sharp corner at $X = 0$ and so is not differentiable at that point. The Dirichelet function $Y = $ (1 when X is rational; 0 when X is irrational) is also not continuous – it is not differentiable anywhere. To be continuous, it is not enough to have a first derivative, that first derivative must also be a differentiable function. That deformations are continuously differentiable is important, because it means that the function must extend between landmarks – it cannot be defined only at certain discrete points and disappear in the regions between them.

If a function blows up (becomes infinite or non-differentiable) between points, we cannot use it to interpolate values between them. This is important because we are using the thin-plate spline as an interpolation function, inferring what happens between landmarks from data at given anatomical points. If it is unreasonable to interpolate, it is unreasonable to use the thin-plate spline for that purpose. It is also unreasonable to interpolate between far distant landmarks, just as it is unreasonable to extrapolate a linear regression far beyond the range of the observed data. If our landmarks are far apart, we have too few data to draw conclusions about what happens between them. For example, in Figure 5.2 we are assuming that the changes in regions between post-cranial landmarks can be inferred from landmarks on the dorsal and ventral periphery. That assumption can be questioned, because if we actually had more landmarks in that region we might find abrupt changes – small regions where the grid dramatically compresses or expands. We are simply assuming that no such localized changes occur.

Another case in which it would be inappropriate to think of shape change as a deformation is when there is change concentrated at a single landmark. That is equivalent to a function with an abrupt change, which violates the assumption of continuity. Such discontinuities can be detected as displacement of one shape coordinate against a background of invariant points. That pattern may be rare, but one close to it has actually been found in data (Myers et al., 1996). In that study, prairie deer mice (*Peromyscus maniculatus bairdii*) fed different diets were found to have skulls that differ only in the location of the tips of the incisors relative to the other skull landmarks. This is an extreme case of a Pinocchio effect (as discussed in Chapter 3). Such highly local changes should be ruled out before any deformation-based method is applied; if such highly localized change is found, it is better to rely on shape coordinates.

There is one other case in which a deformation-based approach might be unwise; when the interpolation spans a large amount of extra-organismal space – that is, when it is interpolating the changes over regions of "tissue" outside the organism. This can happen when landmarks are located at tips of long structures, or on structures that extend far laterally. Normally this is not a serious problem because we can simply avoid interpreting the changes in regions between those landmarks, except to say (perhaps) that the long bony structures are relatively elongated or reoriented more laterally. However, this can be a problem when multiple landmarks are located at tips of long structures and no other landmarks serve to pin down what is happening to the regions between them. It is possible to analyze the changes in relative position and length of those tips using shape coordinates, but it may not be wise to draw a grid interpolating changes at those tips to regions between them – there is no organismal tissue there.

If we do not have one of the special cases described above – that is, if we do not have evidence that some landmarks are largely independent of the others – then we can apply an interpolation function to understand changes between landmarks. Because the interpolation function is continuously differentiable, relative displacements of landmarks can be used to calculate the displacement of any location on the organism. These inferred displacements between landmarks can be illustrated using a variety of graphical styles; Figure 5.2 demonstrates the one most often used, a deformed grid in the style of D'Arcy Thompson (1992).

THE PHYSICAL METAPHOR

The mathematical basis for drawing the picture of the deformed grid is a metaphor – the bending of an idealized steel plate (Bookstein, 1989). According to this metaphor, displacements of landmarks in the X, Y plane (the plane in which we have drawn them in Figure 5.1) are visualized as if they were transferred to the Z-coordinate of an infinite, uniform and infinitely thin steel plate. That is, instead of depicting a landmark as displaced in some direction within the plane of this page, it is visualized as if it were displaced in the third dimension (out of this page).

The metal plate is constrained by little stalks that weld the landmarks in one shape to the landmarks in the other. This is difficult to draw because the imagery is inherently three-dimensional, so imagine two plates and place a configuration of landmarks on each. Now, put one plate above the other, and construct little stalks that attach a landmark on one plate to its homologue on the other plate. If a landmark in one shape is displaced a long distance relative to the other landmarks, construct a long stalk. Thus, when the landmark is displaced a long distance in one direction (such as far anteriorly), the stalk is long; conversely, when displaced only a short distance, the stalk is short. Therefore the stalks are of uneven lengths, and that unevenness means that one plate cannot be flat. The conformation that plate takes is determined by the relative heights of the stalks, and by the distances between them on the plate.

In some cases, the plate simply tilts or rotates (it does not actually bend); in other cases the plate must actually bend, such as when a point in the middle is elevated higher than four surrounding points. That bending may be gentle or quite sharp. For real steel plates, the conformation of the plate tends to minimize the magnitude of bending over the whole plate (as well as the physical energy required to produce that bending). Here we use the expression *tends* to minimize the magnitude and energy of bending, because real steel plates may have flaws, and the situation is not a pure case of work against elasticity. In the ideal case, the bending energy depends solely on the distance between the points and the relative heights of the stalks, and the total amplitude of bending. If we consider two different deformed plates, both describing the same total overall amount of change (the same set of stalk heights) but one with the stalks proportionately closer together, the one that is bent between the more closely spaced points requires more energy than the one that is bent between more widely spaced points.

The bending energy depends on the spacing of the stalks because it is a function of the rate of change in the slope of the plate – i.e. whether the slope of the surface increases rapidly or slowly. In these terms, more energy is required when the slope of the surface changes at a higher rate (for the same net amplitude of change). Imagine a tall stalk surrounded by short ones, which induces a steep slope in the curvature of the plate. The steepness of that slope is proportional to the function being minimized – the rate of change in slope of the surface – and, thus, the function being minimized is a function of the second derivative (the slope of the surface is the first derivative) integrated over the whole surface of the plate. It can also be termed the integral of the quadratic variation over the plate.

To return from ideal plates to the analysis of a deformation, we now project the changes that were visualized as if in the Z-direction back into the X, Y plane (the plane of our landmark data). The idea of bending that had a physical meaning when we were talking about changes in the Z-direction is now reinterpreted as "spatially local information". This interpretation may not be intuitively obvious, but consider what a relatively rapid increase in slope means – that there are contrasting displacements of closely spaced points. When closely spaced points change in opposite directions it requires more energy to bend the plate between them; so there is an inverse relationship between the spatial scale of the change and its metaphorical bending energy. Minimization of bending energy is equivalent to minimization of spatially localized information.

It is always possible to envision changes as highly local by assuming that the plate flattens out immediately after rising, then rises again just at the next stalk, then flattens again, then rises again, etc. The argument against doing so is that this would be the most unparsimonious interpretation possible. By minimizing bending energy, we obtain a more parsimonious description of the change. We do not assume highly localized change unless the data demand doing so.

Uniform and Non-Uniform Components of a Deformation

Some transformations require no bending energy at all; these are equivalent to tilting or rotating the plate. These are often called *affine* or *uniform* transformations, meaning that they leave parallel lines parallel. The terms "affine" and "uniform" are both used to describe the same component of a deformation; "affine" is favored by mathematicians, but "uniform" appears more often in the geometric morphometric literature. Consequently, we will use "uniform" for this component and "non-uniform" for its complement. In our example (see Figure 5.2), if the entire fish simply elongates relative to its depth without any disproportionate lengthening of one region relative to another, it is a uniform elongation. Uniform elongation is equivalent to uniform narrowing, as should be recalled from our discussion of shape variables in previous chapters. Because it is uniform, meaning that the same change occurs everywhere, we need only one descriptor for the change of the whole organism. In contrast, the *non-uniform* or *non-affine* deformations (which involve the metaphorical bending) have regionally differentiated effects.

A deformation can be broken down into uniform and non-uniform components, as in Figure 5.3. Most real biological transformations will have both uniform and non-uniform components. These components are computed separately, so we describe them separately (first the uniform, then the non-uniform), but it is important to bear in mind that a complete description, and an accurate illustration, requires specifying *all* the components.

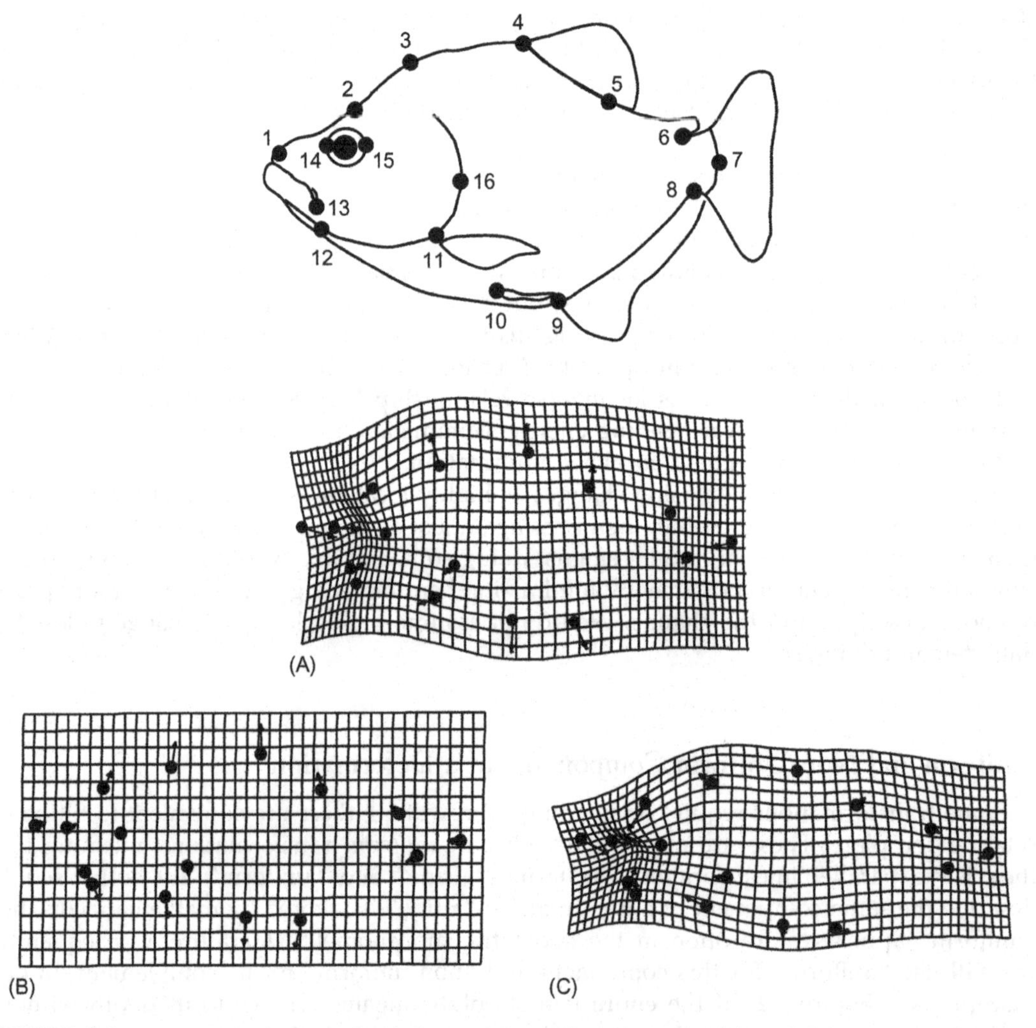

FIGURE 5.3 Ontogenetic change in body shape of *S. gouldingi*, depicting: (A) total deformation and its two components; (B) uniform component; (C) non-uniform component.

Uniform (Affine) Components

There are six distinct types of uniform deformations for landmarks in two dimensions, and they are independent of each other (meaning that they are mutually orthogonal). Figure 5.4 shows these six operations carried out on a square configuration of landmarks. The first four are the familiar ones that do not alter shape: translation along two perpendicular axes (Figure 5.4A,B), scaling (Figure 5.4C) and rotation (Figure 5.4D). These are all used in superimposing shapes. The other two uniform deformations do alter shape: compression/dilation (Figure 5.4E) and shear (Figure 5.4F). Compression/dilation refers to the case in which one direction has expanded (the vertical or *Y*-direction in Figure 5.4E) while the other has contracted (the horizontal or *X*-direction). Shearing refers to translating landmarks along one axis by a distance proportional to their location along the other axis.

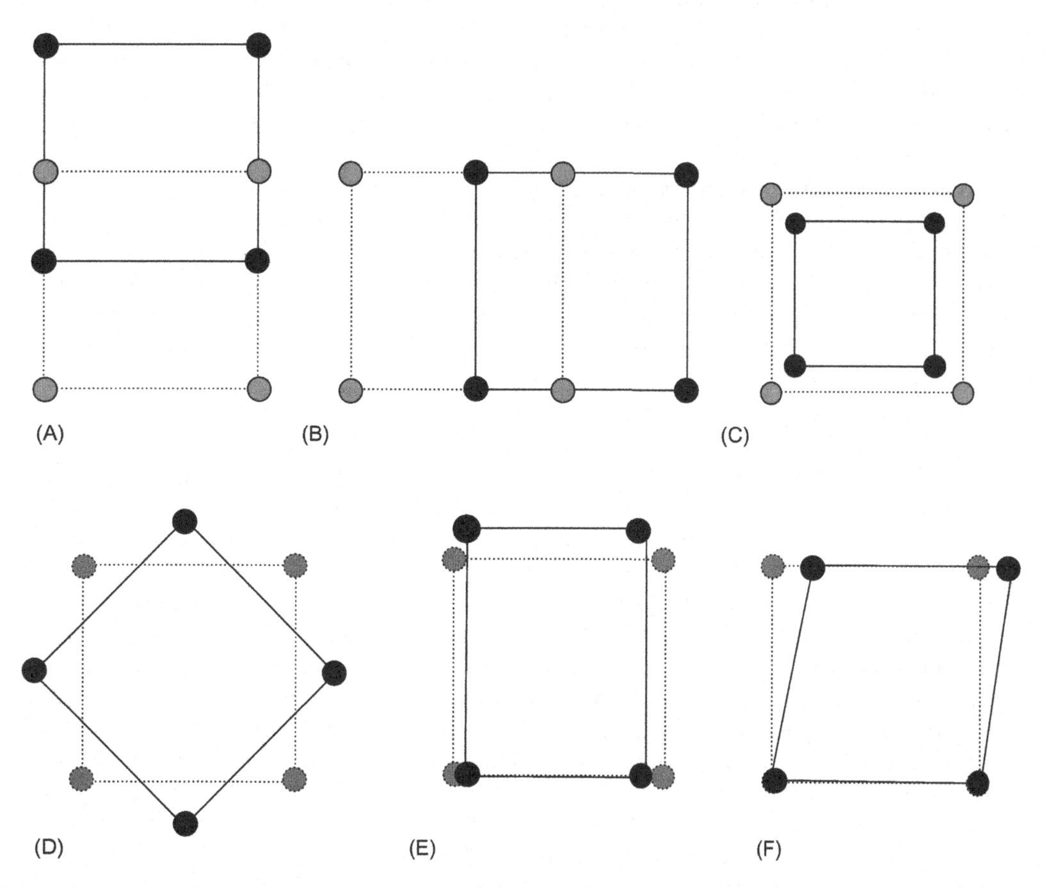

FIGURE 5.4 The six uniform (affine) transformations: (A) translation along the vertical axis; (B) translation along the horizontal axis; (C) scaling; (D) rotation; (E) compression/dilation; (F) shearing. The original (or reference) square is shown with dotted lines, while the deformed shape is shown with solid lines.

Because compression/dilation and shear alter shape whereas translation, rotation and scaling do not, it is common to talk about the two that alter shape without mentioning the ones that do not. All of them need to be accounted for, so we will refer to compression/dilation and shear as the *explicit uniform deformations* or *explicit uniform terms* because they are the ones explicitly tracked. We will refer to the others as the *implicit uniform deformations* or *implicit uniform terms.* They are implicit because they can be mathematically determined from the superimposition method used, the explicit uniform components, and the non-uniform components of a deformation – they are the translation, rotation and scaling that must have been carried out. Both explicit and implicit uniform terms are needed, in addition to the non-uniform terms, to draw the deformation correctly.

Each deformation has an inverse. Applying the inverse of a deformation is equivalent to traveling backwards along the path that was taken until we arrive back at the starting point. We can think of the deformation in terms of a $2K$-dimensional vector (i.e. two dimensions per landmark). There would be a vector at each landmark indicating the direction in which that particular landmark will be mapped under the deformation (although there are only $2K-4$ independent dimensions). In the inverse of the deformation, the directions of the arrows would be reversed. The inverse of a translation is the same magnitude of translation in the opposite direction (negative X instead of positive X). Similarly, we can represent rotation as an angular displacement so its inverse is a *negative* angular displacement (counterclockwise instead of clockwise). Scaling is slightly different because it involves multiplication (whereas translations and rotations could be treated as additions). Scaling is multiplication by a factor F; its inverse is multiplication by the inverse of F $(1/F)$. Unfortunately, the algebraic descriptions of the last two deformations and their inverses are not quite as simple (as we will see below). Graphically, we can see that the inverse of compression/dilation involves a reversal of which axis is compressed and which is dilated, and that the inverse of a shear is a shear of the same amount along the same axis in the opposite direction.

Several different approaches exist to calculating orthogonal axes to represent the two uniform components of shape difference found in a data set. The first approach, as developed by Bookstein (1996) is to determine the pattern of shape change at each landmark of the reference after the two operations of shear and compression/dilation, followed by partial Procrustes superimposition. A derivation of this approach is shown in the Appendix. The other approach is to use the thin-plate spline method described below to partition the shape variation in the data into the affine (uniform) and non-affine components (Rohlf and Slice, 1990; Rohlf and Bookstein, 2003). The affine portion can then be subjected to a singular value decomposition (akin to a principal components analysis) to develop two orthogonal basis vectors (in two dimensions) which span the space of possible affine changes in shape. The uniform shape terms found via singular value decomposition will be linear combinations of shear and compression/dilation. The approach based on singular value decomposition is readily adapted to three-dimensional data as well. The choice of whether to use the affine terms as per Bookstein (1996) or based on singular value decomposition is not particularly important, they are simply slightly different basis sets of the uniform terms. Since the terms should not be interpreted independently of one another in any case, the choice between the two is not noticeable in the final statistical analysis.

Decomposing the Non-Uniform (Non-Affine) Component

The non-uniform part of a deformation differs from the uniform in that it does not leave the sides of a square parallel. However, like the uniform part, the non-uniform can be further decomposed into a set of orthogonal components. The decomposition of the non-uniform deformation is based on the thin-plate spline interpolation function, and produces components called *partial warps*. We first describe an intuitive introduction to partial warps, then a more mathematical one.

AN INTUITIVE INTRODUCTION TO PARTIAL WARPS

The non-uniform component describes changes that have a location and spatial extent on the organism; they are not the same everywhere. They describe spatially graded phenomena such as anteroposterior growth gradients, and more highly localized changes such as the elongation of the snout relative to the eye. The notion of spatial scale is central to the analysis, so we need an intuitive notion of spatial scale. In general (but imprecise) terms, a change at small spatial scale is one confined to a small region of an organism. To refine that idea, and develop a firmer grasp of the concept, we show several components at progressively smaller spatial scales (Figure 5.5).

Figure 5.5A shows a component at large spatial scale that, while broadly distributed, is not the same everywhere (so it is not uniform). The particular example shown in Figure 5.5A is the elongation of the mid-body relative to the more cranial and caudal regions. A more localized change, confined to the posterior region of the body, is shown in Figure 5.5B – a shortening of the region between the dorsal and adipose fins relative to the dorsal fin and caudal peduncle. Because more distant landmarks are not involved in the change, it is more localized than the one shown in Figure 5.5A. Another localized change is shown in Figure 5.5C, this time confined to the cranial region. This is a shortening of the postorbital region relative to the regions just anterior and posterior.

The components we have described above and depicted in Figure 5.5 are partial warps, but to draw them we had to specify their orientation (we drew them as oriented along the anteroposterior body axis). That orientation is not actually specified by the partial warps themselves; rather, it is provided by a two-dimensional vector, the *partial warp scores*. There is one two-dimensional vector per partial warp. These scores express the contribution that each partial warp makes to the total deformation. The scores have an X- and Y-component, and indicate the direction of the partial warp. The idea of direction or orientation should be familiar from previous chapters. In Figure 5.6 we show one partial warp (that depicted in Figure 5.6B) multiplied by three different vectors. It may be easiest to see the directions by looking at the orientation of the vectors at landmarks. Figure 5.6A shows the partial warp oriented horizontally, which in our case corresponds to the X-direction, so the coefficient of the X-component is large and that of the Y-component is negligible. In contrast, Figure 5.6B shows the vector with a negligible X-component and a large Y-component. Figure 5.6C shows the vector with X- and Y-components of equal magnitudes.

We have described partial warps one at a time, but a complete description (and interpretation) requires combining them all. Taken separately, partial warps are purely

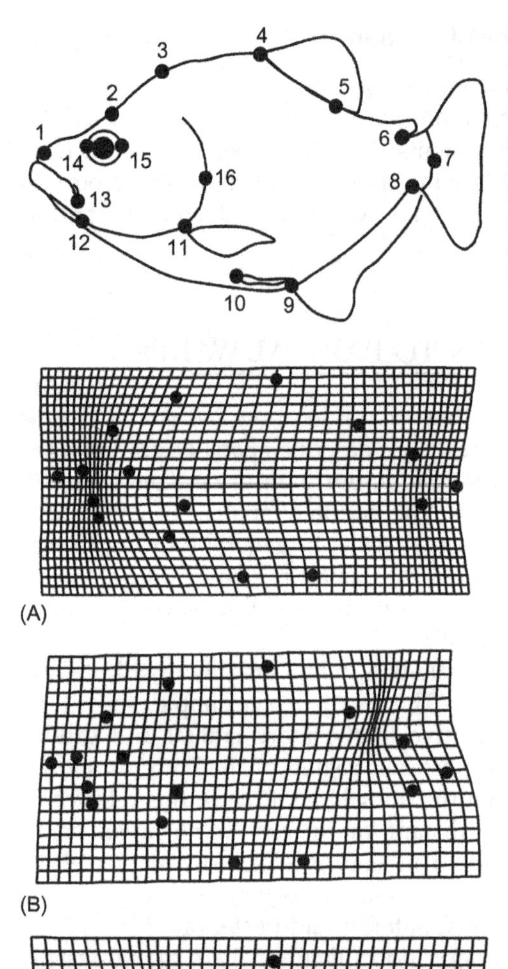

FIGURE 5.5 Three components of the non-uniform deformation, called partial warps. (A) Partial warp at large spatial scale, depicting an expansion of the mid-body relative to the head and caudal body; (B) partial warp at moderate to small spatial scale, depicting a contraction of the region between dorsal and adipose fins relative to the length of the dorsal fin and caudal peduncle; (C) another partial warp at moderate to small spatial scale, depicting a shortening of the postorbital region relative to the preorbital head and anterior post-cranial body.

geometric constructs – a function of the location and spacing of the landmarks of the reference form. They are obtained by a geometric decomposition of the landmarks of the reference form (as explained in detail in the next section). Although they provide a basis for the tangent space, they cannot be interpreted except in these abstract terms – we cannot say, for example, that one part of the change in the ontogeny of the fish is a shortening of the region between dorsal and adipose fins relative to the dorsal fin and caudal peduncle.

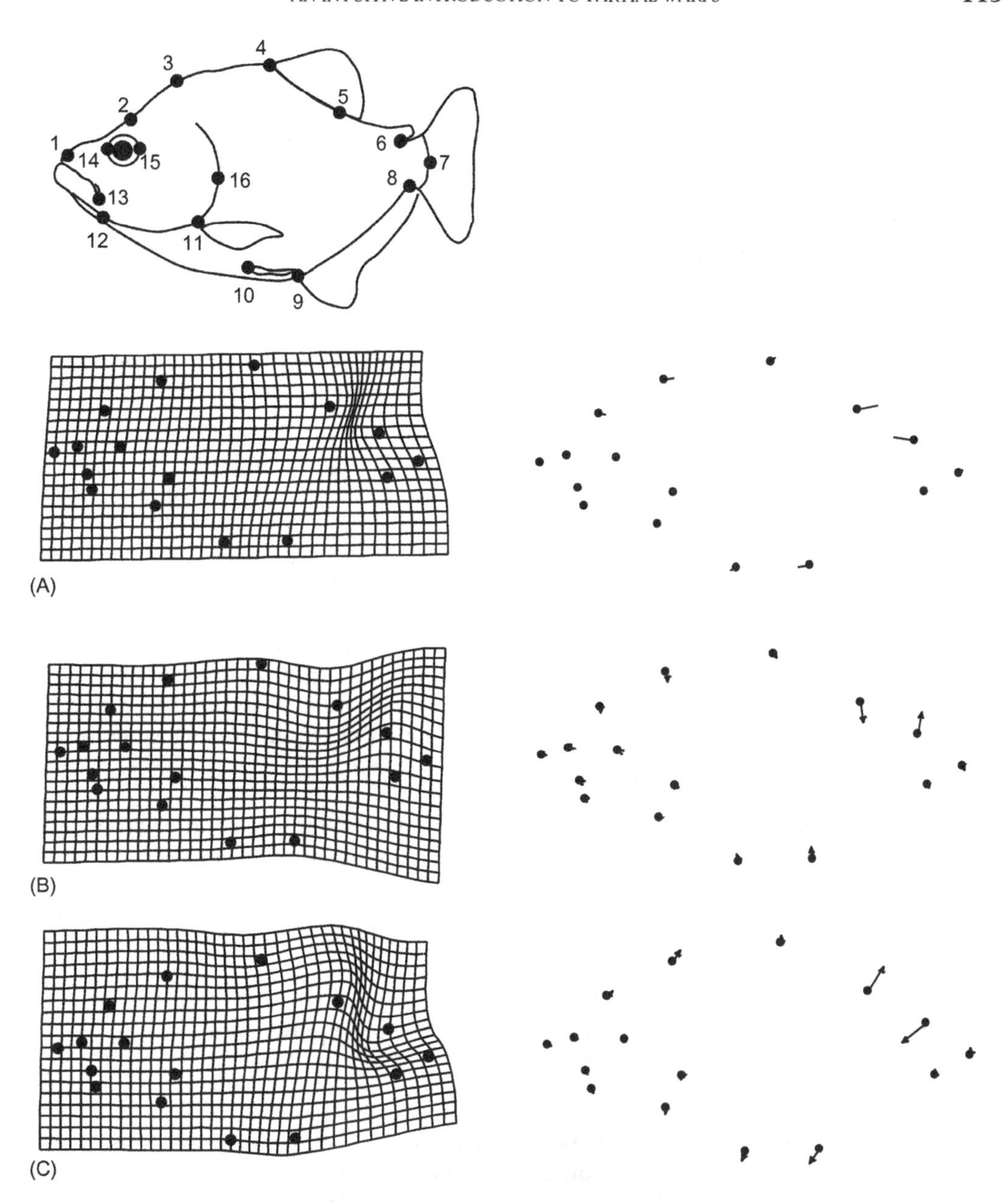

FIGURE 5.6 One partial warp, oriented in three directions. (A) Along the *X*-axis; (B) along the *Y*-axis; (C) equally along *X*- and *Y*-axes. Due to the orientation of our landmark coordinates, the *X*-direction corresponds to the anteroposterior axis, and the *Y*-direction corresponds to the dorsoventral axis. To make it easier to see the direction in which the partial warps are oriented, we also display them by vectors of relative landmark displacements.

That is a component of the deformation, not a component of an ontogeny. Only by looking at the total deformation can we say where change occurs.

To summarize our intuitive presentation of spatial scale, we repeat our major points. First, any non-uniform deformation can be decomposed into a series of components (partial warps) at progressively smaller spatial scales. Each component describes a pattern of relative landmark displacements, based on the spacing and location of landmarks in the reference form. Each partial warp is multiplied by a two-dimensional vector (the partial warp scores) that measures the contribution made by the partial warp (in each direction) to the total deformation. We now present a more technical introduction to the thin-plate spline.

AN ALGEBRAIC INTRODUCTION TO PARTIAL WARPS

Algebraically, partial warps are obtained by eigenanalysis of the *bending energy matrix*. Eigenanalysis may be familiar from a quite different context, for example, principal components analysis, where it is used to extract eigenvectors (PCs) of the variance–covariance matrix of measurements. The exact same mathematics is involved in calculating the partial warps; the difference lies in the matrix being analyzed. Rather than extracting eigenvectors of a variance–covariance matrix, we instead extract them from the bending-energy matrix. (We will discuss eigenanalysis further, in context of principal components analysis in Chapter 6; here we focus on the derivation of the bending energy matrix.)

The idea behind the thin-plate spline is that it will approximate the observed deformation by a linear combination of a function that is the smoothest available and that fully describes the observed deformation. The function satisfying that pair of requirements has the form:

$$Z(X,Y) = -U(R) = -R^2 \ln R^2 \tag{5.1}$$

where R is the distance between a pair of landmarks in the reference configuration (scaled to unit centroid size). This particular function satisfies the biharmonic equation:

$$\Delta^2 \mathrm{U} = \left(\frac{\mathrm{d}^2}{\mathrm{d}x^2} + \frac{\mathrm{d}^2}{\mathrm{d}y^2}\right)^2 \mathrm{U}\,(R) \propto \delta_{(0,0)} \tag{5.2}$$

where $\delta_{(0,0)}$ is the generalized function, or delta function, which is defined to be zero everywhere except at $X = 0$, $Y = 0$, with the seemingly odd requirement that:

$$\int (\delta_{(0,0)} \mathrm{d}x\,\mathrm{d}y) = 1 \tag{5.3}$$

The delta function is oddly behaved, but mathematically tremendously useful, as it has useful normalization properties. It is sometimes called a functional, rather than a function.

U is said to be the fundamental solution of the biharmonic equation, which is the equation for the shape of a thin steel plate lifted to a height Z (X, Y) above the (X, Y)-plane. This is because the bending energy (BE) of the steel plate at a point (X, Y) is given by:

$$\left(\frac{\mathrm{d}^2\mathrm{U}}{\mathrm{d}x^2}\right)^2 + 2\left(\frac{\mathrm{d}^2\mathrm{U}}{\mathrm{d}x\,\mathrm{d}y}\right)^2 + \left(\frac{\mathrm{d}^2\mathrm{U}}{\mathrm{d}y^2}\right)^2 = -BE(X,Y) \tag{5.4}$$

and the total bending energy of the entire plate is:

$$\int (BE(X, Y)\mathrm{d}x\ \mathrm{d}y) \tag{5.5}$$

which is the bending energy at each point integrated over the entire surface. The choice of U (R) minimizes this total bending energy.

For biological purposes, we do not really care about the bending energy of a steel plate. Rather, we care about the connection between bending energy and the curvature of the plate (and their connection to spatial scale). Minimizing bending energy minimizes the curvature of the plate, so when we fit a linear combination of the U (R) function to our data, we are fitting a function that minimizes the amount of curvature needed to model the observed deformations.

Suppose we want a linear combination of U (R) values, centered on each of the K landmarks of our reference form (because we are describing a deformation, we are talking about changes relative to a reference). We need to describe deformations in the X and Y directions, so we form the following linear combinations:

$$fx(X, Y) = A_{X1} + A_{XX}X + A_{XY}Y + \sum_{i=1}^{K} W_{Xi}U(X - X_i,\ Y - Y_i) \tag{5.6}$$

$$f_Y(X, Y) = A_{Y1} + A_{YX}X + A_{YY}Y + \sum_{i=1}^{K} W_{Yi}U(X - X_i,\ Y - Y_i) \tag{5.7}$$

where f_X (X, Y) and f_Y (X, Y) are the spline functions that describe the deformations along the X- and Y-directions relative to the reference form, and W_{Xi} and W_{Yi} are weights of the functions $\mathbf{U}$ ($X - X_i$, $Y - Y_i$), centered on the landmark locations of the reference (X_i, Y_i). The A terms describe uniform (or affine) deformations of the target, using what is known as the six-component uniform model. We need to include those A terms at this stage, but will discard them later in favor of the two uniform components discussed in the Appendix (Equations 5A.23 and 5A.38).

Fitting the functions to the observed deformations is a standard problem in systems of linear equations; we can thus cast the problem into matrix form. We form a $(K+3) \times 2$ matrix $\mathbf{V}$ of the observed deformations at each of the K landmarks, where the deformation at the ith landmark is denoted (X'_i, Y'_i):

$$V = \begin{bmatrix} X'_1 & Y'_1 \\ X'_2 & Y'_2 \\ \vdots & \vdots \\ X'_K & Y'_K \\ 0 & 0 \\ 0 & 0 \\ 0 & 0 \end{bmatrix} = \begin{bmatrix} f_X(X_1, Y_1) & f_Y(X_1, Y_1) \\ f_X(X_2, Y_2) & f_Y(X_2, Y_2) \\ \vdots & \vdots \\ f_X(X_K, Y_K) & f_Y(X_K, Y_K) \\ 0 & 0 \\ 0 & 0 \\ 0 & 0 \end{bmatrix} = \mathrm{LW} \tag{5.8}$$

where **LW** is the product of two matrices **L** and **W**. **L** is the $(K+3)$ $(K+3)$ matrix:

$$\mathrm{L} = \begin{bmatrix} \mathrm{U}(0) & \mathrm{U}(R_{1,2}) & \mathrm{U}(R_{1,3}) & \cdots & \mathrm{U}(R_{1,K}) & 1 & X_1 & Y_1 \\ \mathrm{U}(R_{2,1}) & \mathrm{U}(0) & \mathrm{U}(R_{2,3}) & \cdots & \mathrm{U}(R_{2,K}) & 1 & X_2 & Y_2 \\ U(R_{3,1}) & \mathrm{U}(R_{3,2}) & \mathrm{U}(0) & \cdots & \mathrm{U}(R_{3,K}) & 1 & X_3 & Y_3 \\ \vdots & \vdots & \vdots & \ddots & \vdots & \vdots & \vdots & \vdots \\ U(R_{K,1}) & U(R_{K,2}) & \mathrm{U}(R_{K,3}) & \cdots & U(R_{K,K}) & 1 & X_K & X_K \\ 1 & 1 & 1 & \cdots & 1 & 0 & 0 & 0 \\ X_1 & X_1 & X_3 & \cdots & X_K & 0 & 0 & 0 \\ Y_1 & Y_2 & Y_3 & \cdots & Y_K & 0 & 0 & 0 \end{bmatrix} \tag{5.9}$$

in which $\mathrm{U}(R)$ is the function appearing in Equations 5.6 and 5.7 evaluated at each landmark location (X_i, Y_i). **W** is the $(K+3) \times 2$ matrix of weights and uniform terms appearing in Equations 5.6 and 5.7:

$$\mathrm{W} = \begin{bmatrix} W_{X1} & W_{Y1} \\ W_{X2} & W_{Y2} \\ \vdots & \vdots \\ W_{XK} & W_{YK} \\ A_{X1} & A_{Y1} \\ A_{XX} & A_{YX} \\ A_{XY} & A_{YY} \end{bmatrix} \tag{5.10}$$

So we have the equation:

$$\mathbf{V} = \mathbf{LW} \tag{5.11}$$

in which **L** and **W** are the matrices just described. We wish to solve for **W**, the matrix of coefficients in our spline model, which gives us:

$$\mathbf{W} = \mathbf{L}^{-1}\mathbf{V} \tag{5.12}$$

We can use the weights in the matrix **W** in conjunction with the spline functions in Equations 5.6 and 5.7 to interpolate the observed deformation at the landmarks over the entire specimen. However, it turns out that we can make some further use of the matrix $\mathbf{L}^{-1}$. This matrix is $(K+3)$ by $(K+3)$; if we take the first K rows and the first K columns of $\mathbf{L}^{-1}$, we can form $\mathbf{L}_\mathbf{K}^{-1}$, which is called the *bending energy matrix*.

The bending energy matrix can be rearranged into a series of eigenvectors $\mathbf{E}_\mathrm{i}$, and eigenvalues, λ_i, such that:

$$\mathbf{L}_\mathbf{K}^{-1}\mathbf{E}_\mathrm{i} = \lambda_i\mathbf{E}_\mathrm{i} \tag{5.13}$$

The eigenvectors $\mathbf{E}_\mathrm{i}$ have the usual properties of eigenvectors and, consequently, they are a basis (or a set of coordinate axes) of a space. In this case, the eigenvectors are the basis of the Euclidean space tangent to shape space at the reference shape. This means that we can express the non-affine part of our matrix of observed deformations **V** as a linear combination of the eigenvectors of the bending energy matrix. The eigenvalues are the bending energies required to effect a change (of a given amount of shape difference, i.e. a unit of Procrustes distance) at that spatial scale.

Three of the eigenvalues of the bending energy matrix are zero, corresponding to the components with no bending (with *X*- and *Y*-coefficients, these eigenvectors account for the six uniform components of the deformation). The remaining $K-3$ eigenvectors are the explicitly localized components of a deformation. These eigenvectors are called the *partial warps*; the vector multipliers of the partial warps are called the *partial warp scores* (following Slice et al., 1996). They are "partial" because they describe part of a deformation. We should note that Bookstein (1991) called the eigenvectors of the bending energy matrix *principal* warps, analogous to principal components. By "partial warp", he meant the vector multiple of a principal warp. Slice and colleagues use the term *principal warp* to refer to a partial warp interpreted as a bent surface of the thin-plate spline, and because the latter terminology has become standard, we use it here.

As evident in the definition of $\mathbf{L}_K^{-1}$, only one matrix of landmarks enters into the calculation of bending energy; the coordinates of the form usually called the reference or starting form. Thus, the eigenvectors that give us a coordinate system for shape analyses are a function of one single form. This may be highly counterintuitive, because more familiar eigenvectors, such as principal components, are functions of an observed variance–covariance matrix. They are functions of variation (or differences) among observed forms. That is not the case for the eigenvectors of the bending energy matrix. The eigenvalues of the bending energy are the bending energies that would be required to modify a given shape by a single unit of shape difference at each spatial scale. Thus, the partial warps are not themselves features of shape change, they are simply a coordinate system or basis for the space in which we analyze shape change.

The "*A*" coefficients in Equation 5.10 describe the uniform deformation of the shape. There are six of these coefficients, which is enough to describe the six components of the uniform deformation of shape. However, we know that the reference and the target do not differ by rotation, rescaling or translation, because those differences were removed by the superimposition process. Consequently, we do not need six parameters to describe the uniform component of the deformation, only the two components derived in the Appendix.

By convention, partial (or principal) warps are numbered from the lowest to highest bending energy; the one with the highest number corresponds to the one with greatest bending energy. The two uniform components are sometimes called the zeroth principal warp. Thinking of the uniform components in those terms is useful because it emphasizes that the uniform components cannot be viewed separately from the non-uniform ones. Including the uniform terms also completes the tally of shape variables. The $K-3$ partial warps contribute $2K-6$ scores; adding the two uniform scores brings the count up to $2K-4$.

DECOMPOSING THE DEFORMATION OF THREE-DIMENSIONAL DATA

As in the two-dimensional case, the difference between three-dimensional configurations of landmarks can be described as a deformation of one shape (reference) into the other (target). This deformation can be decomposed into uniform and non-uniform parts (or affine and non-affine). The non-uniform part can be further decomposed into $3(K-4)$

independent components. The uniform part can be further decomposed into twelve independent components; but only five of these change shape.

The numbers of uniform and non-uniform components can be explained if we consider the possible deformations of the simplest three-dimensional shape, a tetrahedron of four landmarks. All deformations of a tetrahedron, like all deformations of a triangle, must be uniform; only when a fifth point is added can we detect non-uniform transformations (i.e. transformations that differ between regions of the tetrahedron). With just four landmarks a deformation can have twelve components, all of them uniform. Seven of the uniform components do not change shape – they are the ones removed by superimposition – which leaves five uniform components that do change shape. With each additional landmark beyond the fourth, there are three possible non-uniform components of deformation (because there are three directions in which that point might move relative to the others), hence $3(K-4)$.

The components of the non-uniform part of a three-dimensional deformation are defined in nearly the same terms as the components of the non-uniform part of a two-dimensional deformation. Again, we use the thin-plate spline model to describe the deformation at any point in space as f_X, f_Y and f_Z, which describe the X-, Y- and Z-components of the deformation:

$$\begin{aligned} f_X(X,\ Y,\ Z) &= A_{X1} + A_{XX}X + A_{XY}Y + A_{XZ}Z + \sum_{i=1}^{K} W_{Xi}U(X - X_i,\ Y - Y_i,\ Z - Z_i) \\ f_Y(X,\ Y,\ Z) &= A_{Y1} + A_{YX}X + A_{YY}Y + A_{YZ}Z + \sum_{i=1}^{K} W_{Yi}U(X - X_i,\ Y - Y_i,\ Z - Z_i) \\ f_Z(X,\ Y,\ Z) &= A_{Z1} + A_{ZX}X + A_{ZY}Y + A_{ZZ}Z + \sum_{i=1}^{K} W_{Zi}U(X - X_i,\ Y - Y_i,\ Z - Z_i) \end{aligned} \tag{5.14}$$

where $U(X-X_i, Y-Y_i, Z-Z_i)$ is a function of the interlandmark distances given by:

$$R_i = \sqrt{(X-X_i)^2 + (Y-Y_i)^2 + (Z-Z_i)^2} \tag{5.15}$$

Again, we have more columns to accommodate the third dimension. The more substantive difference is that $U = |R|$ in contrast to the two-dimensional case in which $U = R^2 \ln R^2$. As in the two-dimensional case, the next steps are to solve for the spline coefficients (the values of A and W) and the eigenvectors of the bending energy matrix (the partial warps).

In both the two-dimensional and three-dimensional cases, the thin-plate spline is only used to solve for the non-uniform components of the deformation; a different approach is taken to solve for the uniform components. Bookstein (1996) shows that the approach he developed to construct a pair of basis vectors for the uniform part of a two-dimensional deformation can be extended to the three-dimensional case. This approach yields three pairs of vectors describing shear and compression/dilation in each of the three two-dimensional planes (*XY*, *YZ* and *XZ*). But remember, there are only five possible shape variables for the uniform part; therefore, the six vectors are not all completely independent. In fact, the problem lies in the three compression/dilation vectors; these three vectors actually describe a two-dimensional space. Bookstein suggests several methods to rectify this problem by constructing an orthonormal basis for this subspace (the current IMP software uses the Gram-Schmidt technique (following Axler, 1996)). These two vectors, combined with the three shear vectors, provide an orthonormal basis for the entire uniform subspace. More recently, Rohlf and Bookstein (Rohlf and Bookstein, 2003) have presented two other methods, both using an SVD to compute an orthonormal basis for the entire uniform subspace (without dividing it into shear and compression/dilation subspaces). The methods differ in how they extract the uniform variation from the total variation. In one, a technique used to compute residuals from a regression is used to compute the uniform component as the residuals from the non-uniform (as the difference between the total deformation and the non-uniform part). In the other, a technique used by Rohlf and Slice (1990) to compute the uniform component directly from superimposed two-dimensional coordinates is extended to three-dimensional coordinates. The new methods differ from that proposed by Bookstein (1996) only in the simplicity of the algorithms; all lead to the same conclusions regarding the differences among populations of shapes.

The result of the completed decomposition (of both uniform and non-uniform components) is an orthonormal basis for the Euclidean space that is tangent to the shape space at the location of the reference shape. Every configuration of landmarks in a data set can be described as a deformation of the reference shape; and that deformation is represented by the full set of scores on the five uniform components and $3(K-4)$ non-uniform components. These scores preserve Procrustes distances and express shape differences as scores on the same number of orthogonal axes as there are dimensions of the shape space (which is equal to the number of statistical degrees of freedom). Consequently, these scores can be used in standard multivariate analyses, assuming that there are no semilandmarks with their reduced dimensionality.

USING THE THIN-PLATE SPLINE TO VISUALIZE SHAPE CHANGE

The combination of the uniform and non-uniform components completely describes any shape change. The set of partial warp scores (including scores on the uniform component) can be used in any conventional statistical analysis and, like the coordinates obtained by GLS, the sum of their squares equals the squared Procrustes distance from the reference. Moreover, like Bookstein's shape coordinates, they have the correct degrees of freedom. Thus, we can use partial warps in any statistical procedure, such as regression, and diagram the results as a deformation.

Interpreting Changes Depicted by the Thin-Plate Spline

Interpretations should be presented in terms of the total deformation, not by detailing the separate uniform and non-uniform components (or the more finely subdivided components of them). Just as we cannot talk about individual landmarks as if they were separately moved, we cannot talk about components of the total deformation as if they were separate parts of the whole. It is important to remember that the changes depicted are based on an interpolation function – we do not actually know what occurs between landmarks, and while the thin-plate spline is a smooth interpolation function, and widely used, other interpolation functions are also available. If we have sparsely sampled some regions of the body, we cannot assume that the spline provides a realistic picture of their changes; there might be many highly localized changes that cannot be detected in the absence of closely spaced landmarks. All we can say is that our data do not require any more localized changes.

We cannot show an example of a biological transformation depicted by the thin-plate spline until we have results to show, so we will borrow examples from later chapters to discuss the description of shape change using the thin-plate spline. In Figure 5.7 we depict the ontogenetic changes in body shape of two species of piranhas: *S. gouldingi* (Figure 5.7A), which we used earlier in this chapter, and *Pygopristis denticulata* (Figure 5.7B). In both species, the head (as a whole) grows less rapidly than the middle of the body, and the eye grows far more slowly than the head. In neither species does the

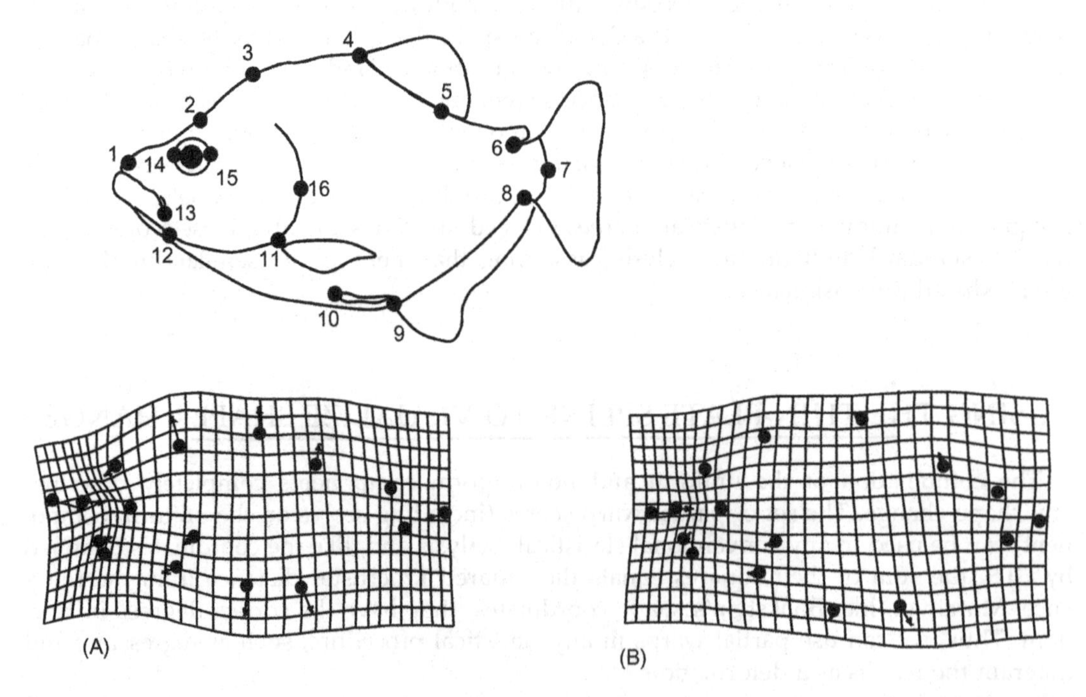

FIGURE 5.7 Ontogenetic shape change for two species of piranhas: (A) *Serrasalmus gouldingi*; (B) *Pygopristis denticulata*.

shortening of the eye result solely from the generally lower cranial growth rates; rather, there is an abrupt (and localized) deceleration of growth rates in the orbital region. However, that does not, by itself, fully account for the apparent contraction of the grid in the head, especially in *S. gouldingi*. Part of the relative shortening of the head, supraorbitally, results from the displacement of the landmark at the epiphyseal bar (landmark 2) towards the anterior landmark of the eye (landmark 14). Suborbitally, the apparent shortening of the head results from the displacement of the posterior jaw landmark (landmark 13) towards the posterior eye landmark (landmark 15), as well as from the more general shortening of the snout and eye. These two species also differ in the ontogeny of posterior body shape. In *S. gouldingi*, the caudal peduncle (the region bounded by landmarks 6, 7, and 8) appears to contract, but no change appears to be localized there – the posterior body generally shortens (as does the head). Growth rates appear to decrease, moving posteriorly from the mid-body to the tail. Because the caudal peduncle is the most posterior part of the body, the growth rates are lowest there. In *P. denticulata*, growth rates decrease more slowly, and most of the change in the posterior body seems to result from the posterior displacement (and relative shortening) of the anal fin. That increases the distance between the pelvic and anal fins (which expands the grid between them), but because that is not a part of the general expansion of the mid-body (it is limited to the ventral region between the fins), the change is ventrally localized. Due to the sparse sampling of landmarks in the middle of the body, there is no abrupt contraction or expansion of the grid such as we see in the head. Sparse sampling of that region makes it difficult to detect localized changes because we cannot show what happens between landmarks when we have not sampled them (quoting Gertrude Stein, "there is no there there").

USING BENDING ENERGY TO SUPERIMPOSE SEMILANDMARKS

In the earlier discussion of semilandmarks, we discussed how semilandmarks could be "slid" along curves to minimize the perpendicular distance between the specimens, and thus the Procrustes distance between the specimens. This was a distance minimizing approach, but it is also possible to use the thin-plate spline to slide landmarks to produce an optimally smooth (non-localized) difference between semilandmarks on two specimens. In this approach, developed by Green (1996) and Bookstein (1997), the first step is a conventional Procrustes superimposition (treating landmarks and semilandmarks as equivalent) to compute a mean configuration and align the targets to it. This is followed by moving the semilandmarks of each target to minimize the bending energy of the thin-plate spline describing the deformation of the reference to that target. The semilandmarks are not free to move in any direction; each is confined to "slide" along the line tangent to the curve at that semilandmark (Figure 5.8). The shape of the curve is not actually known, so the tangent is estimated as the line parallel to the segment connecting adjacent landmarks or semilandmarks (Figure 5.9). After sliding, the superimposition is recomputed; if the new mean configuration differs from the previous mean, the sliding and superimposition are reiterated until they converge on a solution. The justification for this sliding technique is that differences in relative positions of semilandmarks along the curve cannot be informative because this spacing was defined arbitrarily (i.e. extrinsically). Thus, sliding to

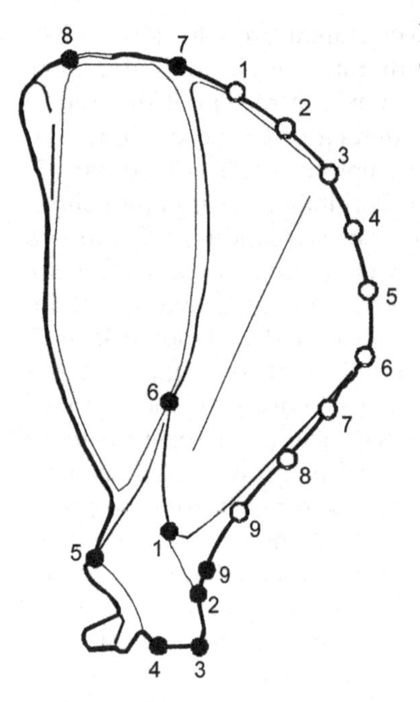

FIGURE 5.8 Semilandmarks on the edge of the squirrel scapula are constrained to slide along tangents to that curve in the reference.

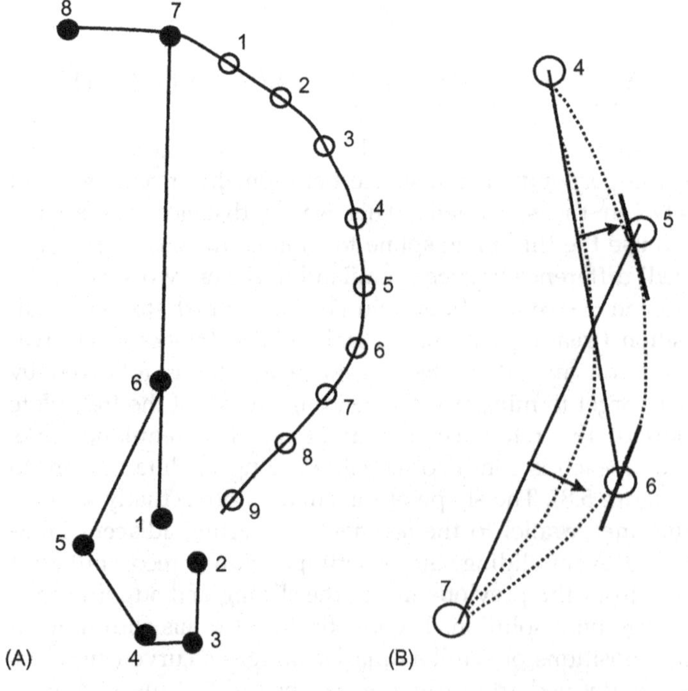

FIGURE 5.9 Tangents estimated as lines parallel to the segment connecting adjacent landmarks or semilandmarks.

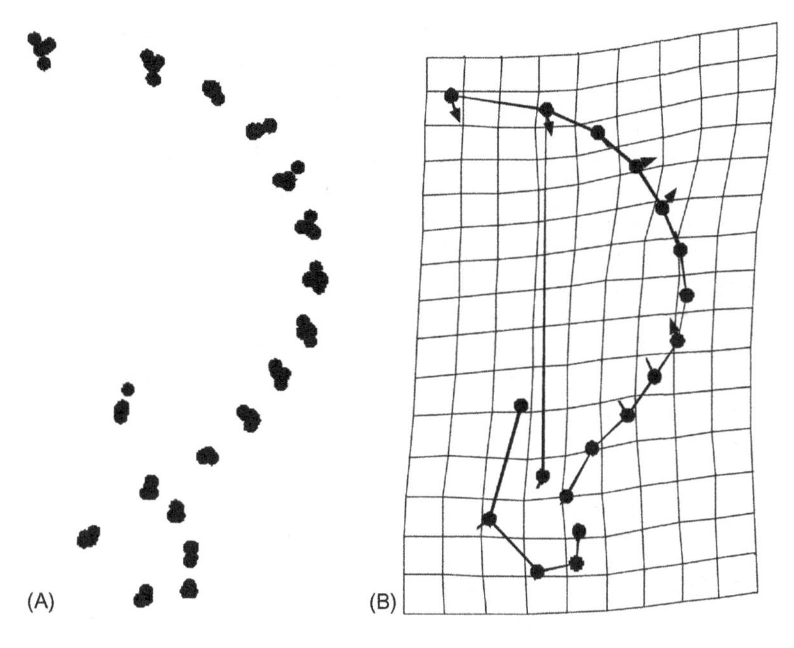

FIGURE 5.10 (A) Scapular landmarks superimposed after sliding to minimize bending energy. Several semilandmarks have ellipses of variation that imply displacements along the anterior edge, particularly the ones in the ventral half. (B) PC1 showing coordinated displacements toward the dorsal end.

minimize the bending energy of the deformation adjusts the spacing of the semilandmarks to minimize the implication that there are shape changes due to differences in that spacing.

Figure 5.10A shows the same data set used earlier, superimposed after sliding to minimize bending energy. Several semilandmarks have ellipses of variation that imply displacements along the anterior edge, particularly the ones in the ventral half. PC1 indicates that these semilandmarks undergo correlated displacements toward the dorsal end as the anterodorsal corner is squared out (Figure 5.10B). Thus, there is little change in the positions of these semilandmarks *relative to each other*, and therefore little localized change along the anterior edge. Most of the localized change in the anterior edge occurs at the corner. As before, the change in shape of the anterior edge is inferred to be the dominant component of shape change. Displacements of the landmarks are generally slight; the exceptions are the ventral displacements of the most dorsal landmarks, which are involved in the general flattening of the dorsal edge and squaring of the anterodorsal corner.

The principal disadvantage of this approach is that the semilandmarks are in new positions relative to the landmarks and the other semilandmarks. However, this may not be the devastating flaw that it seems to be. The underlying premise of sliding is that semilandmarks are not equal to landmarks. As pointed out above, semilandmarks do not represent the same amount of independent information as landmarks because semilandmarks are constrained to lie along the curve at arbitrary intervals. Put another way, if moving semilandmarks does not alter the information about the shape of the curve, then the configuration of landmarks and semilandmarks after sliding might be considered to have the same shape as the configuration before sliding.

The semilandmark method based on bending energy produces smooth differences between curves on different specimens, the distance minimizing method produces specimens with the minimum possible distance between them. There is no clear consensus at this point about which approach is preferable. However, it does appear that the distance minimizing approach will often be more statistically conservative. If we want to show that the mean shape of two groups is different, then the distance minimizing alignment would produce specimens at the minimum possible distance, effectively biasing a statistical test against finding any differences. So, if there is enough difference remaining in the two groups to reject the null hypothesis despite the possible bias of the alignment procedure, one can be reasonably sure the differences in mean shape are not an artifact of the alignment procedure. Any possible bias in the distance biased alignment is certainly against the result we want to show. Bending energy alignment does produce smoother looking differences, but also appears to increase the variance within the data (Sheets et al., 2004, 2006).

APPENDIX

Calculating the Shear and Compression/Dilation Terms

Here we present the mathematical derivation of formulae for calculating the uniform components of a deformation that changes shape. Unlike the formulae for computing the non-uniform part of a shape change, which have been stable over the last decade, the formulae for computing the uniform part have changed repeatedly. Over the last several years, the uniform component has been computed using the formulae presented by Bookstein (1996). The ones based on the Procrustes distance are the ones we present here. We begin with a conceptual framework for Bookstein's derivation of the current formulae; then follow that with the full mathematical details.

Conceptual Framework

The goal of this derivation is to find a unit vector that describes the direction of deformation at each landmark due to shearing or compression/dilation, followed by a Procrustes generalized least squares (GLS) superimposition of the deformed shape back onto the original (undeformed) one. This represents what we measure in data: a deformation followed by a superimposition operation. Thus, both mappings must be taken into account. When we are done, we will have a set of unit vectors that describe the deformation under shearing or under compression/dilation. We can then take the dot product of the observed deformation with the unit vectors to obtain the component of the observed deformation lying along the shear or compression/dilation vectors. These are what we have been calling the explicit uniform components of the deformation.

Notice that we are taking a verbal description of the situation, turning the verbal statement into two mathematical operations or mappings (shear or compression/dilation, followed by the superimposition), then using those mappings to determine the direction of the vectors describing the deformation. That allows us to calculate components of any deformation along those desired directions. What might not be obvious yet is that vectors describing the uniform deformations depend on only one form – the one that we are modeling as deformed, which we will call the reference form (the other is the target). This terminology should be familiar – the reference form is the same one that we discussed in Chapter 3. If you do not wish to read further, you do not need to. You can return directly to the section on decomposing the non-uniform (non-affine) component.

Although we now have a general idea of the procedure, there are still a few ideas that need to be added. The first is the idea of complex number notation for landmark locations, which is often used in mathematical derivations (see Dryden and Mardia, 1998, for example). Consider a landmark configuration consisting of K landmarks in two dimensions, which we will call **Z**, the reference form. Mathematically, we will say:

$$Z = \{Z_j, Z_j = (X_j, Y_j)\}_{j=1}^{K} \tag{5A.1}$$

which means that **Z** is a set of K pairs of landmark positions Z_j, or (X_j, Y_j). It is a useful mathematical shortcut to think of Z_j as being a complex number $Z_j = X_j + iY_j$, where i is the square root of minus one. Complex number notation is often used in texts on the statistics of shape, so understanding this approach is useful.

The next idea is to require that the reference form be rotated to a principal axis alignment, so that $\Sigma_j X_j Y_j = 0$, which will later simplify the mathematics (but may pose problems for aligning specimens in some software, discussed below). The summation Σ_j is from $j = 1$ to $j = K$, and all the summations in the derivation are likewise over all K landmarks. We are also going to assume that the reference has a centroid size of one, so that $\Sigma_j (X_j^2 + Y_j^2) = 1$, and a centroid position of (0, 0), so that $\Sigma_j X_j = 0$ and $\Sigma_j Y_j = 0$.

Mathematical Derivations

Let us consider the two functions of interest: shear, which we will call $\mathbf{S}_1(\lambda)$, and compression/dilation, which we will call $\mathbf{S}_2(\lambda)$ (λ describes the magnitude of the mapping). We will be taking the limit as $\lambda \to 0$ at the end of this derivation, so terms including λ^2 will be discarded. The mappings from a reference form **Z** to a target form **Z′** under these operations are as follows:

$$S_1(\lambda): Z \to Z', Z' = \{Z'_j = (X_j + \lambda Y_j,\ Y_j)\}_{j=1}^{K} \tag{5A.2}$$

$$S_2(\lambda): Z \to Z', Z' = \{Z'_j = (X_j, Y_j + \lambda Y_j)\}_{j=1}^{K} \tag{5A.3}$$

You can probably convince yourself that $\mathbf{S}_1$ describes a shear; the X-coordinates of each point are displaced a distance proportional to their Y-axis position relative to the centroid. Similarly, you should be able to recognize that $\mathbf{S}_2$ describes an expansion of the landmarks along the Y-axis. We do not need to worry about modeling the contraction along the X-axis, even though it must also be occurring, because the Procrustes GLS superimposition will take care of that by requiring that the centroid size be fixed at one.

If **Z** and **Z′** are both centered (i.e. have a centroid position of zero), then the Procrustes superimposition may be approximated as the multiplication of **Z′** by the complex factor $\mathbf{P}_{\mathbf{Z}'}$, where:

$$P_{Z'} = \frac{Z\overline{Z'}}{Z'Z'} \tag{5A.4}$$

and the expression $\overline{Z'}$ refers to the complex conjugate of the complex vector $\mathbf{Z}'$ representing the landmark configuration after the compression/dilation. The Procrustes superimposition of $\mathbf{Z}'$ on $\mathbf{Z}$ is thus $\mathbf{P}_{z'}\mathbf{Z}'$. To get the vectors that describe the uniform deformation, we just subtract the starting position $\mathbf{Z}$ from $\mathbf{P}_{z'}\mathbf{Z}'$ and then divide through by the magnitude of the deformation λ, yielding $(\mathbf{P}_{z'}\mathbf{Z}' - \mathrm{Z})/\lambda$ as the set of vectors describing the deformation.

Further Derivation of the Uniform Components

To find $\mathbf{P}_{Z'}$ for the $\mathbf{S}_2(\lambda)$ mapping (compression/dilation), we note that the numerator of $\mathbf{P}_{z'}$ is:

$$Z\overline{Z'} = \sum_j (X_j + iY_j) \times (X_j - i(Y_j + \lambda Y_j)) \tag{5A.5}$$

which expands to:

$$= \sum_j (X_j^2 - iX_jY_j - iX_j\lambda Y_j + iX_jY_j - (iY_j)^2 - (iY_j)^2\lambda) \tag{5A.6}$$

Because $i^2 = -1$ and the products of X_jY_j sum to zero (under the alignment specified earlier), we can simplify this to:

$$= \sum_j (X_j^2 + Y_j^2 + Y_j^2\lambda) \tag{5A.7}$$

Now add the constraint that $\sum_j (X_j^2 + Y_j^2) = 1$ because we scaled the reference to unit centroid size, and we have:

$$Z\overline{Z'} = 1 + \lambda \sum_j Y_j^2 \tag{5A.8}$$

Now we simplify the denominator of $\mathbf{P}_{z'}$:

$$Z'\overline{Z'} = \sum_j (X_j + iY_j(1 + \lambda)) \times (X_j - iY_j(1 + \lambda)) \tag{5A.9}$$

$$= \sum_j (X_j^2 + Y_j^2(1+\lambda)^2) = \sum_j X_j^2 + Y_j^2(1 + 2\lambda + \lambda^2) \tag{5A.10}$$

$$= \sum_j X_j^2 + Y_j^2 + 2Y_j^2\lambda + Y_j^2\lambda^2 \tag{5A.11}$$

As mentioned before, $\sum_j (X_j^2 + Y_j^2) = 1$, and terms including λ^2 can be discarded in the limit of small λ, so that:

$$Z'\overline{Z'} \cong 1 + 2\lambda \sum_j Y_j^2 \tag{5A.12}$$

This leaves us with:

$$Pz' = \frac{Z\overline{Z'}}{Z'\overline{Z'}} = \frac{(1 + \lambda\sum_j Y_j^2)}{(1 + 2\lambda\sum_j Y_j^2)} \tag{5A.13}$$

We can now expand the term $1/(1+2\lambda\sum_j Y_j^2)$ as $1-2\lambda\sum_j Y_j^2$, keeping only first order terms in λ for this power series expansion. This gives us:

$$\mathrm{P}z' = \frac{Z\overline{Z'}}{Z'\overline{Z'}} \cong \left(1+\lambda\sum_j Y_j^2\right)\left(1-2\lambda\sum_j Y_j^2\right) \cong 1-\lambda\sum_j Y_j^2 \tag{5A.14}$$

to first order in λ.

Now we can calculate the landmark coordinates after the operation of the compression/dilation ($S_2(\lambda)$) and Procrustes superimposition (which is just a multiplication by $\mathrm{P}_{\mathbf{Z}'}$, since $\mathbf{Z}'$ is already centered):

$$\mathrm{P}z'Z' = \left(1-\lambda\sum_j Y_j^2\right)\times Z' = \left\{Z_j = \left(X_j\left(1-\lambda\sum_j Y_j^2\right)\right), \left((Y_j+\lambda Y_j)\left(1-\lambda\sum_j Y_j^2\right)\right)\right\}_{j=1}^{K} \tag{5A.15}$$

The vector describing the displacement from $\mathbf{Z}$ to $\mathbf{P}_{z'}\mathbf{Z}'$ is then:

$$\mathrm{P}z'Z' - Z = \left\{\left(\left(X_j\left(1-\lambda\sum_j Y_j^2\right)-X_j\right), \left((Y_j+\lambda Y_j)\left(1-\lambda\sum_j Y_j^2\right)-Y_j\right)\right)\right\}_{j=1}^{K} \tag{5A.16}$$

$$= \left\{\left(\left(-X_j\lambda\sum_j Y_j^2\right), \left(\lambda Y_j-\lambda Y_j\sum_j Y_j^2-\lambda^2 Y_j\sum_j Y_j^2\right)\right)\right\}_{j=1}^{K} \tag{5A.17}$$

Noting that $\lambda^2 \cong 0$, we can simplify this to:

$$= \left\{\left(\left(-X_j\lambda\sum_j Y_j^2\right), \left(\lambda Y_j-\lambda Y_j\sum_j Y_j^2\right)\right)\right\}_{j=1}^{K} \tag{5A.18}$$

$$= \lambda\left\{\left(X_j\left(-\sum_j Y_j^2\right), Y_j\left(1-\sum_j Y_j^2\right)\right)\right\}_{j=1}^{K} \tag{5A.19}$$

We now define $\gamma = \Sigma_j\, Y_j^2$ and $\alpha = 1-\Sigma_j\, Y_j^2 = \Sigma_j\, X_j^2$, so that $\gamma+\alpha = 1$. After making these substitutions and dividing through by λ, we have:

$$V_2 = \frac{(\mathrm{P}z'-Z')}{\lambda} = \{(-\gamma X_j, \alpha Y_j)\}_{j=1}^{K} \tag{5A.20}$$

which is the vector of the displacements at each landmark point (X_j, Y_j) produced by the mapping $\mathbf{S_2}$ per unit of λ. All we need to do now is to normalize this set so that the length of the vector is one.

The magnitude of this vector is:

$$\sqrt{\sum_j (\gamma^2 X_j^2 + \alpha^2 Y_j^2)} \tag{5A.21}$$

Using the definitions of α and γ to rearrange this and simplify it, we get:

$$= \sqrt{\gamma^2 \sum_j X_j^2 + \alpha^2 \sum_j Y_j^2} = \sqrt{\gamma^2\alpha + \alpha^2\gamma} = \sqrt{\alpha\gamma(\alpha+\gamma)} = \sqrt{\alpha\gamma} \tag{5A.22}$$

So if we normalize $\mathbf{V_2}$, we get:

$$V_2' = \frac{V_2}{\sqrt{\alpha\gamma}} = \left\{\left(-\frac{\gamma}{\sqrt{\alpha\gamma}}X_j, \frac{\alpha}{\sqrt{\alpha\gamma}}Y_j\right)\right\}_{j=1}^{K} \tag{5A.23}$$

$$= \left\{\left(-\sqrt{\frac{\gamma}{\alpha}}X_j, \sqrt{\frac{\alpha}{\gamma}}Y_j\right)\right\}_{j=1}^{K} \tag{5A.24}$$

which is now a unit vector describing a compression/dilation operation followed by Procrustes superimposition.

Similarly, we start with a shearing operation, $\mathbf{S_1}(\lambda)$, and corresponding Procrustes superimposition, $\mathbf{P_{Z'}}$, to find the unit vector corresponding to these operations. First we need to find $\mathbf{P_{Z'}}$ for the $\mathbf{S_1}(\lambda)$ mapping:

$$Z\overline{Z'} = \sum_j (X_j + iY_j) \times (X_j + \lambda Y_j - iY_i) \tag{5A.25}$$

$$= \sum_j (X_j^2 + Y_j^2 + X_jY_j\lambda + iY_i^2\lambda) \tag{5A.26}$$

As before, $\sum_j (X_j^2 + Y_j^2) = 1, \sum_j X_jY_j = 0$ and $\sum_j Y_j^2 = \gamma$; thus:

$$Z\overline{Z'} = 1 + i\gamma\lambda \tag{5A.27}$$

Also:

$$Z'\overline{Z'} = \sum_j (X_j + \lambda Y_j + iY_j) \times (X_j + \lambda Y_j - iY_j) \tag{5A.28}$$

$$\sum_j (X_j + \lambda Y_j)^2 + Y_j^2 = \sum_j (X_j^2 + 2\lambda X_jY_j + \lambda^2 Y^2 + Y_j^2) = 1 \tag{5A.29}$$

Therefore:

$$Pz' = \frac{Z\overline{Z'}}{Z'\overline{Z'}} = \frac{Z\overline{Z'}}{1} = 1 + i\gamma\lambda \tag{5A.30}$$

Now we can simplify:

$$V_1 = \frac{Pz'Z' - Z}{\lambda} = \frac{(1 + i\gamma\lambda)(X_j + \lambda Y_j + iY_j) - (X_j + iY_j)}{\lambda} \tag{5A.31}$$

$$= \frac{X_j + \lambda Y_j + iY_j + i\gamma\lambda X_j + i\gamma\lambda^2 Y_j + i^2\gamma\lambda Y_j - X_j - iY_j}{\lambda} \tag{5A.32}$$

$$= \frac{\lambda Y_j + i\gamma\lambda X_j - \gamma\lambda Y_j}{\lambda} = Y_j + i\gamma X_j - \gamma Y_j \tag{5A.33}$$

This leads to the series of coordinate pairs:

$$= (Y_j(1 - \gamma), \gamma X_j) \tag{5A.34}$$

or

$$V_1 = (\alpha Y_j,\ \gamma X_j) \tag{5A.35}$$

The magnitude of this vector is:

$$\sqrt{\sum_j (\alpha^2 Y_j^2 + \gamma^2 X_j^2)} = \sqrt{\alpha^2 \sum_j Y_j^2 + \gamma^2 \sum_j X_j^2} = \sqrt{\alpha^2\gamma + \gamma^2\alpha} \tag{5A.36}$$

$$= \sqrt{\alpha\gamma(\alpha + \gamma)} = \sqrt{\alpha\gamma} \tag{5A.37}$$

so the unit vector $\mathbf{V}_1'$ obtained by normalizing V_1 is:

$$V_1' = \left\{\left(\frac{\alpha Y_1}{\sqrt{\alpha Y}}, \frac{\gamma X_j}{\sqrt{\alpha\gamma}}\right)\right\}_{j=1}^{K} = \left\{\left(\sqrt{\frac{\alpha}{\gamma}} Y_j, \sqrt{\frac{\gamma}{\alpha}} X_j\right)\right\}_{j=1}^{K} \tag{5A.38}$$

which may now be used to determine the shear component of the uniform deformation.

Some software packages will give you α and γ as used in the calculation of the uniform component, others may give you the unit vectors instead. The expressions are for coordinates of the unit vectors for shear and compression/dilation for a reference form rotated to principal axis orientation. It turns out to be straightforward to rotate them to unit vectors to match any reference orientation preferred by a researcher, although some programs may not offer this option, meaning that the reference may be oddly oriented by the software.

Calculating Uniform Components Based on Other Superimpositions

The approach taken in the above derivation was to determine the unit vectors that would result from a shear or compression/dilation of a reference form, followed by Procrustes superimposition back onto the reference form. It is also possible to determine the unit vectors produced by a shear or compression/dilation of a reference, followed by sliding baseline registration (SBR) or a two-point registration that yields Bookstein coordinates (BC). These unit vectors and specimens can then be used in SBR or BC to calculate the uniform components of the deformation, just as we did with those in Procrustes superimposition. Estimates of the explicit uniform components under SBR are identical to those derived from the Procrustes-based method presented here. This is not surprising, since the Procrustes superimposition differs from SBR only in the implicit uniform deformations (assuming that the Procrustes superimposition, like SBR, is performed with centroid size set to one, two superimpositions differ only in the rotation and translation terms). Thus, a deformation displayed by a Procrustes superimposition shows the same change in *shape* as the deformation displayed by SBR – the differences between them are due to the implicit deformations, and do not alter shape. Deformations shown by BC differ from those in Procrustes superimposition in scale as well as rotation and translation, but these are still implicit uniform terms. Likewise, RFTRA differs from the other superimpositions only in the implicit uniform terms.

References

Axler, S. (1996). *Linear algebra done right*. New York: Springer-Verlag.

Bookstein, F. L. (1989). Principal warps: Thin-plate splines and the decomposition of deformations. *IEEE Transactions on Pattern Analysis and Machine Intelligence*, *11*, 567–585.

Bookstein, F. L. (1991). *Morphometric tools for landmark data: Geometry and biology*. Cambridge: Cambridge University Press.

Bookstein, F. L. (1996). Standard formula for the uniform shape component in landmark data. In L. F. Marcus, M. Corti, A. Loy, G. J. P. Naylor, & D. E. Slice (Eds.), *Advances in morphometrics* (pp. 53–168). New York: Plenum.

Bookstein, F. L. (1997). Landmark methods for forms without landmarks: Morphometrics of group differences in outline shape. *Medical Image Analysis*, *1*, 97–118.

Dryden, I. L., & Mardia, K. V. (1998). *Statistical shape analysis*. New York: John Wiley & Sons.

Green, W. D. K. (1996). The thin-plate spline and images with curving features. In K. V. Mardia, C. A. Gill, & I. L. Dryden (Eds.), *Image fusion and shape variability* (pp. 79–87). Leeds: University of Leeds Press.

Myers, P., Lundrigan, B. L., Gillespie, B. W., & Zelditch, M. L. (1996). Phenotypic plasticity in skull and dental morphology in the prairie deer mouse (*Peromyscus maniculatus bairdii*). *Journal of Morphology*, *229*, 229–237.

Rohlf, F. J., & Bookstein, F. L. (2003). Computing the uniform component of shape variation. *Systematic Biology*, *52*, 66–69.

Rohlf, F. J., & Slice, D. E. (1990). Extensions of the Procrustes method for the optimal superimposition of landmarks. *Systematic Zoology*, *39*, 40–59.

Sheets, H. D., Covino, K. M., Panasiewicz, J. M., & Morris, S. R. (2006). Comparison of geometric morphometric outline methods in the discrimination of age-related differences in feather shape. *Frontiers in Zoology*, *3*, 15.

Sheets, H. D., Kim, K., & Mitchell, C. E. (2004). A combined landmark and outline-based approach to ontogenetic shape change in the Ordovician trilobite *Triarthrus becki*. In A. M. T. Elewa (Ed.), *Morphometrics: Applications in biology and paleontology* (pp. 67–82). New York: Springer.

Slice, D. E., Bookstein, F. L., Marcus, L. F., & Rohlf, F. J. (1996). A glossary for geometric morphometrics. In L. F. Marcus, M. Corti, A. Loy, G. J. P. Naylor, & D. E. Slice (Eds.), *Advances in morphometrics* (pp. 531–551). New York: Plenum.

Thompson, D. A. W. (1992). *On growth and form: The complete revised edition* (2nd ed.). New York: Dover reprint of 1942.

PART 2

ANALYZING SHAPE VARIABLES

CHAPTER

6

Ordination Methods

In this chapter, we discuss two methods for describing the diversity of shapes in a sample: principal components analysis (PCA) and canonical variates analysis (CVA). Our discussion of these methods draws heavily on expositions presented by Morrison (1990), Chatfield and Collins (1980) and Campbell and Atchley (1981). Both methods produce ordinations that simplify descriptions, or provide tools for exploratory data analysis. These ordinations are descriptions of the data, not tests of hypotheses. However, CVA may also be used as multigroup discriminant function, in which the rate of correct assignment of specimens to groups based on shape is used to support specific hypotheses related to the ability to assign individuals to different species (Nolte and Sheets., 2005; Costa et al., 2008; Van Bocxlaer and Schultheiß, 2010; Williams et al., 2012) or as a diagnostic tool (Menesatti et al., 2008, Yee et al., 2009). PCA is a tool for simplifying descriptions of variation among individuals, whereas CVA is used for simplifying descriptions of differences between groups. Both analyses produce new sets of variables that are linear combinations of the original variables. They also produce scores for individuals on those variables, and these can be plotted and used to inspect patterns visually. Because the scores order specimens along the new variables, the methods are called "ordination methods". It is hoped that the ordering provides insight into patterns in the data, perhaps revealing patterns that are convenient for addressing biological questions. The most important difference between PCA and CVA is that PCA constructs variables that can be used to examine variation among individuals within a sample, whereas CVA constructs variables to describe the relative positions of groups (or subsets of individuals) in the sample.

We discuss PCA and CVA in the same chapter because they serve a similar purpose, and because the mathematical transformations performed in them are similar. We describe PCA first because it is somewhat simpler, and because it provides a foundation for understanding the transformations performed in CVA and in other related methods. We begin the description of PCA with some simple graphical examples, and then present a more formal exposition of the mathematical mechanics of PCA. This is followed by a presentation of an analysis of a real biological data set. The description of CVA follows a similar outline; the only difference is that we begin with a discussion of groups and grouping

DOI: http://dx.doi.org/10.1016/B978-0-12-386903-6.00006-X

variables. CVA requires that the individuals be grouped, because the objective of the method is to analyze the relative positions of those groups. Consequently, the sample must be divided into groups before the analysis begins. The description of differences between groups is optimized relative to the variation within those groups. That optimization requires a few more computational steps than PCA, but none of the steps in CVA introduce new mathematical concepts. CVA will be just a new application of ideas you have already encountered in the discussion of PCA.

As we will see, optimization of between groups differences with respect to within group variation has implications for the relative positions of group means and the distances between them, which are discussed in detail by Mitteroecker and Bookstein (2011). In the third section, we discuss an alternative method suggested by them in which PCA is used to analyze differences between group means without altering their positions or distance. This Between Groups PCA (BGPCA) may be more appropriate than CVA for some particular exploratory applications.

PRINCIPAL COMPONENTS ANALYSIS

Geometric shape variables are neither biologically nor statistically independent. For example, the shape variables produced by the thin-plate spline describe variation in overlapping regions of an organism or structure. Because the regions overlap, they are under the influence of the same processes that produce variation; and therefore we expect them to be correlated. Even when they do not describe overlapping regions, morphometric variables (both geometric and traditional) are expected to be correlated because they describe features of the organism that are functionally, developmentally or genetically linked. Their patterns of variation and covariation are often complex and difficult to interpret. The purpose of PCA is to simplify those patterns and make them easier to interpret by replacing the original variables with new ones (principal components, PCs) that are linear combinations of the original variables and independent of each other.

One might wonder why it would be a worthwhile exercise to take simple variables that covary with each other and replace them with complex variables that do not covary. Part of the value of this exercise arises from the fact that the new complex variable is a function of the covariances among the original variables. It thus provides some insight into the covariances among variables, which can direct future research into the identity of the causal factors underlying those covariances. Another useful purpose served by PCA is that most of the variation in the sample usually can be described with only a few PCs. Again this is useful, because it simplifies and clarifies what needs to be explained. Another important benefit of PCA is that the presentation of results is simplified. It is much easier to produce and explain plots of the three PCs that explain 90% of the variation than it is to plot separately and explain the variation on each of 30 original variables.

An indirect benefit of PCA that is useful (but often misused) is that it simplifies the description of differences among individuals. Clusters of individuals are often more apparent in plots of PCs than in plots of the original variables. Finding such clusters can be quite valuable, but those clusters do not represent evidence of statistically distinct entities. Legitimate methods for testing the hypothesis that *a priori* groups are statistically significantly different will be presented in later chapters.

Geometric Description of PCA

Figure 6.1A shows the simple case in which there are two observed traits, $\mathbf{X}_1$ and $\mathbf{X}_2$. These traits might be two distance measurements or the coordinates of a single landmark in a two-dimensional shape analysis. Each point in the scatter plot represents the paired values observed for a single specimen. We expect that the values of each trait are normally distributed, and we expect that one trait is more variable than the other. In this case, $\mathbf{X}_1$ has a larger range of observed values and a higher variance than $\mathbf{X}_2$. In addition, the values of $\mathbf{X}_1$ and $\mathbf{X}_2$ are not independent; higher values of one are associated with higher values of the other. This distribution of values can be summarized by an ellipse that is tilted in the $\mathbf{X}_1$, $\mathbf{X}_2$ coordinate plane (Figure 6.1B). PCA solves for the axes of this ellipse, and uses those axes to describe the positions of individuals within that ellipse.

The first step of PCA is to find the direction through the scatter that describes the largest proportion of the total variance. This direction, the long axis of the ellipse, is the first principal component (PC1). In an idealized case like that shown in Figure 6.1A, the line we seek is approximately the line through the two cases that have extreme values on both variables. Real data rarely have such convenient distributions, so we need a criterion that has more general utility. If we want to maximize the variance that the first axis describes, then we also want to minimize the variance that it does not describe – in other words, we want to minimize the sum of the squared distances of points away from the line (Figure 6.1C). (*Note*: the distances that are minimized by PCA are not the distances minimized in conventional least-squares regression analysis – see Chapter 8.)

The next step is to describe the variation that is not described by PC1. When there are only two original variables, this is a trivial step; all of the variation that is not described by the first axis of the ellipse is completely described by the second axis. So, let us consider briefly the case in which there are three observed traits: $\mathbf{X}_1$, $\mathbf{X}_2$ and $\mathbf{X}_3$. This situation is unlikely to arise in optimally superimposed landmark data, but it illustrates a generalization that can be applied to more realistic situations. As in the previous example, all traits are normally distributed and no trait is independent of the others. In addition, $\mathbf{X}_1$ has the largest variance and $\mathbf{X}_3$ has the smallest variance. A three-dimensional model of this distribution would look like a partially flattened blimp or watermelon (Figure 6.2A). Again PC1 is the direction in which the sample has the largest variance (the long axis of the watermelon), but now a single line perpendicular to PC1 is not sufficient to describe the remaining variance. If we cut the watermelon in half perpendicular to PC1, the cross-section is another ellipse (Figure 6.2B). The individuals in the section (the seeds in the watermelon) lie in various directions around the central point, which is where PC1 passes through the section. Thus, the next step of the PCA is to describe the distribution of data points around PC1, not just for the central cross-section, but also for the entire length of the watermelon.

To describe the variation that is not represented by PC1, we need to map, or project, all of the points onto the central cross-section (Figure 6.2C). Imagine standing the halved watermelon on the cut end and instantly vaporizing the pulp so that all of the seeds drop

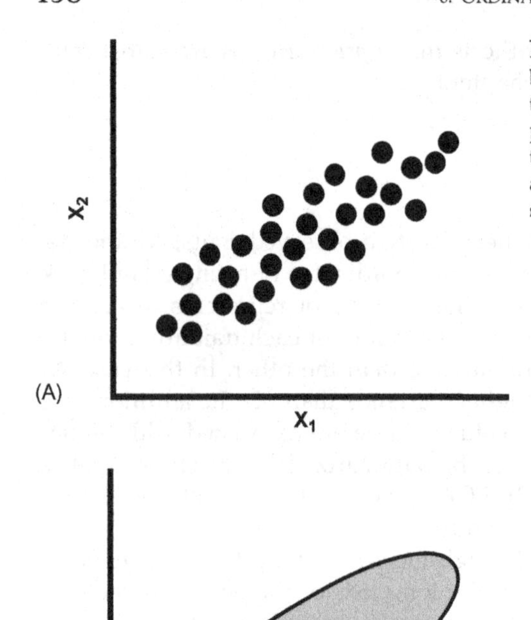

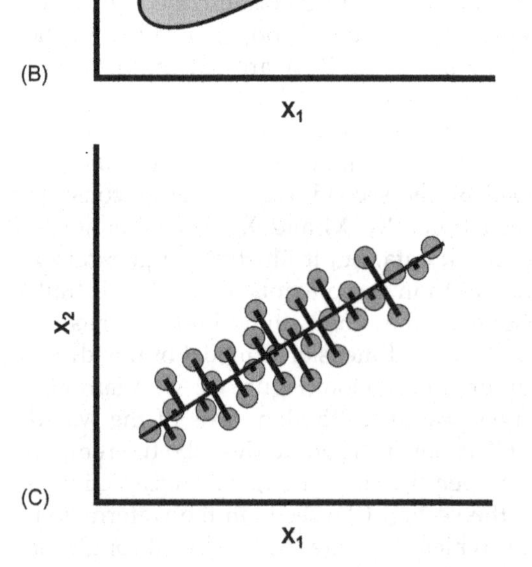

FIGURE 6.1 Graphical representation of the problem to be solved by PCA. (A) Scatter plot of individuals scored on two traits, $\mathbf{X_1}$ and $\mathbf{X_2}$; (B) an ellipse enclosing the scatter of points shown in part (A); (C) a line through the scatter and the perpendicular distances of the individuals from that axis. The goal of PCA is to find the line that minimizes the sum of those squared distances.

vertically onto a sheet of wax paper, then repeating the process with the other half of the watermelon and the other side of the paper. The result of this mapping is a two-dimensional elliptical distribution similar to the first example. This ellipse represents the variance that is not described by PC1. Thus, the next step of the three-dimensional PCA is

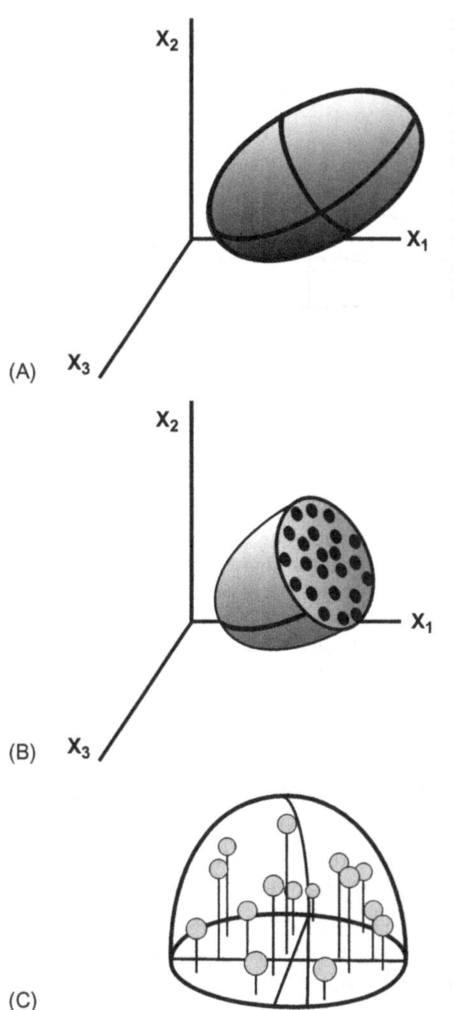

FIGURE 6.2 Graphical representation of PCA on three original variables (X_1, X_2, X_3). (A) The distribution of individual specimens on the three original axes is summarized by a three-dimensional ellipsoid; (B) the three-dimensional ellipsoid is cut by a plane passing through the sample centroid and perpendicular to the longest axis (PC1) at its midpoint, showing the distribution of individuals around the longest axis in the plane of the section; (C) the upper half of the ellipsoid in B has been rotated so that the cross-section is in the horizontal plane. Perpendicular projections of all individuals (from both halves) onto this plane are used to solve for the second and third PCs.

the first step of the two-dimensional PCA – namely, solving for the long axis of a two-dimensional ellipse, as outlined above. In the three-dimensional case, the long axis of the two-dimensional ellipse will be PC2. The short axis of this ellipse will be PC3, and will complete the description of the distribution of seeds in the watermelon. By logical extension, we can consider *N* variables measured on some set of individuals to represent an *N*-dimensional ellipsoid. The PCs of this data set will be the *N* axes of the ellipsoid.

After the variation in the original variables has been redescribed in terms of the PCs, we want to know the positions of the individual specimens relative to these new axes (Figure 6.3). As shown in Figure 6.3A, the values we want are determined by the orthogonal projections of the specimen onto the PCs. These new distances are called principal component *scores*. Because the PCs intersect at the sample mean, the values of the scores represent the distances of the specimen from the mean in the directions of the PCs.

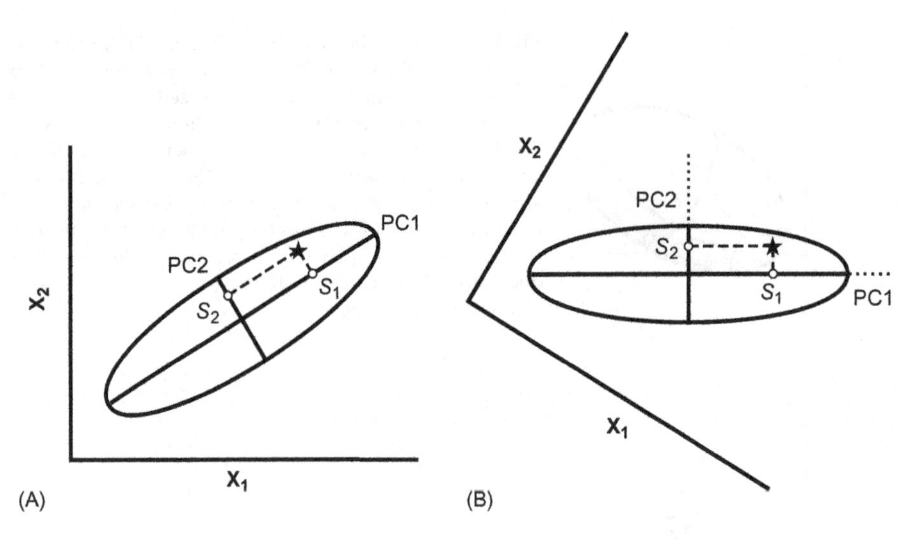

FIGURE 6.3 Graphical interpretation of PC scores. (A) The star is the location of an individual in the sample. Perpendiculars from the star to PCs indicate the location of the star with respect to those axes. The distances of points S_1 and S_2 from the sample centroid (intersection of PC1 and PC2) are the scores of the star on PC1 and PC2. (B) The figure in part A has been rotated so that PCs are aligned with the edges of the page. The PCs will now be used as the reference axes of a new coordinate system; the scores on these axes are the location of the individual in the new system. The relationships of the PC axes to the original axes has not changed, nor has the position of the star relative to either set of axes.

In effect, we are rotating and translating the ellipse into a more convenient orientation so we can use the PCs as the basis for a new coordinate system (Figure 6.3B). The PCs are the axes of that system. All this does is allow us to view the data from a different perspective; the positions of the data points relative to each other have not changed.

As suggested by Figure 6.4, we could compute an individual's score on a PC from the values of the original variables that were observed for that individual and the cosines of the angles between the original variables and the PCs. In our simple two-dimensional case, the new scores, Y, could be calculated as:

$$Y_1 = A_1X_1 + A_2X_2 \tag{6.1}$$

where A_1 and A_2 are the cosines of the angles α_1 and α_2 and the values of individuals on X_1 and X_2 are the differences between them and the mean, not the observed values of those variables.

It is important to bear in mind for our algebraic discussion that Equation 6.1 represents a straight line in a two-dimensional space. Later we will see equations that are expansions of this general form and represent straight lines in spaces of higher dimensionality. So, in case the form of Equation 6.1 is unfamiliar, the next few equations illustrate the simple conversion of this equation into a more familiar form. First, we rearrange the terms to solve for X_2:

$$Y_1 - A_1X_1 = A_2X_2 \tag{6.2}$$

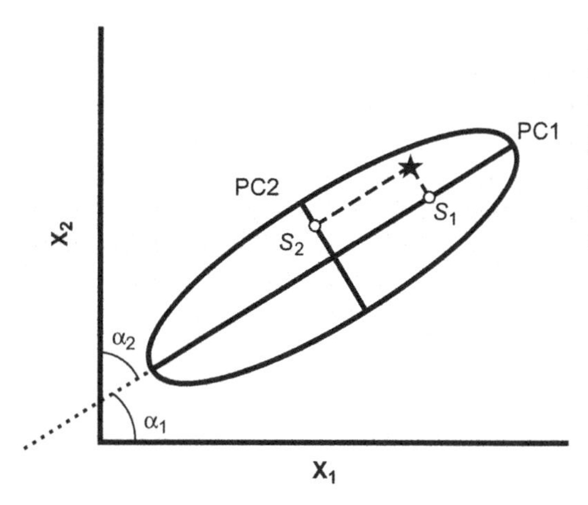

FIGURE 6.4 Graphical interpretation of PC scores, continued. The angles α_1 and α_2 indicate the relationship of PC1 to the original axes $\mathbf{X_1}$ and $\mathbf{X_2}$. Thus, S_1 can be computed from the coordinates of the star on $\mathbf{X_1}$ and $\mathbf{X_2}$ and the cosines of the angles between PC1 and the original axes. S_2 can be computed from the coordinates of the star on $\mathbf{X_1}$ and $\mathbf{X_2}$ and the cosines of the angles between PC2 and the original axes.

$$A_2X_2 = -A_1X_1 + Y_1 \tag{6.3}$$

$$X_2 = \frac{-A_1X_1}{A_2} + \frac{Y_1}{A_2} \tag{6.4}$$

Then we make two substitutions ($M = -A_1/A_2$ and $B = Y_1/A_2$) to produce:

$$X_2 = MX_1 + B \tag{6.5}$$

Thus, the formula for the PC is, indeed, the formula for a straight line.

Algebraic Description of PCA

We begin this description of PCA by repeating the starting conditions and the constraints we want to impose on the new variable. We have a set of observations of P traits on N individuals, where P is the number of shape variables (not the number of landmarks). The data comprise P variances and $P(P-1)/2$ covariances in the sample. We want to compute a new set of P variables (PCs) with variances that sum to the same total as that computed from the variances and covariances of the original variables, and we also want the covariances of all the PCs to be zero. In addition, we want PC1 to describe the largest possible portion of variance, and we want each subsequent PC to describe the largest possible portion of the variation that was not described by the preceding components.

The full set of observations can be written as the matrix **X**:

$$\mathbf{X} = \begin{bmatrix} X_{11} & X_{12} & X_{13} & \cdots & X_{1P} \\ X_{21} & X_{22} & X_{23} & \cdots & X_{2P} \\ X_{31} & X_{32} & X_{33} & \cdots & X_{3P} \\ \vdots & \vdots & \vdots & \ddots & \vdots \\ X_{N1} & X_{N2} & X_{N3} & \cdots & X_{NP} \end{bmatrix} \tag{6.6}$$

where X_{NP} is the value of the Pth coordinate in the Nth individual. We can also think of this as a P-dimensional space with N points plotted in that space – just a multidimensional version of the simplistic examples presented in the previous section.

Our problem is to replace the original variables ($\mathbf{X_1}, \mathbf{X_2}, \mathbf{X_3}, \ldots \mathbf{X_P}$), which are the columns of the data matrix, with a new set of variables ($\mathbf{Y_1}, \mathbf{Y_2}, \mathbf{Y_3}, \ldots \mathbf{Y_P}$), the PCs that meet the constraints outlined in the first paragraph of this section. Each PC will be a straight line through the original P-dimensional space, so we can write each $\mathbf{Y_j}$ as a linear combination of the original variables:

$$\mathbf{Y_j} = A_{1j}\mathbf{X_1} + A_{2j}\mathbf{X_2} + \cdots + A_{Pj}\mathbf{X_P} \tag{6.7}$$

which can be expressed in matrix notation as:

$$\mathbf{Y_j} = \mathbf{A_j^T X} \tag{6.8}$$

where $\mathbf{A_j^T}$ is a vector of constants $\{A_{1j}, A_{2j}, A_{3j} \ldots A_{Pj}\}$. (The notation $\mathbf{A_j^T}$ refers to the *transpose*, or row form, of the column matrix $\mathbf{A_j}$.) All this means is that the new values of the individuals, their PC scores, will be computed by multiplying their original values (listed in matrix $\mathbf{X}$) by the appropriate values of $\mathbf{A_j^T}$ and summing the appropriate combinations of multiples. Now we can see that our problem is to find the values of $\mathbf{A_j^T}$ that satisfy the constraints outlined above.

The first constraint we will address is the requirement that the total variance is not changed. Variance is the sum of the squared distances of individuals from the mean, so this is equivalent to requiring that distances in the new coordinate system are the same as distances in the original coordinate system. The total variance of a sample is given by the sample variance–covariance matrix $\mathbf{S}$:

$$\mathbf{S} = \begin{bmatrix} s_{11} & s_{12} & s_{13} & \cdots & s_{1P} \\ s_{21} & s_{22} & s_{23} & \cdots & s_{2P} \\ s_{31} & s_{32} & s_{33} & \cdots & s_{3P} \\ \vdots & \vdots & \vdots & \ddots & \vdots \\ s_{P1} & s_{P2} & s_{P3} & \cdots & s_{PP} \end{bmatrix} \tag{6.9}$$

in which s_{ii} is the sample variance observed in variable $\mathbf{X_i}$, and s_{ij} (which is equal to s_{ji}) is the sample covariance observed in variables $\mathbf{X_i}$ and $\mathbf{X_j}$.

We can meet the requirement that the total variance is unchanged by requiring that each PC is a vector of length one. If we multiply matrix $\mathbf{X}$ by a vector of constants as indicated in Equation 6.8, the variance of the resulting vector $\mathbf{Y_j}$ will be:

$$\mathrm{Var}(\mathbf{Y_j}) = \mathrm{Var}(\mathbf{A_j^T}X) = \mathbf{A_j^T S A_j} \tag{6.10}$$

Thus, the constraint that variance is unchanged can be formally stated as the requirement that the inner product or dot product of each vector $\mathbf{A_j^T}$ with itself must be one:

$$\mathbf{A_j^T A_j} = 1 = \sum_{k=1}^{P} A_{kj}^2 \tag{6.11}$$

This means that the sum of the squared coefficients will be equal to one for each PC. Substituting Equation 6.11 into Equation 6.10 yields $\mathrm{Var}(\mathbf{Y_j}) = \mathbf{S}$, demonstrating that the constraint has been met.

The next constraint is the requirement that principal component axes have covariances of zero. This means that the axes must be *orthogonal*. More formally stated, this constraint is the requirement that the dot product of any two axes must be zero. For the first two PCs, the constraint is expressed as:

$$\mathbf{A_1^T A_2} = 0 = \sum A_{1i}A_{2i} \tag{6.12}$$

The general requirement that the products of corresponding coefficients must be zero for any pair of PCs is expressed as:

$$\mathbf{A_i^T A_j} = 0 \tag{6.13}$$

The requirements imposed by Equations 6.11 and 6.13 indicate that we are solving for an *orthonormal basis*. A basis is the smallest number of vectors necessary to describe a vector space (a matrix). An orthogonal basis is one in which each vector is orthogonal to every other, so that a change in the value of one does not necessarily imply a change in the value of another – in other words, all the variables are independent, or have zero covariance (Equation 6.13) in an orthogonal basis. An orthonormal basis is an orthogonal basis in which each axis has the same unit length. This very particular kind of normality was imposed by the first requirement (Equation 6.11). In an orthonormal basis, a distance or difference of one unit on one axis is equivalent to a difference of one unit on every other axis; consecutive steps of one unit on any two axes would describe two sides of a square.

So far, we have defined important relationships among the values of $\mathbf{A_i^T}$. There is an infinite number of possible orthonormal bases that we could construct to describe the original data. The third constraint imposed above defines the relationship of the new basis vectors to the original vector space of the data. Specifically, this constraint is the requirement that the variance of PC1 is maximized, and that the variance of each subsequent component is maximized within the first two constraints.

We begin with the variance of PC1. From Equation 6.10 we know that:

$$\mathrm{Var}(\mathbf{Y_1}) = \mathrm{Var}(\mathbf{A_1^T X}) = \mathbf{A_1^T S A_1} \tag{6.14}$$

The matrix $\mathbf{S}$ can be reduced to:

$$\Lambda = \begin{bmatrix} \lambda_1 & 0 & 0 & \cdots & 0 \\ 0 & \lambda_2 & 0 & \cdots & 0 \\ 0 & 0 & \lambda_3 & \cdots & 0 \\ \vdots & \vdots & \vdots & \ddots & \vdots \\ 0 & 0 & 0 & \cdots & \lambda_p \end{bmatrix} \tag{6.15}$$

where each λ_i is an *eigenvalue*, a number that is a solution of the *characteristic equation*:

$$\det(\mathbf{S} - \lambda_i \mathbf{I}) = 0 \tag{6.16}$$

In the characteristic equation, $\mathbf{I}$ is the $P \times P$ identity matrix:

$$\mathbf{I} = \begin{bmatrix} 1 & 0 & 0 & \cdots & 0 \\ 0 & 1 & 0 & \cdots & 0 \\ 0 & 0 & 1 & \cdots & 0 \\ \vdots & \vdots & \vdots & \ddots & \vdots \\ 0 & 0 & 0 & \cdots & 1 \end{bmatrix} \quad (6.17)$$

If each original variable in the data matrix $\mathbf{X}$ has a unique variance (cannot be replaced by a linear combination of the other variables), then each λ_i has a unique value greater than zero. Furthermore, the sum of the eigenvalues is equal to the total variation in the original data.

For each eigenvalue, there is a corresponding vector $\mathbf{A_i}$, called an *eigenvector*, such that:

$$\mathbf{SA_i} = \lambda_i \mathbf{A_i} \quad (6.18)$$

This must be true, because we have already required:

$$\det(\mathbf{S} - \lambda_i \mathbf{I}) = \mathbf{0} \quad (6.19)$$

Therefore:

$$(\mathbf{S} - \lambda_i \mathbf{I})\mathbf{A_i} = \mathbf{0} \quad (6.20)$$

which can be rearranged to:

$$\mathbf{SA_i} = \lambda_i \mathbf{A_i} \quad (6.21)$$

Thus, the eigenvectors are a new set of variables with variances equal to their eigenvalues and covariances equal to zero. Because the covariances are zero, the eigenvectors satisfy the constraint of orthogonality. Eigenvectors usually do not meet the constraint of normality ($\mathbf{A_i^T A_i} = 1$), but this can be corrected simply by rescaling. Accordingly, the rescaled eigenvectors are the PCs, which comprise an orthonormal basis for the variance–covariance matrix $\mathbf{S}$.

All that remains is to order the eigenvectors so that the eigenvalues are in sequence from largest to smallest. We can now show that the variance of PC1 is the first and largest eigenvalue. From Equation 6.10 we have $\mathrm{Var}(\mathbf{Y_j}) = \mathbf{A_j^T S A_j}$, and from Equation 6.18 we have $\mathbf{SA_i} = \lambda_i \mathbf{A_i}$. Putting these together, we get:

$$\mathrm{Var}(\mathbf{Y_1}) = \mathbf{A_1^T} \lambda_1 \mathbf{A_1} \quad (6.22)$$

We can rearrange this to:

$$\mathrm{Var}(\mathbf{Y_1}) = \lambda_1 \mathbf{A_1^T A_1} \quad (6.23)$$

which simplifies to λ_1 because we have already imposed the constraint that ($\mathbf{A_1^T A_i} = 1$).

A Formal Proof That Principal Components are Eigenvectors of the Variance–Covariance Matrix

This is the derivation as presented by Morrison (1990). Let us suppose that we have a set of measures or coordinates $\mathbf{X} = (\mathbf{X}_1, \mathbf{X}_2, \mathbf{X}_3 \ldots \mathbf{X}_P)$, and we want to find the vector $\mathbf{A}_1 = (A_{11}, A_{21}, A_{31} \ldots A_{P1})$ such that:

$$\mathbf{Y}_1 = A_{11}\mathbf{X}_1 + A_{21}\mathbf{X}_2 + A_{31}\mathbf{X}_3 + \cdots + A_{P1}\mathbf{X}_P \tag{6.24}$$

We would like to maximize the variance of $\mathbf{Y}_1$:

$$s^2_{Y_1} = \sum_{i=1}^{P}\sum_{j=1}^{P} A_{i1}A_{j1}s_{ij} \tag{6.25}$$

where s_{ij} is the element on the ith row and jth column of the variance–covariance matrix $\mathbf{S}$ of the observed specimens. We can write the variance of $\mathbf{Y}_1$ in matrix form as:

$$s^2_{Y_1} = \mathbf{A}_1^{\mathbf{T}}\mathbf{S}\mathbf{A}_1 \tag{6.26}$$

Now we seek to maximize $s^2_{Y_1}$ subject to the constraint that $\mathbf{A}_1$ has a magnitude of one, which means that $(\mathbf{A}_1^{\mathbf{T}}\mathbf{A}_1 = 1)$. To do this, we introduce a term called a Lagrange multiplier λ_1, and use it to form the expression:

$$s^2_{Y_1} + \lambda_1(1 - \mathbf{A}_1^{\mathbf{T}}\mathbf{A}_1) \tag{6.27}$$

which we seek to maximize with respect to $\mathbf{A}_1$. Therefore, we take this new expression for the variance of $\mathbf{Y}_1$ and set its partial derivative with respect to $\mathbf{A}_1$ to zero:

$$\frac{\partial}{\partial A_1}\left\{s^2_{Y_1} + \lambda_1(1 - \mathbf{A}_1^{\mathbf{T}}\mathbf{A}_1)\right\} = 0 \tag{6.28}$$

Using Equation 6.26, we can expand the expression for the partial derivative to:

$$\frac{\partial}{\partial A_1}\{\mathbf{A}_1^{\mathbf{T}}\mathbf{S}\mathbf{A}_1 + \lambda_1(1 - \mathbf{A}_1^{\mathbf{T}}\mathbf{A}_1)\} = 0 \tag{6.29}$$

which we now simplify to:

$$2(\mathbf{S} - \lambda_1\mathbf{I})\mathbf{A}_1 = 0 \tag{6.30}$$

where $\mathbf{I}$ is the $P \times P$ identity matrix. Because $\mathbf{A}_1$ cannot be zero, Equation 6.30 is a vector multiple of Equation 6.16, the characteristic equation. In Equation 6.30, λ_1 is the eigenvalue and $\mathbf{A}_1$ is the corresponding eigenvector.

Given Equation 6.30, we can also state that:

$$(\mathbf{S} - \lambda_1\mathbf{I})\mathbf{A}_1 = 0 \tag{6.31}$$

This can be rearranged as:

$$\mathbf{S}\mathbf{A}_1 - \lambda_1\mathbf{I}\mathbf{A}_1 = 0 \tag{6.32}$$

and simplified to:

$$\mathbf{SA_1} - \lambda_1 \mathbf{A_1} = 0 \tag{6.33}$$

and further rearranged so that:

$$\mathbf{SA_1} = \lambda_1 \mathbf{A_1} \tag{6.34}$$

This leads to the following substitutions and rearrangements of Equation (6.26):

$$s^2_{Y_1} = \mathbf{A_1^T S A_1} = \mathbf{A_1^T} \lambda_1 \mathbf{A_1} = \lambda_1 \mathbf{A_1^T A_1} = \lambda_1 \tag{6.35}$$

Thus, the eigenvalue λ_1 is the variance of $\mathbf{Y_1}$.

Interpretation of Results

As we stated above, PCA is nothing more than a rotation of the original data; it is simply a descriptive tool. The utility of PCA lies in the fact that many (if not all) of the features measured in a study will exhibit covariances because they interact during, and are influenced by, common processes. Below, we use an analysis of jaw shape in a population of tree squirrels to demonstrate how PCA can be used to reveal relationships among traits.

Fifteen landmarks were digitized on the lower jaws of 31 squirrels (Figure 6.5). These landmarks capture information about the positions of the cheek teeth (2–5), the incisor (1, 14 and 13), muscle attachment areas (6, 9–12, 15) and the articulation surface of the jaw joint (7 and 8). The 31 specimens include 23 adults and 8 juveniles (individuals lacking one or more of the adult teeth).

Figure 6.6 shows the landmark configurations of all 31 specimens, after partial Procrustes superimposition. This plot does not tell us much beyond the fact that there is shape variation in the sample. We can infer from the areas of the scatters for individual landmarks that there is not much variation in the relative positions of the cheek teeth. In contrast, many of the ventral landmarks have noticeably larger scatters, suggesting that their positions relative to the teeth are more variable.

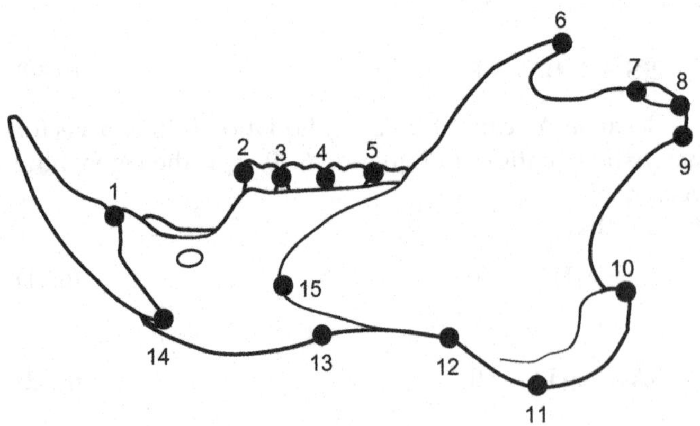

FIGURE 6.5 Outline drawing of the lower jaw of the fox squirrel, *Sciurus niger*, showing the locations of 15 landmarks.

To obtain more precise information about the pattern of shape variation, the 31 sets of landmark coordinates are converted into shape variables (see Chapter 3 for review), and these shape variables are subjected to PCA. The 15 landmarks yield 26 shape variables, so there are 26 PCs, and 26 scores for each specimen (its score on each component). The output from PCA consists of the list of coefficients describing the PCs, the variance of each component, each component's percentage of the total variance, and the scores of each specimen on each component.

As shown in Table 6.1, each PC has progressively less variance. Many of the components represent such small proportions of the total variance that it is reasonable to ask whether they describe anything biologically meaningful. One common rule of thumb is to interpret only those components that represent more than 5% of the variance. In the squirrel jaw example, PCs 1 through 5 meet this criterion. They account for a total of 80.4% of the variance in the sample, leaving 19.6% undescribed. This may seem like a large proportion of the variance to omit from further analysis, but it is doubtful that any one of the remaining 21 components describes a meaningful amount of variance.

The similarity in magnitudes of variances described by most components can be seen in a *scree* plot, in which the variance, or percentage of the total variance, is plotted against the ordinal number of the PC (Figure 6.7). In this example, there is a large difference between the variances of the first two PCs, and much smaller differences between successive pairs of components. This difference is reflected in the scatter-plot of scores in the two axes (Figure 6.8); the range of scores is much larger on PC1 than PC2, indicating that PC1 accounts for a much larger portion of the total variance. If two components have similar variances (e.g. if the distribution of scores in Figure 6.8 were closer to circular), then we have grounds to question whether either of them can be attributed to a distinct causal factor. Thus, an alternative rule of thumb is to find the inflection point on the scree plot and interpret only those components to the left of the inflection point (where the variance of each component is distinct from the variance of the following component). The main

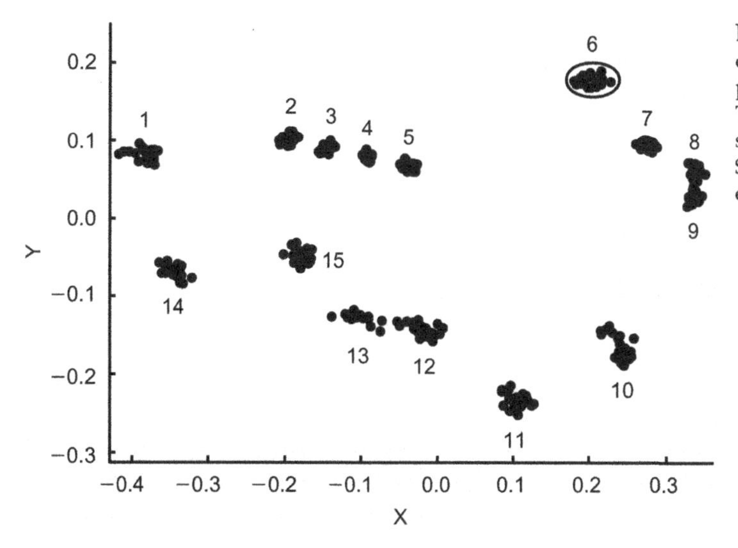

FIGURE 6.6 Plot of landmark coordinates of 31 *S. niger* jaws after partial Procrustes superimposition. The locations of landmark 6 in all 31 specimens are enclosed by an ellipse. Similar ellipses could be drawn for each landmark.

TABLE 6.1 Eigenvalues From PCA of Squirrel Jaws

PC	Eigenvalues	% of Total Variance
1	1.13×10^{-3}	51.56
2	2.15×10^{-4}	9.83
3	1.64×10^{-4}	7.49
4	1.36×10^{-4}	6.22
5	1.16×10^{-4}	5.32
6	9.52×10^{-5}	4.36
7	7.18×10^{-5}	3.28
8	5.45×10^{-5}	2.49
9	4.49×10^{-5}	2.05
10	3.58×10^{-5}	1.64
11	3.25×10^{-5}	1.49
12	2.36×10^{-5}	1.08
13	1.79×10^{-5}	0.82
14	1.37×10^{-5}	0.63
15	9.83×10^{-6}	0.45
16	9.31×10^{-6}	0.43
17	6.87×10^{-6}	0.31
18	3.72×10^{-6}	0.17
19	3.06×10^{-6}	0.14
20	2.17×10^{-6}	0.10
21	1.66×10^{-6}	0.08
22	7.04×10^{-7}	0.03
23	5.37×10^{-7}	0.02
24	3.62×10^{-7}	0.02
25	1.15×10^{-7}	0.01
26	5.02×10^{-8}	<0.01

difficulty with applying this rule is that scree plots often do not have inflection points that are as obvious as the one in Figure 6.7.

Fortunately, there is a more rigorous approach to testing whether two successive PCs have distinct variances. This is an application of a test developed by Anderson (1958) and discussed in Morrison (1990). The null hypothesis is that some sets of R consecutive

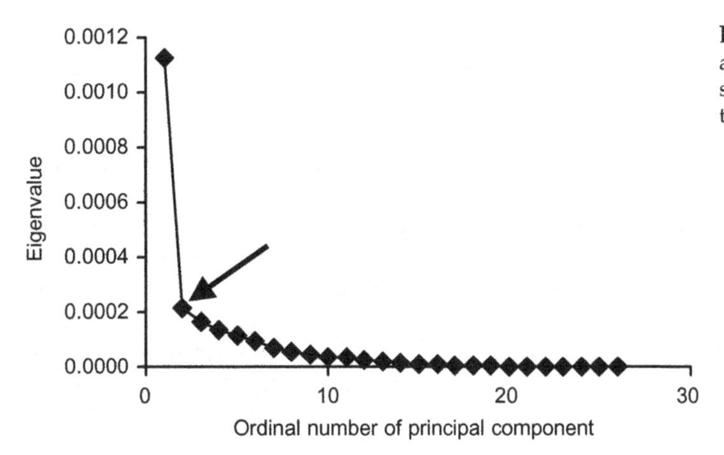

FIGURE 6.7 Scree plot of the variance described by each PC for the squirrel jaw data set. Arrow indicates the inflection point.

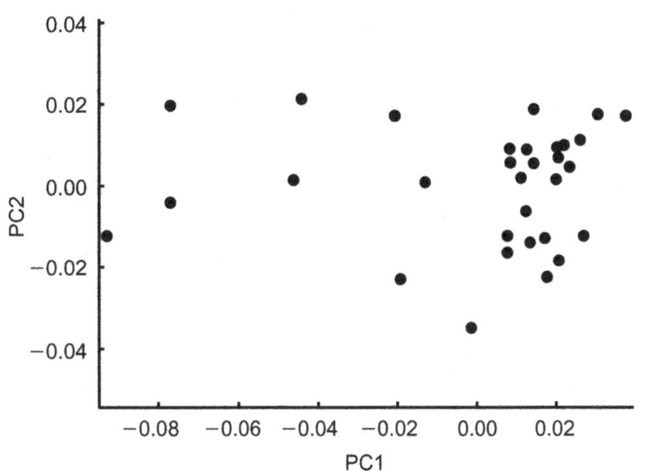

FIGURE 6.8 Scatter plot showing scores on the first two PCs for the sample of 31 squirrel jaws shown in Figure 6.6.

eigenvalues are equal to each other. In other words, the variation described by these components cannot be distinguished from random variation. The eigenvalues are numbered from $Q+1$ to $Q+R$, where Q is a function of P (the total number of eigenvalues) and R (the number of the particular components of interest) such that $Q=P-R$. Anderson (1958) derived a χ^2 statistic based on the likelihood-ratio criterion to test the hypothesis that the $Q+1$ eigenvalue is not distinct from the higher numbered eigenvalues:

$$\chi^2 = -N \sum_{j=Q+1}^{N} \ln \lambda_j + NR \ln\left(\frac{\sum_{j=Q+1}^{N} \lambda_j}{R}\right) \tag{6.36}$$

where N is the sample size minus one. When N is large, the degrees for freedom are $(\frac{1}{2}\ R(R+1)-1)$ ($d.f.=2$ when $R=2$). In the special case where $Q+R=P$, the test evaluates whether variation in the last R eigenvectors is spherical. To test two successive eigenvalues, R is set to 2. For the squirrel jaw example, comparison of the first two eigenvalues yields $\chi^2 = 19.12$, which has a p-value less than 0.0001. Comparison of the second and third eigenvalues yields $\chi^2 = 0.55$, which has a p-value of 0.76. Thus, PC1 is the only one with a distinct eigenvalue, and the only one that can be regarded as biologically meaningful.

If you use several software packages to run PCAs, you may occasionally find the results differ in signs for the PCs (when that happens, the scores for individuals on those axes also differ by a sign). Reversed axes and scores can be disconcerting, but there is no need to worry – the sign of a PC is arbitrary. If $\mathbf{A_1}$ is an eigenvector corresponding to λ_1, then so is $-\mathbf{A_1}$. If we change the sign on $\mathbf{A_1}$, then the score of the jth specimen on the first axis will also change sign; $\mathbf{Y_j} = \mathbf{A_1^T X_j}$ so the product $\mathbf{Y_1 A_1}$ does not change sign. In other words, the eigenvectors $\mathbf{A_1}$ and $-\mathbf{A_1}$ are simply mirror images. The choice of sign has no effect on the interpretations of this component, and no effect on the computation of the subsequent component (a vector orthogonal to $\mathbf{A_1}$ will also be orthogonal to $-\mathbf{A_1}$).

To this point we have not discussed how to interpret the pattern of variation represented by a PC. That rests on the coefficients of the PC, which express the relationship between the PC and the original variables. Because our original variables are shape variables, we can generate a picture of shape variation along any PC by multiplying the original shape variables by the coefficients of the PC and summing them. Figure 6.9 shows the result of that computation for PC1 of shape variation in the sample of squirrel jaws.

We should note that many of the studies applying PCA to geometric data call the method "relative warps analysis" (RWA). PCA and RWA are not exactly equivalent, because the components of variance extracted by RWA are sometimes weighted by bending energy (originally, RWA was an analysis of components of variation relative to bending energy, hence the term "relative" in the name of the method). When variation is not weighted by bending energy, RWA is PCA. We prefer the more familiar term.

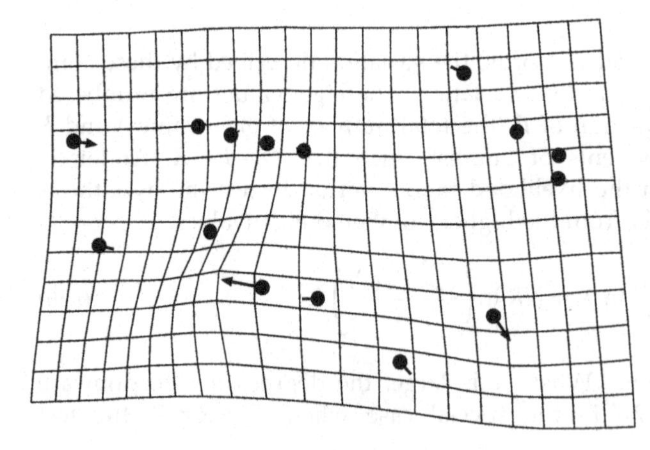

FIGURE 6.9 Pattern of shape change along PC1 for the 31 squirrel jaws shown in Figure 6.6. Circles indicate the locations of the landmarks in the mean shape of the sample; arrows indicate the changes in the relative positions of the landmarks as the score on PC1 increases. The deformed grid illustrates the thin-plate spline interpolation over the entire form.

CANONICAL VARIATES ANALYSIS

The purpose of CVA is to simplify the description of differences among *groups* and to form mathematical functions which may be used to assign specimens to groups, acting as a multiaxis discriminant function. For example, CVA could be used to describe differences in mandible shape among queens, soldiers and workers in a colony of ants, and to determine the rate at which individual ants could be correctly assigned to one of these classes. It could also be used to describe differences in soldier morphology among colonies, species, or more inclusive categories. If individuals in a study can be sorted into mutually exclusive sets, CVA can be used to describe the differences among those sets. CVA does not provide a test of differences in the mean forms, that test is best performed using the General Linear Model framework (see Chapters 8 and 9), which encompasses the familiar MANOVA methods. CVA does allow estimation of the rate at which specimens may be effectively sorted or assigned as members of *a priori* groups, which implies both differences in the mean shape and also some degree of non-overlap in the distributions of traits. This analysis of assignments or classifications makes CVA a useful complement to tests of difference in mean form.

There are many similarities between CVA and PCA. Like PCA, CVA constructs a new coordinate system (the canonical variates, CVs) and determines the scores on those axes for all individuals in a study. Also, the CVs are linear combinations of the original variables and are constrained to be mutually orthogonal. However, whereas PCA is used to describe differences among individuals, CVA is used to describe differences among group means. In this sense, CVA is analogous to a PCA of the group means. Another difference between CVA and PCA is that CVA uses the patterns of within-group variation to scale the axes of the new coordinate system. Because of this rescaling, CVs are not simply rotations of the original coordinate system, and distances in CV space are not equal to distances in the original coordinate system. (This is where the analogy breaks down.) As a result of the rescaling, CV1 is the direction in which groups are most effectively discriminated, which is not necessarily the direction in which the group means are most different.

An important difference between PCA and CVA is that distances computed along the CVA axes are not equivalent to Procrustes distances and, thus, some care must be taken in interpreting positions in morphospaces defined by CVA axes. The number of CVA axes that can appear in an analysis is also limited to the number of distinct groups present minus one (assuming this is smaller than the degrees of freedom per individual), another difference relative to PCA, in which the number of meaningful axes is controlled by the degrees of freedom per measured individual.

Groups and Grouping Variables

A group is a set of individuals that share a particular state of a discontinuous trait. Examples of groups include sexes, color morphs, species, and supraspecific categories like guilds. The groups analyzed by CVA must be mutually exclusive, meaning that they cannot comprise nested or intersecting sets. In other words, the groups differ in the values of a categorical variable, which is sometimes called a "qualitative trait" or a "grouping variable". The important characteristic of these variables is that they are not measured nor arrayed in a sequence; they do not have intrinsic numerical values, nor do they have an inherent order or sequence.

Sometimes, features that can be scored on a continuous graded scale are treated as categorical variables. For example, the proportions of meat and vegetation in an animal's diet can be quantified and scored along a continuum. Nevertheless, it is a common practice to sort diets into a small number of categories (e.g. carnivore, herbivore, omnivore). Other traits that might be treated in a similar fashion include geographic location and age. There are several reasons for treating these kinds of traits as categorical variables. One is a lack of sufficient information to justify or support a more finely graded analysis – for example, a researcher may not have precise data on the proportions of food items in the diets of all species or individuals in a study. Another reason for treating a quantifiable trait as a categorical variable is that the investigator may not want to impose a hypothesis of ordering on the data, which is often a consideration when groups are not dispersed along a single straight line. Similarly, the investigator may not want to assume that all steps are of equal value (e.g. ontogenies often can be divided into discrete instars or age classes based on sequences of developmental events, but the sequentially numbered steps may represent different amounts of time or ontogenetic change). Under these circumstances, a quantifiable trait may be treated as a categorical variable and CVA would then be used to describe differences among the groups delineated by distinct states. However, the user should be aware that taking this approach also limits the inferences that can be drawn from the result – for example, an observation that age classes can be differentiated does not necessarily imply the kind of monotonic progression from age to age that can be inferred from a regression.

Geometric Description of CVA

To develop a geometric intuition for CVA, we return to the metaphor of a slightly flattened watermelon. In PCA, we described the positions of seeds within the watermelon by finding its greatest dimensions. In CVA, we are not interested in the positions of seeds in the watermelon; instead, we want to describe the positions of the watermelons in the field (centroids of the ellipses in Figure 6.10). If all we want to know is the location of each

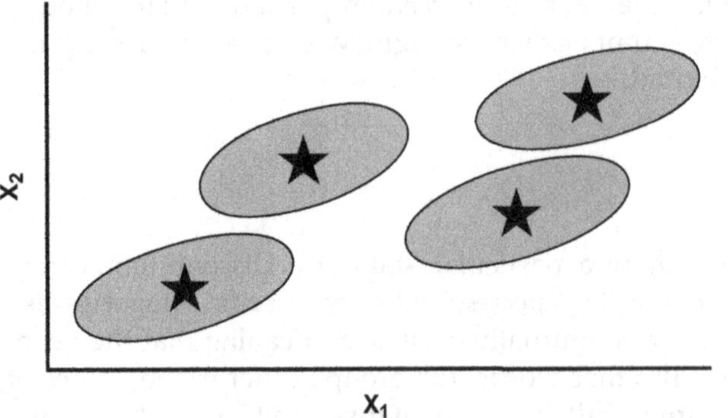

FIGURE 6.10 Ellipses of variation in two dimensions (X_1 and X_2) for four sample populations. Stars indicate locations of the means of each sample.

melon, we could simply plot each melon's centroid; however, suppose we want to find the direction in which it is easiest to walk across the field without stepping on any of the melons (perhaps we want to spread fertilizer in the field). To solve this problem, we want to find the direction in which the melons are farthest apart. This requires that we know the averages of the shapes and orientations of all the melons, not just the position of each melon's centroid.

Similarly, CVA begins with a PCA of the pooled (averaged) within-group variances. This gives us a new coordinate system in which we can describe the position of each group. In our field, we begin by defining a new coordinate system that would be aligned with the axes of the average melon (Figure 6.11).

Now we can see that the melons overlap more in the direction defined by the long axis of the average melon. To take this into account, we rescale this axis proportionate to the elongation of the average melon. In effect, we distort our plot of the field until the average melon looks circular rather than elliptical (Figure 6.12).

Now we can solve for the direction in which melons tend to be farthest apart in the rescaled space by performing a PCA on the group centroids. The axes produced by this last computation are the CVs (Figure 6.13A). The scores of individuals on the CVs are the projections of the individuals onto these new coordinate axes (Figure 6.13B).

Because computation of the CVs involves a rescaling, interpretation of CV scores can be complex. If we undo the rescaling and rotation that were used to solve for the CVs (Figure 6.14), we see that each CV is a linear combination of the original variables. However, we also see that the CVs are not orthogonal axes in the original coordinate space. Furthermore, distances on CVs are not equivalent to distances in the original space.

Note that, in this example, there are more groups than variables in the original data set. In such cases, the number of CVs will be equal to the number of variables. Most studies will have fewer groups than variables and, in these cases, the number of CVs will be one less than the number of groups. If there are three groups in a study, the differences among

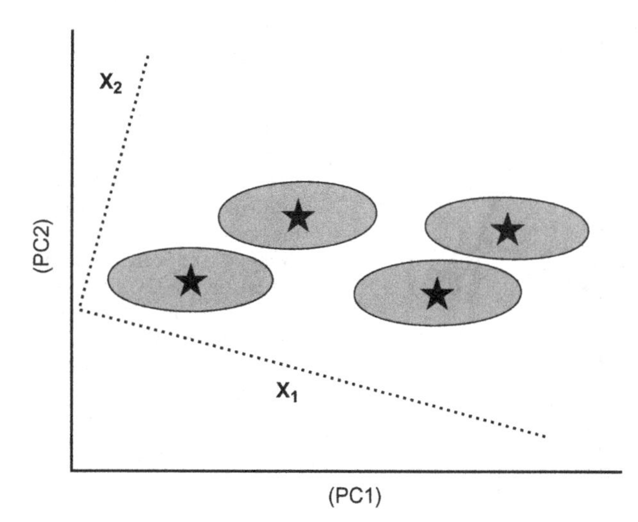

FIGURE 6.11 Graphical representation of the first step of CVA. The entire data set is rotated to a new coordinate system that is aligned with the PCs of the pooled variances. At this stage the relative positions of the four samples (and the individuals within groups) have not changed. The original coordinate system (Figure 6.10) is shown in the dotted lines. The axes of the new coordinate system are labeled in parentheses because we have not specified the location of the average sample, only the directions of its variances.

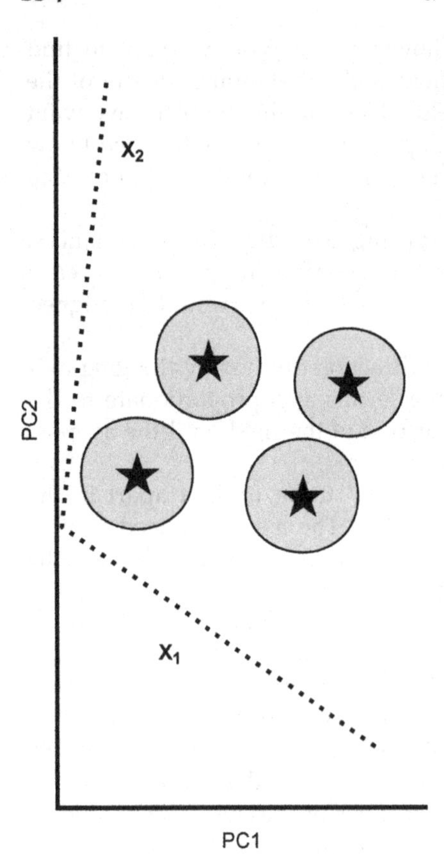

FIGURE 6.12 Graphical representation of the second step of CVA. The new coordinate system (solid lines) is rescaled in proportion to the pooled within-group variances in the original space. Variation within samples will be circular in the new space if the original variances were all identical. Note that the axes of the original coordinate system (dotted) are not orthogonal in the new space. Furthermore, distances in the new space are not equivalent to distances in the original space (Figure 6.10).

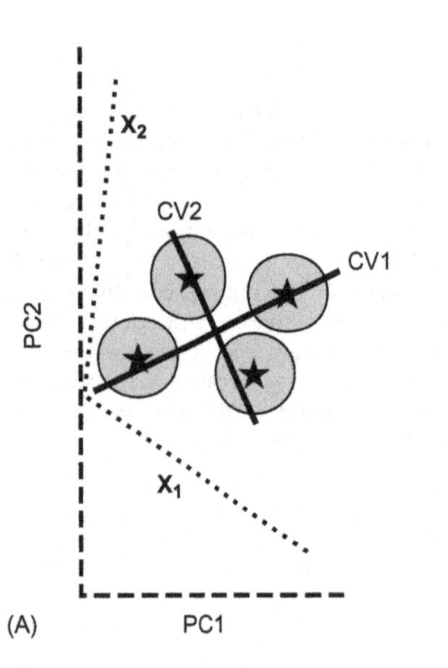

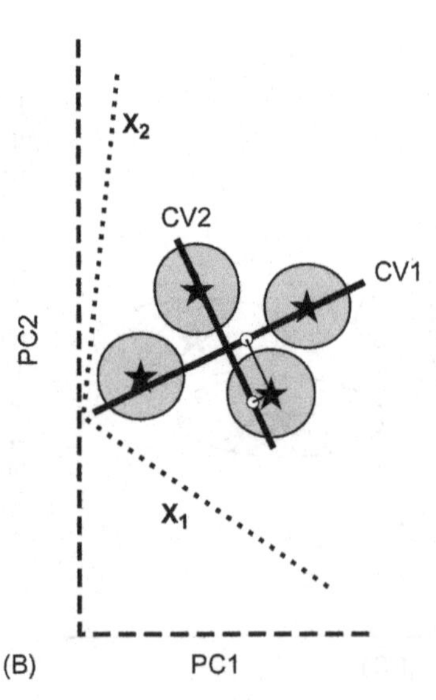

FIGURE 6.13 Graphical representation of the final steps in CVA. (A) CV1 is the direction through the rescaled space (outer, dashed axes) in which the group means are most different; CV2 is the direction orthogonal to CV1 in which the group means are most different. (B) Scores of individuals in the CV space are their projections onto the CVs. Circles represent the scores of one of the sample means.

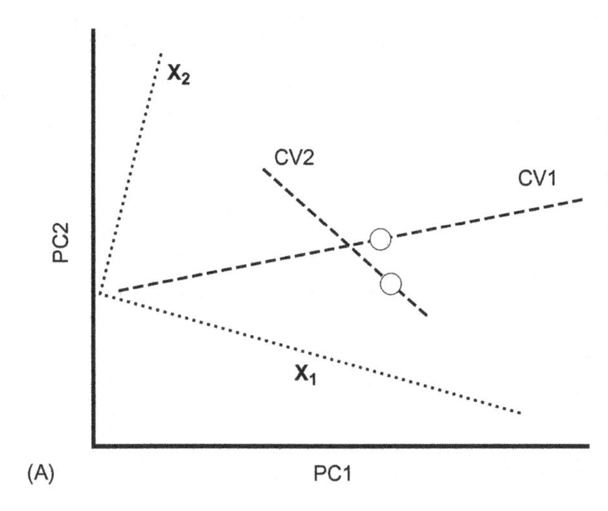

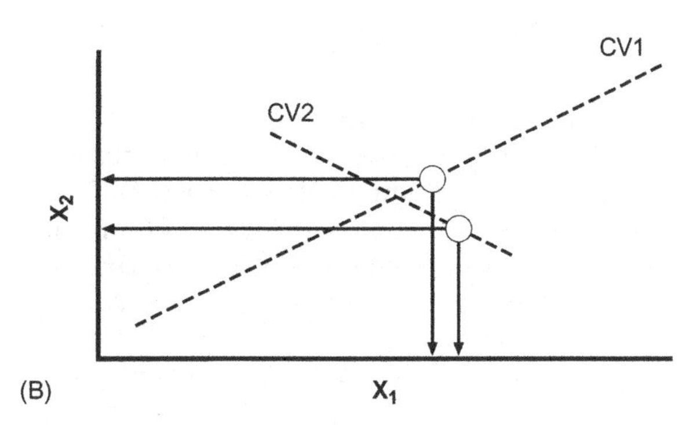

FIGURE 6.14 Interpretation of CV scores in terms of the original axes. (A) Rescaling the axes has been reversed, restoring orthogonality of $\mathbf{X_1}$ and $\mathbf{X_2}$. White circles represent the scores of one individual on each CV. (B) Rotation of the original axes is reversed, restoring the original orientation. Arrows show projections of the CV scores onto the original axes; each CV score represents a combination of scores on the original axes.

them can be summarized as a plane defined by two vectors, whether the original data included three variables or 300.

Algebraic Description

In CVA, as in PCA, we begin with a set of measures or coordinates $\mathbf{X} = (\mathbf{X_1}, \mathbf{X_2}, \mathbf{X_3} \ldots \mathbf{X_P})$, and we want to find the vector $\mathbf{A_1} = (A_{11}, A_{21}, A_{31} \ldots A_{P1})$ such that:

$$\mathbf{Y_1} = A_{11}\mathbf{X_1} + A_{21}\mathbf{X_2} + A_{31}\mathbf{X_3} + \cdots + A_{P1}\mathbf{X_P} \tag{6.37}$$

In PCA, we solved for the eigenvalues and eigenvectors of the variance–covariance matrix **S**. In CVA, we are concerned with the ratio of two variance–covariance matrices: one is the pooled within-groups variance–covariance matrix, $\mathbf{S_W}$, which represents the deviations of individuals from their respective group means; the other is the between-groups variance–covariance matrix, $\mathbf{S_B}$, which represents the portion of the total variance (deviations from the grand mean) not explained by $\mathbf{S_W}$. In other words, $\mathbf{S_W}$ represents differences within groups, and $\mathbf{S_B}$ represents differences between the groups. So, in CVA we want to find the $\mathbf{Y_1}$ that maximizes the ratio of between-group variance to within-group variance. The within-group variance of $\mathbf{Y_1}$ is:

$$s^2_{Y_1\text{within}} = \mathbf{A_1^T S_W A_1} \tag{6.38}$$

and the between-group variance of $\mathbf{Y_1}$ is:

$$s^2_{Y_1\text{between}} = \mathbf{A_1^T S_B A_1} \tag{6.39}$$

The form of these expressions should be familiar from our discussion of PCA. As before, we use the Lagrange multiplier λ_1 to form the expression:

$$\left(\frac{s^2_{Y_1\text{between}}}{s^2_{Y_1\text{within}}}\right) - \lambda_1 = \mathbf{A_1^T A_1} \tag{6.40}$$

then make the substitutions indicated by Equations 6.38 and 6.39 to form:

$$\left(\frac{\mathbf{A_1^T S_B A_1}}{\mathbf{A_1^T S_W A_1}}\right) - \lambda_1(1 - \mathbf{A_1^T A_1}) \tag{6.41}$$

This is the expression we will maximize relative to $\mathbf{A_1}$, under the constraint that $\mathbf{A_1^T A_1} = 1$. Taking the partial derivative of this expression again yields a characteristic equation that can be solved for the eigenvalues and corresponding eigenvectors of $\mathbf{S_W^{-1} S_B}$. Note that the matrix inversion in Equation 6.41 does require that the pooled within group variance–covariance matrix be invertible. This, in turn, requires that the variance–covariance be of full-rank, meaning that the number of variables must equal the number of degrees of freedom in the system, which poses some challenges, particularly when working with semilandmarks. The number of CVA axes appearing is also limited to a maximum of the number of groups minus one.

Interpretation of Results

To help interpret the CVA result, and to illustrate the effect of the rescaling step on the result, we first show PCA on the data set that will be used in the CVA example. This data set is composed of 15 landmarks on the lower jaws of 119 squirrels from three geographic areas (Figure 6.15). For each landmark, the cloud of points overlaps broadly, suggesting similarly broad overlaps in the distribution of the whole shapes. Closer examination shows there are slight differences in the relative positions of some landmarks – circles predominate at one end of the clusters, triangles at the other. This is most evident for landmark 13, which is more anterior in the western Michigan sample and more posterior in the southern sample. The scores on PC1 (Figure 6.16A) show there are differences between the distributions of mandible shape in the three samples, and that each sample varies considerably along this axis. As shown by the deformation grid (Figure 6.16B), that difference primarily consists of shifts in the relative position of landmark 13, as surmised from the superimposed shapes, but there is also expansion of the angular process (landmarks 10–12) and contraction of the space between the molars and the tips of the coronoid and

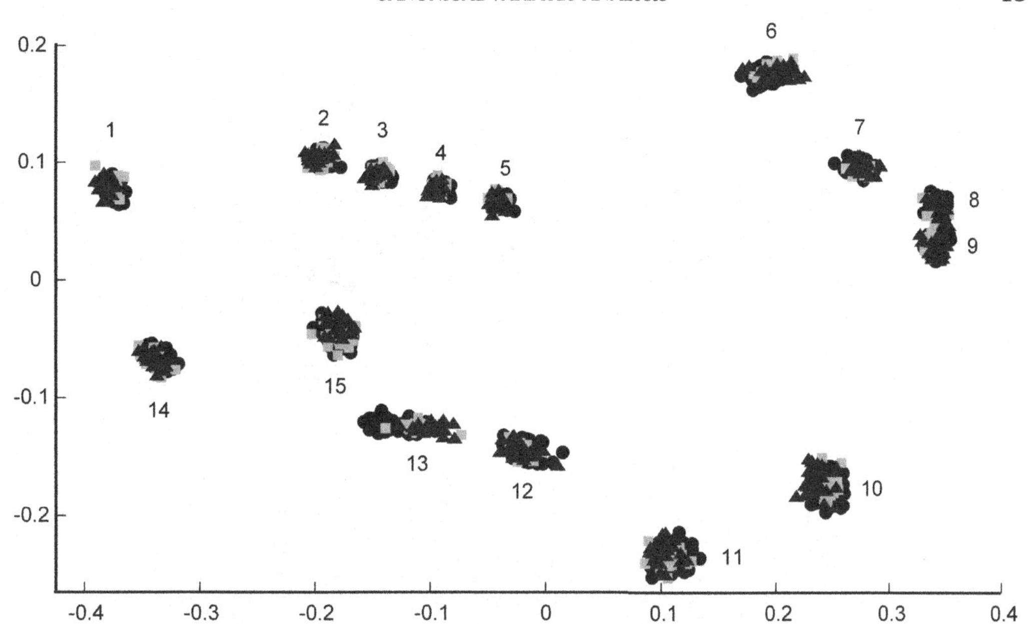

FIGURE 6.15 Landmark coordinates, in partial Procrustes superimposition, for 119 squirrels from three geographic samples. Circles = western Michigan, gray squares = eastern Michigan, triangles = southern states.

condyloid processes (between landmarks 5 and 6–9). This dimension is highly variable within samples, such that the variation is quite large relative to the differences between means. Consequently, PC1 is unlikely to meet the criterion for an efficient discriminator even though the differences on this axis might be statistically significant. Another axis that has relatively less variation compared to the differences between means would be a better discriminator.

Results of CVA will look different from those of PCA for two crucial reasons. First, CVA is describing differences between groups, and the direction in which group means are most different is not necessarily the direction in which individuals are most different. Second, CVA does not simply rotate the original data to the axes that maximize the group differences, it finds the axes that optimize between-group differences *relative* to within-group variation and, in general, these axes will be different directions from the ones that maximize between-group differences. In addition, optimization also involves rescaling such that the new axes are scaled differently from the original axes and scaled differently from each other. Consequently, distances and relative positions in CV space can be quite different from distances and positions in the original data, and interpretations of results can be counterintuitive.

Because there are only 3 sample groups, there can be only 2 CVs. The plot of scores on those two axes (Figure 6.17) shows that it was possible to find two axes of differentiation, that the 3 means are not collinear. The distributions also show more circular distributions with much less overlap than was seen in the PC scores. Thus, different combinations of

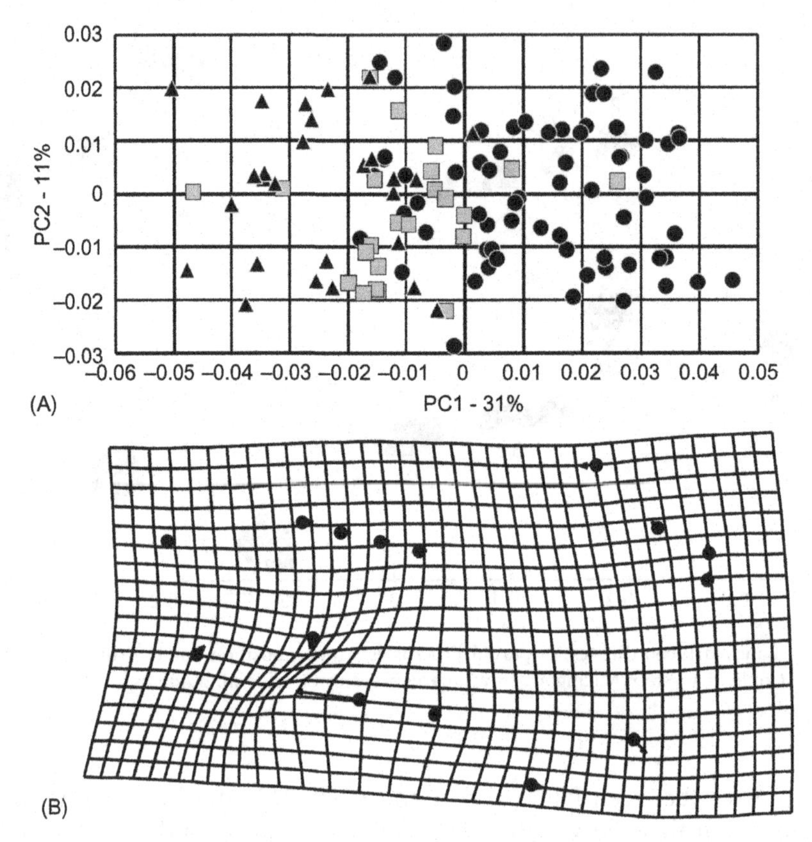

FIGURE 6.16 Shape variation in three geographic samples of squirrel jaws analyzed by PCA. (A) Scores on the first 2 axes (circles = western Michigan, squares = eastern Michigan, triangles = southern states). (B) Deformation showing shape change association with increasing scores on PC1.

the shape variables were found by CVA to be more effective discriminators of these samples than were the PCs.

As we did with the PCs, we multiply the original shape variables by the coefficients of the CVs and sum them. This produces a series of vectors of relative landmark displacement that illustrates the shape differentiation represented by the CVs. As shown in Figure 6.18A, the amount of the shape difference described by CV1 of this data set is imperceptible. When the deformation is exaggerated to visualize the pattern, it can be seen that differences in the relative heights of the teeth are the most useful trait for discriminating among the groups. Relative tooth heights are not an efficient discriminator because the differences between groups are large, but because the variation within groups is even smaller than the differences between groups. Consequently, a biologically insignificant feature is determined to be diagnostically important. Figure 6.18C illustrates all of the other shape differences that are correlated with CV1. These changes, which may be helpful for diagnosing group members, include a portion of the shape differences that were detected using PCA. This figure demonstrates another important point to bear in mind when using

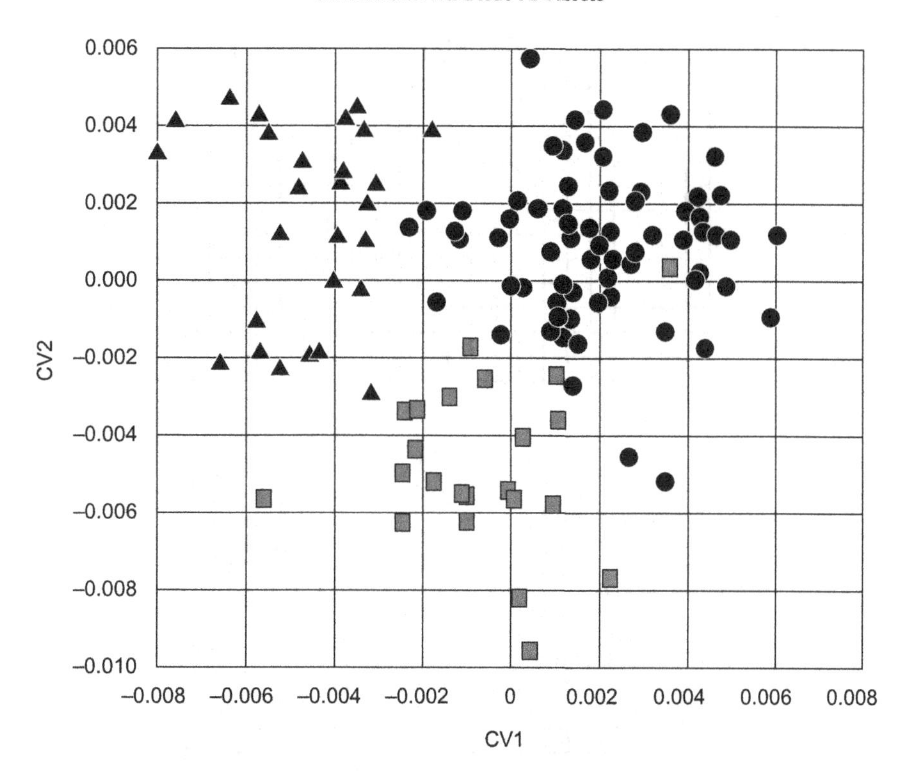

FIGURE 6.17 CV scores for the three geographic samples of squirrel jaws. Circles = western Michigan, gray squares = eastern Michigan, triangles = southern states.

CVA: the CV is not a complete description of the difference between groups, it is just that part of what differs between groups that varies least within groups.

Like PCA, CVA will compute a set of axes under the specified constraints, regardless of whether the differences between groups are statistically significant. The optimal discriminator in a data set need not be a statistically significant discriminator, and a statistically significant discriminator may not turn out to be particularly effective in assigning specimens to groups. To determine how many CVs are effective discriminators, we employ Bartlett's (1947) test for differences in the value of Wilk's lambda (Λ). Wilk's Λ is the within-groups sum of squares divided by the total sum of squares (within-plus between-groups):

$$\Lambda = \frac{\det(\mathbf{W})}{\det(\mathbf{T})} = \frac{\det(\mathbf{W})}{\det(\mathbf{W}+\mathbf{B})} \tag{6.42}$$

where det is the determinant of the matrix. Conveniently, Λ can be computed as the product of the eigenvalues of $\mathbf{W}(\mathbf{W}+\mathbf{B})^{-1}$. Bartlett's test uses the following formula to estimate a χ^2 test statistic:

$$\chi^2 = -(W-(P-\mathrm{B}+1)/2)\ln \Lambda \tag{6.43}$$

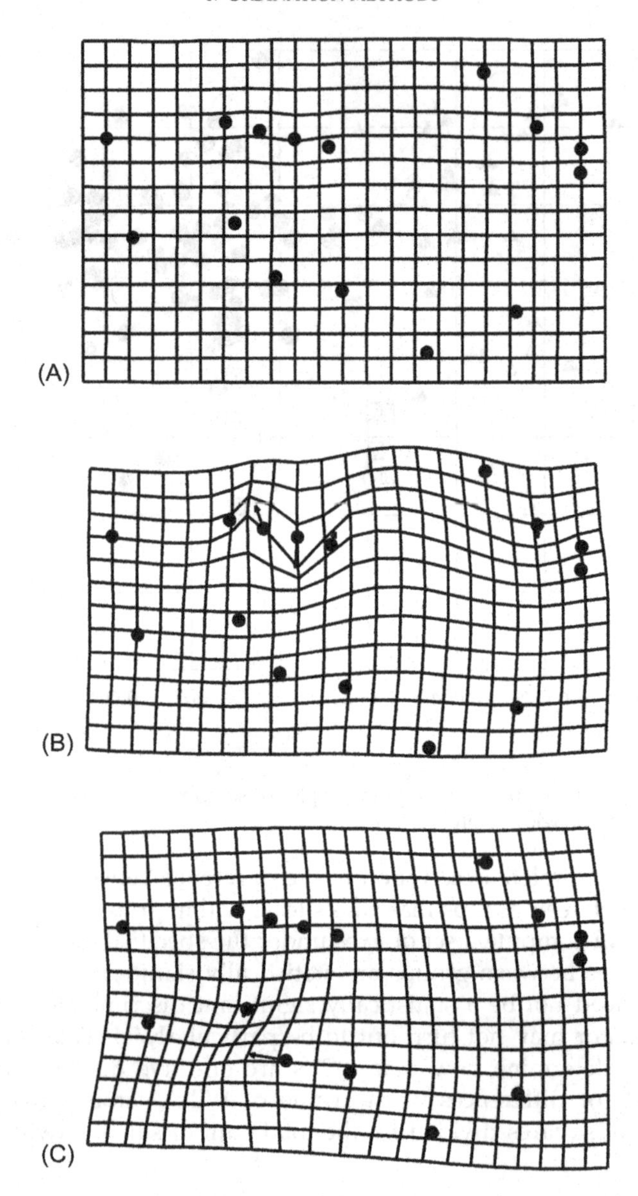

FIGURE 6.18 Shape differentiation associated with CV1 for the three geographic samples of squirrel jaws. (A) Transformation of the reference shape to the shape corresponding to a score of 0.01 on CV1, reflecting the actual magnitude of difference between the means of the western Michigan and southern samples. (B) Transformation of the reference shape to the shape corresponding to a score of 0.1 on CV1, reflecting exaggeration by a factor of 10. (C) Deformation representing all of the shape change correlated with CV1, for an individual with a score of 0.1 on CV1.

In this expression, P is the number of variables, $W = N - B - 1$ (where N is the total number of individuals) and $B = G - 1$ (where G is the number of groups). The degrees of freedom are determined by the product of P and B.

The testing procedure begins by computing the estimated χ^2 in which Λ is the product of the eigenvalues of all CVs. If this value is significantly greater than expected for the given degrees of freedom, it is safe to infer there are statistically significant differences among the groups. In the squirrel jaw example, there are three groups and 26 shape variables, and the maximum possible number of meaningful CVs is two. Bartlett's test on both CVs yields a χ^2 of 206.6, with 52 degrees of freedom, for a p-value less than 0.000001. This result indicates that at least some of the groups in the study can be discriminated using scores on these two CVs.

We do not yet know whether both CVs contribute to discrimination of the groups, so the next step is to remove the eigenvalue for the first CV (the most efficient discriminator) and repeat the test. Reducing the number of CVs reduces the number of groups that can be discriminated, which reduces B by 1 and the degrees of freedom by P. These changes produce a χ^2 of 83.5 with 26 degrees of freedom for a p-value that is still less than 0.000001. Thus, the second CV also contributes to discriminating among the groups.

In general, the test is reiterated using the remaining R $(=B - i)$ eigenvectors until $R = 0$ (all eigenvalues have been removed) or some set of R remaining eigenvectors fails the test. If R goes to zero, the analysis will have shown that some groups can be discriminated on the CV that is the least efficient discriminator. If a set of R eigenvectors fails the test, then only the first $B - R$ CVs contribute to discriminating among the groups. Note that the test cannot be taken to indicate that all groups can be discriminated, and it does not indicate which groups can be discriminated.

One simple approach to assessing the utility of the CVs for discriminating among groups can also be evaluated using the Mahalanobis distances of specimens from the group mean. The means are computed using the *a priori* group assignments. The Mahalanobis distance between a specimen **X** and the mean **M** of a group, is given by:

$$\mathbf{D} = \sqrt{(\mathbf{X}-\mathbf{M})^{\mathrm{T}}\mathbf{S}^{-1}(\mathbf{X}-\mathbf{M})} \tag{6.44}$$

where $\mathbf{S}^{-1}$ is the inverse of the variance–covariance matrix of the CV scores of the specimens. The predicted group membership of each specimen based on the scores is determined by assigning each specimen to the group whose mean is closest (under the Mahalanobis distance) to the specimen. All of the CVs that pass Bartlett's test, and only those CVs, are used to compute the Mahalanobis distances and assign specimens to groups.

When specimens are assigned to a group using CV axes estimated using the same data set, the resulting rate of correct specimen assignment to groups are referred to as a *resubstitution rate* of assignment. Resubstitution rates involve a certain degree of circularity, in that the same data was used to create the discriminant functions (CV axes) and to assess the performance of those functions. This process leads to some level of over-fitting of the model to the data, and an overestimate of the effectiveness of the CVA. A number of approaches have been developed to produce more reasonable estimates of the *actual error rate* for a classification method (Knoke, 1986; Schiavo and Hand, 2000). The *actual error rate* is the rate of error for the particular assignment method given the data, a rate contingent on our given data; there is also an *expected error rate* which would be the rate we would achieve given another data set similar to the one we have collected.

The simplest approach to estimate the actual error rate (given a particular data set) is to use a cross-validation or jackknife procedure (Efron, 1983; Efron and Tibshirani, 1995; Van Bocxlaer and Schultheiß, 2010). In these procedures, the data set is divided into two subsets, a training set and a test set. The test set may be as small as one specimen, or as much as 50% of the data set, with the training set consisting of all remaining data. The CVA is then fit to the training set, and used to assign the members of the training set to a group. This process is repeated for a large number of possible variations of the test set, and the rate of correct specimen assignment is computed over all the test sets employed. This yields a cross-validation (or jackknife, when the test set has $n = 1$) rate of correct specimen assignment, which is a better predictor of the overall effectiveness of the method than the resubstitution rate. Most modern software will have some form of jackknife or cross-validation method available.

The difference between the resubstitution rate and the cross-validation rate of assignment can be substantial. The plots of CVA scores produced by most software systems are resubstitution rates, and so the patterns produced by these plots must be viewed with substantial caution, as they may overstate the effectiveness of the method. Nicely separated groups on a CVA plot may not translate into effective cross-validation rates, or a statistically reliable method. It is not unusual for a CVA to indicate that one or more of the CV axes produced were statistically significant, but to produce assignment rates no better than expected by chance. For this reason, it seems wise to require both statistically significant CVA axes and a cross-validation rate of correct specimen assignment that is substantially better than chance.

It turns out to be relatively straightforward to compute the random rate of correct specimen assignment that can be achieved via random sorting. A biased random allocation of specimens to groups proportional to the number of individuals in the group will yield the highest random rate. If we have a total of N specimens distributed among m groups, and n_i members in the ith group, then the maximum rate of random assignment of specimens to groups is given by:

$$\textit{Random Rate} = \sum_{i=1}^{m} \frac{n_i^2}{N^2} \tag{6.45}$$

This is achieved by a random rule of assigning each specimen to a group with a random probability n_i/N. In such a situation, Equation 6.45 is the expected rate of correct specimen assignments. If the group sizes are all equal, this rate will simply be $1/m$, but if the group sizes are unequal, the random rate will be higher than that.

In studies focusing on classification rate, rather than morphospace analysis, incorporating size into a CVA may help improve the classification rate. In such situations, it is possible simply to include the log of centroid size as an additional column in the data matrix. As discussed in Chapter 4, the new space, sometimes called Procrustes Form Space, would not be a shape space. Another possible approach would be to use the Procrustes Size Preserving methods discussed in Chapter 14.

TABLE 6.2 CVA Classification Table for 119 Squirrel Jaws

A priori Assignments	*A posteriori* Assignments			
	Western Michigan	Eastern Michigan	Southern States	Total
Western Michigan	62	4	3	69
Eastern Michigan	1	22	0	23
Southern states	0	1	26	27

The *a priori* classifications are based on the geographic localities where specimens were collected. The *a posteriori* assignments are based on Mahalanobis distances of individuals from the means of the *a priori* groups. Total is the total number of specimens in each geographic sample. Thus, 62 specimens in the western Michigan sample were correctly classified using Mahalanobis distance, and 7 were misclassified as members of one of the other geographic samples.

TABLE 6.3 CVA Classification Table for 119 Squirrel Jaws, using a Jackknife Cross-Validation Analysis

A priori Assignments	*A posteriori* Assignments			
	Western Michigan	Eastern Michigan	Southern states	Total
Western Michigan	56	6	7	69
Eastern Michigan	4	18	1	23
Southern states	1	5	21	27

Re-analysis of classification rates for the three geographic samples of squirrel jaws analyzed in Table 6.2. The overall rate of correct classifications has dropped from 92% in the resubstitution rates to 80% in the jackknife.

Table 6.2 shows the resubstitution classification results based on CVA of the three samples of squirrel jaw discussed earlier. As shown in the first row, 62 of the 69 western Michigan squirrels have jaws that are closer to the mean of their sample than to the mean of another sample, based on the Mahalanobis distance. In contrast, only one specimen from each of the other samples is farther from the mean of its own sample than it is from the mean of another sample. Like the plot in Figure 6.17, this result contributes to the general impression that the members of these three groups can usually be discriminated. Using a jackknife cross-validation test, we can see that the samples may not be quite so distinct (Table 6.3). The rate of correct classification drops by a little more than 10%. It is still clear that the three groups can usually be discriminated, but it is also more apparent that a substantial proportion of individuals on the fringes of the shape distributions are apt to be misclassified based on shape alone.

CVA of Rank-Deficient Data

One issue that arises in working with CVA is the need to invert an estimated, pooled variance–covariance matrix, which requires that the degrees of freedom in the matrix be greater than the number of variables in the matrix. This is a problem at small sample sizes, or when using semilandmarks (because each semilandmarks is represented by two variables but has only one degree of freedom). The simplest approach to this problem is to use PCA as a dimensionality reduction tool, and carry out the CVA on some subset of the PC scores, rather than on the actual data. The issue that then arises is how to determine the number of PC axes to use. If all PC axes with non-zero eigenvalues are included, then there is no loss of information, but most software has some level of rounding error present, which can make it difficult to determine if small eigenvalues are zero or not. Some workers will simply use a set of PC axes comprising 95 or 99% of the variance. One study showed decreased overfitting of the CVA when fewer PC axes were used (Sheets et al. 2006). The Between Groups PCA (Mitteroecker and Bookstein, 2011) may be another viable approach to this issue.

A radically different approach to working with rank-deficient data is to use a machine learning approach to specimen classification such as the Weka system (Witten and Frank, 2005). These approaches attempt to build computer-based classification rule systems, without reference to parametric statistical models. The performance of these methods is typically assessed using cross-validation methods, and these machine learning methods appear to do as well as CVA, at least with some data sets (Van Bocxlaer and Schultheiß, 2010).

BETWEEN GROUPS PRINCIPAL COMPONENTS ANALYSIS

To avoid the problems engendered by the rescaling step in CVA, Mitteroecker and Bookstein (2011) suggest Between Groups Principal Components Analysis (BGPCA). This method analyzes differences between means without regard to the magnitude or pattern of within group variation. It is, simply, a PCA of the means. Absence of the rescaling step eliminates distortion of the distances between the means. It also obviates concerns about the artificiality of discriminators based on trivial distinctions, although this may not be advantageous in all applications. Restricting the analysis to the subspace defined by the means reduces the dimensionality of the space from the number of variables ($2k-4$, for a set of two-dimensional landmarks) to a maximum of 1 less than the number of groups. BGPCA finds the principal components of this subspace.

Figure 6.19A shows the mean shapes of the jaw in the three squirrel samples analyzed previously. Again, the position of landmark 13 differs greatly between the mean shapes; differences in other regions are also apparent. The first PC of this subspace is approximately the axis between the two most different means – those of the western Michigan and southern samples (Figure 6.19B). The second axis, as in other PCAs, is constrained to be orthogonal to the first. In this example, it describes the divergence of the eastern Michigan sample from the axis connecting the other two means. Were the differences between means less neatly balanced, or if there were more groups, the axes might not be so conveniently aligned with the means. As might be predicted from the shape comparison in Figure 6.19A, the deformation illustrating the shape differences associated with PC1 scores shows that the mean shapes differ primarily in the relative position of landmark 13, but also in shape of the angular process and space between the other posterior processes and the molars (Figure 6.19C). In this particular data set, the mean shapes happen to have differed primarily in the direction of highest within-sample variation. This might not have

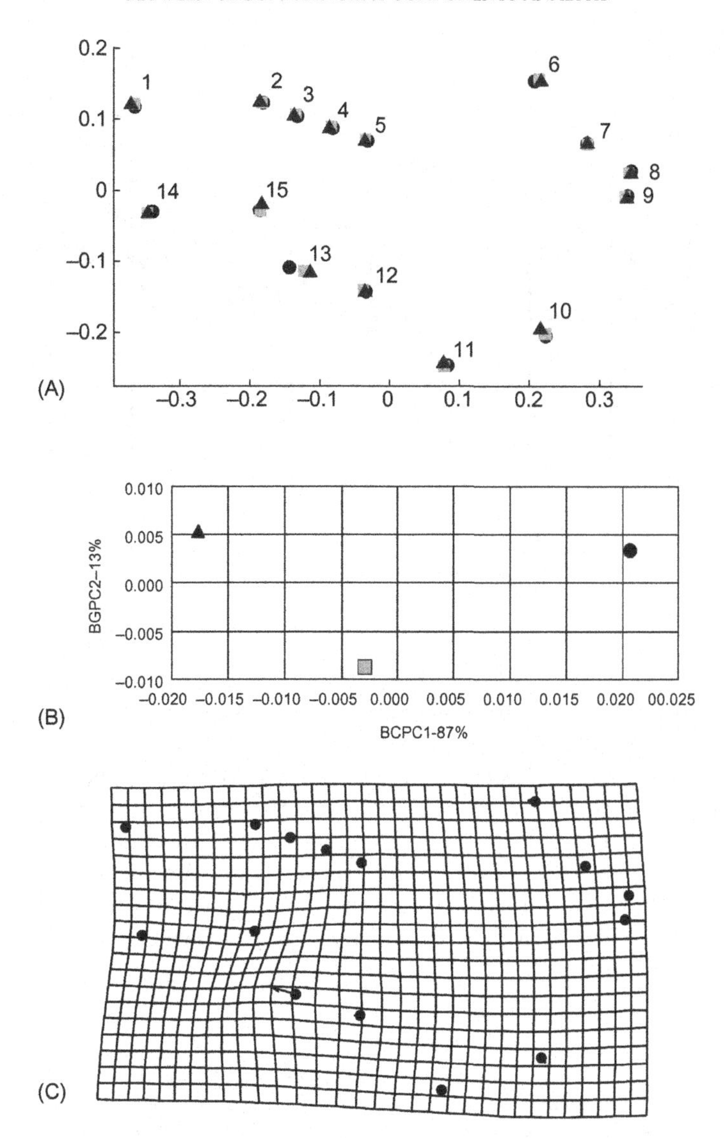

FIGURE 6.19 Differences in mean shape of squirrel jaws from the three geographic samples, analyzed by BGPCA. (A) Mean shapes. (B) Scores. (C) Deformation showing the shape difference between means of the western and southern samples along BGPC1.

been the case if the samples were more distantly related or if they were collected from more distinctive habitats.

BGPCA explicitly disregards the issue of discrimination, however, if one has an interest in evaluating discrimination, or has other reasons for examining variation along these axes, the scores of individuals on these axes can be obtained easily by multiplying the data matrix (superimposed landmark coordinates) by the eigenvectors of the BGPCA. The

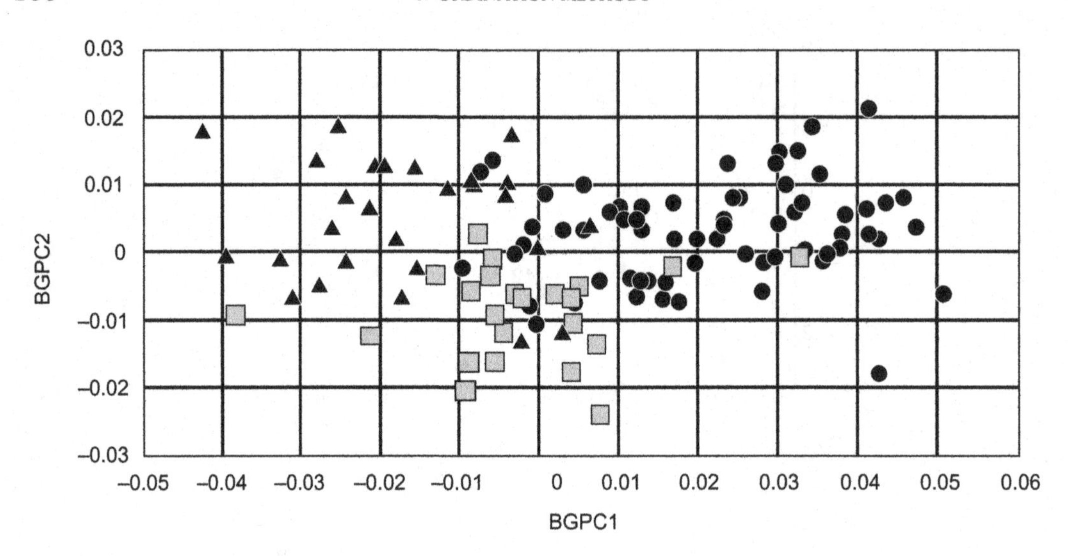

FIGURE 6.20 Scores of all 119 individuals projected onto BGPCs.

result of that computation for the 119 squirrel jaws is shown in Figure 6.20. It is important to remember that this is the distribution of shapes in a subspace that was defined by the positions of the means in a larger space. In the case of this example, the scores in the original 26-dimensional space have been projected onto a two-dimensional plane. The PCs of the original space accounted for only 42% of the variation; the plane of means likely accounts for less than that. It is important to remember that all variation orthogonal to this subspace has been excluded. It is only the differences between means that remain undistorted by this projection. The utility of examining scores of individuals on these axes will depend on the scientific merits of focusing subsequent analyses on this subspace.

References

Anderson, T. W. (1958). *An introduction to multivariate analysis*. New York: Wiley.

Bartlett, M. S. (1947). Multivariate analysis. *Journal of the Royal Statistical Society, Series B, 9*, 176–197.

Campbell, N. A., & Atchley, W. R. (1981). The geometry of canonical variates analysis. *Systematic Zoology, 30*, 268–280.

Chatfield, C., & Collins, A. J. (1980). *Introduction to multivariate analysis*. London: Chapman & Hall.

Costa, C., Aguzzi, J., Menesatti, P., Antonucci, F., Rimatori, V., & Mattoccia, M. (2008). Shape analysis of different populations of clams in relation to their geographical structure. *Journal of Zoology, 276*, 71–80.

Efron, B. (1983). Estimating the error rate of a prediction rule: Improvement on cross-validation. *American Statistician, 37*, 36–48.

Efron, B., & Tibshirani, R. (1995). *Cross-validation and the bootstrap: Estimating the error rate of a prediction rule*. Canada: (Technical Report) Department of Statistics, University of Toronto.

Knoke, J. D. (1986). The robust estimation of classification error rates. *Computers and Mathematics with Applications, 2*, 253–260.

Menesatti, P., Costa, C., & Paglia, G., et al. (2008). Shape-based methodology for multivariate discrimination among Italian hazelnut cultivars. *Biosystems Engineering, 4*, 417–424.

Mitteroecker, P., & Bookstein, F. (2011). Linear discrimination, ordination, and the visualization of selection gradients in modern morphometrics. *Evolutionary Biology, 38*, 100–114.

Morrison, D. F. (1990). *Multivariate statistical methods.* New York: McGraw Hill.

Nolte, A. W., & Sheets, H. D. (2005). Shape based assignment tests suggest transgressive phenotypes in natural sculpin hybrids (Teleostei, Scorpaeniformes, Cottidae). *Frontiers in Zoology, 2*, 11 <http://www.frontiersinzoology.com/content/2/1/11>

Schiavo, R. A., & Hand, D. J. (2000). Ten more years of error rate research. *International Statistical Review, 68*, 295–310.

Sheets, H. D., Corvino, K. M., Panasiewicz, J. M., & Morris, S. R. (2006). Comparison of geometric morphometric outline methods in the discrimination of age-related differences in feather shape. *Frontiers in Zoology, 3*, 15 <http://www.frontiersinzoology.com/content/3/1/115>

Williams, S. T., Hall, A., & Kuklinski, P. (2012). Unraveling cryptic diversity in the Indo-West Pacific gastropod genus *Lunella* (Turbinidae) using elliptic Fourier analysis. *American Malacological Bulletin, 30*, 189–206.

Witten, I. H., & Frank, E. (2005). *Data mining: Practical machine learning tools and techniques.* San Francisco: Morgan Kaufmann.

Van Bocxlaer, B., & Schultheiß, R. (2010). Comparison of morphometrics techniques for shapes with few homologous landmarks based on machine-learning approaches to biological discrimination. *Paleobiology, 36*, 497–515.

Yee, W. L., Chapman, P. S., Sheets, H. D., & Unruh, T. R. (2009). Analysis of body measurements and wing shape to discriminate *Rhagoletis pomonella* and *Rhagoletis zephyria* (Diptera: Tephritidae) in Washington State. *Annals of the Entomological Society of America, 102*, 1013–1028.

7

Partial Least Squares Analysis

Partial Least Squares (PLS) is a method for exploring patterns of covariation between two (and potentially more) blocks of variables. It can be used to study covariation between shape and environmental variables, such as between nasal cavity morphology and temperature and aridity (Noback et al., 2011) or between shape and a collection of climatic, geographic and biotic variables, e.g. vegetation type and human density (Monteiro et al., 2003). PLS can also be used to analyze the relationship between form and function, such as the relationship between horn morphology and fighting behavior in bovids (Lundrigan, 1996) or between morphology and disease status (Lowe et al., 1997; Bookstein et al., 2002). As well as being useful for analyzing the relationship between shape and non-shape variables, PLS can also be used to analyze the covariances between two or more blocks of shape variables. That ability to examine relationships between two or more blocks of shape variables makes PLS useful for synthesizing information about three-dimensional morphologies from two two-dimensional views (e.g. Rohlf and Corti, 2000) or for relating the shapes of functionally interacting parts such as the maxillary and mandibular dentitions (Sheets et al., in press). Because PLS can be used to examine the covariance between blocks of shape variables, the method can be used to examine morphological integration and modularity (Bookstein et al., 2003; Klingenberg et al., 2003; Bastir and Rosas, 2004; Bastir et al., 2005; Mitteroecker and Bookstein, 2007).

PLS, like the ordination methods discussed in the previous chapter, reduces the dimensionality of the data (of both blocks), yielding axes that explain the covariance between blocks, ordered from the pair that explains the maximal covariance to the pair that explains the least, all of which are mutually orthogonal. It also gives scores on those axes along with the proportion of the total covariance between blocks explained by that pair of axes and the correlation between the scores for each pair of axes. It obviously differs from Principal Components Analysis (PCA) in that it examines the covariance between blocks rather than the variance within a block. It also differs from other methods that examine the relationship between sets of variables, for reasons that will be discussed in more detail below. An important feature of PLS is that the variables within the blocks need not be independent of each other. For example, in the study relating nasal cavity

Geometric Morphometrics for Biologists
DOI: http://dx.doi.org/10.1016/B978-0-12-386903-6.00007-1

morphology to temperature and aridity, the measures of temperature included mean yearly temperature, coldest monthly temperature and warmest monthly temperature, and the measures of aridity included mean yearly vapor pressure, lowest monthly vapor pressure and highest monthly vapor pressure (Noback et al., 2011). Additionally, the number of variables can greatly exceed the number of cases, making PLS useful for discriminating between groups when there are far more shape variables than specimens (Barker and Rayens, 2003; Sheets et al., 2006; Mitteroecker and Bookstein, 2011). This makes PLS particularly useful for geometric morphometric studies because it is common to have far more shape variables than specimens in studies, especially when the data comprise semilandmarks as well as landmarks.

PLS is probably unfamiliar to many biologists even though it has been used extensively in the social sciences (see Wold, 1966; Bookstein, 1982; Jöreskog and Wold, 1982), in clinical studies (e.g. Sampson et al., 1989; Streissguth et al., 1993; Lowe et al., 1997), economics (Fornell and Bookstein, 1982) and chemistry (Kemsley, 1996; Barker and Rayens, 2003). Thus, a large part of this chapter discusses similarities and differences between PLS and more familiar methods, including regression, Principal Components Analysis (PCA) and Canonical Correlation Analysis (CCA). Because of the potentially wide range of applications of this method, it is important to understand how it is related to the other methods that address similar questions. Before discussing the method in more detail, we note that one approach to PLS employs a mathematical technique not yet introduced in this text: Singular Value Decomposition (SVD). SVD is related to the more familiar decomposition by eigenanalysis, which is used to extract principal components from the variance–covariance matrix and also the partial warps from the bending-energy matrix. SVD offers a more general approach, one that is needed because in PLS we are decomposing matrices that are not square and symmetric, meaning that the number of rows and columns need not be equal and the first row is not also the first column. There are other approaches to PLS analysis available as well (Wold, 1966; Streissguth et al., 1993; Bookstein et al., 2003). Because PLS uses SVD, the vectors generated by PLS are often called *Singular Axes* (SAs); in some geometric morphometric studies they are also called *Singular Warps*.

ANALYZING COVARIANCES BETWEEN BLOCKS AND SIGNIFICANCE TESTING

We begin the analysis with two blocks of data that are measured on the same individuals, such as a block of shape variables and another of ecological variables. Given these blocks, we might first wish to determine whether the two blocks covary. To that end, we wish to quantify the covariance between the two blocks and to determine whether that covariance exceeds what we might obtain by chance. We thus have a one set of p_1 variables, $\mathbf{Y}_1$, our first block, which will be a set of shape variables, and a second set of p_2 variables $\mathbf{Y}_2$, which could also be shape data or another set of measurements. We can now compute the variance–covariance matrix, $\mathbf{R}$, which can be thought of as comprising the variance–covariance matrices *within* blocks $\mathbf{Y}_1$ and $\mathbf{Y}_2$ ($\mathbf{R}_1$ and $\mathbf{R}_2$, respectively) with dimensions $p_1 \times p_1$ and $p_2 \times p_2$ respectively and the covariance matrix between the two blocks $\mathbf{R}_{12}$ (with dimension p_1 by p_2), giving

$$\mathbf{R} = \begin{bmatrix} \mathbf{R}_1 & \mathbf{R}_{12} \\ \mathbf{R}_{12}^t & \mathbf{R}_2 \end{bmatrix} \tag{7.1}$$

where $\mathbf{R}_{12}^t$ is the transpose of $\mathbf{R}_{12}$. The variance–covariance of the combined data can thus be thought of as having three distinct parts, the first two of which are the variance–covariance matrices within each block ($\mathbf{R}_1$, $\mathbf{R}_2$) and the third, which is the covariance between the two blocks ($\mathbf{R}_{12}$).

The covariance between the two blocks can be quantified by Escoufier's coefficient (Escoufier, 1973), which is a multivariate extension of the ordinary univariate correlation. That coefficient is given by the expression:

$$\mathrm{RV} = \frac{trace(R_{12}R_{12}^t)}{\sqrt{trace(R_1R_1^t)trace(R_2R_2^t)}} \tag{7.2}$$

The numerator is the summed squared covariances between the two sets of variables and the denominator is the square root of the product of the summed squared variances within each block. Escoffier's coefficient thus ranges from 0 (no covariance) to 1 (complete covariance). The statistical significance of **RV** can be tested by randomizing the order of observed values (the rows of the matrix $\mathbf{Y}_1$, for example) and recomputing the value of the coefficient for the permuted version of $\mathbf{Y}_1$ and $\mathbf{Y}_2$ (Klingenberg, 2009). Note that this will alter the covariance between blocks, but not the variance within each. If the observed value of **RV** lies outside the confidence interval of values obtained by the permutations, for some chosen α, then the observed **RV** (and therefore the covariance between the two blocks) is statistically significant.

MATHEMATICAL DETAILS OF TWO BLOCK PLS

Given that matrix, $\mathbf{Y}_1$, of p_1 variables measured on n specimens, and the other block, $\mathbf{Y}_2$, of p_2 variables, we compute the variance–covariance matrix, $\mathbf{R}$, as discussed above, which comprises the within-block variance–covariance matrices of blocks $\mathbf{Y}_1$ and $\mathbf{Y}_2$ ($\mathbf{R}_1$ and $\mathbf{R}_2$, respectively) and the covariance matrix between the two blocks $\mathbf{R}_{12}$ (as shown in Equation 7.1). We then perform a singular value decomposition (SVD) of $\mathbf{R}_{12}$:

$$\mathrm{R}_{12} = USV^t \tag{7.3}$$

S is a $p_1 \times p_2$ diagonal matrix whose entries are the P_{min} singular values, λ_I (there are as many singular values as there are variables in the *smaller* block, P_{min}). The matrices **U** and **V** have dimensions $p_1 \times p_{min}$ and $p_2 \times p_{min}$, respectively; their columns are the Singular Axes (SAs). The first columns of **U** and **V** comprise the paired SAs corresponding to the first singular value λ_1, just as the first Principal Component (PC1) is the axis corresponding to the first eigenvalue of the variance–covariance matrix. The SAs are ordered by decreasing singular values, just as PCs are ordered by decreasing eigenvalues. Scores on

SAs are calculated just like scores on PCs, i.e. by multiplying the data for each specimen by the SA (i.e. taking the dot product between an SA and the data for a specimen). Scores are calculated for each block separately.

The fraction of the total covariance of the two blocks expressed by the *i*th pair of singular axes is given by:

$$\frac{\lambda_I^2}{\sum_{j=1}^{P_{min}} \lambda_j^2} \tag{7.4}$$

Whether a singular value is larger than we would expect from randomly related blocks is determined by comparing the observed singular value to the distribution produced by randomly permuting the covariance structure between blocks. In such a permutation test, the vectors of observations, each representing a specimen in the first block, are randomly associated with vectors of observations from the second, thereby randomizing the covariance structure between blocks without altering the variance–covariance structure within the blocks. If the observed singular value lies outside the 95% confidence interval obtained from the permuted data sets, the observed SA is judged to be statistically significant. The correlation between the scores on the two blocks on the *i*th SA is also a measure of the statistical significance of the axis, and this correlation also may be tested via a permutation test in exactly the same manner.

Three Block PLS

It is possible to extend PLS to more than two blocks of data, such as when we have three blocks of data $\mathbf{Y}_1$, $\mathbf{Y}_2$ and $\mathbf{Y}_3$, and seek the linear combinations of variables within each block (expressed as vectors $\mathbf{U}_1$, $\mathbf{U}_2$, $\mathbf{U}_3$) which produce the greatest covariation of scores ($s_1 = \mathbf{Y}_1\mathbf{U}_1$, $s_2 = \mathbf{Y}_2\mathbf{U}_2$, $s_3 = \mathbf{Y}_3\mathbf{U}_3$). The vectors $\mathbf{U}_1$, $\mathbf{U}_2$, $\mathbf{U}_3$ are the singular axes of such a system, and s_1, s_2 and s_3 are the scores along the singular axes. It is then possible to compute the correlation between each pair of scores r_{1-2}, r_{1-3} and r_{2-3}. As discussed by Bookstein and colleagues (2003), there is no standard approach to a multiblock PLS; several are possible based on different properties of the PLS method. Bookstein and colleagues follow the approach taken by Streissguth and colleagues (1993), which uses an iterative method to estimate $\mathbf{U}_1$, $\mathbf{U}_2$ and $\mathbf{U}_3$.

In this approach, one starts with an arbitrary choice of the first axes $\mathbf{U}_1$, $\mathbf{U}_2$ and $\mathbf{U}_3$, using random values or $1/n^{0.5}$ if there are n entries in the vector $\mathbf{U}$. The following set of steps (Bookstein et al., 2003) is then iterated:

(a) Compute the scores (s_1, s_2)

$$s_1 = \mathbf{Y}_1\mathbf{U}_1 \tag{7.5a}$$

$$s_2 = \mathbf{Y}_2\mathbf{U}_2 \tag{7.5b}$$

$$s_3 = \mathbf{Y}_3\mathbf{U}_3 \tag{7.5c}$$

(b) Normalize the scores to unit variance

$$s_1 = s_1/(\text{standard deviation}(s_1)) \quad (7.6a)$$

$$s_2 = s_2/(\text{standard deviation}(s_2)) \quad (7.6b)$$

$$s_3 = s_3/(\text{standard deviation}(s_3)) \quad (7.6c)$$

(c) Compute the correlations between the scores

$$r_{1-2} = \text{correlation of } s_1 \text{ and } s_2 \quad (7.7a)$$

$$r_{1-3} = \text{correlation of } s_1 \text{ and } s_3 \quad (7.7b)$$

$$r_{2-3} = \text{correlation of } s_2 \text{ and } s_3 \quad (7.7c)$$

(d) Update the estimates of the **U** vectors

$$\mathbf{U}_1 = \mathbf{Y}_1^t(r_{1-2}s_2 + r_{1-3}s_3) \quad (7.8a)$$

$$\mathbf{U}_2 = \mathbf{Y}_2^t(r_{1-2}s_1 + r_{2-3}s_3) \quad (7.8b)$$

$$\mathbf{U}_3 = \mathbf{Y}_3^t(r_{1-3}s_1 + r_{2-3}s_2) \quad (7.8c)$$

The sequence of steps is then repeated, using the **U** vectors at the end of each iteration as inputs into the next. This is repeated until the changes in **U** (or correlations r) do not change within some desired tolerance level. It is then possible to test the significance of the observed correlations using permutations.

USING PLS TO COMPARE PATTERNS OF COVARIANCE BETWEEN BLOCKS ACROSS GROUPS

PLS is usually used to examine patterns of covariances between blocks of variables measured in a single sample, but it can also be used to compare those covariances between samples, as in a comparative analysis of geographic variation. Such comparisons rely on the same logic (and methods) used in comparative analyses of regression equations or PCs because in all of these we are asking if the biologically corresponding vectors point in the same direction. To answer that question, we can compute the angle between comparable SAs, then test it statistically (using, for example, a bootstrapping procedure). In a similar fashion, we can also compare SAs to PCs, asking whether the major dimension of covariance *between* blocks is equivalent to the dominant dimension of variation *within* blocks. For example, when our data come from an ontogenetic series, the major dimension of variance within each block is likely to be the ontogenetic vector, and the major dimension of the covariance between blocks may be the developmental covariance between the two blocks. Comparing SAs to PCs can be especially useful for understanding causes of variance when PLS indicates a significant relationship between morphology and some collection of environmental variables. That same relationship between morphology and the environment may also explain the dominant axis of morphological variation.

COMPARING PLS TO OTHER METHODS

PLS resembles several other methods, including simple and multiple regression, principal components analysis (PCA), and canonical correlation analysis (CCA). PLS can also be used to discriminate between groups, making it an alternative to discriminant function analysis (DFA) and canonical variates analysis (CVA). The relationship between methods is complicated because PLS can be approached from multiple perspectives, but we focus on PLS as solving one particular sort of eigenstructure problem and having the constraints on the directions of the SAs noted above, i.e. that they be mutually orthogonal. Below we briefly compare methods so that you can decide which is most useful for your purposes.

PLS Compared to Multiple Regression

Both PLS and multiple regression can examine the relationship between two multivariate sets of variables, but they differ in two important respects. First, and most importantly, PLS does not require that the variables in either block be uncorrelated with each other, and works most effectively when they are not, whereas multiple regression has difficulty determining the variance explained by highly correlated predictive variables. In PLS, the correlations among the variables are thought to reflect their joint response to underlying (unobserved) variables, often called "latent variables" (a concept frequently used in PLS). To estimate a latent variable, it is important to have multiple observed variables because their correlations are explained by their dependence on the latent variable. For example, to measure the latent variable "climate" we would use multiple observed climatic variables (e.g. maximum monthly temperature, minimum monthly temperature, maximum monthly precipitation, minimum monthly precipitation, seasonality, etc.). The correlations among them are explained by "climate". Rather than exploring the structure of these measurements within a block to extract that latent variable, PLS seeks the combination of the climate variables that maximally covary with the other block of variables – the linear combination of climate variables most relevant for explaining the other block. These coefficients are called *saliences* because they indicate which variables in one block are most relevant (salient) for explaining covariation with the other block. The ability to find that combination is enhanced by having multiple correlated observed variables.

In striking contrast, the coefficients of a multiple regression express the dependence of the dependent variables (e.g. shape) on one independent variable, *with all others held constant*. Consequently, correlations between the independent variables are a problem for the method. When the independent variables are correlated with each other, most of the variance in the dependent variables will be associated with one independent variable, the one first entered into the model, leaving little to be explained by the others, as discussed in the chapter on General Linear Models. Even though *all* the independent variables might affect the dependent variables, only one might be accorded a high weight, making the others appear to have trivial explanatory power. That is because they are explaining the *residual* variance, i.e. the variance not already explained by the one with the large coefficient. Multiple regression is thus poorly suited to cases in which the independent variables are correlated with each other. In contrast, PLS is specifically intended for the case in which multiple variables (within each block) are measuring the same factor. The problem posed by correlated independent variables in a multiple regression is sometimes solved by conducting a preliminary PCA to obtain uncorrelated variables (the PCs), then regressing the dependent variable (e.g. shape) on the PCs. As a result, the construction of independent variables is determined only by the covariances among them, without considering their relationship to the dependent variables. In PLS, the axes for both blocks are determined by the covariances between the blocks, which can yield axes that need not correspond to the PCs within blocks.

Another difference is that regression typically casts one set of variables as dependent on the other, whereas PLS treats them symmetrically. That is, PLS does not assume that one set of variables is independent and the other dependent. Both sets are treated as jointly (and linearly) related to the same underlying causes. What makes the symmetry of the method important is that (Model 1) regression is based on a model that assumes that the independent variable is controlled and therefore all of its variation is explained by the experimental manipulation; it is measured without unexplained variation ("error"). Hence all the unexplained variance in the data is ascribed to the dependent variable. No such model underlies PLS and so no error is ascribed to any variables (in either block). For this reason alone, we would not expect to obtain the same coefficients from PLS as we obtain from regression.

There is, however, a form of PLS more comparable to regression, PLS–Regression (Wold et al., 2001). This method uses the basic machinery of PLS, the SVD of the cross-block covariance matrix, R (Equation 7.1) as the initial step in the procedure. That first step yields the pair of linear combinations, SA1, for the two blocks, plus the scores for the paired SA1s. Then, instead of regressing the first block, **Y**, on the second, **X**, **Y** is regressed on the vector obtained from the scores for **X** (which may be normalized or otherwise weighted). Further details on this method are beyond the scope of this chapter; there are several algorithms for the procedure as well as several methods for obtaining the vector of scores for the X block (see Mevic and Wehrens, 2007). When the variables in the X block (i.e. the predictors) are all uncorrelated, PLS–Regression will be equivalent to ordinary least squares linear regression on those variables (Wold et al., 2001).

PLS Compared to PCA

PLS and PCA resemble each other in one important respect: both reduce the dimensionality of the data by extracting a set of mutually orthogonal axes. As you recall from Chapter 6, PCs are extracted from a variance–covariance matrix (by eigenanalysis), producing a set of mutually orthogonal dimensions (eigenvectors), ordered according to the amount of variance each one explains. Similarly, PLS decomposes a matrix into mutually orthogonal axes, ordered according to the amount of *covariance* between blocks explained by each one. The most obvious difference is that PCA examines variation within a single block of variables whereas PLS examines the covariation between blocks. Consequently, one of the primary differences between PCs and SAs is that SAs, unlike PCs, come in pairs. For each singular value there is a pair of axes that, taken together, accounts for the patterns of covariances between blocks. But despite this obvious difference, both PLS and

PCA impose a similar constraint on the analysis: both define axes to be mutually orthogonal. Just as PC2 is defined to be orthogonal to PC1, SV2 is defined to be orthogonal to SV1. This is important when biological factors are not orthogonal, which may be the general rule. Even though the axes (both PCs and SAs) provide a useful, simplified space in which to explore patterns in the data, the axes themselves, beyond the first, need not correspond to any biological factors. It is likely that PC1 and SA1 have a biological interpretation when they account for a very large proportion of the variance or covariance, but the remaining axes are, by definition, constrained to be orthogonal to them, making their interpretation more dubious. This same issue arises when using PCA for explanatory or even comparative purposes (see Rohlf and Corti, 2000; Houle et al., 2002; Angielczyk and Sheets, 2007).

Another important similarity between the methods, which also should inspire a cautious approach to interpreting results, is that PLS extracts *linear* combinations of variables (like PCA) but the relationship between blocks may be non-linear. In such cases, the first dimension may represent the dominant linear trend, and others represent orthogonal deviations from linearity. Thus, we would need to interpret SV1 together with SV2 to understand the relationship between the two blocks, recognizing that a single non-linear factor accounts for both. Of course, the issue of linearity is also important whether we are analyzing the data by PCA/PLS, by regression, or by the method discussed in the following section, CCA. However, most workers recognize that linearity is an important assumption of regression; non-linearity might not seem so important in studies using PCA or PLS because neither method is explicitly based on a linear model so the impact of non-linear relationships among variables might not seem to violate assumptions of the method.

Unlike the situation for PCA, there is no analytic statistical test of the significance of SAs, meaning that there is no analytic test for the difference in length between SA1 and SA2, and so forth. However, as mentioned above, resampling-based approaches can be applied to test the hypothesis that SA1 (and succeeding SAs) explain more covariance than expected by chance. A permutation test, discussed by Rohlf and Corti (2000), determines whether the singular values are larger than could be produced by a random permutation of associations among variables between blocks (keeping within-block associations intact).

PLS Compared to Canonical Correlation Analysis

Canonical correlation analysis examines the correlation between blocks of variables and it closely resembles multiple regression although, as in the case of PLS, CCA treats both blocks symmetrically. CCA thus differs from multiple regression and resembles PLS in that neither block is construed as comprising a block of causal variables with the other comprising the responses. One important difference between CCA and PLS is the quantity maximized by the two procedures. CCA seeks pairs of axes (canonical axes) that are maximally correlated *with each other*. That is, CCA seeks an axis, a linear combination of variables, from one block that is maximally correlated with a linear combination of variables from other block. In contrast, PLS seeks axes that maximally account for the covariance between *blocks* (for a more detailed comparison between CCA and PLS, see Rohlf and

Corti, 2000). Another important difference is that the coefficients produced by a CCA are interpreted like partial regression coefficients, meaning that (as discussed above), each coefficient indicates the contribution made by an independent variable *when all others are held constant*. In this way, CCA, like multiple regression, is not well-suited to analyses in which the variables within a block are correlated. As discussed above, constructing axes that are mutually uncorrelated with each other within each block by a preliminary PCA need not yield the same axes as those constructed by maximizing the covariances between the two blocks.

PLS compared to Canonical Variates/Discriminant Analysis

PLS might not seem comparable to a CVA because PLS examines the relationship between two blocks of variables whereas CVA discriminates between groups. However, canonical correlation analysis and canonical variates analysis are closely related techniques and PLS is related to both. PLS has been used to discriminate between groups, such as between types of dementia (Gottfries et al., 1995) and even between years of a vintage port wine (Ortiz et al., 1996). When used for purposes of discrimination, one block of variables consists of codes that indicate an individual's membership in a group. The resultant scores can then be inspected to assess the separation between groups (Barker and Rayens, 2003; Mitteroecker and Bookstein, 2011). The procedure for discrimination by PLS is equivalent to the method introduced in the last chapter, the Between-group PCA, when shape is one block and (normalized) codes for the groups are the other (Mitteroecker and Bookstein, 2011). This approach is particularly useful when the number of variables greatly exceeds the number of individuals; under these conditions, the results of a CVA can be what looks like very large differences between groups even when the "groups" are random samples from a single population.

APPLICATIONS OF PLS

PLS can be used to address a large range of biological questions about the relationship between shape and other variables. To exemplify some of these applications, we consider two cases in which both blocks comprise shape data, but the questions asked about the relationships between the blocks differ and some of the methodological details also differ. We also consider several applications that relate shape to other variables, surveying several studies to show the diversity (and treatment) of those non-shape variables.

Using PLS as an Exploratory Tool to Characterize a Population: The Anterior Human Dentition

The forensic discipline of bitemark analysis has come under scrutiny (Rothwell, 1995; Pretty and Sweet, 2001; Bowers, 2006; Pretty, 2006; NAS, 2009) due to a number of criminal convictions based on bitemark analysis that were later overturned based on DNA evidence (Bowers, 2006). Forensic identification of post-mortem victims based on the examination of dental records that incorporate detailed information about the entire dentition are effective and non-controversial but, in bitemark analysis, only the incisal surfaces of the six anterior mandibular and maxillary teeth (the biting dentition) typically leave an impression. In assessing the effectiveness of bitemark evidence, Sheets et al. (in press) examined the patterns of both variation in the positions of the incisal surfaces of the six anterior teeth, and covariation in tooth position using 1099 three-dimensional scans of human dentitions.

These dentitions were digitized by a commercial laboratory as part of the process of constructing occlusal mouth guards for patients drawn from a random sample of these private clinical patients in the USA (all identifying personal information was stripped from the records used, and IRB approval was obtained for this study). One of the goals of this study was simply to document systematically the patterns of variation and covariation in this relatively large sample. This simple documentation of the patterns was a task specifically called for in the 2009 report on forensic sciences (NAS, 2009), and is helpful in assessing the strength of evidence linking a suspect dentition to a bitemark, particularly with the substantial distortion of the impression produced by skin (Bush et al., 2011). This was a "sample of convenience", and some care is warranted in attempting to extend these results to various human subpopulations.

On each scanned three-dimensional dentition, ten semilandmarks were placed along the incisal surface of each anterior tooth (Figure 7.1). The data were then analyzed using both PCA (Figures 7.2 and 7.3) and PLS (Figure 7.4). The results of the PCA are shown as paired views, both frontal and occlusal, because there is substantial interest in the degree

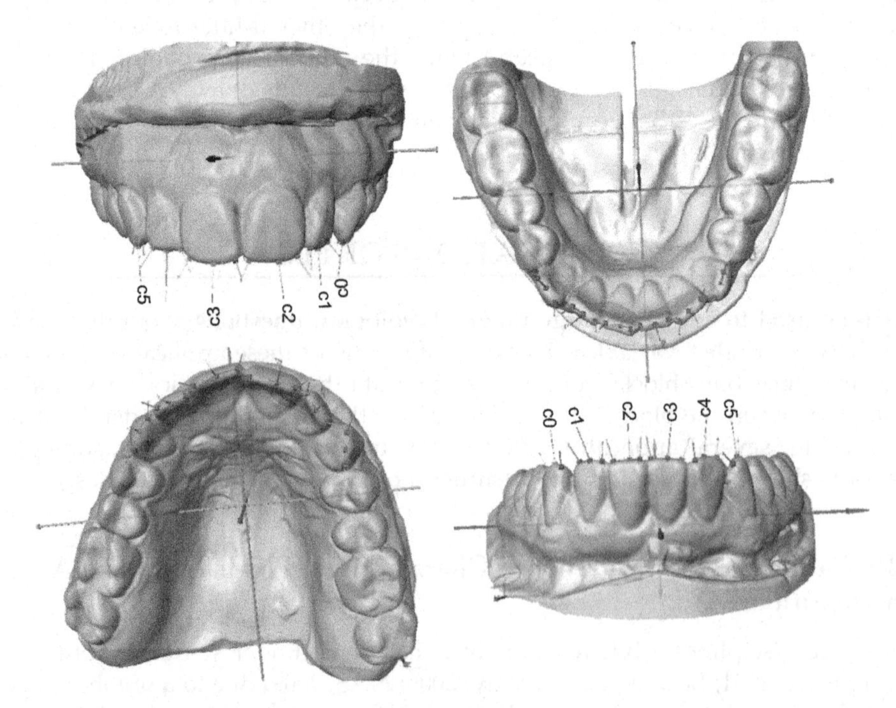

FIGURE 7.1 Occlusal and frontal views of three-dimensional scans of a cast made of a human dentition. Semilandmarks were placed along the occlusal surfaces of the six anterior teeth in the maxilla and mandible.

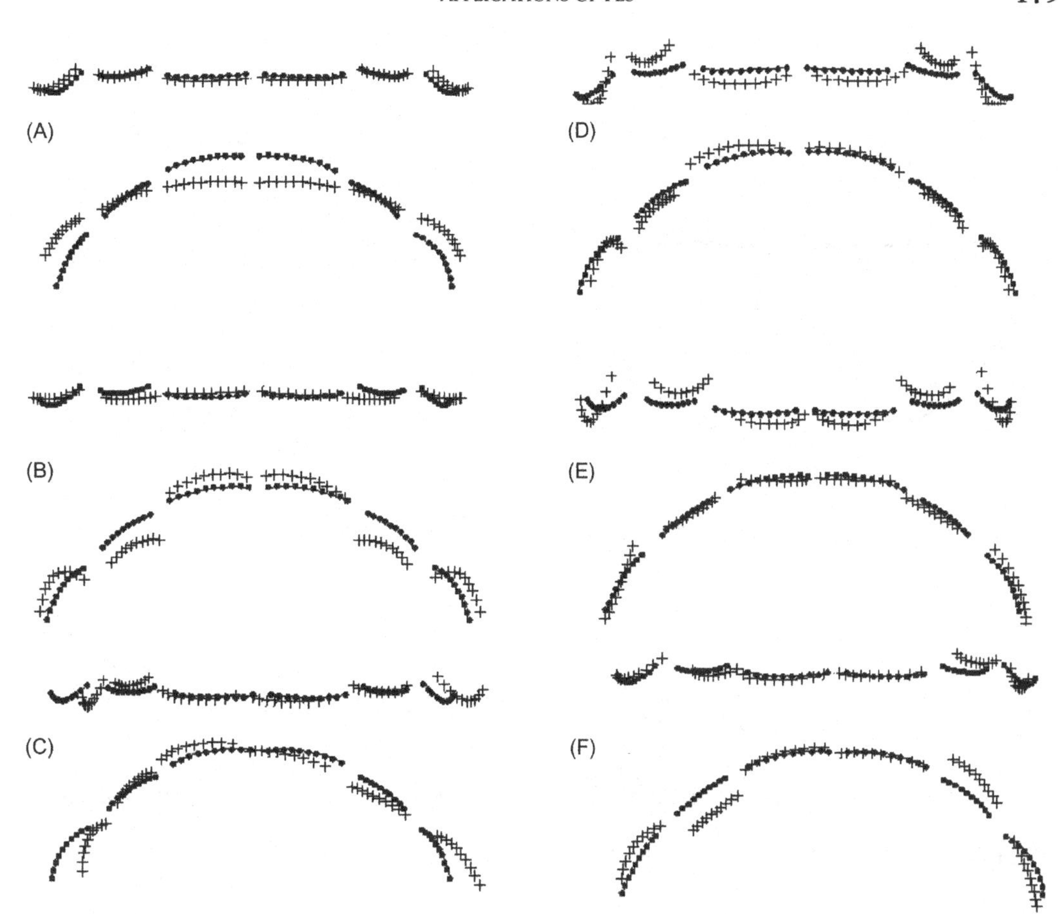

FIGURE 7.2 PCs 1–6 (A–F) of the maxillary dentition in frontal and occlusal views (see Figure 7.1). The data for the PC axis variation are shown as floating gray crosses. The average dentition over the entire set is shown as a series of black dots connected by solid lines along the incisal edges. The crosses indicate the pattern of differences for specimens with positive scores along the axis; specimens with negative scores would have the reversed pattern. It should be noted that the data shown in Figures 7.2 and 7.3 are three-dimensional in nature and that rotation in three-dimensional space permits a better understanding of the shape variance.

of variance present in the vertical direction in the human dentition and the impact of this variance on the resultant bitemarks. PLS results are shown only in occlusal view. The first 11 SAs were judged to be significant based on permutation test, but the first four explain a total of 81.9% of the covariation. Thus, the remaining seven axes are statistically significant, but explain little of the covariation.

Clearly identifiable patterns emerge from both PCA and PLS (Tables 7.1 and 7.2), with the predominant feature in both being the contrast between relatively wide versus relatively narrow arches. Not surprisingly, these show a strong covariance between upper and lower dentitions. The second PC and SA are also highly similar, showing the relative displacement of the central and lateral incisors. Because this sample was obtained from

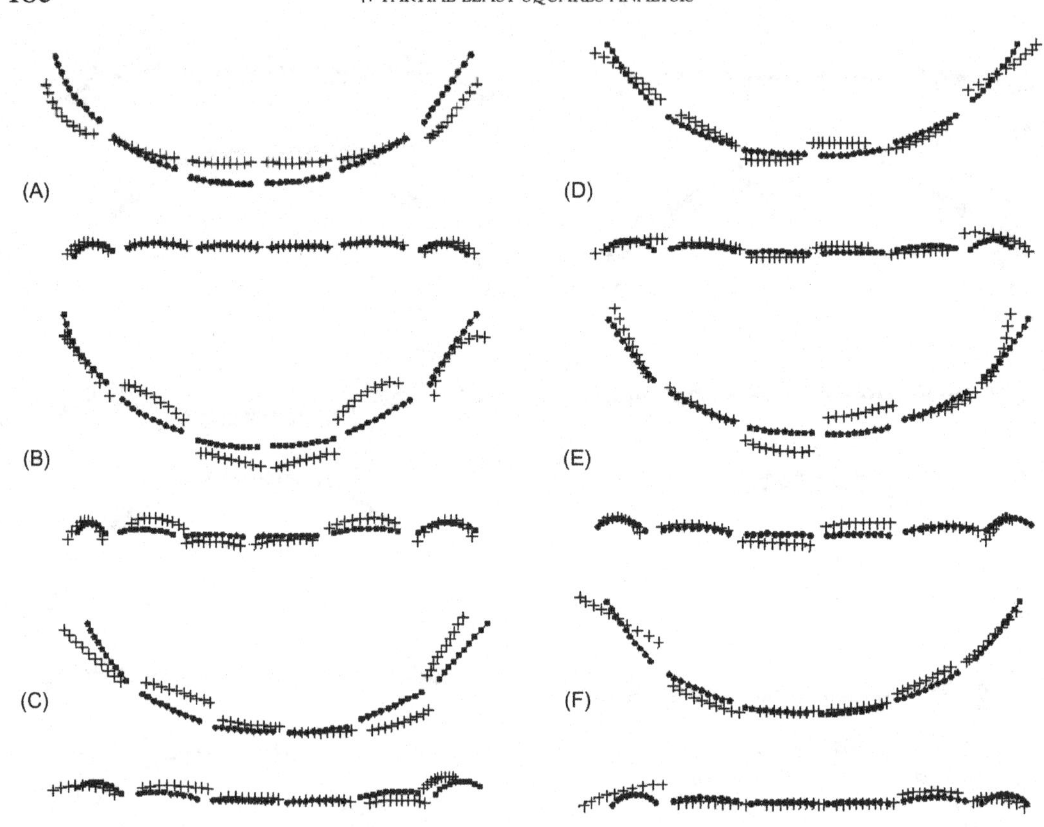

FIGURE 7.3 PCs 1–6 (A–F) for the mandibular dentition, in occlusal and frontal view.

patients being fitted for occlusal mouth guards in private dental practices in the USA, it is probably biased toward individuals with relatively high levels of dental care.

Using PLS to Examine Morphological Integration and Modularity

One of the most promising applications of PLS is in studies of morphological integration and modularity (subjects covered in more depth in Chapter 12). Numerous studies, especially of primates, use PLS for that purpose (e.g. Bookstein et al., 2003; Bastir and Rosas, 2004, 2005, 2006; Bastir et al., 2005, 2007, 2008; Mitteroecker and Bookstein, 2007, 2008; Laffont et al., 2009; Gkantidis and Halazonetis, 2011). Here we focus on the covariance between two parts of the rodent mandible. The rodent mandible has become a favored model system for studies of morphological integration and modularity, and the dominant hypothesis is that there are two modules, one comprising the tooth-bearing region, the other the muscle-bearing region (e.g. Cheverud et al., 1991; Mezey et al., 2000; Klingenberg et al., 2003). This hypothesis is shown in Figure 7.5. It may be evident that it is difficult to define these two regions precisely because the incisor extends well into the muscle-bearing region and the dominant muscles for biting and chewing, the lateral,

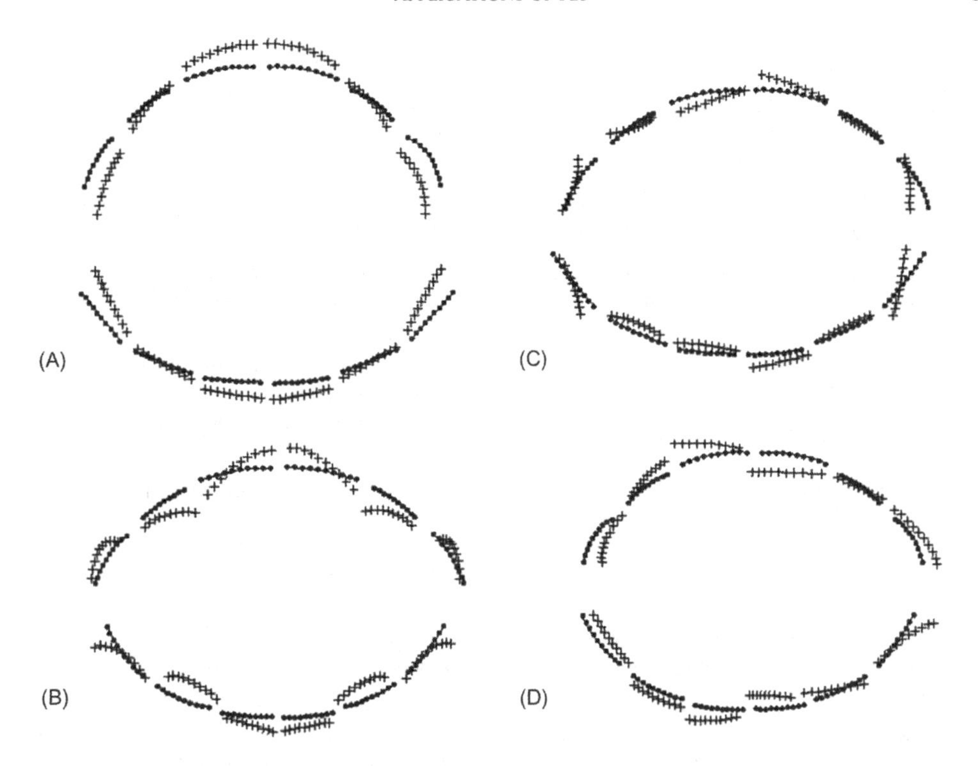

FIGURE 7.4 The first four SAs (A–D), showing patterns of covariation in the maxillary and mandibular dentition (occlusal view with maxilla at top).

medial and superficial masseters, extend well into the tooth-bearing region. Nevertheless, the two parts are commonly divided along the line shown in Figure 7.5, and are hypothesized to be both developmental and functional modules.

The mandible is a single skeletal element and it is measured as a single configuration of landmarks (and semilandmarks). To analyze its integration we can either retain the information about the interconnections between the two parts within the whole, or we can treat each module as a single configuration of its own and analyze the relationship between the two configurations (Klingenberg, 2009). The major methodological distinction between the two approaches is the first approach involves superimposing the entire configuration, whereas the second involves superimposing each block separately. The major conceptual distinction between the two approaches, as discussed by Klingenberg is that the first explicitly considers information about the connection of the subsets, which is important when some of the covariation between subsets arises from variation in their connections rather than from simultaneous variation within the two subsets. The second approach ignores the anatomical connection of the two subsets, including information about the relative sizes and positioning of the parts, focusing on the covariation due to joint changes of shape within each subset.

In either case, our first objective is to measure the covariance between the two putative modules. Because the hypothesis of modularity predicts that the two parts will *not* be

TABLE 7.1 Variance Explained By Each PC Axis of the Human Dentition Data Set, With a Description of the Pattern Implied By the Axis

PCA Axis	Percent of Variance	Pattern Implied By the Axis
1	Mandible: 54.0 Maxillary: 39.8	Arch width-positive scores indicate a relatively wide arch, negative scores indicate a relatively narrow arch, in both maxillary and mandible
2	Mandible: 8.8 Maxillary: 12.6	Positive scores indicate a labial displacement of both central incisors, relative to a lingual displacement of both lateral incisors, with a slight labial shift of the canines in the maxillary
3	Mandible: 6.7 Maxillary: 7	This axis implies a pattern of left–right asymmetry in both maxillary and mandible, although the specific details of the asymmetry differ. In the mandible, there is a "bulge" of all teeth to one side, whereas in the mandible, all 4 incisors shift in a line relative to the canines
4	Mandible: 4.2 Maxillary: 5.2	Asymmetry in the location of the central incisors in the mandible. One shifted lingually and the other labially, while the adjacent lateral incisors shift in the opposite directions. In the maxillary, this is a different pattern, a labial shift of one central incisor with an accompanying lingual shift of the lateral incisor and canine
5	Mandible: 3.9 Maxillary: 4.8	In the mandible, this axis implies opposing lateral and lingual shifts of the central incisors, while in the maxillary, this axis describes outward shifts of both canines
6	Mandible: 3.6 Maxillary: 3.8	Opposing lingual–labial shifts of the canines and lateral incisors appear in the maxillary, while asymmetric lingual–labial shifts appear in the two lateral incisors of the maxillary, with some changes in the orientation of one canine

Note that the descriptions are stated in terms of positive scores, negative scores simply reverse the pattern.

TABLE 7.2 Covariance Explained by Each PLS Axis of the Human Dentition Data Set, with a Description of the Pattern Implied by the Axis

PLS Axis	Percent of Covariance	Pattern Implied By the Axis
1	65.4	A contrast between wide and narrow arches
2	7.2	Strong labial displacement of both central incisors of both the upper and lower dentition.
3	5	Strong left-right asymmetry pattern
4	4.3	Opposing tilts of the central incisors and lateral incisors of both maxillary and mandible, and minor amounts of rotation of the canines

correlated, we then need to determine whether the observed covariance is lower than we would obtain between two randomly partitioned subsets of landmarks that have the same number of landmarks as the *a priori* partitions. Our second objective is to describe the dominant axes of covariation between the two modules should they covary.

Using the first approach, we obtain an RV coefficient which, as you recall, is analogous to an R^2, of 0.506. For an R^2, this is a very high value; not surprisingly, the null hypothesis

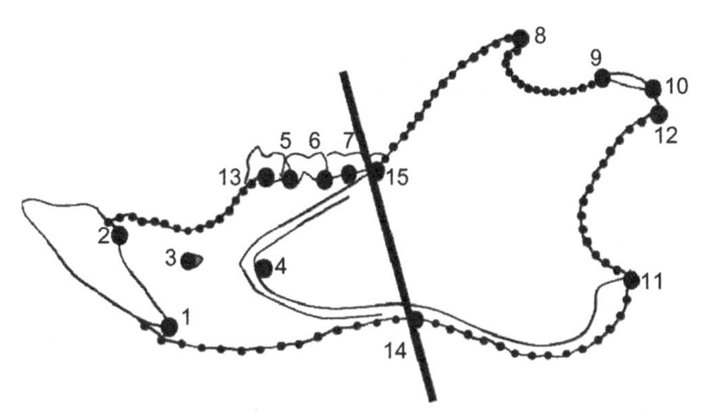

FIGURE 7.5 Subdivision of the mandible into two putative functional and developmental modules.

of independence between blocks is rejected ($P < 0.001$) by a permutation test. SA1 explains 46.6% of the covariance between blocks and the correlation between scores is 0.93. The dominant axis of covariation between the two blocks is shown in Figure 7.6A, which can be compared to the dominant axis of variation (PC1) of the mandible (Figure 7.6B). Doing the analysis the second way isolates the covariation due specifically to joint changes in shape *within* each subset, produces a lower RV of just 0.302, which is still a moderately high value for an R^2, and this too is statistically significant ($P < 0.001$). SA1, however, accounts for just 40.1% of the covariance and the correlation between the scores is 0.74. The pictures (Figure 7.7) show the dominant axis of covariation between the two blocks, the paired SA1s.

Using PLS to Relate Shape to Ecological Factors

Several studies have used PLS to analyze the relationship between shape and environmental factors, both abiotic and biotic (Fadda and Corti, 1998; Ruber and Adams, 2001; Monteiro et al., 2003; Arif et al., 2007; Pulcini et al., 2008; Fornell et al., 2010; McGuire, 2010; Noback et al., 2011). We single out three of them to show how such analyses can be conducted, focusing on the environmental variables.

Cichlid Body Shape and the Biotic Environment: The Relationship Between Body Shape and Trophic Morphology

Ruber and Adams (2001) investigated the relationship between body shape and trophic morphology in Lake Tanganyika cichlids. These cichlids are noted for their morphological diversity, and phylogenetic studies have revealed extensive convergence in trophic specializations, especially the specializations of their dentitions. Ruber and Adams asked whether other body shape covaries with trophic morphology, specifically, with the specialized dentitions. They sampled body shape by 14 landmarks on the external morphology of these fishes from 17 populations from four lineages, which were divided into phylogeographic clades. The landmarks comprise the first block of data, the second consists of four morphological variables that previous studies had shown to be correlated with trophic morphology: (1) gape width; (2) interorbital width; (3) oral tooth counts of all erupted teeth on the premaxillary; and (4) dental bones. Thus, the morphological variables are

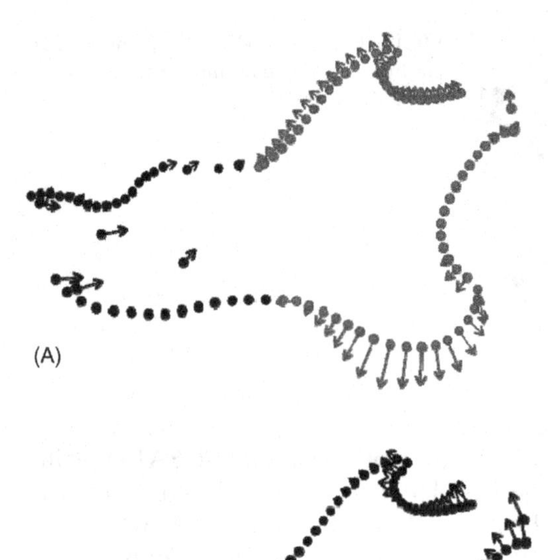

FIGURE 7.6 SA1 for the two subdivisions of the mandible, analyzed as two parts of a whole. (A) SA for the two mandibular blocks (Block 1 is shown in black, Block 2 in gray); (B) PC1 for the mandible as a whole.

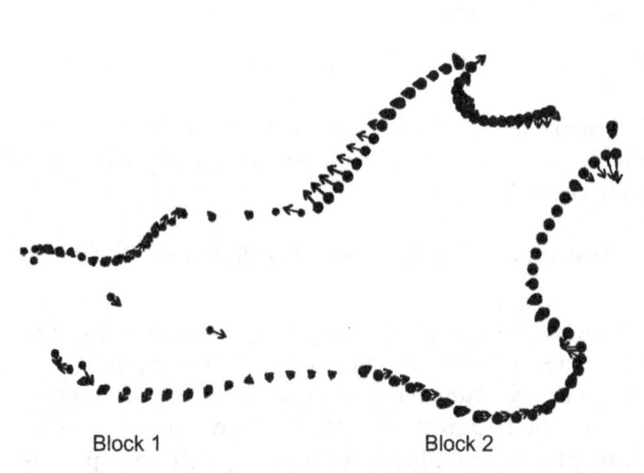

FIGURE 7.7 SA1 for the two subdivisions of the mandible, analyzed as two separate parts.

proxies for the "trophic environment". To determine whether these trophic variables are related to shape, Ruber and Adams used PLS and found that specimens with low tooth counts and a small gape have relatively elongated heads and a forward directed snout whereas those with high tooth counts and a large gape have deep heads and a ventrally directed snout. Using phylogenetic generalized least squares (PGLS) analysis, Ruber and

Adams found a statistically significant relationship between body shape and trophic morphology, taking phylogeny into account.

Human Nasal Cavity Morphology and the Climatic Environment: Temperature and Vapor Pressure

The functional morphology of the nasal cavity morphology was investigated by Noback and colleagues (2011). The nasal cavity functions to humidify and warm the air before it reaches the lungs, and the question posed by this study is whether humans that live in more demanding environments (with lower temperatures and higher aridity) have nasal cavities that enhance turbulence and air–wall contacts to improve the conditioning of the air. Noback and colleagues recorded three-dimensional coordinates of 21 landmarks on the bony morphology of the nasal cavity of 100 individuals, from five climatic zones (selecting specimens to represent indigenous populations of the area rather than those affected by modern western lifestyles or health care). They examined the relationship between nasal cavity shape and indicators of temperature and aridity, which include monthly observations of temperature and vapor pressure (obtained from KNMI Climate Explorer) from 1901 to 2006 for the geographical locations (or region) of each individual cranium. The data on climate may not represent the climatic conditions over the past thousands of years during which the populations diversified but it is the most comprehensive and detailed database on climate available. The climate variables include (1) mean yearly temperature, (2) coldest monthly temperature, (3) warmest monthly temperature, (4) mean yearly vapor pressure, (5) lowest monthly vapor pressure and (6) highest monthly vapor pressure. The shape data comprise the means for each of the 10 populations.

SA1 of the environmental block is a general climatic factor, with all climatic variables loading positively and nearly equally on that axis, with the two extremes being the coldest and driest versus warmest and most humid. SA2 for the environmental block is a contrast between temperature and vapor pressure (i.e. warm and dry versus cool and humid). SA1 explains 94.3% of the covariation between nasal cavity shape and climate, yielding a correlation of 0.77 between the scores on the paired SA1. SA2 explains only 5.1% of the covariation between shape and climate, yielding a correlation of 0.55 between the two blocks. Although the covariances explained by both SA1 and SA2 are statistically significant based on permutation tests, only SA1 is interpreted because it explains vastly more of the covariance between the blocks. The shape changes associated with climate are found in the nasal aperture, upper nasal cavity and nasopharynx. Populations from cool dry climates differ from those from warm humid climates by relatively higher and narrower nasal aperture, a relatively high and narrow upper nasal cavity and a shortening of the nasopharynx with an increase in the relative length of the posterior cavum length. The authors conclude that the morphology of the bony nasal cavity appears to be associated mainly with temperature and the nasopharynx with humidity, with the changes in shape being consistent with increased contact between air and mucosal tissue in cold, dry climates (increasing turbulence during inspiration and increasing the surface area to volume ratio in the upper nasal cavity). They also recognize that the shape differences are modest and that populations overlap, perhaps suggesting that the shape of the nasal cavity represents a compromise and/or that nasal cavity morphology lacks extreme specializations that would reduce the versatility of a generalist, mobile species.

Environmental Correlates of Geographic Variation in Skull and Mandible Shape of a Rodent, the Punaré Rat (Thrichomys apereoides)

In an investigation of the environmental sources of geographic variation in an echimyid rodent sampled from Ceará, Paraíba, Pernambuco, Alagoas, Bahia, Goiás and Minas Gerais in Brazil, Monteiro and colleagues (2003) examined the relationship between skull and mandibular shape and a collection of environmental variables, including geographic, climatic and biotic environmental variables. In this region, increases in latitude and longitude are associated with increases in altitude and rainfall and decreases in mean annual temperature and human density; moreover, the vegetation changes from the arid caatingas to the cerrado savannas. The environmental block comprises geographic variables (1) latitude and (2), longitude and (3) altitude plus the climatic variables (1) mean temperature and (2) rainfall, and two biotic variables, (1) human population density and (2) vegetation type (a categorical variable). The environmental data were obtained from the Enciclopédia dos Municípios Brasileiros. The skull was measured in three views (dorsal, ventral and lateral) and the mandible was measured in lateral view.

SA1 explained from 69% to 90% of the covariation between shape and the environment. When summarizing the results, we focus on the two views for which the association between geographic variation shape and the environment is statistically significant. For the ventral view, the environmental SA1 represents a contrast between latitude, longitude, altitude, versus vegetation type and human density. With increasing scores (towards the south), snouts shorten and narrow relative to the skull, jugals become relatively shorter and the tympanic bullae and foramen magnum reduce, plus there are more localized changes in basicranial bones. The distribution of the scores for shape SA1 on the environmental SA1 suggests an environmental gradient from north to south, but shape appears to follow that gradient only from north to the most northern of the southern populations. The southern populations show increasing scores on the environmental SA1 but the scores are nearly constant on the shape SA1 (with the exception of one population from Goiás). For the lateral view, the environmental SA1 represents a contrast between latitude, longitude, altitude and rainfall versus temperature, vegetation type and human density. The associated shape changes include a relative shortening of the snout and brain case, a decrease in relative length of the jugal plus a general dorsovental shallowing of the skull. The scores on the paired SA1 suggest two clusters of populations, northern and southern; the northern populations vary little along either the shape or environmental SA1 whereas the southern populations, which have higher scores on the shape SA1, vary along the environmental SA1 but not along shape SA1.

References

Angielczyk, K. D., & Sheets, H. D. (2007). Investigation of simulated tectonic deformation in fossils using geometric morphometrics. *Paleobiology, 33*, 125–148.

Arif, S., Adams, D. C., & Wicknick, J. A. (2007). Bioclimatic modelling, morphology, and behaviour reveal alternative mechanisms regulating the distributions of two parapatric salamander species. *Evolutionary Ecology Research, 9*, 843–854.

Barker, M., & Rayens, W. (2003). Partial least squares for discrimination. *Journal of Chemometrics, 17*, 166–173.

Bastir, M., O'Higgins, P., & Rosas, A. (2007). Facial ontogeny in Neanderthals and modern humans. *Proceedings of the Royal Society B-Biological Sciences, 274*, 1125–1132.

Bastir, M., & Rosas, A. (2004). Comparative ontogeny in humans and chimpanzees: Similarities, differences and paradoxes in postnatal growth and development of the skull. *Annals of Anatomy-Anatomischer Anzeiger, 186*, 503–509.

Bastir, M., & Rosas, A. (2005). Hierarchical nature of morphological integration and modularity in the human posterior face. *American Journal of Physical Anthropology, 128*, 26–34.

Bastir, M., & Rosas, A. (2006). Correlated variation between the lateral basicranium and the face: A geometric morphometric study in different human groups. *Archives of Oral Biology, 51*, 814–824.

Bastir, M., Rosas, A., & Sheets, H. D. (2005). The morphological integration of the hominoid skull: A partial least squares and PC analysis with morphogenetic implications for European mid-pleistocene mandibles. In D. Slice (Ed.), *Modern morphometrics in physical anthropology* (pp. 265–284). New York: Kluwever Academic/Plenum Publishers.

Bastir, M., Sobral, P. G., Kuroe, K., & Rosas, A. (2008). Human craniofacial sphericity: A simultaneous analysis of frontal and lateral cephalograms of a Japanese population using geometric morphometrics and partial least squares analysis. *Archives of Oral Biology, 53*, 295–303.

Bookstein, F. L. (1982). The geometric meaning of soft modeling, with some generalizations. In K. G. Jöreskog, & H. Wold (Eds.), *Systems under indirect observation: causality, structure, prediction* (pp. 55–74). New York: North Holland Publishing Co.

Bookstein, F. L., Gunz, P., & Ingeborg, H., et al. (2003). Cranial integration in Homo: Singular warps analysis of the midsagittal plane in ontogeny and evolution. *Journal of Human Evolution, 44*, 167–187.

Bookstein, F. L., Streissguth, A. P., Sampson, P. D., Connor, P. D., & Barr, H. M. (2002). Corpus callosum shape and neuropsychological deficits in adult males with heavy fetal alcohol exposure. *Neuroimage, 15*, 233–251.

Bowers, C. M. (2006). Problem-based analysis of bitemark misidentifications: The role of DNA. *Forensic Science International, 159S*, S104–109.

Bush, M. A., Bush, P. J., & Sheets, H. D. (2011). A study of multiple bitemarks inflicted in human skin by a single dentition using geometric morphometrics. *Forensic Science International, 211*, 1–8.

Cheverud, J. M., Hartman, S. E., Richtsmeier, J. T., & Atchley, W. R. (1991). A quantitative genetic analysis of localized morphology in mandibles of inbred mice using finite-element scaling analysis. *Journal of Craniofacial Genetics and Developmental Biology, 11*, 122–137.

Escoufier, Y. (1973). Le traitement des variables vectorielles. *Biometrics, 29*, 751–760.

Fadda, C., & Corti, M. (1998). Geographic variation of Arvicanthis (Rodentia, Muridae) in the Nile Valley. *Zeitschrift für Saugetierkunde-International Journal of Mammalian Biology, 63*, 104–113.

Fornell, C., & Bookstein, F. (1982). Two structural equation models: LISREL and PLS applied to consumer exit-voice theory. *Journal of Marketing Research, 19*, 440–452.

Fornell, R., Cordeiro-Estrela, P., & De Freitas, T. R. O. (2010). Skull shape and size variation in Ctenomys minutus (Rodentia: Ctenomyidae) in geographical, chromosomal polymorphism, and environmental contexts. *Biological Journal of the Linnean Society, 101*, 705–720.

Gkantidis, N., & Halazonetis, D. J. (2011). Morphological integration between the cranial base and the face in children and adults. *Journal of Anatomy, 218*, 426–438.

Gottfries, J., Blennow, K., Wallin, A., & Gottfries, C. G. (1995). Diagnosis of dementias using partial least squares discriminant analysis. *Dementia, 6*, 83–88.

Houle, D., Mezey, J., & Galpern, P. (2002). Interpretation of the results of common principal components analysis. *Evolution, 56*, 433–440.

Jöreskog, K. G., & Wold, H. (1982). *Systems under direct observation: Causality–structure–prediction.* North Holland Publishing Co.

Kemsley, E. K. (1996). Discriminant analysis of high-dimensional data: A comparison of principal components analysis and partial least squares data reduction methods. *Chemometrics and Intelligent Laboratory Systems, 33*, 47–61.

Klingenberg, C. P. (2009). Morphometric integration and modularity in configurations of landmarks: Tools for evaluating a priori hypotheses. *Evolution & Development, 11*, 405–421.

Klingenberg, C. P., Mebus, K., & Auffray, J. C. (2003). Developmental integration in a complex morphological structure: How distinct are the modules in the mouse mandible? *Evolution & Development, 5*, 522–531.

Laffont, R., Renvoise, E., Navarro, N., Alibert, P., & Montuire, S. (2009). Morphological modularity and assessment of developmental processes within the vole dental row (Microtus arvalis, Arvicolinae, Rodentia). *Evolution & Development, 11*, 302–311.

Lowe, A. A., Özbeck, M. M., Miyamoto, K., & Fleetham, J. A. (1997). Cephalometric and demographic characteristics of obstructive sleep apnea: An evaluation with partial least squares analysis. *The Angle Orthodontist*, *67*, 143–154.

Lundrigan, B. (1996). Morphology of horns and fighting behavior in the family Bovidae. *Journal of Mammalogy*, *77*, 462–475.

McGuire, J. L. (2010). Geometric morphometrics of vole (Microtus californicus) dentition as a new paleoclimate proxy: Shape change along geographic and climatic clines. *Quaternary International*, *212*, 198–205.

Mevic, B. -H., & Wehrens, R. (2007). The pls package: Principal component and partial least squares regression in R. *Journal of Statistical Software*, *18*, 1–23.

Mezey, J. G., Cheverud, J. M., & Wagner, G. P. (2000). Is the genotype-phenotype map modular?: A statistical approach using mouse quantitative trait loci data. *Genetics*, *156*, 305–311.

Mitteroecker, P., & Bookstein, F. (2007). The conceptual and statistical relationship between modularity and morphological integration. *Systematic Biology*, *56*, 818–836.

Mitteroecker, P., & Bookstein, F. (2008). The evolutionary role of modularity and integration in the hominoid cranium. *Evolution*, *62*, 943–958.

Mitteroecker, P., & Bookstein, F. (2011). Linear discrimination, ordination, and the visualization of selection gradients in modern morphometrics. *Evolutionary Biology*, *38*, 100–114.

Monteiro, L. R., Duarte, L. C., & dos Reis, S. F. (2003). Environmental correlates of geographical variation in skull and mandible shape of the punare rat *Thrichomys apereoides* (Rodentia: Echimyidae). *Journal of Zoology*, *261*, 47–57.

Noback, M. L., Harvati, K., & Spoor, F. (2011). Climate-related variation of the human nasal cavity. *American Journal of Physical Anthropology*, *145*, 599–614.

Ortiz, M., Sarabia, L., Symington, C., Santamaria, F., & Iniguez, M. (1996). Analysis of ageing and typification of vintage ports by partial least squares and soft independent modeling class analogy. *Analyst*, 1009–1013.

Pretty, I. A. (2006). The barriers to achieving an evidence base for bitemark analysis. *Forensic Science International, Suppl. 1*, S110–S120.

Pretty, I. A., & Sweet, D. (2001). The scientific basis for human bitemark analyses: A critical review. *Science and Justice*, *41*, 85–92.

Pulcini, D., Costa, C., Aguzzi, J., & Cataudella, S. (2008). Light and shape: A contribution to demonstrate morphological differences in diurnal and nocturnal teleosts. *Journal of Morphology*, *269*, 375–385.

Rohlf, F. J., & Corti, M. (2000). Use of two-block partial least-squares to study covariation in shape. *Systematic Biology*, *49*, 740–753.

Rothwell, B. R. (1995). Bitemarks in forensic dentistry: A review of legal, scientific issues. *Journal of the American Dental Association*, *126*, 223–232.

Ruber, L., & Adams, D. C. (2001). Evolutionary convergence of body shape and trophic morphology in cichlids from Lake Tanganyika. *Journal of Evolutionary Biology*, *14*, 325–332.

Sampson, P. D., Streissguth, A. P., Barr, H. M., & Bookstein, F. L. (1989). Neurobehavioral effects of prenatal alcohol: Part II. Partial least squares analysis. *Neurotoxicology and Teratology*, *11*, 477–491.

Sheets, H. D., Bush, P. J., & Bush, M. A. Patterns of variation and match rates of the anterior biting dentition: Characteristics of a database of 3D scanned dentition. *Journal of Forensic Science* (in press).

Sheets, H. D., Covino, K. M., Panasiewicz, J. M., & Morris, S. R. (2006). Comparison of geometric morphometric outline methods in the discrimination of age-related differences in feather shape. *Frontiers in Zoology*, *3*, 15.

Streissguth, A. P., Bookstein, F. L., Sampson, P. D., & Barr, H. M. (1993). *The enduring effects of prenatal alcohol exposure on child development: Birth through seven years, a partial least squares solution.* Ann Arbor, MI: University of Michigan Press.

Committee on Identifying the Needs of the Forensic Sciences Community and Commissioned by the National Research Council. Strengthening Forensic Science in the United States: A Path Forward. 2009. The National Academies Press http://www.nap.edu/openbook.php?record_id=12589.

Wold, H. (1966). Estimation of principal components and related models by iterative least squares. In P. R. Krishnaiaah (Ed.), *Multivariate analysis* (pp. 391–420). New York: Academic Press.

Wold, S., Sjöström, M., & Erickson, L. (2001). PLS-regression: A basic tool of chemometrics. *Chemometrics and Intelligent Laboratory Systems*, *58*, 109–130.

CHAPTER

8

Statistics

Organisms vary for reasons beyond our control and often beyond our understanding. Variation is of obvious biological importance because evolution could not occur without it, but variation is also a source of frustration for biologists, as evident in what has been termed the Harvard Law of Biology: "under the most carefully controlled conditions, biological material does whatever it damn well pleases" (quoted by Ellen Larsen [2005, p. 115], in a book entirely devoted to the subject of variation [Hallgrímsson and Hall, 2005]). Because organisms, even when reared under carefully controlled conditions, vary in the outcome of development, we cannot assume that all those outcomes are due to the treatment that we applied experimentally. The problem of interpreting experimental outcomes is obviously much more difficult when nature did the experimenting, not us. Given that there will always be variation that we cannot explain, we cannot safely ascribe all experimental outcomes, whether the experiments are controlled or natural, to the treatments. This inexplicable variation is the "error" term in statistical analyses – any variation that we *can* explain is not error so long as the factor explaining it is included in our statistical model.

Variation further complicates drawing inferences about the experimental results because we rarely, if ever, measure every single individual in the population of interest. Almost always we instead draw a sample from that population and hope to infer something about the population from the sample. For example, if our experiment is run in the laboratory, we are rarely asking questions about the response of our particular laboratory population to the specific treatment that we applied. Should we give some mice liquid diets and others regular laboratory pellets and measure their jaws to see the impact of dietary consistency on their jaws, we are not asking whether these particular groups of mice differ. Rather, we want to know if mice, more generally, will differ in their jaw morphologies because of differences in dietary consistency. Similarly, when we analyze natural populations, we are rarely interested solely in the specific organisms that we measure, we want to generalize from those samples to the population as a whole. For example, we do not ask whether adult chipmunks (*Tamias alpinus*), collected between 1911 and 1919, whose skulls are contained in the mammal collection of the Museum of Vertebrate Zoology, vary in jaw shape because they vary in size, or if these particular chipmunks are

Geometric Morphometrics for Biologists
DOI: http://dx.doi.org/10.1016/B978-0-12-386903-6.00008-3

sexually dimorphic in jaw shape. Instead, we want to generalize from that sample to *T. alpinus*. That concern for generality motivates statistical analysis because we would not need statistical tests if we cared only about the particular organisms that we've observed. We could easily determine if their means differ in whatever variables interest us – we'd just measure those individuals, calculate the mean of each sample and look at the numbers. It is precisely because we want to make inferences about the populations from which the samples were drawn that we need statistical methods of inference.

This chapter presents an introduction to formulating and testing hypotheses. We will focus on two simple hypotheses: (1) adult *T. alpinus* vary in shape because they vary in size; (2) adult *T. alpinus* are sexually dimorphic in shape. In the next chapter, we will test the more complex hypothesis that adult *T. alpinus* are sexually dimorphic in shape, controlling for size, i.e. they differ in shape when compared at the same size as well as other complex hypotheses. In this chapter, we also restrict ourselves to balanced designs, meaning that our sample sizes are equal in all groups which, in the case of an analysis of sexual dimorphism, a balanced design means that we have equal numbers of males and females. In the next chapter, we consider unbalanced designs, i.e. the case in which groups differ in sample sizes.

In general outline, the first step in any statistical analysis is to turn the biological hypothesis into a formal statistical model. Then the coefficients of that model are estimated, and the model is tested for its statistical significance. To explain these steps, we will consider our first example, the biological hypothesis that chipmunk jaw shapes vary because of variation in size. An important distinction between that biological hypothesis and our statistical model is that our mathematical model says nothing about causality. Instead, the model says that we can *predict* one variable (shape) from another (size). Based on the good fit of our model to the data, we might conclude that size predicts shape and, in light of that, we might be tempted to conclude that size *explains* shape. However, even if the model fits well, size might not be a cause of shape for at least two reasons. First, size is not a process. In the context of developmental biology, we can explain size in terms of the proliferation of cells that add tissue to a structure. Because growth rates vary over the organism, cell proliferation (in conjunction with cell death, cell differentiation, deposition of an extracellular matrix, etc.) produces changes in shape. In this context, saying that size "explains" shape does not mean that size itself causes shape; rather, it means that we are using "size" as shorthand for all those developmental processes that jointly alter size and shape. Second, even if size predicts shape, we cannot infer that it actually causes shape because we have not manipulated size and determined that those manipulations affect shape. If the model fits the data, what we have demonstrated is that the relationship between size and shape is predicted by a particular mathematical model.

We begin with the formulation of the model for the simple bivariate case in which we have one *dependent variable* (Y) and one *independent variable* (X), each of which is measured on N individuals. The model is the equation of a straight line, hence the term "linear regression". We are fitting the equation of a straight line to the data to find the coefficients that best predict shape from values of the independent variable (e.g. size). More specifically, we are trying to find the best estimates of the coefficients m and b of the equation:

$$Y_i = mX_i + b + \varepsilon_i \tag{8.1}$$

where Y_i is the dependent variable measured for the ith specimen, m is the slope of the line, b is the Y-intercept of the line, and ε_i is "error" (the variation in Y not explained by X). Our objective is to estimate m and b and then to determine whether they are statistically different from zero. This is the model for any hypothesis in which the predictor is a continuous variable. Size exemplifies a continuous variable because there is always a size between any two others. In the case of categorical factors, such as the other factor that we will consider (sex), we cannot find values between any two of them. There are no values between "male" and "female".

When the assumption of linearity holds, our statistical analysis can tell us if Y is only weakly dependent on X – meaning that knowledge about X does not enable us to predict Y. It is also possible that the relationship of the two variables is statistically significant, but that m is such a small number that the effect of X on Y is biologically trivial. It may be a *statistically significant* relationship, in that it is stronger than expected by chance, but it might not be *biologically significant*. Recognizing this distinction is important, because statistical significance is a matter of sample size and the power of a test. With very large samples, or very powerful tests, we might have little difficulty rejecting the null hypothesis. However, if X accounts for very little of the variation in Y, X provides little biological insight into Y. We therefore need to pay as much attention to the explanatory power of X and to the magnitude of its impact on Y as to the statistical results. The fraction of the variance in Y explained by X (and the model) provides the needed information about explanatory power.

As mentioned above, *when the assumption of linearity holds*, our statistical analysis can tell us whether we can predict Y from X. The reason for emphasizing this assumption is that a strong but non-linear relationship might look like a weak linear one. Consequently, we might end up rejecting our biological model because the statistical analysis suggests a weak relationship between variables, but the relation is actually strong but not linear. Fortunately, in some cases of a non-linear relationship between the variables, it is easy to transform the independent variable to make the relationship linear. For example, a number of studies of ontogenetic allometry use the logarithm of centroid size, rather than centroid size itself, as the independent variable. That transformation is useful when most of the shape change occurs over small values of X, such as when most shape change occurs early in ontogeny (as it often does). We should note that it does not matter whether the logarithm is taken to base 10 (log) or base e (ln) because these differ only by a constant, i.e. $\log(X) = \log(e)\ \ln(X) = 0.4329\ \ln(X)$. In other cases, other transformations of X (such as other trigonometric functions) might do a better job of linearizing the relationship between variables.

In a moment, we will present the equations that provide the best estimates of m and b, but to explain why they are considered "best" we first need to consider how that decision could be made, in general. The standard approach for deriving the best estimator is to choose an *error function*. By minimizing that error, we find the optimal values for the parameters. A least squares analysis, as the term suggests, uses the sum of squared residuals as the error function, so that is the function minimized. We then express the relationship between that error term and the regression model:

$$\sum_{i=1}^{N} \varepsilon_i^2 = \sum_{i=1}^{N} (y_i - mx_i - b)^2 \tag{8.2}$$

where $x_i = X_i - <X>$ (the difference between an observed value of X_i and its expected value $<X>$, which is the sample mean) and $y_i = Y_i - <Y>$ (the difference between an observed value of Y_i and its expected value $<Y>$), x_i and y_i are called *centered* versions of the original variables. Thus, we are summing residuals, or deviations from expected values, over all N individuals in a population. By minimizing this function, we will obtain the best estimates for m and b.

To find the values of m and b that minimize the sum of squared residuals, we set the derivative to zero (for both m and b). As you recall from calculus, the derivative of a function is zero at the maximum and minimum. We then solve for m and b. Using this optimization method, the equation for the slope, m, can be written as:

$$m = \frac{\sum xy}{\sum x^2} \tag{8.3}$$

which is the sum of the products of the deviations divided by the sum of the squared deviations of the X values (each sum is taken over all individuals). In other words, the slope is the ratio of the deviations of Y to the corresponding deviations of X. When the corresponding deviations are identical, the slope is one; when the deviations of Y are a consistent multiple of the deviations of X, the slope will be that multiple.

Substituting the $X_i - <X>$ for x_i and $Y_i - <Y>$ for y_i allows us to compute m directly from the observed values. The sum of the products can be written as:

$$\sum xy = \sum (X_i - <X>)(Y_i - <Y>) \tag{8.4}$$

which can be simplified to:

$$N\sum X_i Y_i - \sum X_i \sum Y_i \tag{8.5}$$

After applying a similar substitution and simplification to the sum of the squared deviations, we can write:

$$m = \frac{\left(N\sum_{i=1}^{N} X_i Y_i\right) - \left(\sum_{i=1}^{N} X_i \sum_{i=1}^{N} Y_i\right)}{\left(N\sum_{i=1}^{N} X_i^2\right) - \left(\sum_{i=1}^{N} X_i\right)^2} \tag{8.6}$$

Now that we have an expression for the slope, we can solve for the intercept, b, and complete the equation for the regression. When $b = 0$, $<Y> = m<X>$, so we can calculate b from the observed values, X_i and Y_i, and the sample size, N:

$$b = <Y> - m<X> = \frac{\sum_{i=1}^{N} Y_i - m\sum_{i=1}^{N} X_i}{N} \tag{8.7}$$

In addition to an estimate of the value of m, we will also need measures of the uncertainty of that estimate. These measures will be used to test whether m is significantly different from zero (because if we cannot say that, we cannot claim that Y depends on X), and to test whether the value of m differs between samples (whether the relationship between X and Y is different).

Before we derive the measures of uncertainty, it will be useful to introduce some shorthand notation. The sums of squares of the deviations x_i and y_i will be:

$$s_{xx} = \sum_{i=1}^{N} x_i^2 \tag{8.8}$$

and

$$s_{yy} = \sum_{i=1}^{N} y_i^2 \tag{8.9}$$

Similarly, the sum of the products of the deviations will be:

$$s_{xy} = \sum_{i=1}^{N} x_i y_i \tag{8.10}$$

In testing whether the regression is significant, it is important to keep in mind that we are asking whether the relationship between X and Y explains a significant proportion of the variance in Y. If we knew the values of the error terms, ε_i, we could compute their variance and use those estimates to determine the proportion of variance in Y that is explained by the regression of Y on X. More often than not, ε_i are unknown, so we need a different approach.

What we can do is to compute an F-ratio from the information that we have. F is a ratio of variances (or mean squared deviations) that are sums of squared deviations divided by the appropriate degrees of freedom for the terms in the model. The degrees of freedom of the model are simply equal to the number of estimated or fitted parameters in the model. The ratio of the sum of squared deviations explained by the regression is S^2_{XY}/S_{XX}. This has one degree of freedom, so the proportion of the variance explained is also S^2_{XY}/S_{XX}. Recall that the slope is s_{XY}/s_{XX}, so the explained variance can also be written as $m \cdot s_{XY}$. The unexplained or residual sum of squared deviations is $s_{YY} - m \cdot s_{XY}$, which has $N-2$ degrees of freedom, so the unexplained variance is $(s_{YY} - m \cdot s_{XY})/(N-2)$. F is the explained variance divided by the unexplained, so F is $(N-2)m \cdot s_{XY}/(s_{YY} - m \cdot s_{XY})$ with 1 and $N-2$ degrees of freedom. The corresponding p-value indicates the likelihood that such a high F is due to chance, meaning that such a large proportion of the variance in Y explained by the regression of Y on X is due to chance.

THE CORRELATION COEFFICIENT

The correlation coefficient (r), which ranges from minus one to one, expresses the strength of the linear relationship between X and Y. Its squared value (r^2), which ranges from zero to one, indicates the fraction of the variance in Y that is explained by X. The expression for r^2 is:

$$r^2 = \frac{s^2_{XY}}{s_{XX} s_{YY}} \tag{8.11}$$

It is common to regard high r^2 values as indicating high explanatory power of the model. However, even high values of r^2 need not be statistically significantly greater than zero. For that reason we need to test the statistical significance of r^2, which we can do (assuming normality of the residuals) using the expression:

$$z = \frac{1}{2}\ln\left[\frac{(1+r)}{(1-r)}\right] \tag{8.12}$$

which is a normally distributed variable, with variance equal to $1/(N-3)$, where N is the sample size (see the derivation in Freund and Walpole, 1980), a calculation that assumes that the residuals are independent and normally distributed. So, based on an analytic model of the distribution of r values, we can test whether or not the variance explained by the model is larger than we expected by chance.

The other approach to testing the significance of an observed r value is to use a permutation test of the significance of the regression, an approach which dates back to Fisher (1935). The null hypothesis we would like to disprove can be stated as:

> $\mathbf{H_0}$: The variance explained by this model for this particular data set is no greater than might occur by chance, meaning that there is no association between the X and Y values that differs from what we might expect to occur randomly.

This hypothesis contains a statement about the *exchangeability* (Anderson, 2001b) of the X and Y variables in our data set, namely, that the relationship between X and Y is exchangeable. That is because, if the null hypothesis ($\mathbf{H_0}$) is true, if we randomly shuffled the X_i and Y_i values to create new pairings, permuting the original data, we would expect the model to fit the permuted data as well as it fits the original data. Because the relationship between X and Y is exchangeable under $\mathbf{H_0}$, if $\mathbf{H_0}$ is true, the model should have the same predictive power for the permuted data as it did for the original data. That allows us to state a basis for rejecting the null hypothesis: if we form a large number of permuted data sets, we can determine how many of them have as large an r value as the original data set did. If only 3% of the permuted data sets have as large an r value as the original data set does, we can use this observed 3% rate to claim that there is only a 3% chance that the observed r values could have arisen from a randomly permuted set of data. Permutation methods are discussed in more detail later (see the Appendix of this chapter and the discussion of permutations in the next chapter). But it is important to note that the permutation method used here does assume that the residuals are independent of one another, just as the analytic model did. The permutation assumes that the residuals also came from the same distribution, but does not require that the distribution be normal, a difference from the analytic model discussed earlier.

MULTIVARIATE REGRESSION

To apply this theory to shape we need to extend it to the multivariate case. Our dependent variable, for the case of two-dimensional data consisting solely of landmarks, is a vector with $2K-4$ components. That number will need to be adjusted for three-dimensional data and for data consisting of landmarks plus semilandmarks, but for the remainder of this discussion we will mention only the two-dimensional landmark case. The adjustments are straightforward except that, in the case of landmark-only data, the dimensionality of the data equals the number of partial warps (including the two uniform components), which will not be the case for data that include semilandmarks as well as landmarks. That distinction, however, is not important when using permutation or bootstrap methods based on Procrustes distances, as we will see later. Regression models can be framed in terms of partial warp scores or principal component scores or coordinates of landmarks because all the mathematics involved is linear so a rotation of the data will not alter the answers, so long as the mathematics is done correctly.

To regress shape on an independent (scalar) variable, we regress the shape data on the independent variable. For example, suppose we have P partial warp and uniform components, which we can write as a row vector $\{Y_1, Y_2, Y_3, \ldots Y_P\}$. Then the (linear) model for the regression of that vector on a scalar (X) is:

$$\{Y_1, Y_2, Y_3, \ldots Y_P\} = \{m_1, m_2, m_3, \ldots m_P\}X + \{b_1, b_2, b_3, \ldots b_P\} + \{\varepsilon_1, \varepsilon_2, \varepsilon_3, \ldots \varepsilon_P\} \tag{8.13}$$

where $\{m_1, m_2, m_3, \ldots m_P\}$, $\{b_1, b_2, b_3, \ldots b_P\}$ and $\{\varepsilon_1, \varepsilon_2, \varepsilon_3, \ldots \varepsilon_P\}$ are vectors of slope and intercept coefficients and residuals, respectively. Although this expression looks far more complicated than the one for a bivariate regression, it actually is not. In fact, we can determine the ith component of the slope and intercept terms using the same m_i and b_i values that minimize the residuals in the corresponding bivariate model. Each observation Y is now a vector, as are the slope, intercept and each of the errors.

Estimating slope and intercept coefficients is no more complex in the multivariate case than it was in the bivariate case. But in one important respect, the analysis actually is more complex – checking the assumption of linearity. There are at least two ways to check this assumption for multivariate data, although neither is ideal. One is to look at the relationship between each individual component of shape and the independent variable, such as by regressing each partial warp on size. If one or more exhibits a strong and highly non-linear relationship, such as shown in Figure 8.1A, then it is unlikely that shape and size are linearly related. This method for checking linearity is not ideal because it falls back on inspecting multiple bivariate regressions when it is multivariate linearity that really matters. Another approach is to estimate the Procrustes distance between each specimen and the shape at the lowest value on the independent variable. Regressing that distance on the independent variable may show if *that* relationship is non-linear (as in Figure 8.1B). If it is not, it is unlikely that shape and size are linearly related. This method is again not ideal, because the Procrustes distance measures only the magnitude of the difference between each specimen and the reference, not its direction. Two specimens that differ a great deal from each other in shape may be equally distant from the reference. Despite the deficiencies of these two less than ideal methods, we can use them to check whether it is unlikely that shape is linearly related to size. The results shown in Figure 8.1 both indicate a non-linear relationship of shape and centroid size, and both suggest that shape might be linearly related to the log of centroid size. That linear relationship to the log of centroid size is suggested by the shape of the curves because they depict a very rapid change in shape relative to size over the smaller values of size. So we can try a log

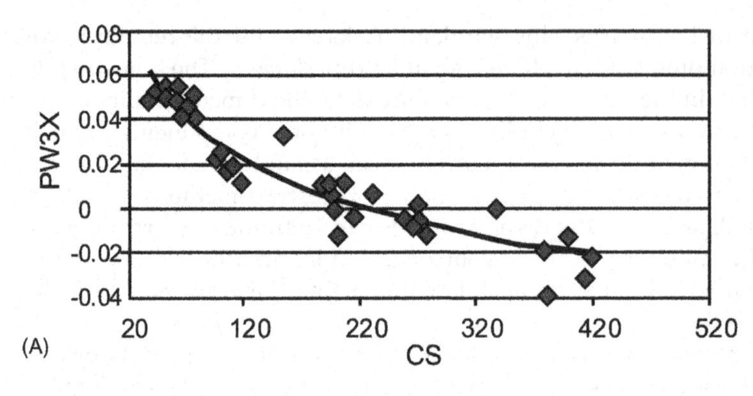

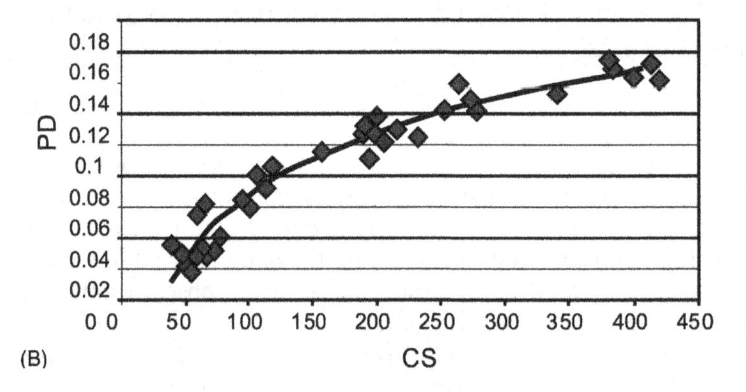

FIGURE 8.1 Checking the assumption of a linear relationship between shape and the independent variable: (A) using a single variable plotted on centroid size; (B) using the Procrustes distance of each specimen from the shape having the smallest size, plotted on centroid size.

transform of centroid size, repeating the two analyses to check for linearity (Figure 8.2). Both plots now suggest a nearly linear relationship between shape and log centroid size. Thus, we would use log centroid size as our independent variable.

To this point, we have talked about the assumption of linearity as that is usually stated in context of bivariate regression. However, multivariate studies make another assumption of linearity, which is that all components of the dependent variable be linearly related to the independent variable. In other words, we are assuming that all the components of shape are linearly related to each other, as we have assumed they are all linearly related to the independent variable. This assumption will not hold if some components of shape are linearly related to the independent variable but some others are non-linearly related to it. The components of shape cannot be linearly related to each other if different ones fit differently shaped curves. Because this departure from the assumption of non-linearity is specific to multivariate data, it may not be intuitively obvious what the assumption means. What it means is that the slope of the relationship between shape and the independent variable is constant – the values $\{m_1, m_2, m_3, \ldots m_P\}$ are not functions of the independent variable.

In some cases, such as in studies of ontogeny, the shape variable correlated with age may change from age to age. If that is the case, we cannot model the ontogeny of shape by a single vector of slope coefficients because that vector would change with time. The ontogenetic trajectory of shape is then a curving path in shape space, not a straight line. The assumption of multivariate linearity can be checked in two ways although, again, neither

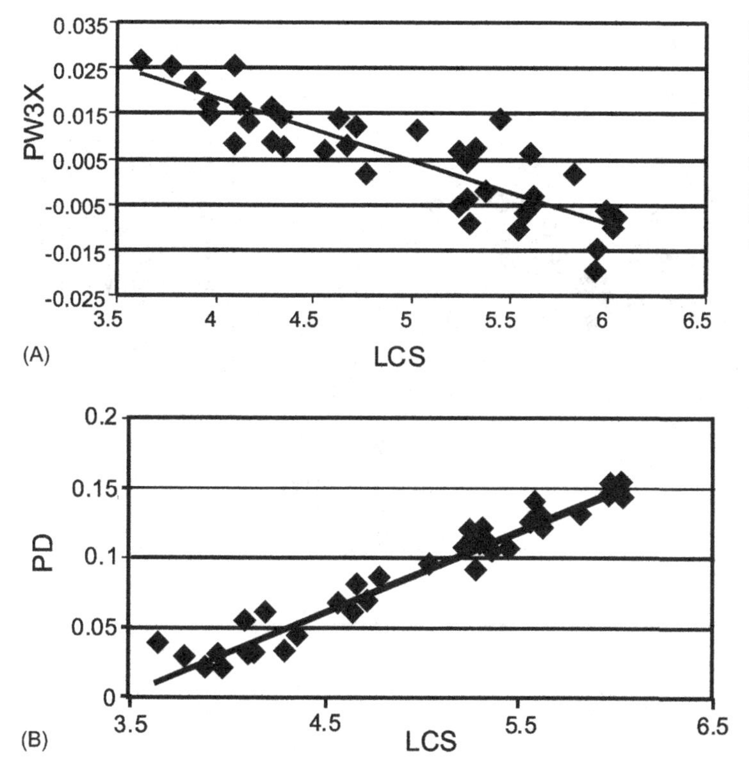

FIGURE 8.2 Checking the assumption of a linear relationship between shape and the independent variable: (A) using a single variable plotted on ln centroid size; (B) using the Procrustes distance of each specimen from the shape having the smallest size, plotted on ln centroid size.

method is ideal. One is to conduct a principal components analysis (PCA) of the data, and check for a statistical relationship between multiple PCs and the independent variable (Figure 8.3). In the example shown in Figure 8.3A, there is a substantial deviation from linearity – not only is PC1 correlated with age (which is expected) but PC2 and PC3 also are, with PC2 and PC3 describing the deviations from the linear trend represented by PC1. The assumption can also be checked by regressing several shape variables on each other (Figure 8.3B) because, if the relationship among these variables is non-linear, we must reject the assumption of multivariate linearity.

When shape data violate the assumption of multivariate linearity, there is no easy way to transform them. They are not individual variables that can be individually transformed because all of them, taken together, represent a single variable – shape. If we log transform some of the components, we thereby alter the meaning of "shape". Also, whenever the dependent variable is transformed, the error structure of the data is also affected. That is not the case when the independent variable is transformed, because that variable is presumed to be measured without error. The non-linear dynamics of the shape variable are not just a nuisance, they are biologically interesting but they do complicate statistical analyses.

Presuming that the assumption of linearity actually is met, we can go forward with the analysis and test the hypothesis that our independent variable predicts shape. The classic analytic approach represents the variance (which includes total variance, the variance explained by the model and the residual) by variance–covariance matrices instead of

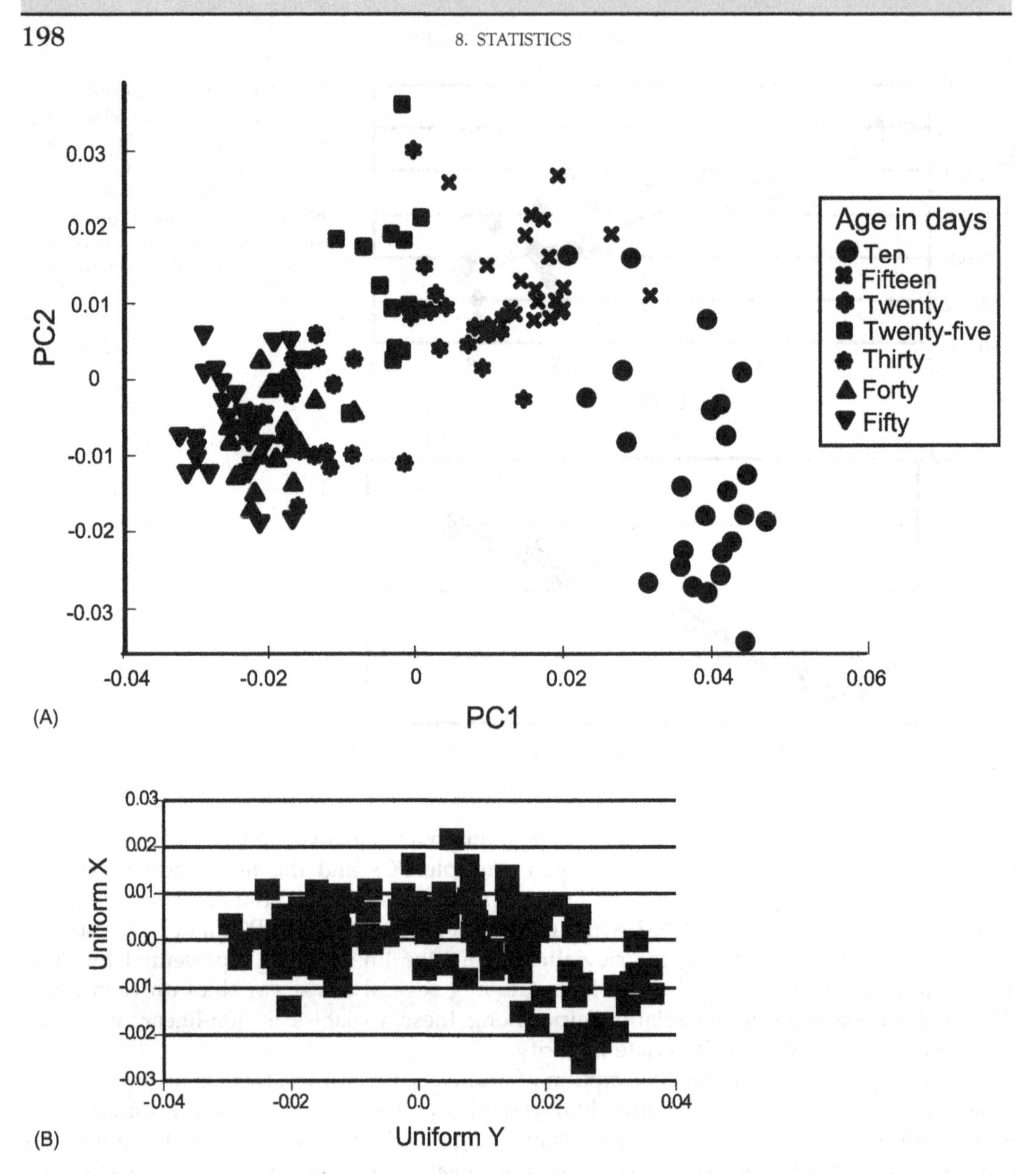

FIGURE 8.3 Checking the assumption of multivariate linearity of the dependent variable: (A) using principal components analysis; (B) using two shape variables on each other (the two uniform components).

scalar values. Because we are dealing with a multivariate system, the test for the significance of the regression is different from the bivariate case. The classical, analytic multivariate statistical approach uses Wilks' Lambda, Λ, which is:

$$\Lambda = \frac{det(\Sigma_P)}{det(\Sigma)} \tag{8.14}$$

where Σ_P is the variance–covariance matrix of the predicted values of Y at a value of X in the data set, *det* is the determinant of the matrix, and Σ is the variance–covariance matrix for the original set of variables (e.g. partial warp scores or principal component scores or any other complete set of scores for our data). Wilk's lambda can also be computed as a function of the eigenvalues of the inverse of the variance–covariance matrix of the residual $(\Sigma_R)^{-1}$ times the variance–covariance matrix of the predicted values (Σ_P). Approximations are available to convert the Λ value into an F-statistic or a χ^2 value. Several other conventional multivariate test criteria can also be used, including Roy's maximum root test, Pillai's trace, Hotelling–Lawley trace) (Rencher, 1995), all of which give the same results when there is only one independent variable and the sample size is large. When the sample size is small (relative to the number of landmarks), the authors' experience is that Wilk's Λ can substantially overestimate the variance explained by the regression model, perhaps due to difficulties associated with estimating variance–covariance matrices at small sample sizes. The other analytic tests could be expected to share this behavior. It is not clear what sample size must be used to obtain consistent results but the general rule of thumb is that there should be five times as many observations as estimated parameters. The number of parameters rises rapidly for landmark data, especially for three-dimensional landmark data, and even more so for semilandmark data, making sample size a substantial concern. For that reason, results of these analytic tests should be viewed with caution unless sample sizes are large relative to the number of parameters.

DISTANCE-BASED METHODS OF HYPOTHESIS TESTING

There is an alternative approach to testing statistical hypotheses, which uses variances expressed in units of Procrustes distance rather than variance–covariance matrices of the partial warp scores or other variables (Goodall, 1991). By this approach, we use the Procrustes distance between each individual's observed shape and its expected value given that individual's value on the independent variable. The summed squared Procrustes distances give a measure of the variance in shape that is *not* explained by X. Thus, it is a measure of the residual, i.e. the variance not explained by the regression, because the distances being squared and summed are the deviations *from* the regression, hence they are not explained by the model. This distance-based approach has the advantage of expressing deviations in terms of the familiar (and meaningful) units of Procrustes distance and it also has the large advantage of expressing variance as a simple squared Procrustes distance – a scalar rather than a matrix. Using distance metrics in statistics has proven quite useful in other contexts as well (e.g. Anderson, 2001a,b).

The generalized form of this test is an F-ratio of the variance explained by the regression model relative to that not explained by the regression model, in which the variance is expressed as a summed square Procrustes distance. Goodall's (1991) original derivation (to be discussed later), was an F-test for the difference in the means of the two groups relative to the variance within each one. The test does make restrictive assumptions about the variance at each landmark, specifically, that it is normally, independently and identically distributed at each landmark. Rather than use a test that depends on this restrictive model, we can instead use permutation tests based on the concept of exchangeability of the independent variable, as discussed above in the context of a simple bivariate regression. Under the null hypothesis that the dependent variable is uncorrelated with the independent variable, the significance of the observed *F*-ratio (the mean summed squared Procrustes distance explained by the model relative to the mean square residual expressed in the same units) is tested by comparing that observed value to the distribution of *F*-ratios produced via random permutation (discussed in greater detail, below). If the observed *F*-ratio is extreme relative to that obtained from the permuted values (at some desired alpha level), then the null hypothesis (i.e. that *X* is not a linear predictor of *Y*) may be rejected at that alpha level.

Examples: Testing the Null Hypothesis that *X* is Not a Linear Predictor of Y

We test the null hypothesis that *X* is not a linear predictor of *Y*, using centroid size as our *X* variable, for a sample comprising an ontogenetic series of *Serrasalmus gouldingi*, a data set measured at 16 landmarks (the case shown above [see Figures 8.1, 8.2] when we checked the assumption of linearity). The null hypothesis is thus that shape is unrelated to size over ontogeny, i.e. that growth is isometric. If we can reject that null hypothesis, we can say that shape is a function of size. We also test this same null hypothesis for a second case, a sample of adult alpine chipmunks, *T. alpinus*, captured between 1911 and 1919 from an elevational transect through the Sierra Nevada mountains near Yosemite. This second example differs from the first in that the sample comprises solely adults and also because the data comprise 85 semilandmarks as well as 15 landmarks.

In the analysis of *S. gouldingi*, a regression of the full set of partial warps on the natural log of centroid size yields a value of Wilk's Λ of 0.006785 corresponding to an *F*-statistic of 47.05 with 28 and 9 degrees of freedom ($P = 4.46 \times 10^{-7}$). Thus, it is highly improbable that the null hypothesis is true; we would therefore reject it in favor of the alternative, which is that shape is allometric, meaning that it changes as a function of size. However, the sample size is fairly small ($N = 38$) and, as noted above, there is good reason to worry about the reliability of the Wilk's Λ statistic when the sample size is small. We could instead use Goodall's *F*-test, which determines the proportion of the shape variation that is not predicted by size, summing the squared Procrustes distances between the observed and expected shape for each individual, given its size. From that sum, we conclude that 27.66% of the shape variance is *not* explained by the regression. Thus, 100% − 27.66% = 72.34% of the shape variance *is* explained by size. We obtain an *F*-ratio of 94.199, with 28 and 1008 degrees of freedom (for numerator and denominator, respectively). From the tabulated values of *F*, we would conclude that the probability of obtaining such an extreme value when the null hypothesis is true is less than 0.00001. Considering the dubious assumptions of this test, we can use permutations to determine the significance of the *F*-ratio and, not surprisingly, only 0.1% of the *F*-ratios obtained by permutations are as large as the one obtained from the data. Thus, we can reject the null hypothesis that shape is unrelated to size over ontogeny of this species. The magnitude of this effect can be appreciated visually by the depiction of the regression as a deformation (Figure 8.4). That magnitude can be quantified by the Procrustes distance between the shapes at the lowest and highest values of the independent variable, which is 0.21.

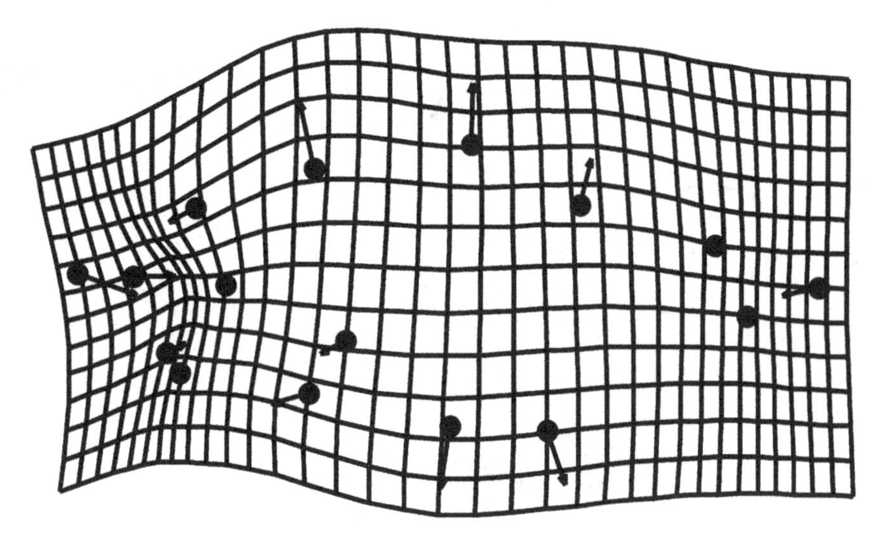

FIGURE 8.4 The ontogenetic allometry of *Serrasalmus gouldingi*, depicted as a deformation.

In our second example, the null hypothesis is less obviously dubious because all individuals are adults and they are nearly the same size; centroid size in this sample ranges from 69.54 to 76.24, with a coefficient of variation of just 2.1%. We first test the null hypothesis that there is no relationship between size and shape for the 15 landmarks, which makes it possible to use the multivariate tests (e.g. Wilks' Λ) as well as Goodall's F. We then test the null hypothesis of no relationship between size and shape using the full data set of 15 landmarks and 85 semilandmarks, which cannot be tested multivariately because the number of coordinates (200) vastly exceeds our sample size (104). For the data restricted to the 15 landmarks, we obtain a value for Wilks' Λ of 0.5567, which yields an approximate F-ratio of 2.38, with 26 and 77 degrees of freedom (for numerator and denominator, respectively). The probability of obtaining an F-ratio this large, with those degrees of freedom, is 0.002. Using Goodall's F-test, the percent unexplained by the regression is 97.61, and the F-ratio is 2.49, with 26 and 2652 degrees of freedom; the probability of obtaining an F-ratio this large, with these degrees of freedom, is less than 0.005. Because the assumptions are so dubious, we use permutations to determine the significance of F. In this case, 0.2% of the permuted values exceed the observed one, so we again conclude that shape is allometric rather than isometric. The effect, as shown in Figure 8.5A, is subtle so we exaggerate it 10-fold to make it visible. Given that these 15 landmarks provide so little information about mandibular shape, especially the complex curvature of the jaw, we redo this analysis including the 85 semilandmarks. Using Goodall's F-test, we determine that size does not explain 96.67% of the variation in shape and thus that it does explain 3.33% of the variation. Goodall's F for this case is 3.51, with 196 and 19992 degrees of freedom; the probability of obtaining an F-ratio this large, with these degrees of freedom, is less than 0.005. Using permutations instead to determine the significance of F, we find that 0.1% of the values for F obtained by permutation equal or exceed the observed one, thus we again conclude that size has a statistically significant, if relatively small, impact on

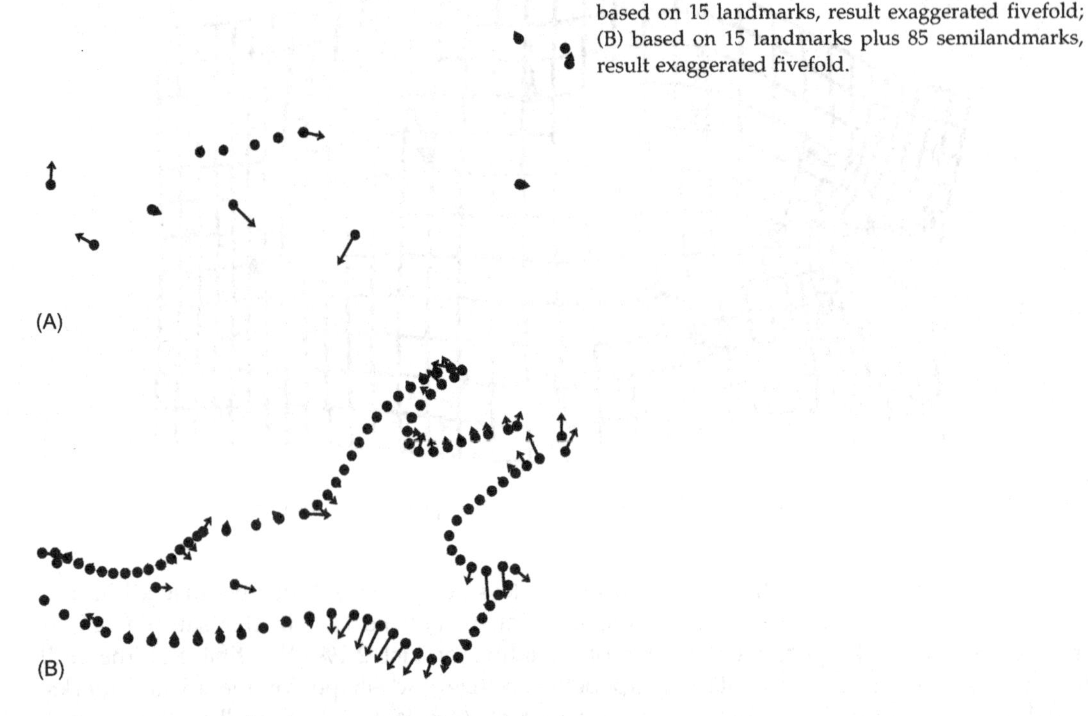

FIGURE 8.5 Allometry of *Tamias alpinus* (A) based on 15 landmarks, result exaggerated fivefold; (B) based on 15 landmarks plus 85 semilandmarks, result exaggerated fivefold.

shape. Depicting that effect (Figure 8.5B), magnified fivefold, shows that the impact of size is particularly pronounced on the curvature of the angular process, the depth of the molar alveolus and the width of the condyloid process.

COMPARING TWO MEANS

A classic problem in statistics is to determine whether or not the mean value of a measurement (or a set of one-dimensional measurements) in one group is different from that of another group. These groups are distinguished because they differ in some discrete, categorical variable, such as a treatment that we applied to them, or some property such as sex. In such cases, we cannot subdivide the classifying variable; there is no treatment between the ones that we apply, nor is there a value for sex between male and female. In some cases, the classifying variable is not necessarily discrete and categorical, but it is treated as if it were. For example, populations from different localities might be treated as categorically distinct even though there are geographical coordinates between any two sites. Similarly, populations sampled at two different times might be treated as categorically distinct even though there is a date in between any two others.

The comparison between two groups that differ categorically can be expressed as a mathematical model of the form:

$$Y_i = A(i) + B + \varepsilon_i \tag{8.15}$$

where Yi is our response variable and $A(i)$ denotes the group to which specimen i belongs, B is the mean value of Y (computed over both the groups) and ε_i is the error term (i.e. the residual). The response variable is often called a "factor" and the values on the factor are its "levels". For example, if we want to compare the adult height of males and females, the two values for A terms would represent the two levels (male, female) on the factor "sex". The mean value of the first level (group 1) is $A(1) + B$, and that of the second level is $A(2) + B$. Thus, the values for the two levels ($A(i)$) are deviations from the mean, and the difference between the two group means is given by $A(2)-A(1)$. Notice how similar this is to the regression model above; not surprisingly, we can use regression to test for the difference in mean. We do this by "dummy coding" our groups and then regressing on the dummy coded predictors. The dummy codes for this simple case, with two groups and equal sample sizes, would consist of two codes, 1 for the members of the first group and −1 for the members of the second. This pair of codes makes the mean value over both groups zero and the groups deviate by equal amounts from the mean. We can thus test the null hypothesis that *the* deviation from the mean (the coefficients for Ai) is no greater than zero. In the following chapter, we'll see that this test for a difference in means is a special case of what is known as the General Linear Model.

Testing the Difference Between Mean Shapes of Two Groups

The classical multivariate approach to testing for differences in the means of two groups makes use of the null hypothesis H_0: $A(1) = A(2) = 0$, meaning that the two groups have the same mean. This has been tested using Hotelling's T^2 (a multivariate form of the common t-test) which is not conceptually different from the F-test used to test regression models. To calculate Hotelling's T^2, one first computes the pooled within group variance−covariance matrix:

$$\Sigma_p = \Sigma_r = (n_1\Sigma_1 + n_2\Sigma_2)/(n_1 + n_2 - 2) \tag{8.16}$$

where n_1 and n_2 are the sample sizes of the two groups, and Σ_1 and Σ_2 are the within-group variance−covariance matrices. This pooled within-group variance−covariance matrix is also the variance−covariance matrix of the residuals, or the errors, because this is the part of the variance not explained by the factor of the model. From this pooled estimate, we can calculate a squared Mahalanobis distance between the means of the two groups (Dryden and Mardia, 1998):

$$D^2 = (A(2) - A(1))^T \sum\nolimits_r^{-1} (A(2) - A(1)) \tag{8.17}$$

From this we can compute an F-ratio

$$F = \frac{n_1 n_2 (n_1 + n_2 - Q - 1)}{(n_1 + n_2)(n_1 + n_2 - 2)Q} D^2 \tag{8.18}$$

where Q is the degrees of freedom per specimen Y_i. The resulting statistic has an F distribution with Q degrees of freedom in the numerator and $n_1 + n_2 - Q - 1$ in the denominator. While this expression looks complex, the important aspect to realize is that it is essentially a ratio of the variance explained by the model, which is proportional to squared difference in the means, and the residual or error variance, which is the variance not explained by the model.

There are some limitations to Hotelling's T^2 due to the fact that it uses the variance–covariance matrix when constructing the test statistic. As discussed above in the context of the multivariate tests for regression models, using the variance–covariance matrix requires that the matrix Σ_r be inverted and thus this matrix must be of full rank. When working with landmark data, the variance–covariance matrix of the coordinates is not of full rank because we have $2K$ (or $3K$) coordinates but only $Q = 2K - 4$, $3K - 7$ or $2K + L - 4$ (etc.) degrees of freedom. To obtain a matrix of full rank for landmarks, we could use Bookstein shape coordinates or partial warp scores or the appropriate number of principal component scores. Semilandmarks pose more of a challenge not only because of the rank of the variance–covariance matrix but also because we will rarely have sample sizes four or five times the number of coordinates of semilandmarks. Another problem is that the test presupposes that the two samples have the same variance–covariance matrix.

Fortunately, we can use Goodall's *F*-test (Goodall, 1991), as discussed above in context of regression. In the case of a comparison between two means, Goodall's *F*-test is given by:

$$F = \frac{n_1 + n_2 - 2}{n_1^{-1} + n_2^{-1}} \frac{D_{1-2}^2}{\left(\sum D_1^2 + \sum D_2^2\right)} \tag{8.19}$$

where D_{1-2}^2 is the squared Procrustes distance between the two means, and the summed terms on the bottom are the squared Procrustes distances of the specimens within each group around the mean of their group. The n_i values are again the sample sizes of the two groups, and Q is the degrees of freedom in the measurements. This *F*-value is approximately distributed as *F* distribution with $df_{numerator} = Q$, and $df_{denominator} = (n_1 + n_2 - 2)Q$. Clearly, this *F*-ratio is again a ratio between the variance explained by the model and the unexplained variance.

In the derivation of this *F*-test, the assumption is that variation is isotropic normal scatter at each landmark, meaning that all landmarks vary equally and independently, which is not typically the case for actual data. Goodall's *F*-test is thought to be fairly robust to violations of the underlying assumptions but, just as we used resampling-based tests to test the significance of Goodall's *F* in the case of regression, we will do so when comparing group means. The *F*-ratio is first computed exactly as in Equation 8.19, but to test the significance of *F*, we will either permute members of the groups (sampling without replacement) or bootstrap the data (sampling with replacement). A more thorough explanation of these and other randomization procedures is in the Appendix to this chapter. These resampling procedures are used to determine how often an *F*-value as large or larger than the observed one is obtained over many resamplings, providing an estimate of the *p*-value of the observed *F*-ratio.

Testing the Null Hypothesis that Chipmunk Jaw Shape is Not Sexually Dimorphic

To exemplify the comparison between two means, we will test the null hypothesis that alpine chipmunk jaws are not sexually dimorphic, i.e. that there is no difference between the shapes of male and female jaws. For the data restricted to the 15 landmarks, we obtain a value of Wilks' Λ of 0.6964, with 26 and 77 degrees of freedom (for numerator and denominator, respectively, yielding an approximate F-ratio of 1.29. The probability of obtaining an F-ratio this large, with those degrees of freedom, when the null hypothesis is true, is 0.1946. We thus would not reject the null hypothesis that mandibular shape is not sexually dimorphism. Using Goodall's F-test, the percent of the variance unexplained by the regression is 96.18, so 3.82% of the variance *is* explained by the regression. The F-ratio is 4.06, with 26 and 2652 degrees of freedom, and the probability of obtaining an F-ratio this large, with these degrees of freedom, is less than 0.005. Only 0.1% of permutations yield an F-ratio this large or larger, so we can reject the null hypothesis that males and females do not differ in shape. The effect, shown in Figure 8.6A, is exaggerated fivefold because it is subtle, being most pronounced on the coronoid process, depth of the ramus and (perhaps) curvature of the incisor. Given that these 15 landmarks provide so little information about mandibular shape, especially the complex curvature of the jaw, we redo this analysis including the 85 semilandmarks. Using Goodall's F-test, we determine that sex does not explain 95.59% of the variation in shape and thus that it *does* explain 4.41% of the variation. Goodall's F for this case is 4.71, with 196 and 19992 degrees of freedom; the probability of obtaining an F-ratio this large, with these degrees of freedom, is less than 0.005. Using permutations instead to determine the significance of F, we find that 0.1% of the values for F obtained by permutation equal or exceed the observed one, thus we again conclude that sex has a statistically significant, if relatively small, impact on shape. Depicting

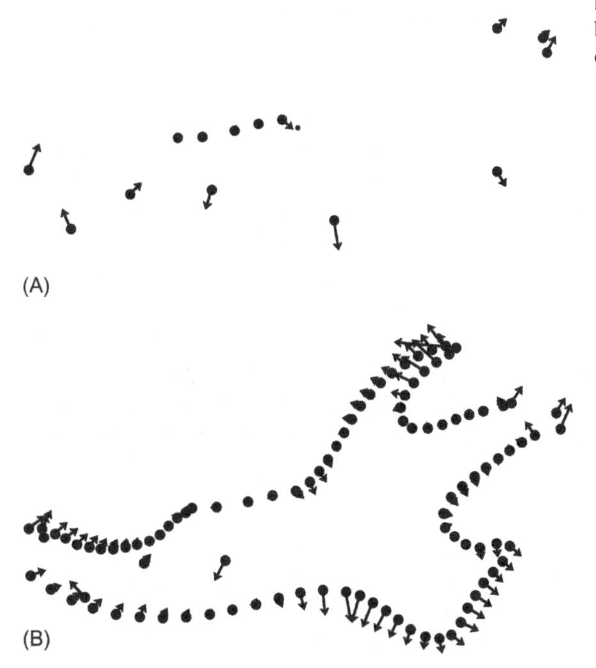

FIGURE 8.6 Sexual dimorphism of *T. alpinus* (A) based on 15 landmarks, exaggerated 10-fold; (B) based on 15 landmarks plus 85 semilandmarks, exaggerated 10-fold.

that effect (Figure 8.6B), magnified fivefold, shows that the impact of size is particularly pronounced on the orientation and width of the coronoid process, curvature and length of the angular process, position of the masseteric fossa, and curvature of the incisor alveolus.

ONE-WAY ANOVA/MANOVA

Having considered a case in which there are only two groups, we now extend the analysis to three or more. The categorical variable in the analysis of variance (ANOVA) is typically called a factor or, in the case of experimental studies, a treatment; the different values it takes are called the levels of the factor. In the case above, there were only two levels of the factor sex. We will now extend the analysis to factors that have more than two levels, although in this chapter we still consider only one factor. The factor could be species, with one level per species, so for nine species we would have nine levels of the factor. In an experimental case, we might have three treatments, such as three diets (liquid, soft, hard), in which case we have three levels of the treatment. The question we typically ask in such cases is whether the factor influences the dependent variable. Classically, this is called a single factor analysis of variance (ANOVA) or multivariate analysis of variance (MANOVA), or a one-way ANOVA/MANOVA. This is also a simple example of a General Linear Model, which encompasses a wide range of analyses, including regression, ANOVA, MANOVA (with one or more than factors), and Analysis of Covariance (ANCOVA/MANCOVA), which contain a mixture of categorical factors and continuous variables (covariates), plus a number of other models. A lot of details about the nature of factors, and the design of experiments, become critical once we have more than a single factor. These are the subject of the next chapter. In this one, we consider only a simple one factor (one-way) ANOVA/MANCOVA.

In the discussion of regression and of Goodall's *F*-test, we have seen how to characterize the variance in a data set using Procrustes distances, and to measure the variance explained by a model. In both linear regression and the pairwise comparison of means, we saw that a permutation test could be used to determine the significance of an *F*-ratio based on summed squared Procrustes distances. There are other approaches to charactering variance that use what is called the sum of squares and cross products matrices (SSCP, which are linearly related to the variance–covariance matrices) and there are analytic statistical tests available based on SSCP, which we discuss in the next chapter. In many ways, it is much easier to use and explain Procrustes distances and permutation tests than to work with SSCP matrices. We thus begin by discussing a univariate ANOVA as a way of introducing the basic ideas of the method, and to explain how sums of squares are formed. The extension of these ideas and explanation to Procrustes distances and to permutation tests is then relatively straightforward.

Univariate ANOVA With One Factor

In this section, we examine the development of a general linear model (GLM) for a univariate variable (*Y*) that is hypothesized to depend on a single fixed factor (*A*). We will show one approach to calculating variances based on sums of squares, using the estimated means of the specimens in each level of factor *A*. This approach is conceptually easy to understand, but it is not in the matrix notation we will need to use in the next chapter, and it is not the approach used in most computer-based calculations. Most modern approaches to calculating the sums of squares use matrix algebra and the differences between the sums of squares are explained by models expressed in terms of design matrices. The simple summation methods presented below are easier to understand at an introductory level, but difficult to scale up to larger problems and are probably more prone to rounding errors. Researchers interested in programming their own GLM methods will need to consult more advanced texts to develop a complete understanding of these approaches (Anderson, 2001a, b, 2006; Anderson and Robinson, 2001; Rencher and Schaalje, 2008, for starters).

Suppose that we have a univariate dependent variable *Y*, which depends on a factor *A*, which has *J* distinct levels and n_j specimens per level. We will not require that there be equal numbers of specimens in each level (also called a cell) at this point. However, we will require that *Y* be centered, i.e. its mean value is zero, thereby removing one degree of freedom. For the *i*th ($i = 1$ to n_j) specimen in cell j ($j = 1$ to J) we have the model

$$Y_{ij} = \alpha_j + \varepsilon_{ij} \tag{8.20}$$

where α_j is the contribution of the *j*th level of the factor to the value, and ε_{ij} is the error. Notice that we require that the mean value of the residual terms ε_{ij} be zero, with variance σ_e^2, and also that the mean value of *Y* be zero (because *Y* is centered). Consequently, the $n_j\alpha_j$ terms summed over all the cells must also equal zero,

$$\sum_{j=1}^{J} n_j\alpha_j = 0 \tag{8.21}$$

As a result, there are *J* values for α_j but only $(J - 1)$ of them are independent because the constraint that they sum to zero removes one degree of freedom. Some authors include a mean value of *Y*, $\overline{Y}$ or μ in the expression, rather than requiring that *Y* be centered, so you may see the form

$$Y_{ij} = \mu + \alpha_j + \varepsilon_{ij} \tag{8.22}$$

We can now look at variance partitioning, i.e. splitting the variation into the portion explained by the factor and that left unexplained, which is typically called the residual or error term, just as in linear regression. We need to do this variance partitioning to understand how to form *F*-ratios in the context of a one-way ANOVA. We will do this by first looking at the summed square values around the mean value ($\overline{Y}$), then splitting that into two terms, the first being the scatter about the mean ($\overline{Y}_j$) of each level of factor *A*, and the other being the scatter of the mean values of each level about the total mean. The total sum of squares is given by:

$$SS_{total} = \sum_{j=1}^{J}\sum_{i=1}^{n_j}(Y_{ij} - \overline{Y})^2 = \sum_{j=1}^{J}\sum_{i=1}^{k}(Y_{ij} - \overline{Y}_j)^2 + \sum_{j=1}^{J} n_j\overline{Y}_j^2 \tag{8.23}$$

Note that $\overline{Y}$ must be zero, due to the requirement that Y be centered. We have included the $\overline{Y}$ term here so that our expression for the sums of squares (SS) will be consistent with the standard presentations of these ideas.

The first term in the expression above represents the error and the second term is the sum of squares (SS) due to A, or what is sometimes called the between-group sum of squares:

$$SS_{error} = \sum_{j=1}^{J}\sum_{i=1}^{k}(Y_{ij} - \overline{Y}_j)^2 \tag{8.24}$$

The SS_{error} has an expected value of the degrees of freedom multiplied by σ_e^2. The degrees of freedom in the error term are given by:

$$df_{error} = \sum_{j=1}^{J} n_j - J \tag{8.25}$$

Therefore, the expected mean square error (or residual) is estimated as:

$$MS_{error} = \frac{SS_{error}}{df_{error}} \tag{8.26}$$

which has an expected value of $EMS_{error} = \sigma_e^2$. Whereas the expected mean square (EMS) is calculated from the model, the mean square error (MS_E) term is calculated from the data. If the model describes the data, then MS_E and EMS should be similar, differing only by relatively minor random variation.

The sum of squares contributed by factor A is:

$$SS_A = \sum_{j=1}^{J} n_j \overline{Y}_j^2 \tag{8.27}$$

The mean square value for factor A (MS_A) is $SS_A/(J-1)$ because the degrees of freedom, df_A are $(J-1)$. The expected mean square (EMS_A) has two sources, one being the pooled variance of the error terms across groups, the other being the squared effects of the factor values:

$$EMS_A = \sigma_e^2 + \sum_{j=1}^{J} n_j \frac{\alpha_j^2}{(J-1)} \tag{8.28}$$

The null hypothesis we want to test is that factor A does not contribute to the value of Y. Since we have required that the mean value of Y be zero (when we centered it), the null hypothesis is that all α_j values are equal, and are, in fact, equal to zero. Under these conditions, the expected mean square value of EMS_A is simply σ_e^2.

The F-ratio of the variance explained by the model to the variance of the errors (or the residual) is expected to be one if the null hypothesis is true. Thus, we can compute the F-ratio based on the data as:

$$F = \frac{MS_A}{MS_{error}} = \frac{SS_A/df_A}{SS_{error}/df_{error}} \tag{8.29}$$

which will follow an F-distribution with degrees of freedom df_A, df_{error}. Working with univariate data, we can simply look up this value in a table of F-values to find the associated p-value. Notice that the F-ratio is the variance explained by the model divided by the unexplained variance, just as when we used an F-ratio to determine the variance explained by a regression model, or used Goodall's F-test to compare the mean shapes of two groups. Because the expected mean term in the numerator is equal to the denominator plus one additional term, the F-value will be larger than one if the additional term:

$$\sum_{j=1}^{J} n_j \frac{\alpha_j^2}{(J-1)} \tag{8.30}$$

is not zero. The null hypothesis is that this term actually *is* zero so, under the null hypothesis, the F-ratio will be one so larger F values indicate lower probabilities that the null hypothesis is true. If the null hypothesis is rejected, we may then interpret the MS_A term as the variance explained by the factor A, and compare it to the unexplained variance estimate, MS_{error}.

EXTENSION OF THE UNIVARIATE ANOVA TO MULTIVARIATE SHAPE DATA

There is a simple approach to extending the single factor ANOVA to shape data (the more complex approaches are discussed in the next chapter, never fear). Because we have a well-understood measure of differences in shape, i.e. the Procrustes distance, sums of squared Procrustes distances may be used to characterize variance in data sets as we discussed already in the context of regression and the comparison of two groups. To extend the single factor univariate ANOVA to multiple groups, we simply replace all the summed square differences in the equations above by summed square Procrustes distances around the means of each level of A, and about the overall mean shape (Klingenberg and McIntyre, 1998; Rohlf, 2009). Variance partitioning proceeds exactly as discussed in the univariate case. The resulting ratio:

$$F = \frac{MS_A}{MS_{error}} = \frac{SS_A/df_A}{SS_{error}/df_{error}} \tag{8.31}$$

is referred to as a Generalized Goodall's F-test (Rohlf, 2009) or a pseudo-F-test (McArdle and Anderson, 2001).

As in the case of linear regression and the pairwise comparison of means, we need not rely on an analytic model of the distribution of the Generalized Goodall's F-statistic. Instead, we can use a permutation approach to test the null hypothesis that the variance explained by the factor is due to a random association between specimens and group levels. To test this null hypothesis, we permute the group labels assigned to each specimen, randomly associating each specimen with a label. We then compute the F-ratio for each permuted data set and the distribution of F values obtained over many permutations can then be used to test the observed F-value at any desired α level. To reject the null hypothesis at some α value, less than α of the permutation sets must equal or exceed the observed *F*-value. So, for example, to reject the null hypothesis at α = 0.05, less than 5% of the *F*-ratios obtained from the permuted data can equal or exceed the observed one.

APPENDIX: AN OVERVIEW OF RANDOMIZATION AND MONTE CARLO METHODS

In this Appendix, we aim to present the principles underlying randomization methods in a coherent fashion. More complete discussions of the topics presented in this chapter can be found in the texts by Efron and Tibshirani (1993) and Manly (1997), and the work of Anderson and colleagues (Anderson, 2001a,b, 2006; Anderson and Robinson 2001; Anderson and Ter Braak, 2003). We discuss four classes of methods, including the bootstrap, jackknife, and permutation tests, and Monte Carlo simulations. To illustrate these methods, we focus on a few univariate statistical tests. The extension to multivariate statistics is not difficult but, by examining applications to univariate statistics, it may be easier to acquire an intuitive understanding of how these methods work. The basic ideas appeared in the work of R.A. Fisher in the 1930s, but the ideas and techniques were neither developed extensively nor used widely until recently. Perhaps the best summary is contained in the title of Efron's (1979) paper, *Computers and the theory of statistics, thinking the unthinkable.* The approach he outlined was indeed unthinkable prior to the advent of computers, and could not be used widely until computers became fast and inexpensive enough to be generally available to researchers, which accounts for the long time lag between the development of the ideas and their widespread application. These methods are computationally intensive because they replace the complex analytic mathematical methods of classical statistics by an extensive use of randomization and repeated calculations. The enormous number of calculations required by these methods makes them unthinkable without inexpensive (and fast) computers.

Resampling Statistics

Classical statistics relies on algebraic derivations of formulae based on a limited number of well-studied distributions, particularly the normal (Gaussian), *F*-, gamma, chi-square, uniform, and Poisson distributions. To see how resampling-based methods can provide an alternative, we will work through one simple example.

Suppose X is a set of 31 observations of a length:

$$X = \{2, 2, 3, 4, 2, 5, 3, 2, 6, 2, 3, 4, 6, 2, 1, 4, 3, 7, 2, 3, 4, 4, 5, 8, 5, 2, 1, 3, 4, 4, 3\} \quad (8A.1)$$

In this case, $N = 31$. We can compute the mean (denoted $<X>$ for "the expectation of X") by:

$$<X> = \sum_{i=1}^{N} \frac{X_i}{N} \quad (8A.2)$$

where X_i is the ith element in the list. In our example, $<X> = 3.52$. Of course, we also need to quantify our uncertainty in this value. If we assume that the distribution of X fits the model of a normal distribution, then the standard error of the mean is given by the standard deviation σ divided by the square root N (the number of observed individuals). The standard deviation is:

$$\sigma = \left(\frac{\sum_{i=1}^{N} (X_0 - <X>)^2}{N-1} \right)^{1/2} \tag{8A.3}$$

so the standard error of the mean (SEM) is:

$$\text{SEM} = \frac{\sigma}{\sqrt{N}} = \left(\frac{\sum_{i=1}^{N} (X_0 - <X>)^2}{N(N-1)} \right)^{1/2} \tag{8A.4}$$

For our example, $\sigma = 1.69$ and $N = 31$, so $\text{SEM} = 1.69/(31)^{1/2} = 0.304$.

The 95% confidence interval for the mean, *assuming a normal distribution*, ranges from $<X> -1.96(\text{SEM})$ to $<X> +1.96(\text{SEM})$ because, for a normal or Gaussian distribution, 95% of the values in the distribution lie within 1.96 standard deviations of the mean. So, for our example, $1.96\ \text{SEM} = 0.304$ and $<X> = 3.52$, so the 95% confidence interval is from 2.92 to 4.12. Suppose that we want to claim that the average body length of this population is greater than 3.0 cm. Again, using the normal distribution, we can calculate that the chance of the mean being less than or equal to 3.0 is 0.049%, so we can reject the hypothesis that the mean is less than or equal to 3.0 at a 5% confidence level, meaning that we accept a 5% chance of rejecting the null model when it was true (Type I error).

What difficulties arise in this example? First, we have assumed that the distribution is normal. This is important even though statistics based on the normal distribution are known to be robust to violations of the assumptions of normality. Nevertheless, as the distribution departs further from normality, larger errors appear in the results, leading to increased error rates. The validity of the normal distribution for our example has not been determined. Is that assumption reasonable? If the distribution is normal, 1.9% of the measurements will be less than or equal to zero (that is the expectation under the model). Does that pose a problem? Yes, because we are measuring lengths, and *none* can be less than zero, under any circumstances – in fact, the lower bound may be substantially larger than zero (due to physiological constraints on the size of the organism). So we *know* that our distribution must deviate from the normal distribution, at least with respect to the expectation that the mean will be zero. Perhaps that deviation has only a small effect on our estimate of SEM, but we are relying on the reputation of the normal distribution as a robust estimator to reassure ourselves about that. We really do not know what effect that lower bound has on our statistical inferences. We could of course transform our values, subtracting the mean from all of them, for example. And there are other distribution models besides the normal, or we could use other transformations (taking the natural log for example), in an effort to arrive at normally distributed variables.

The other difficulty we face is the lack of an exact formula for the standard error of many statistics, or of functions of statistics that we might want to work with. Suppose we want to know the standard error in the median of the distribution. We can calculate the median of our measurements of X, which equals 3.0, but can we actually conclude that the median of the population is greater than 2.0? We do not really know the range of values that the median might take on for this distribution, and the normal model provides no estimate of the uncertainty in the median. The standard deviation and variance of populations are also of tremendous biological interest, but how do we estimate the range of values for these statistics?

Resampling-Based Methods

Having noted that we can face serious difficulties when we assume a normal distribution and rely on the theory based on it, we now examine methods that allow us to make statistical inferences without assuming any distribution.

The Bootstrap

We begin with the bootstrap because it is probably the easiest to understand. It was not the first computer-based statistical method developed; in fact it is one of the more recent (it was developed from jackknife and permutation methods). The term "bootstrapping" comes from the novel *Baron Münchausen's Narrative of his Marvelous Travels and Campaigns in Russia*, by Rudolph Erich Raspé (1785), in which the Baron falls to the bottom of a deep lake. He cannot figure out what to do until, at the last moment, he thinks to pull himself up by his own bootstraps. This describes, fairly accurately, the approach used in a bootstrap procedure: the observed data themselves are used as a basis for resampling. We will approximate the unknown statistical distribution from which the data were drawn by (randomly) resampling our data.

A bootstrap set is a set of data of the same sample size as the original data set, whose elements are *randomly drawn with replacement* from our original set of observations. To draw them randomly (with replacement) from a set of N elements, a uniformly distributed random number from 1 to N is generated by a random number generator. The corresponding element from the original set of observations then forms the first element in the bootstrap set. For example, given our 31 observations, we will construct a sample that also has 31 observations. The number provided by the random number generator is 8, so we take the value of the eighth individual of our sample as the first value in the bootstrap set. This procedure is repeated N times. Note that a single value from the original data set may appear multiple times in a bootstrap set because we are sampling *with replacement*, meaning that we do not remove an individual from the sample after we have placed its value in the bootstrap set. As a result, some values might not appear at all in the bootstrap set.

To see how a bootstrap set is formed, we consider an abstract, symbolic example. Suppose $\mathbf{C}$ contains five values:

$$\mathbf{C} = \{C_1, C_2, C_3, C_4, C_5\} \tag{8A.5}$$

To form a bootstrap version of $\mathbf{C}$, we generate a list of five random numbers, each independently chosen and ranging from 1 to 5 (because $N = 5$):

$$L = \{5\ \ 2\ \ 4\ \ 3\ \ 5\} \tag{8A.6}$$

The numbers in L are the ordinal positions of the elements of $\mathbf{C}$ so $\mathbf{C}_{\mathbf{Bootstrap}}$ contains the corresponding values of $\mathbf{C}$ (e.g. $L_1 = 5$, so it corresponds to the fifth element of $\mathbf{C}$, which is C_5). Thus:

$$\mathbf{C}_{\mathbf{Bootstrap}} = \{C_5, C_2, C_4, C_3, C_5\} \tag{8A.7}$$

Note that C_5 appears twice in this bootstrap set whereas C_1 does not appear even once.

Returning to the numerical example presented earlier:

$$\mathbf{X} = \{2,\ 2,\ 3,\ 4,\ 2,\ 5,\ 3,\ 2,\ 6,\ 2,\ 3,\ 4,\ 6,\ 2,\ 1,\ 4,\ 3,\ 7,\ 2,\ 3,\ 4,\ 4,\ 5,\ 8,\ 5,\ 2,\ 1,\ 3,\ 4,\ 4,\ 3\} \tag{8A.8}$$

To form a bootstrap set, $\mathbf{X}_{\mathbf{Boot}}$, from $\mathbf{X}$, we generate the list, $\mathbf{B}$, of 31 random numbers:

$$\mathbf{B} = \{30,\ 8,\ 19,\ 16,\ 28,\ 24,\ 15,\ 1,\ 26,\ 14,\ 20,\ 25,\ 29,\ 23,\ 6,\ 13,\ 29,\ 13,\ 28,\ 2,\ 11,\ 26,\ 1,\ 5,\ 7,\ 7,\ 19,\ 9,\ 7,\ 1\} \tag{8A.9}$$

We then select the elements of $\mathbf{X}$ corresponding to those ordinal values:

$$X_{\text{Boot}} = \{4,\ 2,\ 2,\ 4,\ 3,\ 8,\ 1,\ 2,\ 2,\ 2,\ 3,\ 5,\ 4,\ 5,\ 5,\ 6,\ 4,\ 4,\ 6,\ 3,\ 2,\ 3,\ 2,\ 2,\ 2,\ 3,\ 3,\ 2,\ 6,\ 3,\ 2\} \tag{8A.10}$$

We can now calculate the mean, standard deviation and median of $\mathbf{X}_{\mathbf{Boot}}$: $<\mathbf{X}_{\mathbf{Boot}}> = 3.39$, $\sigma_{X_{\text{Boot}}} = 1.62$, and median($\mathbf{X}_{\mathbf{Boot}}$) = 3. These values are slightly different from those of the original distribution, $<X> = 3.52$; $\sigma = 1.69$, and median ($\mathbf{X}$) = 3.0. To arrive at an estimate of the confidence intervals for these statistics, we will compute a large number ($N_{\text{Bootstrap}}$) of bootstrap sets. We will then determine the 95% confidence interval over the $N_{\text{Bootstrap}}$ sets, forming a *bootstrap estimate* of the confidence intervals on the mean, standard deviation and the median. If we generate 200 bootstrap sets based on $\mathbf{X}$, we find that the 95% confidence interval for the mean is 3.00 to 4.10; for the standard deviation the confidence interval is 1.23 to 2.10, and for the median it is 3.00 to 4.00. The normal model predicted a 95% confidence interval for the mean, 2.91 to 4.12, so the two methods approximately agree. They appear to differ at the lower boundary (at small lengths), which is where we expect departures from the normal distribution, for the reasons discussed earlier.

The approach outlined here may be extended to virtually any statistic and to any function, univariate or multivariate. For example, we can use it to perform *t*-tests, which are used to compare the means of two samples. It is possible that the difference in numerical values of two means is due solely to an arbitrary division of one group into two. Because of the variation within the population, drawing two samples from it can result in two samples that differ numerically in their means.

Let us look again at our sample of 31 measured lengths:

$$X = \{2,\ 2,\ 3,\ 4,\ 2,\ 5,\ 3,\ 2,\ 6,\ 2,\ 3,\ 4,\ 6,\ 2,\ 1,\ 4,\ 3,\ 7,\ 2,\ 3,\ 4,\ 4,\ 5,\ 8,\ 5,\ 2,\ 1,\ 3,\ 4,\ 4,\ 3\} \tag{8A.11}$$

and consider a second group of 18 lengths:

$$Y = \{2,\ 2,\ 3,\ 2,\ 4,\ 2,\ 3,\ 2,\ 8,\ 9,\ 2,\ 9,\ 3,\ 2,\ 3,\ 3,\ 3,\ 9\} \tag{8A.12}$$

Using the normal model, we find that $<X> = 3.52$, $\sigma_X = 1.69$, and $<Y> = 3.94$ and $\sigma_Y = 2.71$. To test whether the means are different, we find the probability of statistic t:

$$t = \frac{(<Y> - <X>)}{\sqrt{\left(\frac{\sigma_X^2(N_X - 1) + \sigma_Y^2(N_Y - 1)}{N_X + N_Y - 2}\right)\left(\frac{N_X + N_Y}{N_X N_Y}\right)}} \tag{8A.13}$$

with degrees of freedom equal to $(N_X + N_Y - 2)$. For relatively large values of N_X and N_Y, the t-value will be normally distributed with a mean of zero and a standard deviation of one, *provided that* the null hypothesis of equal means is true. If the absolute value of t exceeds 1.96, we may conclude that, under the normal model, there is only a 5% chance of the mean values being that different by chance. We can thus reject the null hypothesis at a 5% level of confidence.

The problem is that the list of lengths contained in **Y** is highly non-normal. Most values are close to 3, but there are several around 8 or 9, so **Y** appears to be rather bimodal. Also, in a normal distribution with a mean of $<Y> = 3.94$ and a standard deviation of $\sigma_Y = 2.71$, we would expect that 7.3% of the measured lengths would be less than zero. So, the distribution of **Y** departs substantially from normality, more so than does the distribution of **X**.

To form a bootstrap version of the t-test, we *simulate the null hypothesis we wish to reject*. This simple principle is the key to understanding how to form your own bootstrap tests when asking novel statistical questions. The null hypothesis of the t-test is that the means of the two groups are equal, which we can also phrase as the hypothesis that the two groups in question came from a single underlying distribution that was arbitrarily subdivided into two groups. If this were the case, any difference between the means would arise simply by chance. So, to test this hypothesis, we assume that the null hypothesis is true – i.e. that **X** and **Y** were drawn from the same population. This means that under the hypothesis, specimens are *exchangeable* between the two groups (Anderson, 2001b). Therefore, we merge the two sets of observations (**X** and **Y**) into a common pool of specimens (**Z**) and draw (with replacement) two bootstrap sets from **Z**, one of size N_X and one of size N_Y, and compute the differences in means between the two bootstrap sets. This is repeated $N_{Bootstrap}$ times. We can then determine the number of times in which the difference between the means of paired bootstrap sets exceeds the observed difference between the means of **X** and **Y**. Expressed as a proportion of the total, we get an estimate of the probability that the observed difference is due to chance; i.e. if the difference between means of pairs of bootstrap samples exceeds the observed differences in 5% (or fewer) of the total number of iterations, we can reject the null hypothesis that the means are equal. This is simply another way of phrasing the statement that the observed difference is statistically significant at a 5% confidence level if the observed difference between means exceeds the 95th percentile of differences between means of the bootstrap sets.

A symbolic example of this merging and subsequent formation of two bootstrap sets may help to develop an understanding of how the test operates. Suppose we have a set **C** of five elements, and a set **D** of four elements:

$$\mathbf{C} = \{C_1, C_2, C_3, C_4, C_5\} \tag{8A.14}$$

$$\mathbf{D} = \{D_1, D_2, D_3, D_4\} \tag{8A.15}$$

The merged set, **M**, would have nine elements:

$$\mathbf{M} = \{C_1, C_2, C_3, C_4, C_5, D_1, D_2, D_3, D_4\} \tag{8A.16}$$

To draw two bootstrap sets out of **M**, we would form a list of five random integers (because there are five elements in **C**), and the elements in **M** corresponding to this list would be the elements in the bootstrap version of **C**:

$$L_1 = \{7\ \ 5\ \ 1\ \ 8\ \ 5\} \tag{8A.17}$$

$$\mathbf{C}_{\mathbf{Bootstrap}} = \{D_2, C_5, C_1, D_3, C_5\} \tag{8A.18}$$

Note that two elements in $\mathbf{C}_{\mathbf{Bootstrap}}$ come from **D**. A second list of four integers is used to form a bootstrap version of **D**:

$$L_2 = \{2\ \ 4\ \ 9\ \ 9\} \tag{8A.19}$$

$$\mathbf{D}_{\mathbf{Bootstrap}} = \{C_2, C_4, D_4, D_4\} \tag{8A.20}$$

The approach we used to produce the bootstrap versions of **C** and **D** reflects the null hypothesis that **C** and **D** come from a common underlying distribution. The elements of **C** and **D** are thus interchangeable.

The difference between means of the bootstrapped versions of **C** and **D** can be determined by many repetitions, developing a bootstrap estimate of the distribution of the differences between means produced by the null hypothesis (given the data). When we carry out this bootstrap *t*-test on our numerical example, sets **X** and **Y**, we find that 268 of 1000 bootstrap sets (26.8%) have a difference between means as large or larger than that between the means of **X** and **Y**. Thus, we cannot reject the null hypothesis that these samples were drawn from populations with equal means, the difference between them being due solely to chance. Using a *t*-test based on the normal distribution, we would have rejected that null hypothesis. Because both samples appear to have non-normal distributions, it seems reasonable to attribute the difference between results to violating the assumption of normality.

Permutation Tests

Permutation tests pre-date bootstrap tests, having been introduced by R.A. Fisher in the 1930s as a basis for supporting the ideas of the Student's *t*-test rather than as a tool for computation. With the advent of computers, permutation methods could be used profitably for statistical inference. Permutation tests operate in much the same manner as bootstrap tests, but differ in that they resample groups *without* replacement. This makes permutation tests suitable for hypothesis testing, but not for the estimation of confidence intervals (Efron and Tibshirani, 1993; Good, 1994; Manly, 1997).

Again, we can look at a simple, abstract example of how a permutation set is formed to get a sense of how the approach works, and how it differs from the bootstrap. Consider two data sets **C** and **D**:

$$\mathbf{C} = \{C_1, C_2, C_3, C_4, C_5\} \tag{8A.21}$$

$$\mathbf{D} = \{D_1, D_2, D_3, D_4\} \tag{8A.22}$$

with sample sizes of five and four respectively. We form the merged set **M** of nine elements:

$$\mathbf{M} = \{C_1, C_2, C_3, C_4, C_5, D_1, D_2, D_3, D_4\} \quad (8A.23)$$

To produce permuted versions of **C** and **D**, we want to resample **M** without replacement. To do this, write a list of nine integers and randomly permute it to form a list **L**:

$$\mathbf{L} = \{5\ 2\ 6\ 8\ 7\ 3\ 9\ 4\ 1\} \quad (8A.24)$$

The first five values in **L** are the ordinal values of the elements in **M**, placed in the permuted version of **C**:

$$\mathbf{C}_{permutation} = \{C_5, C_2, D_1, D_3, D_2\} \quad (8A.25)$$

The last four values in the list are the ordinal values of the elements in **M** that are placed in the permuted version of **D**:

$$\mathbf{D}_{permutation} = \{C_3, D_4, C_4, C_1\} \quad (8A.26)$$

Note the different way that the permutation sets (Equations 8.25, 8.26) and bootstrap sets (Equations 8.18, 8.20) are constructed from **C** and **D**.

To carry out a permutation test of the hypothesis that the means of the two groups **X** and **Y**, we would first compute the difference between the means of the two groups, which have sample sizes of $N_X = 31$ and $N_Y = 18$. The second step is to merge the two data sets into a single larger one and form a series of paired permutation sets, each drawn from the merged data set. The first permutation set in each pair, containing N_X specimens, is drawn randomly without replacement from the merged set. The second permutation set of the pair contains the remaining N_Y elements of the merged data set. (No element of the original sets appears twice in the paired permutation sets, and none is omitted.) The difference between means of the two permutation sets is then calculated, and repeated for $N_{Permutation}$ sets. The proportion of times in which the difference between the means of the paired permutation sets exceeds that between the original data sets is taken as the probability that the observed value could have arisen by a random splitting of a single underlying distribution.

The permutation test of the difference between the means of sets **C** and **D** indicates that 21.3% of the permuted sets had a difference in means equal to or greater than the observed difference of 0.428, so we cannot reject the null hypothesis that the means are equal at a 5% level of confidence. The permutation test has produced results agreeing with the bootstrap test (in which 26.8% of the bootstrap sets had a difference between means as large or larger than the observed data set).

It is possible to form permutation tests for a wide variety of statistical hypotheses in a manner similar to the bootstrap (see Efron and Tibshirani, 1993; Good, 1994; Manly, 1997) and, in the next chapter, we will see permutations of residuals from the full or reduced model (Anderson and Ter Brakk, 2003). However, there is an important difference between the permutation and bootstrapping approaches due to fundamental differences in how they operate. Permutation tests are not suited to the estimation of confidence intervals because the standard deviation of the estimates of a parameter (such as a mean or median) is not a reliable estimate of the standard error in that parameter. Rather, the permutation test yields an estimate of the range of parameter values possible under the null model simulated by the test. In contrast, the standard deviation of the bootstrap estimates of the same parameter yields a reliable estimate of its standard error because the bootstrap resampling simulates a repetition of the process of selecting specimens from the population (Efron and Tibshirani, 1993). When used for hypothesis testing, both methods tend to give very similar results, so it is difficult (and perhaps unnecessary) to determine which approach is preferable in most cases. To some extent, the choice between them appears to be a matter of preference among writers of software. There are some reasons to think that permutation tests may yield a more exact achieved significance level (ASL) than bootstrap approaches (Efron and Tibshirani, 1993; Good, 1994), but this is at the cost of precluding estimates of confidence intervals (or standard errors) on the statistics involved.

The Jackknife

Jackknife methods (Quenouille, 1949; Tukey, 1958) also preceded bootstrap methods and, to some extent, have been supplanted by them. Jackknife estimates are obtained by resampling such that one element is left out at a time. If there are N specimens in a sample, then it is possible to form N jackknife data sets, each with $N-1$ specimens. If we again look at the set $\mathbf{C}$:

$$\mathbf{C} = \{C_1, C_2, C_3, C_4, C_5\} \quad (8A.27)$$

The five possible jackknife versions of $\mathbf{C}$ are:

$$C_{J1} = \{C_2, C_3, C_4, C_5\} \quad (8A.28)$$

$$C_{J2} = \{C_1, C_3, C_4, C_5\} \quad (8A.29)$$

$$C_{J3} = \{C_1, C_2, C_4, C_5\} \quad (8A.30)$$

$$C_{J4} = \{C_1, C_2, C_3, C_5\} \quad (8A.31)$$

$$C_{J5} = \{C_1, C_2, C_3, C_4\} \quad (8A.32)$$

Jackknife data sets will always be more similar to the original data set than bootstrap sets are because bootstrapping offers a greater variety of ways of resampling the data. The jackknife may be viewed as an approximation to the bootstrap (Efron and Tibshirani, 1993), and it is a good approximation when the changes in the statistic are smooth or linear with respect to changes in the data. The mean is a linear statistic, but the median is not (because the median may change abruptly as observations are added or subtracted from the sample). Therefore, jackknife and bootstrap estimates of the mean will not differ much but estimates of the median may differ considerably.

There are some approaches to combining the bootstrap and the jackknife (see particularly Efron, 1992; Efron and Tibshirani, 1993, Chapter 19, on assessing the error of bootstrap estimates), but otherwise the jackknife appears to offer few advantages over the bootstrap.

Cross-validation testing of models (Manly, 1997) is somewhat similar to jackknife testing. Cross-validation is used to test the performance of predictive models, like regression, or the performance of discriminant function or canonical variates analysis. In these applications, some portion of the data (anywhere from 1 specimen to 50% of the data) is held aside as a test set, while the model or discriminant function is fitted to the remainder of the data, which is designated as the training data. The quality of the fit of the model, or the performance of the discriminant function, is then evaluated on the test data. This approach yields an estimate of the performance of the model, if it were to be used on new data. Cross-validation is particularly helpful in detecting overfitting of models.

Monte Carlo Methods

Monte Carlo methods compare the value of an observed statistic to the range of values expected under a given null hypothesis, assuming a model of the populations involved. Like analytical statistical methods, Monte Carlo methods require making assumptions about the nature of the distribution from which populations are drawn. They then fit parameters of the distributional models to the observed samples. In contrast, analytic statistical approaches use algebraic derivations to estimate the values of statistics (and standard errors in those statistics) based on the nature of the underlying distributions. The distinction is that Monte Carlo approaches generate random data sets based on the parameters and distribution of the model; those random data sets are drawn from model distributions having the same sample size as the original one. The distribution of the statistic of interest (estimated over many computer-generated Monte Carlo sets) is used to estimate the mean and standard deviation of that statistic under the null model and the model distribution used. Monte Carlo methods can be used both for hypothesis testing and for generating confidence intervals.

Monte Carlo methods use numerical simulations to avoid the need for extensive algebraic computations and approximations. It may often be easier to program a Monte Carlo simulation than to determine analytically the distribution of an intricate statistical function, particularly when the statistic is not a linear function. Because it is necessary to assume a model of the distributions of the samples, the Monte Carlo method shares most of the primary weaknesses of analytic statistics; if the observed distribution departs substantially from the model, the Monte Carlo sets will not represent the actual system of interest. One useful feature of the Monte Carlo method is the ability to determine the effect of different distributional models (the ones typically used are the uniform, normal or Gaussian, and Poisson) on the range of values estimated by the Monte Carlo sets. The comparison of observed distributions to those produced by Monte Carlo methods is a powerful approach to hypothesis testing.

For example, if we wish to determine the significance of the observed difference in the means of sets **X** and **Y**:

$$\mathbf{X} = \{2, 2, 3, 4, 2, 5, 3, 2, 6, 2, 3, 4, 6, 2, 1, 4, 3, 7, 2, 3, 4, 4, 5, 8, 5, 2, 1, 3, 4, 4, 3\} \tag{8A.33}$$

$$\mathbf{Y} = \{2, 2, 3, 2, 4, 2, 3, 2, 8, 9, 2, 9, 3, 2, 3, 3, 3, 9\} \tag{8A.34}$$

we will test the null hypothesis that the two sets (**X** and **Y**) came from the same underlying distribution, with the observed difference between them being due to a random assignment of specimens into groups. To form the Monte Carlo set, we will assume that the single underlying distribution is normal. We then estimate the mean and standard deviation of this underlying distribution by merging the data sets into a single group. The mean of the single distribution is 3.67 and the standard deviation is 2.1. To determine the significance of the observed difference in the means of the two groups, we generate a series of paired Monte Carlo sets, one with a sample size $N_X = 31$, one with a sample size $N_Y = 18$, and we determine the difference between the two means. We then determine the proportion of $N_{Monte\ Carlo}$ sets in which the difference between the means of the paired Monte Carlo sets exceeds that observed between the means of the original data sets.

For the sets **X** and **Y** above, the Monte Carlo sets were generated under the assumption that both samples were drawn from the same normal distribution, with a mean of 3.67 and a standard deviation of 2.1 (the mean and standard deviation of the combined data sets). In 480 of 1000 pairs of Monte Carlo sets (48%), the difference between the means of the paired Monte Carlo sets exceeds the observed difference between the means of the original data sets, thus the null hypothesis of a single underlying normal distribution cannot be rejected. It should be noted that the combined data set (of all specimens in **X** and **Y**) is probably not normally distributed, so we might want to repeat the Monte Carlo test using other models of the underlying distribution.

Monte Carlo simulations are particularly useful for testing different hypothetical situations when the underlying distributions are believed to be well known. Monte Carlo methods can be used in cases when bootstrap methods cannot, such as to estimate the effect of increasing the sample size on the estimated variance; Monte Carlo simulations are not limited by the observed sample sizes (as bootstrap methods are).

Example: Resampling Tests and Regression Models

To this point, we have focused on *t*-tests, but computer-based methods are useful for a wide variety of tests. To develop a more general understanding of these methods, we now show how bootstrap and permutation methods can be used in regression analysis. As presented earlier in this chapter, the model for a regression, given N observations for pair of measurements (X_i, Y_i):

$$Y_i = A + BX_i + \varepsilon_i \tag{8A.35}$$

The slope, B, is given by:

$$B = \frac{s_{XY}}{s_{XX}} \tag{8A.36}$$

The intercept, A, is given by:

$$A = <Y> - B<X> \tag{8A.37}$$

where $<X>$ and $<Y>$ are the expected values (means) of the X_i and Y_i values, and

$$s_{XY} = \sum_{i=1}^{N} (X_i - <X>)^2 \tag{8A.38}$$

$$s_{XY} = \sum_{i=1}^{N} (X_i - <X>)(Y_i - <Y>) \tag{8A.39}$$

are the values of A and B which minimized the summed square residuals (ε_i). This sum of squared error terms is:

$$\text{Error} = \sum_{i=1}^{N} (Y_i - A - BX_i)^2 = \sum_{i=1}^{N} (\varepsilon_i)^2 \tag{8A.40}$$

under the assumption that the residuals are independently and identically normally distributed.

To show that there is a statistically significant dependence of Y on X, it is sufficient to show that the confidence interval on the slope excludes zero. This is equivalent to showing that there is a non-zero correlation between Y and X, which may be tested using the squared value of the correlation coefficient (R^2) between X and Y, which indicates the fraction of the variance in the dependent variable (Y) that is explained by the independent variable (X). The expression for R^2 is:

$$R^2 = \frac{s_{XY}^2}{s_{XX} s_{YY}} \tag{8A.41}$$

where

$$s_{YY} = \sum_{i=1}^{N} (Y_i - <Y>)^2 \tag{8A.42}$$

It is very common to interpret high R^2 values as being indicative of high explanatory power in a regression model. There is a method of testing whether an R^2 value is statistically significant (under the assumption of normality of the residuals), by the expression:

$$\frac{1}{2} \ln\left(\frac{1+R}{1-R}\right) \tag{8A.43}$$

which is a normally distributed variable, with variance equal to $1/(N-3)$, where N is the sample size.

The significance of the slope can be assessed by a permutation test. The objective is to determine the range of slopes that could be generated by random permutations of the associations among X and Y values, since the null hypothesis implies that these values are exchangeable. Thus, we again adopt the strategy of assuming that the null hypothesis is true (which, in this case, is that the associations among X and Y values are random). The associations of the X_i values with the Y_i are then randomized, generating a permutation set of paired X and Y values with the same distribution of X and Y values as in the data, but with randomized combinations of X and Y. The regression model is then fitted to each permutation set, and the slope (or correlation coefficient) is calculated. The distribution of the regression slopes (or the correlation coefficients) generated by the permutation sets can be used to determine if the observed regression slope (or correlation coefficient) could have been produced by a random association among X and Y variables. If the observed slope (or correlation coefficient) is outside the 95% confidence interval of the permutation sets, then we can reject the null hypothesis that the slope (or correlation coefficient) does not differ from zero. Note that the permutation test estimates the range of slopes (or correlation coefficients) *produced by the null model*, not by the observed data. Thus we reject the null hypothesis by showing that the observed statistic lies outside the range of the values predicted by the null model.

To carry out a bootstrap test of the significance of the regression line, two approaches are available: one is to bootstrap the *paired observations* (X_i, Y_i); the other is to bootstrap the *residuals* from the regression. Note that we could also use permutation methods, on either the raw data or the residuals. When bootstrapping specimens, we form bootstrap sets by sampling (with replacement) from the paired specimen values (X_i, Y_i) to form a bootstrap set. The regression model is fitted and the slope (or correlation coefficient) is determined for each bootstrap set, forming a bootstrap estimate of the confidence intervals for the slope (or correlation coefficient). This yields a confidence interval on the slope itself, so that if it excludes zero, we can reject a null hypothesis that the regression slope (or correlation) is zero.

The alternative is to bootstrap the residuals, by first determining the residuals to the bootstrap, and the Y values that are predicted by the regression model for each X value:

$$Y_{\text{predicted}} = A + BX \tag{8A.44}$$

Then the residuals are randomly combined with the paired X_i and $Y_{predicted}$ values, both of which are resampled (with replacement). This approach produces a wider variety of possible paired values of X_i and Y_i; it can be thought of as bootstrapping the variable part of the distribution, independently of the portion that is dependent on X. The range of slopes (or correlation coefficients) is determined over many bootstrap sets; if the 95% confidence interval for the slope (or correlation coefficient) excludes zero, we can infer that there is a statistically significant dependence of Y on X at a 5% confidence level.

The discussion of how a permutation test is used to determine the statistical significance of a regression slope serves as a useful illustration of the differences in approach between bootstrap and permutation methods. In the permutation method, the approach is to estimate the confidence interval under the *null model*, given the distribution of observed data. Thus, if the observed statistic is outside the confidence interval of the null, the observed statistic is judged to be significant. In contrast, the bootstrap approach estimates the range of the statistic on the *observed data* (rather than the range under the null). Permutation tests almost always focus on estimating distributions under the assumption that the null model is true, whereas bootstrap methods can be used to estimate the distribution of a statistic either over the observed data or under an assumption that the null is true.

Issues Common to All Resampling Methods

Statistical Power

When evaluating the utility of statistical tests we tend to focus on the rate of Type II errors (i.e. failing to reject the null hypothesis when it is false and the alternative is true).

That is because we can control the rate of Type I error (i.e. falsely rejecting the null hypothesis when it is true). Type I error rates are controlled by setting the *alpha* level of the test and so statistical tests cannot be said to differ in their rates of Type I error. In contrast, they can differ in their rates of Type II error. The *power* of a statistical test is its ability to distinguish between the false null hypothesis and the true alternative, and it is sometimes expressed as 1 minus the rate of Type II error.

Estimating the power of statistical tests turns out to be both difficult, and neglected by many researchers. Some work indicates that permutation, bootstrap and analytic tests have equivalent statistical power when the data meet the requirements of the analytic tests (Hoeffding, 1952; Robinson, 1973; Romano, 1989; Manly, 1997). Edgington (1995) reports higher statistical power for randomization tests when there are violations of the assumptions of the analytic statistical tests. Efron and Tibshirani (1993) present an approach to estimating power, given a specific sample size. The approach offered by Sheets and Mitchell (2001) is to use Monte Carlo methods to estimate the rates of Type II error under several plausible alternatives to the null hypothesis. Despite the attendant difficulty in estimating the statistical power of different tests, randomization-based tests seem to have at least as much statistical power as the more familiar analytical tests.

How Many Repetitions?

Regardless of the method used, the researcher is always faced with the question of how many replications or repetitions should be made. We want a small bias and standard deviation, but it is not clear how many replications are required to achieve this end. The number of independent bootstrap samples that one may form out of N specimens is $(2N-1)!/N!(N-1)$ (Efron and Tibshirani, 1993), which is over 90 000 for $N=10$ specimens. In most cases, even thousands of bootstrap replicates will not come close to exhausting all possible bootstrap sets. Typically, a modest subset of all possible sets is adequate for most statistical questions. Estimates of standard errors can usually be produced using only 100 or fewer bootstrap sets (Efron and Tibshirani, 1993), but reliable estimates of confidence intervals may require using many more. It does not appear that there is complete consensus on this issue (see Efron, 1992; Efron and Tibshirani, 1993; Jackson and Somers, 1989; Manly, 1997), but it does seem that more repetitions are necessary for estimating confidence intervals in that we must estimate a specific percentile point value, than for hypothesis testing (see Manly, 1997) or for estimating of standard errors (Efron and Tibshirani, 1993). If computer time is not an issue, a range of 1000 to 2000 bootstrap tests is recommended for estimating a 95% confidence interval on a parameter (Efron, 1987; Efron and Tibshirani, 1993) and, in light of the very fast computers now generally available, far more than these are feasible. When the time necessary to complete a calculation is a factor, one approach is to increase the sample size steadily until arriving at a value that is stable with respect to further increases in sample size. The stability criterion is perhaps most applicable to hypothesis testing, where we may not need to know the exact confidence level of the observed statistic – only that we can (or cannot) reject the null hypothesis at a 5% confidence level.

Using this sequential approach, we could, for example, run a bootstrap t-test and find that in 100 bootstrap tests the difference in means exceeds the observed difference 40 times (yielding $p=0.40$). It is probably safe to state that we cannot reject the null at a 5% confidence level in light of that result. A repetition of the bootstrap procedure might yield a slightly different confidence level, even changing by several percentage points, but it is highly unlikely to yield $p < 0.05$. Similarly, in such a bootstrap t-test, if the difference in bootstrap means never exceeds the observed difference in means (in 100 bootstrap sets), a single repetition of the bootstrap calculations at 100 bootstrap sets confirms that $p < 0.05$ appears to be reasonable (although the authors would probably use more than 100 repetitions in results intended for publication, just to be cautious). The difficulty arises when the bootstrap estimate of the p-value is very close to the desired confidence level ($p = 0.05$ in this example). In such a case, a large number of bootstrap sets may be warranted.

It is worth remembering that for $N_{Bootstrap}$ sets, the smallest confidence level we could possibly estimate is $1/N_{Bootstrap}$ – e.g. for 1000 bootstraps, the smallest confidence level we could ever hope to estimate is $1/1000 = 0.001$. The estimate of the confidence interval at 0.001, using 1000 bootstrap sets, is essentially based on the value obtained from a single bootstrap set (the one producing the largest or smallest value out of the 1000 sets examined). This suggests that it would be more appropriate to use 10 000 to 20 000 sets to obtain an estimate of the confidence interval at 0.001, so that the estimate is based on the results of 10 to 20 bootstrap sets (the 10 or 20 most extreme values out of the 10 000 or 20 000 total sets). In most cases, it is not necessary to estimate confidence intervals at 0.1% (0.001) because 5% confidence intervals are the standard, and are achievable with lower numbers of bootstraps.

When in doubt about the number of bootstrap sets that should be used to establish a particular confidence interval, the safest approach is to repeat the analysis after doubling the number of bootstrap sets (to determine whether that doubling alters the confidence level). This doubling should be repeated until the estimate stabilizes; the iterative approach may be time-consuming, but it is preferable to a blind reliance on a rule of thumb.

Summary

Randomization tests provide a useful alternative to the more familiar analytical statistical approaches, particularly when the observed distribution departs substantially from the assumptions of analytic models, or when no analytic estimate is available for the confidence interval of a specific statistic needed for the analysis. The performance of these methods appears to be equal to that of analytic methods even though the greater flexibility of these randomization approaches does come at the cost of increased computational time and it will sometimes be necessary to produce specialized software for novel tests).

References

Anderson, M. J. (2001a). A new method for non-parametric multivariate analysis of variance. *Austral Ecology, 26*, 32–46.

Anderson, M. J. (2001b). Permutation tests for univariate or multivariate analysis of variance and regression. *Canadian Journal of Fisheries and Aquatic Sciences, 58*, 626–639.

Anderson, M. J. (2006). Distance-based tests for homogeneity of multivariate dispersions. *Biometrics, 62*, 245–253.

Anderson, M. J., & Robinson, J. (2001). Permutation tests for linear models. *Australian & New Zealand Journal of Statistics, 43*, 75–88.

Anderson, M. J., & Ter Braak, C. J. F. (2003). Permutation tests for multi-factorial analysis of variance. *Journal of Statistical Computation and Simulation, 73*, 85–113.

Dryden, I. L., & Mardia, K. V. (1998). *Statistical shape analysis*. New York: John Wiley & Sons.

Edgington, E. S. (1995). *Randomization tests*. New York: Marcel Dekker.

Efron, B. (1979). Computers and the theory of statistics, thinking the unthinkable. *Society for Industrial and Applied Mathematics Review, 21*, 460–480.

Efron, B. (1987). Better bootstrap confidence intervals. *Journal of the American Statistical Association, 82*, 171–185.

Efron, B. (1992). Jackknife-after-bootstrap standard errors and influence functions. *Journal of the Royal Statistical Society Series B, Methodological, 54*, 83–127.

Efron, B., & Tibshirani, R. J. (1993). *An introduction to the bootstrap*. London: Chapman & Hall.

Fisher, R. A. (1935). *The design of experiments*. Edinburgh: Oliver & Boyd.

Freund, J. E., & Walpole, R. E. (1980). *Mathematical statistics* (3rd ed.). Englewood Cliffs, NJ: Prentice-Hall.

Good, P. (1994). *Permutation tests: A practical guide to resampling methods for testing hypotheses*. New York: Springer-Verlag.

Goodall, C. (1991). Procrustes methods in the statistical analysis of shape. *Journal of the Royal Statistical Society, Series B: Methodological, 53*, 285–339.

Hoeffding, W. (1952). The large-sample power of tests based on permutation of observations. *Annals of Mathematical Statistics, 23*, 169–192.

Jackson, D. A., & Somers, K. M. (1989). Are probability estimates from the permutation models of Mantel's test stable? *Canadian Journal of Zoology, 67*, 766–779.

Klingenberg, C. P., & McIntyre, G. S. (1998). Geometric morphometrics of developmental instability: Analyzing patterns of fluctuating asymmetry with Procrustes methods. *Evolution, 52*, 1363–1375.

Larsen, E. (2005). Developmental origins of variation. In B. Halgrímsson, & B. K. Hall (Eds.), *Variation* (pp. 113–129). Elsevier/Academic Press.

Manly, B. F. (1997). *Randomization, bootstrap and Monte Carlo methods in biology*. London: Chapman & Hall.

McArdle, B. H., & Anderson, M. J. (2001). Fitting multivariate models to community data: A comment on distance-based redundancy analysis. *Ecology, 82*, 290–297.

Quenouille, M. (1949). Approximate tests of correlation in time series. *Journal of the Royal Statistical Society B, 11*, 18–44.

Raspé, R. E. (1785). Baron Münchhausen's narrative of his Marvelous Travels and Campaigns in Russia.

Rencher, A. C. (1995). *Methods of multivariate analysis*. New York: John Wiley & Sons.

Rencher, A. C., & Schaalje, G. B. (2008). *Linear models in statistics*. Hoboken, NJ: Wiley.

Robinson, J. (1973). Large-sample power of permutation tests for randomization models. *Annals of Statistics, 1*, 291–296.

Rohlf, F. J. (2009). tpsRegr 1.37. *Ecology and evolution*. State University of New York at Stony Brook.

Romano, J. P. (1989). Bootstrap and randomization tests of some non-parametric hypotheses. *Annals of Statistics, 17*, 141–159.

Sheets, H. D., & Mitchell, C. E. (2001). Why the null matters: Statistical tests, random walks and evolution. *Genetica, 112*, 105–125.

Tukey, J. W. (1958). Bias and confidence in not quite large samples. (Abstract). *Annals of Mathematical Statistics, 29*, 614.

CHAPTER

9

General Linear Models

In the last chapter, we examined two simple models, one a linear regression and the other a comparison of two means. It is likely obvious that we will need more complex models to answer questions left open by those two analyses. In particular, in our analyses of alpine chipmunk (*Tamias alpinus*) jaw shape, we found that both size and sex have a statistically significant impact. What we now want to know is whether sex has a significant impact on shape, controlling for size, and whether alpine chipmunk jaws are sexually dimorphic in their response to size. To answer these questions we need more complex models but, like the simple models that we already presented, these more complex ones are all examples of *General Linear Models*, which have important features in common. One common feature is that we can test the statistical significance of the models using *F*-tests, and a second common feature is that we can state how much of the variation is explained both by the model and by error by *variance partitioning*. It turns out that a wide range of models can be analyzed using the same general approach, including the models used in Analysis of Variance (ANOVA), Analysis of Covariance (ANCOVA), Multivariate Analysis of Variance (MANOVA), Multivariate Analysis of Covariance (MANCOVA), Multiple Regression (and so forth). This list might not actually seem impressively long, but some of these procedures admit to numerous designs (many of which have names and apparently specialized methods to fit). But, despite their diversity, all can be viewed as cases of the General Linear Model. Because some of the named models have multiple names (owing to the diversity of fields that have used them), we will try to alert you to the multiple terms for a single model or concept.

General Linear Models comprise a broad class of predictive mathematical models employed in statistics which, as is evident from their name, are linear. More specifically, they are *linear in their fitted parameters*. What this means is that any change in the predicted outcome is linearly related to a change in the fitted parameters. For example, if we change a parameter by some value δ, resulting in a change of Δ in the predicted value, then a change of 2δ in the parameter will produce a change of 2Δ in the predicted value. The most common of the general linear models is the univariate linear regression model:

$$Y = mX + b + \varepsilon \tag{9.1}$$

Geometric Morphometrics for Biologists
DOI: http://dx.doi.org/10.1016/B978-0-12-386903-6.00009-5

which is clearly linear in both fitted parameters, *m* and *b*.

A function such as:

$$Y = me^{X} + \varepsilon \tag{9.2}$$

is still linear in the fitted parameter *m*, even if not in the independent variable *X*. Because it is linear in *m*, this expression is an example of a General Linear Model. However, an expression such as:

$$Y = \mathrm{m}e^{\alpha X} + \varepsilon \tag{9.3}$$

is not within the family of General Linear Models because it is not linear in the fitted parameter α.

Working with General Linear Models allows us to use a wide range of models, but we are most likely to be interested in simple extensions of the linear regression model that we discussed in the last chapter. From that starting point, we can build more complex models by incorporating both continuous and additional categorical independent variables. For example, our response variable, *Y*, might depend on categorical factors *A* and *B* plus a continuous variable X, as well as on the unknown sources of the error term ε. This dependence is written as:

$$Y_i = A_i + B_i + A_i \times B_i + \beta_x(A, B, A \times B)X_i + \varepsilon_i \tag{9.4}$$

where Y_i is the dependent variable (either univariate or multivariate) for the *i*th specimen. Y_i can be either a simple number (a scalar), such as centroid size, or a vector of *K* real values for multivariate data, such as shape. *A* and *B* are categorical variables, which are usually termed "factors" and *X* is a continuous variable, which is often termed a "covariate". "$A_i \times B_i$" is known as a "crossed" or "interaction" term, meaning that factor *A*'s impact on Y_i depends on Y_i's value on factor *B*. In this model, the slope term, β_x, is a function of *A*, *B* and the interaction term $A \times B$. In the univariate case, the fitted terms A_i, B_i, $A_i \times B_i$ and $\beta_x(A, B, A \times B)$ are scalars whereas in the multivariate case, they are vectors having as many coefficients as there are variables, and so is the error term ε_i. The covariate X_i, however, is univariate in this case.

Just as we saw in our discussion of ANOVA in the last chapter, the factors are categorical variables such as sex, diet class or species, and the covariates are continuously-valued variables such as size, position along a geographic transect or fitness, etc. Throughout this chapter we will assume that the continuous variables (*Y* and *X*) are centered, i.e. their mean values are zero. When these variables are not centered, the model would include an explicit term for the mean value (and the data would have an additional degree of freedom that is lost in the process of centering). In the case of shape data, *Y* is typically shape, expressed in terms of the difference in shape between each specimen and the mean. Each factor has two or more levels (i.e. distinct values of the factor) corresponding to the number of groups. For example, in the case of sexual dimorphism that we discussed in the last chapter, the factor is sex, which has two levels: "male" and "female". Many factors have more than two levels; habitat, for example, has (among other possibilities): "montane", "mid-elevation", "desert", "lacustrine", "riverine", "marine," etc.

We can use the model presented in Equation 9.4 to ask a series of questions such as: (1) Does shape (Y) depend on either or both of the factors A and B? (2) Is there an interaction between the factors? (3) Does shape depend on a covariate X? (4) Does the slope of Y on X depend on one or both of the factors? (5) Does the slope depend on the interaction between the two factors? As well as asking such "yes" or "no" questions, we can also ask what fraction of the total variation in shape is explained by each term, factor and covariate. The set of techniques collectively used to answer such question is known as General Linear Models (GLM).

GLM include a variety of models, such as the linear regression and two group comparisons of the last chapter, plus techniques such as analysis of variance (ANOVA), analysis of covariance (ANCOVA) and their multivariate equivalents, multivariate analysis of variance (MANOVA), multivariate analysis of covariance (MANCOVA), multiple regression, and a range of other models (that do not all have names). The model described above, which has two factors, A and B and a covariate X, would be called a two-factor ANCOVA (or MANCOVA). Like the family of ANOVA methods, GLM methods use the distribution of sums of squares (or mean squares), which are proportional to the variance contributed by each factor (and covariate) plus the error terms. When the data are univariate, the sums of squares are scalars but when the data are multivariate, they are matrices "sums of squares and cross products" (SSCP) matrices. Whether univariate or multivariate, sums of squares are used to form F-ratios for hypothesis testing and to estimate (and decompose) the variation explained. Whether the data are univariate or multivariate, the error term (ε) represents the residual unexplained variance or "noise". Different methods are required when using SSCP matrices for hypothesis testing, but the fundamental concepts are the same in both the univariate and multivariate case.

This chapter will discuss GLM, beginning with applications to univariate data to lay the foundation, then extending them to multivariate data. We consider both classical statistical models and those tested by permuting the data. We begin with a general overview of factors and experimental design because that design has a major impact on the efficacy of these statistical methods. One of the most important considerations is the number of factors to be tested because the number of interaction terms grows with the number of factors. For example, in a model containing three factors (A, B and C) there are four interaction terms, $A \times B$, $A \times C$, $B \times C$ and $A \times B \times C$. Adding a fourth factor increases the number of interaction terms to 11 and so on. The rapid growth of interaction terms, and the attendant increase in the number of parameters that must be estimated, makes the effective use of complex models a daunting task even though they are powerful.

FACTORS AND EXPERIMENTAL DESIGN

The nature of the factors in a particular experiment, and the manner in which data are collected will both have a major impact on how the data can be analyzed as well as the types of questions that can be addressed effectively. We first explain the distinction between *fixed* and *random* factors, then the distinction between *crossed* and *nested* factors and then the distinction between main effects and interactions. After that, we focus on the distinction between balanced and unbalanced designs because the procedures for analyzing unbalanced designs can be disappointingly uncertain (Searle, 2006). Although permutation methods will be helpful in many applications involving shape data, they share many of the limitations of classical analytic methods. Thus, the impact of unbalanced designs should be considered when designing the study to avoid being forced to confront the problems they pose after the data are collected. It can be very disheartening to contemplate leaving out measured individuals or even factors to balance the data.

Fixed and Random Factors

The distinction between fixed and random factors can be both subtle and difficult to explain because the same factor can be either fixed or random depending on context. In general, when we are interested in the specific factors and can collect data on all relevant levels of the factor, then the factor can be treated as fixed. In contrast, when the factors are not of particular interest in their own right, and the levels that are measured are a random sample of the levels that could have been measured, the factor is treated as random. In general, in the case of a random factor, the factor itself might not be of any special interest but it is one that needs to be taken into account when testing hypotheses about the fixed factors. Random factors are sometimes viewed as "nuisance terms" because they are both uninteresting and contribute to variation in the sample.

To make this distinction more concrete, we can consider two cases in which the same factor is either fixed or random depending on the specifics of the question being asked. In the first case, suppose that we want to test the hypothesis that the shapes of fly wings depend on altitude and sex. We have sampled both sexes at three altitudes, and we have sampled them at different times (although we made sure to sample both sexes, and flies at all altitudes, at the same times). We are not interested in seasonality of shape, but we might nonetheless suspect that the season in which the flies were collected might affect wing shape, and that flies at higher altitudes might be differently affected by season than those at lower altitudes. Thus, even though we are not interested in time, we need to consider the possibility that some of the variation within the sample is temporal. In that case, habitat and sex would be fixed factors, but time would be random. The dependence of shape on time is not a hypothesis that we wish to test, nor have we exhaustively sampled (or controlled) for it. We have merely taken a sample of shapes at several times. We do not particularly care when any sample was collected – we only care that they were collected at different times. As a result, we want to know how much of the variation is due to collecting date, but we do not care whether the flies collected at one particular time differ significantly from those collected at another specific time. In this context, time is merely a nuisance variable because it could explain some of the variance in the sample; including it as a term in the model is important both to avoid ascribing the variation explained by it to one of the fixed factors of interest, and to lessen the unexplained variation in the data (by explaining it). In contrast, say that we do want to know whether time has an impact on wing shape. We may want to know whether high and low altitude flies, of either sex, respond to climate change, and if that response depends on altitude or sex. In this case, time would be a fixed factor. Thus, whether a factor is fixed or random depends on what hypothesis one intends to test, and exactly how data collection is structured. It is not a feature of the variable itself.

Typically, the statistical significance of random factors is not the hypothesis that really interests us. Those factors are measured simply to determine how much variation can be attributed to them. We might also want to quantify the variance explained by fixed factors, and doing so may seem to blur the distinction between the two kinds of factors. The distinction, however, remains important, especially when a study contains both fixed and random factors because the denominator of the F-ratio depends on whether the factor is random or fixed.

Crossed and Nested Factors

In the examples presented above, we measured males and females at all altitudes in all years. At least in principle, any individual could have been allocated to any "treatment" (by nature if not by us). In some cases, all the individuals at one level on one factor are also all at the same level on another. For example, in an experiment on the impact of dietary consistency on mandible morphology, infant deer mice could not be removed from their mothers and raised by different mothers (Myers et al., 1996). Consequently, all siblings ate the same food. That is important because we would expect siblings to be more similar to each other than to unrelated mice not only because they eat the same diet but also because they are genetically related to each other and share the same uterine and nestling environment. There are many families that eat the same diet, but no families eat more than one diet. Thus, within any single level of the diet factor there are many families, but at any single level of the family factor there is only one diet. In this case, family is *nested* within the diet factor. Because variation among families may contribute substantially to the variation in shape, this source of variation needs to be taken into account when assessing the impact of diet on shape. As may be obvious, we are not particularly interested in whether family A differs from family B, or from any other family. All that we care about is whether variation among families contributes to variation in shape, and whether all families respond similarly to diet, making it safe to generalize about the impact of dietary consistency on shape.

To take another sample, suppose that we are rearing fish in a large number of tanks to test for the impact of water temperature on body shape. For each temperature, we have multiple tanks but (obviously) all the fish within any given tank are reared at the same temperature. We want to assess the impact of water temperature on body shape and to ensure that our result is general rather than depending on the particular tank in which the fish were reared. Because the fish within the same tank may have something in common aside from the temperature of rearing, and that common feature might vary across tanks even if water temperature does not vary, that other (unmeasured) factor could contribute to the variation among fish. It might contribute both to the variation among fish reared at the same temperature and to the response of the fish to water temperature. Although we do not care whether the fish in one tank differ, on average, from the fish in any other specific tank, we do care whether variation among tanks contributes to our experimental error and, even more importantly, whether the response to the factor of interest depends on the nested term. Again, the nested factor is also a random factor.

Whether factors are fixed or random is important to the statistical design, as is whether factors are crossed or nested. The distinction between fixed and random is important because the question asked about fixed factors is whether the average shapes differ across those specific levels. In this case, there is no experimental error associated with the selection or measurement of the levels. The levels are not randomly selected hence we do not have to consider the random variation among them. But when the factor is random, the levels we sample are random samples of the factor and it is therefore measured with error (just like the dependent variable is). Hence there is an error term associated with the measurement of the factor, just as there is with any random variable. That adds an additional error term to the model. Even though the numerical values for each level do not depend on whether the factor is fixed or random, there is an additional population parameter to estimate when the factor is random. Whether factors are crossed or nested is important not only because the nested terms are random but also because nesting affects the design of the statistical test; as we discuss in more detail below, when we design a scheme for permuting observations, the individuals being permuted must be equivalent to each other in order to be exchangeable. Rather than exchanging *individuals* as if they were all equivalent to each other, levels of the nested factor are the equivalent units (and therefore the exchangeable ones). For example, rather than permuting all the deer mice as if they each were equivalent to any other, we would permute whole litters.

Main Effects and Interaction Terms

The impact of each factor, considered one at a time, is known as a "main effect". For example, the impact of diet on shape is a main effect of diet. If we have just one factor, that one main effect is all that concerns us (other than error). But when we have two or more factors, both factors will have main effects and it is also possible that the two factors interact. If they interact, the impact of one factor depends on the level of the other. For example, consider those male and female flies that we collected at different altitudes. We might find that females differ from males in their response to altitude. That differential response is the interaction effect – the effect of one factor (altitude) depends on the level of the other (sex). When the interaction term is significant, then we cannot generalize about the effect of the main factor because its effect is conditional on another factor. We could say that altitude's effect on female shape is general (so long as that does not also depend on the impact of time) but we cannot say that altitude's effect on shape is general because it is not – it depends on sex. In general, we would not interpret main effects in the presence of interactions. When the design is unbalanced, meaning that the sample sizes are not equal for all the levels in the analysis, it can be difficult to partition main effects and interactions.

Decomposition of Variance

We often need to decompose the variation of the dependent variable (shape) into the contributions made by the factors, covariate(s), interaction terms and error. This decomposition is done by determining the variation explained by the combinations or subsets of the total model. Differences in the variance explained by the various combinations of independent variables sort out (i.e. decompose) the variance into the portions explained by each term of the model. Or perhaps we should say that they do so under ideal experimental conditions, including an ideal (i.e. balanced) design. Significance testing of each contribution is done using *F*-tests, which are ratios between the variances explained by different terms. As discussed earlier, variance can be characterized using variance–covariance matrices (based on SSCP), or squared Procrustes distances, and there are both analytic and numerical approaches to estimating significances of the observed *F*-ratios, which will be discussed later. The important feature here is that all of this depends on being able to decompose variance.

Balanced and Unbalanced Design

In a balanced design, each possible combination of levels of the various factors has the same sample size. If we lay the factors out along the rows and columns of a matrix, each "cell" in that matrix is a unique combination of levels of all factors. In a balanced design, the number of individuals within all cells is equal. For example, say that we have two factors, handedness and sex. The levels of handedness are right-handed and left-handed, and the levels of sex are male and female. In a balanced design, we would have equal numbers of right- and left-handed individuals and equal numbers of males and females; and furthermore, equal numbers of right-handed females, left-handed females, right-handed males and left-handed males. But we can see that numbers are not equal in all cells of Table 9.1. We have equal numbers of right- and left-handed females but we have an excess of right-handed males. This kind of design is called "unbalanced".

Unbalanced designs present serious problems for statistical analyses because the experimental design, not biology, induces a correlation between factors. In this case, handedness is correlated with sex. If the sex of a particular specimen is female, then the chances are equal that the individual is right- or left-handed but that is not the case for males. For males, the chance of being right handed is 70%. That is a problem for the following reason. Suppose that being right-handed causes a detectable change in shape of the hand (or other part of the limb or even the brain). If we compare the variance explained by sex, without considering handedness, we would see that on average, most men had the shape associated with right-handedness (because 70% of the males are right-handed). In contrast, the average female would not have that shape because half do and half do not. If we were to consider handedness alone, we would see the impact of being right-handed, but it would be difficult to tell if that effect was due to

TABLE 9.1 Number of Individuals Sampled by Sex and Handedness

Sex	**Handedness**	
	Right	**Left**
Female	50	50
Male	70	30

being right-handed or male. Whereas the right-handed sample is biased toward males, the left-handed sample is biased toward females. As a result, it is difficult to conclude that any difference found between right- and left-handed samples is due purely to handedness. Part of the difference could instead be due to sex. In effect, the two factors are confounded by the experimental design, making it difficult to separate the effects of the factors and also to estimate the terms.

Unfortunately, unbalanced designs are far more common in biological studies than balanced ones. In some cases, the design might have been balanced at the outset of the experiment, but it becomes unbalanced when the organisms die or escape or when specimens are damaged in the process of capture or preparation. In field studies, animals may wander away from the study plots or be uncatchable on any given day. In observation of museum collections, the number of specimens that can be used to estimate the levels of a factor depends on the number of undamaged specimens at those levels contained in the collection(s).

It is possible to ensure having a balanced design by restricting the number of individuals in each cell to the number contained in the smallest cell. For example, to obtain a balanced design for our analysis of sexual dimorphism in the last chapter, we removed the excess females to equalize the numbers of males and females. Had we added another factor to the design, such as whether the animals came from the Yosemite area or the southern Sierras, we would have ended up with only 24 animals per cell because there are only 24 males from the southern Sierras. Balancing a design by randomly leaving out specimens is sometimes recommended as an alternative to using an unbalanced design (Underwood, 1997). The reason for recommending a balanced design, even if that means leaving out data, is not that sums of squares cannot be calculated. Rather, it is that the results can be difficult to interpret. Additionally, the fact that the factors are confounded raises problems for the analysis and there is some controversy about the best procedure to use for calculating sums of squares for unbalanced designs (an overview of some procedures is given below, in *Unbalanced Designs and Sums of Squares*).

DESIGN MATRICES

Design matrices play a crucial role in the statistical analysis because, as is obvious from their name, they encode the design of the study. The design matrix contains the information about the number of factors, the levels of each factor, the number of distinct combinations of factors in each interaction term, etc. This matrix is often called the "**X**" matrix because it is represented by **X** in the following expression, which is a strikingly compact equation that applies to a wide range of models, both univariate and multivariate. It may appear to be complex because it is written in terms of matrices, but the pay-off is the ability to formulate virtually any linear model in the following form:

$$\mathbf{Y} = \mathbf{XB} + \varepsilon \tag{9.5}$$

In this expression, **Y** is the centered data matrix for the dependent variable (typically shape in our case). Because the data are centered, the mean of each column is zero and

Y contains the deviations of each individual from the overall mean. **B** is the matrix of coefficients of the model, which will be fitted to the data, **X** is the centered design matrix and ε is the matrix of residuals or error terms. If **Y** is a matrix with N rows (one per specimen) and Q columns, the matrix of residuals, ε, will also have N rows and Q columns. The size of the design matrix, **X,** $C \times N$, depends on the design, i.e. on the number of factors, the number of levels of each factor and the number of distinct combinations of factors in any interaction terms, as well as the number of interaction with covariates. It can also depend on how the model is coded because the number of columns, ignoring interaction terms for the moment, could either equal $G-1$ (where G is the number of groups) or G. It takes $G-1$ columns to specify the design, so using G columns makes the coding scheme redundant (and the **X** matrix is then not invertible). We will therefore focus on design matrices that have $G-1$ columns.

To understand the codes, it is important to remember that we are using regression to analyze categorical factors. We therefore need values for the categorical factors that make the results of the regression interpretable. One coding method is called "dummy coding". According to this method, all individuals are coded as either a zero or one to indicate each individual's level on each categorical factor; including all interaction terms. Which group is coded as zero or one is arbitrary, but the interpretation depends on the codes because the intercept is the mean of the group coded as zero. Usually, the control group is the one coded zero and the null hypothesis is that the means of the other groups do not differ from the mean of the control group. The coefficients for the other groups give the deviations from the control group mean. If there are three groups, it takes two columns to encode a single factor; all individuals belonging to the first group will have ones in the first column and zeros in the second, all individuals belonging to the second group will have zeros in the first column and ones in the second and all individuals belonging to the third group will have zeros in both columns. To obtain the codes for the interaction terms, the columns of codes for the factors are multiplied by each other. Coding can become complex when factors are nested, so the **X** matrices for these more complex designs are discussed later in context of the more complex models.

An alternative coding method is called "effect coding"; according to this method, all individuals are coded as negative one, zero, or positive one. If there are only two groups, the first one is coded as -1, the other as 1, and if the design is balanced, the mean for the column is zero. If there are three groups, the first is coded as -1, the *last* as 1, and the second by 0; in the second column, the first group is coded as 0, the second as -1, and the third as 1. Using this method, the intercept is the grand mean and $\mathbf{X}_1$ is the deviation of the first group from that mean, $\mathbf{X}_2$ is the deviation of the second group from that same mean, etc. So, when testing the statistical significance of the coefficients for **X**, we are testing the null hypothesis that one group does not differ from the grand mean by more than expected by chance. As mentioned above, the codes for interaction terms are obtained by multiplying the columns for the interacting factors.

Coding is more difficult when the design is unbalanced for a reason that may become obvious if you consider that the grand mean will not be zero when there are different numbers of positive and negative ones. The codes will therefore have to be modified to ensure that the grand mean is still zero and that the columns of **X** are mutually orthogonal. One approach is to code the first group as $(N-n_i)/N$ where n_i is the number of individuals in that group and N is the total sample size. So, if there are 10 individuals in the first group, and 100 total, the code for the first group is $(100 - 10)/100$ or 0.9. Then all other individuals will be coded as $-n_i/N$ which, for our example, would yield $-10/100$ or -0.1. As a result, we would have 10 individuals coded 0.9 and 90 coded as -0.1, so the sum of the codes for that column equals zero. In the next column, the n_i term would be the number of individuals in the second group, and so forth. As you might imagine, coding a design matrix can become very tedious, and error-prone, when there are many groups and factors. Fortunately, there are programs that can do this for you. Two of them, the model.matrix function in the stats package in R (R_Development_Core _Team, 2011), and XMatrix (Anderson, 2003), are discussed in the workbook.

THE FORM OF A GENERAL LINEAR MODEL

In general, **X** will have N rows and C columns and will also be a centered matrix. For a simple linear regression model or, as discussed above, for a single factor model with only two levels, **X** would be an N row by one column ($N \times 1$) matrix. The fitted parameters of the model are in a K by C matrix **B**. Different models are expressed by how we form the design matrix **X**.

The value of **B** is then estimated from:

$$\mathbf{B} = (\mathbf{X}'\mathbf{X})^{-1}\mathbf{X}'\mathbf{Y} \tag{9.6}$$

where $\mathbf{X}'$ denotes a matrix transposition, and $\mathbf{X}^{-1}$ indicates a matrix inversions.

The value of the **Y** matrix predicted by the model is:

$$\mathbf{Y}_{\text{pred}} = \mathbf{XB} = \mathbf{HY} \tag{9.7}$$

where $\mathbf{H} = \mathbf{X}(\mathbf{X}'\mathbf{X})^{-1}\mathbf{X}'$ and the residuals may be estimated as:

$$\varepsilon = \mathbf{Y} - \mathbf{Y}_{\text{pred}} = \mathbf{Y}(\mathbf{I} - \mathbf{H}) \tag{9.8}$$

where **I** is the identity matrix. As we have required that the matrices be centered, the original sum of squares and cross products matrix in the data may be calculated as:

$$\mathbf{SS}_{\text{total}} = \mathbf{Y}'\mathbf{Y} \tag{9.9}$$

where $\mathbf{Y}'$ is the transpose of **Y**. If we are dealing with univariate data, this is simply called the sum of squares, a scalar value. For multivariate data, the diagonal of the matrix is the sum of squares for each variable and the off-diagonals are the cross products. If we divided the SS_{total} by $(n - 1)$, the degrees of freedom, we have the variance or the variance–covariance matrix of the data. The residual or error sum of squares may be calculated as:

$$\mathbf{SS}_{\text{error}} = \varepsilon'\varepsilon \tag{9.10}$$

The sum of square matrix explained by any given model may be found as:

$$\mathbf{SS}_{\text{model}} = \mathbf{SS}_{\text{total}} - \mathbf{SS}_{\text{error}} \tag{9.11}$$

In addition to being a very simple looking expression even when the model is complex, the approach based on matrix methods is relatively straightforward to implement in software.

F-TESTS AND MEAN SQUARES

We can write out expressions for the mean squares for the model or submodels (discussed in more detail below), based on the assumption that the model is true. These terms are just the sum of squares divided by the degrees of freedom in the model. It will turn out that the mean square terms of the various subsets of the model are typically related to one another in simple additive ways, meaning that models that differ only by additions of terms will be identical in their predicted mean square (EMS) values for terms common to all the models (at least for balanced designs). So, if we have two models that differ only by one factor, the model that contains that factor would be termed the "full model" and the model lacking it would be known as the "reduced model"; the ratio of the MS values for these two models, is the *F*-ratio for the two models, just as it is when the MS of one factor is compared to error MS. Both *F*-ratios can be assessed for their statistical significance using either analytic or resampling methods, just as in the case of a linear regression model. When the design is unbalanced, the situation is more complex because, depending on the approach taken, either the expressions depend on the sample size in each cell or the sums of squares for the terms will not add up to the total sums of squares. We will therefore defer consideration of unbalanced designs until we have more fully discussed the simpler case of balanced designs.

In the case of univariate data, the sum of squares and the mean squares are simple scalar values, so forming *F*-ratios is simple and obvious. That is not the case for multivariate data. There is a number of approaches to working with the sum of squares and cross products (SSCP) matrices that arise from the matrix methods discussed above. Alternatively, it is possible to use permutation-based approaches based on summed squared Procrustes distances, using the "outer sum" method developed by Anderson and colleagues (Anderson, 2001a,b; Anderson and Robinson, 2001), discussed in more detail below. Rather than tackling the several related topics all at once (i.e. how to calculate mean squares, the relevant *F*-ratios for both balanced and unbalanced designs and the extension of these procedures to multivariate data), we instead start by looking at a few types of models, applying them to univariate data. After laying this foundation, we discuss unbalanced designs. Thereafter, we review a range of experimental designs and give the expected mean squares and *F*-tests. We then extend the models to multivariate data and then explain the permutation tests.

Univariate Data with One Factor

In this section, we examine the development of a GLM for a univariate variable Y that is hypothesized to depend on a single fixed factor A. We show one approach to calculating variances based on sums of squares, using the estimated means of the specimens in each level of the factor A. This approach is conceptually easy to understand, but it is not in the matrix notation presented earlier, and it is not the approach used in most computer-based calculations because most modern approaches to calculating sums of squares are based on matrix algebra. Thus, they use the difference between sums of squares explained by different models expressed in terms of design matrices. The simple summation methods are easier to understand at an introductory level, but harder to scale up to large problems and probably more prone to rounding errors. Researchers interested in programming their own GLM methods will need to consult more advanced texts to develop a complete understanding of these approaches (Rencher, 1995; Searle, 1997, 2006; Anderson, 2001a,b; Rencher and Schaalje, 2008).

Let us suppose that Y has J distinct levels and n_j specimens per cell. We won't require a balanced design at this point but we will require that Y be centered (i.e. the mean value is zero), removing one degree of freedom. For the ith specimen in cell j we have the model

$$Y_{ij} = \alpha_j + \varepsilon_{ij} \tag{9.12}$$

where α_j is the contribution of the jth level of the factor to the value, and ε_{ij} is the error. Notice that we require that the mean value of the residual terms ε_{ij} be zero, with variance σ_e^2, and the mean value of Y is zero (because Y is centered). Consequently, the $n_j\alpha_j$ terms summed over all the cells must also equal zero:

$$\sum_{j=1}^{J} n_j\alpha_j = 0 \tag{9.13}$$

This means that there are J values α_j but only $(J-1)$ are independent because, as mentioned above, constraining the sum to be zero removes one degree of freedom.

We can now look at variance partitioning to understand how to form F-ratios. We start by looking at the summed square values around the mean value, which is $\overline{Y}$, then we split these sums of squares into two terms, one due to the scatter about the mean of each group (level of the factor $\overline{Y}_j$), and the other due to the scatter of the group means about the grand mean. The total sum of squares is:

$$SS_{total} = \sum_{j=1}^{J}\sum_{i=1}^{n_j}(Y_{ij}-\overline{Y})^2 = \sum_{j=1}^{J}\sum_{i=1}^{k}(Y_{ij}-\overline{Y}_j)^2 + \sum_{j=1}^{J} n_j(Y_j-\overline{Y})^2 \tag{9.14}$$

Note that $\overline{Y}$ must be zero; we have included $\overline{Y}$ here so that our expression for the SS will be consistent with the other standard presentations of these ideas.

The first term represents the error and the second is the SS due to A (the between groups of factor sum of squares).

$$SS_{error} = \sum_{j=1}^{J}\sum_{i=1}^{k}(Y_{ij}-\overline{Y}_j)^2 \tag{9.15}$$

The SS_{error} term, also called the residuals, has an expected value equal to the degrees of freedom times σ_e^2. The degrees of freedom in the error term are given by:

$$df_{error} = \sum_{j=1}^{J} n_j - J \tag{9.16}$$

The expected mean square error (or the expected mean square residual) is then estimated as

$$MS_{error} = \frac{SS_{error}}{df_{error}} \tag{9.17}$$

whose expected value (EMS_{error}) is σ_e^2. In contrast to the mean square (MS) term, which is calculated from the data, the expected mean square (EMS) is calculated from the model. If the model does describe the data, the MS and the EMS should be similar, differing only by relatively minor random variation.

The sum of squares for the between groups term, or that contributed by the factor A (when $\overline{Y} = 0$) is

$$SS_A = \sum_{j=1}^{J} n_j \overline{Y}_j^2 \tag{9.18}$$

The mean square value is estimated as $MS_A = SS_A/(J-1)$ because the degrees of freedom are $df_A = (J-1)$. The expected mean square (EMS_A) has two contributions, a pooled variance of the error terms across the groups plus a component representing the squared effects of the factors:

$$EMS_A = \sigma_e^2 + \sum_{j=1}^{J} n_j \frac{\alpha_j^2}{(J-1)} \tag{9.19}$$

The null hypothesis we want to test is that the factor A does not contribute to the value of Y. Having constrained the mean of Y to be zero, the null hypothesis is therefore that the α_j values are all equal and are, in fact, all equal to zero. Under these conditions, the expected mean square value of EMS_A is simply σ_e^2. The F-ratio of the variance explained by the model relative to the error or residual variance is then expected to be 1 should the null hypothesis be true. Consequently, we can compute the F-ratio based on the data as:

$$F = \frac{MS_A}{MS_{error}} = \frac{SS_A/df_A}{SS_{error}/df_{error}} \tag{9.20}$$

which will follow an F-distribution with degrees of freedom df_A, df_{error}. Notice that the F-ratio is the variance explained by the model divided by the unexplained variance, just as it was when we applied F-tests to regression models, or used Goodall's F-test to compare the mean shapes of two groups. Because the expected mean term in the numerator is equal to the denominator plus one additional term, the F-value will be larger than one if the additional term:

$$\sum_{j=1}^{J} n_j \frac{\alpha_j^2}{(J-1)} \tag{9.21}$$

is not zero. A larger F-value would indicate a lower probability that the null hypothesis is true. Rejecting the null hypothesis allows us to interpret the MS_A term as the variance explained by the factor A, and to compare that to the unexplained variance, MS_{error}.

GENERALIZING AND EXTENDING THE SIMPLE UNIVARIATE ANOVA

The simple univariate case with only one factor is only a starting point, but it nevertheless illustrates all the analytic procedures used by GLM methods. In all cases, we would follow the same basic steps, albeit adapting them to multivariate data:

1. *Find the sum of squares* due to each term in the model and use them to determine the variance explained by each term (when possible). When working with multivariate shape data, there are two approaches to finding these sums of squares: (a) the sum of squares and cross products matrix (SSCP) or (b) the summed squared Procrustes distances between specimens. Either can be used as the multivariate equivalent to the simple sum of squares. This can be complex when the design is unbalanced, as discussed below.
2. *Form F-ratios* of different models to test the significance of the model. In the univariate example given above, the numerator model states that factor A influences Y. The denominator model states that Y is just a random value with variance σ_e^2. When we discussed that case above, we referred to the denominator as the MS_{error}, but this is really just a special case. In general, it is preferable to think of the denominator as another possible model because the denominator will not always be MS_{error}. The important idea is that the expected mean square in the numerator differs from the expected mean square denominator only by a single term, the term that represents the factor of interest. We want to test the significance of that factor and to construct the appropriate test we need to derive the expected mean squares for each part of the model. As we consider more complex models with more factors and covariates, we will need an F-ratio for each term. Fortunately, there are many published tables that list the expected mean squares for various models (although usually only for balanced designs). The excellent text by Lorenzen and Anderson (1993), for example, describes how to form sums of squares and F-ratios for a wide range of univariate, balanced designs, which may be adapted with reasonable care to other analyses. For multivariate data, more complex analytic approaches are needed due to the use of sums of squares and cross products (SSCP) matrices.
3. *Test the significance of the F-ratios.* The tests can use sums of squares and cross products matrices or distance based measures of sums of squares. Analytic tests are available for the first and permutation methods are available for both, such as when Procrustes distances are used to estimate sums of squares.

MODELS

Univariate Two Factor Balanced Design

We will now consider a case in which we have two factors, with a univariate dependent variable Y, which depends on two factors A and B (which have p and q levels respectively). As an example of a model of this sort, consider the alpine chipmunks that we analyzed in the last chapter. Because we are considering only univariate models at this point, Y would be "size". We have two factors; the first is "sex", the second is "region" because the chipmunks were sampled in two regions, the Yosemite region and the southern Sierras. So factor A would be "sex" and factor B would be "region". For this case, the full model is:

$$Y = A + B + A \times B + \varepsilon \tag{9.22}$$

A is the effect of sex on size, B is the effect of region on size and $A \times B$ is the interaction term between sex and region, and ε is again the error term with variance σ_e^2 and mean zero (we again center Y). Note that the significance of the interaction term, $A \times B$ is tested according to the contribution that it makes to the total variance in addition to that made by the two *main effects* (i.e. the contributions made by A and B, treated individually). There are a number of different possible designs for this two factor case because either A or B (or both) can be fixed or random. In this example of sex and region, both factors are fixed. But there are models containing two factors with one fixed and the other random; this type of model is said to be *mixed*. A two-factor mixed model is also used for studies of fluctuating asymmetry (i.e. random deviations from bilateral symmetry). There are two main effects, "Individual" and "Side". The first is the variation among individuals, the second is the difference between right and left sides, and the interaction term, "Individual × Side" is the variation among individuals in "sidedness". In this case, "Individual" is a random term – we are randomly sampling individuals and we do not particularly care whether any one individual differs from any other in shape. "Side" is a fixed factor (because we do care about the difference between right and left sides). In this case, it is actually the interaction term that most interests us because our objective is to quantify fluctuating asymmetry. In some cases, both factors are random, a common design used in quantitative-genetic studies where the objective is to decompose the variance into the part explained by variation in genotype and that explained by variation in environment.

When one factor is fixed and the other is random, it becomes possible to consider two different types of constraints on the interaction terms $A \times B$. Remember that Y is centered, and so the sums of all group factors (for both A and B, as in Equation 9.13) are also required to be zero. However, there are two possibilities for the interaction terms $A \times B$, either that the summed interaction terms be zero (as in Equation 9.13, but summed over the interaction terms), which is called the restricted interaction model, or alternatively that there is no restriction on the interaction terms. The restricted model does imply a correlation of the interaction terms, any two interaction terms will not be independent within a given level of B for example (Quinn and Keough, 2002). We follow the recommendations of Quinn and Keough (2003) and Searle et al. (1992) in presenting and using the unrestricted interaction model.

Table 9.2 shows the sums of squares for each of the terms in the model. Again, the sums of squares are shown in a simple summation notation. The F-ratios used to test the significance of each term in the model, for the various combinations of random and fixed factors, are shown in Table 9.3 (note the different denominators for A and B depending on whether the factors are fixed or random, and that the restricted model of interactions is used here), and the calculation of the variance components for each combination is shown in Table 9.4. Results for the analysis of the impact of sex and region on chipmunk jaw size are shown in Table 9.5. Sex does have a significant impact on chipmunk jaw size; females' jaws are larger

TABLE 9.2 Sums of Squares, Degrees of Freedom and Mean Squares for a 2 Way, Balanced Univariate ANOVA

Source	SS	df	MS
A	$SS_A = NJ\sum_{i=1}^{I}(\overline{y}_i - \overline{y})^2$	I − 1	$S_A/(I - 1)$
B	$SS_B = NI\sum_{j=1}^{J}(\overline{y}_j - \overline{y})^2$	J − 1	$SS_B/(J - 1)$
A × B	$SS_{AB} = N\sum_{i=1}^{I}\sum_{j=1}^{J}(\overline{y}_{ij} - \overline{y}_i - \overline{y}_j + \overline{y})^2$	(I − 1)(J − 1)	$SS_{AB}/(I - 1)(J - 1)$
Residual	$SS_{error} = \sum_{k=1}^{N}\sum_{i=1}^{I}\sum_{j=1}^{J}(y_{ijk} - \overline{y}_{ij})^2$	IJ(N − 1)	$SS_{error}/IJ(N - 1)$
Total	$SS_{total} = \sum_{k=1}^{N}\sum_{i=1}^{I}\sum_{j=1}^{J}(y_{ijk} - \overline{y})^2$	IJN − 1	

There are I levels of factor A, and J levels of factor B, with N specimens per cell. The expressions of the form $\overline{y}_i$ indicate means over all cells with level i (or j, or the combination ij), while $\overline{y}$ is the overall mean of all data.

TABLE 9.3 Terms of the F-Ratio for a Univariate Two Factor, Balanced Design

Source	A, B Fixed	A,B Random	A Fixed, B Random
A	MS_A/MS_{error}	MS_A/MS_{AB}	MS_A/MS_{AB}
B	MS_B/MS_{error}	MS_B/MS_{AB}	MS_B/MS_{error}
A × B	MS_{AB}/MS_{error}	MS_{AB}/MS_{error}	MS_{AB}/MS_{error}

TABLE 9.4 Estimates of the Variances Components for a Univariate Two Factor, Balanced Design

Source	A, B Fixed	A,B Random	A Fixed, B Random
A	$(MS_A - MS_{error})/NJ$	$(MS_A - MS_{AB})/NJ$	$(MS_A - MS_{AB})/NJ$
B	$(MS_B - MS_{error})/NI$	$(MS_B - MS_{AB})/NI$	$(MS_B - MS_{error})/NI$
AB	$(MS_{AB} - MS_{error})/N$	$(MS_{AB} - MS_{error})/N$	$(MS_{AB} - MS_{error})/N$
Error	MS_{error}	MS_{error}	MS_{error}

on average than males' jaws (74.18 mm vs 72.88 mm for females and males, respectively). In contrast, region does not have a statistically significant impact on jaw size, nor is there any stastistically significant interaction between sex and region. The analysis of fluctuating asymmetry of chipmunk jaws shows that the interaction term (which is fluctuating asymmetry) is statistically significant although the main effect of "Sides" is not (Table 9.6).

Univariate Two Factor Nested Model

In this case, one of the two factors, B, is nested within factor A, which we denote as $B(A)$. As an example of nesting, we earlier introduced the case of the deer mice that were fed three diets, with all members of each litter being fed the same diet. Thus, in this case,

TABLE 9.5 A Two-Way Fixed Factor Analysis of Variance: The Impact of Sex and Region on Chipmunk Jaw Size

Source	SS	df	MS	*F*	*P*
Sex	45.57	1	45.57	22.357	0.000
Region	4.413	1	4.413	2.165	0.144
Sex × Region	0.113	1	0.113	0.056	0.814
Error	230.323	113	2.038		

TABLE 9.6 A Two-Way Mixed Model Analysis of Variance: Fluctuating Asymmetry of Alpine Chipmunk Jaw Size

Source	SS	df	MS	*F*	*P*
Individual	960.08	93	10.3234	18.103	0.0001
Sides	1.34	1	1.3357	2.342	0.129
Individual × Sides	53.04	93	0.57027	1.512	0.0089
Measurement error	70.91	188	0.37716		

litter is the nested term (it is nested within diet). Because litter is nested within diet, we do not find all values of "litter" for all levels of "diet" – litter 1, for example, is found only at one level of "diet" ("liquid") and similarly, litter 2 is found only at one level of "diet" ("powder"). The model looks like this,

$$Y = A + B(A) + \varepsilon \tag{9.23}$$

There is no interaction term for a nested model because there are no independent contrasts of the levels of B within differing levels of A, hence we have no way to estimate the interaction between the factors. What this means is that we have no information about the impact of a powdered diet on litter 1, nor of a liquid diet on litter 2 and we therefore cannot determine whether litters respond differently to diets. Table 9.7 shows a calculation of the sums of squares, degrees of freedom and expected mean squares for a two-factor, nested design. Table 9.8 lists the F-ratios for A and B for different combinations of fixed and random values of A and B although, in the case of nested models, the nested term is almost invariably random. Table 9.9 shows the results of the analysis of diet's impact on deer mouse jaw size. Diet does have a significant impact on jaw size when tested against the nested term "litter".

Univariate Three Factor Model

In this case, we have distinct factors, A, B, and C. In addition to their main effects, we also have the interaction terms: $A \times B$, $A \times C$, $B \times C$ and $A \times B \times C$. The model looks like this:

$$Y = A + B + C + A \times B + A \times C + B \times C + A \times B \times C + \varepsilon \tag{9.24}$$

TABLE 9.7 Sums of Squares, Degrees of Freedom and Mean Squares for a Two-Way, Balanced, Nested Univariate ANOVA

Source	SS	df	MS
A	$SS_A = NJ\sum_{i=1}^{I}(\bar{y}_i-\bar{y})^2$	$I-1$	$SS_A/(I-1)$
B(A)	$SS_{B(A)} = N\sum_{i=1}^{I}\sum_{j=1}^{J}(\bar{y}_{j(i)}-\bar{y}_i)^2$	$I(J-1)$	$SS_{B(A)}/I(J-1)$
Residual	$SS_{error} = \sum_{k=1}^{N}\sum_{i=1}^{I}\sum_{j=1}^{J}(y_{ijk}-\bar{y}_{j(i)})^2$	$IJ(N-1)$	$SS_{error}/IJ(N-1)$
Total	$SS_{total} = \sum_{k=1}^{N}\sum_{i=1}^{I}\sum_{j=1}^{J}(y_{ijk}-\bar{y})^2$	$IJN-1$	

There are I levels of factor A, and J levels of factor B, which are nested within A, with N specimens per cell. The expressions of the form $\bar{y}_i$ indicate means over all cells with level i (or j, or the combination ij), while $\bar{y}$ is the overall (grand) mean of the data.

TABLE 9.8 F-Ratios and Variance Components for a Two Factor, Nested, Balanced Design

Source	A Fixed, B Random	A,B Fixed	Component
A	$MS_A/MS_{B(A)}$	MS_A/MS_{error}	$(MS_A - MS_{B(A)})/NJ$
B(A)	$MS_{B(A)}/MS_{error}$	$MS_{B(A)}/MS_{error}$	$(MS_{B(A)} - MS_{error})/N$
Error	MS_{error}		

TABLE 9.9 A Two-Way, Nested Analysis of Variance

Source	SS	df	MS	*F*	*P*
Diet	0.01694	2	0.00847	4.68	0.0132
Litter (Diet)	0.10135	56	0.00181	4.203	0.000
Error	0.09774	227	0.00043		

Preliminary results for the impact of dietary consistency on deer mouse jaw size. One factor; "litter" is nested within the other,"diet", because all members of the same litter ate the same food.

A (partial) table showing how to form *F*-ratios for a three factor ANOVA is shown in Table 9.10. For a more complete description of three factor ANOVAs see Quinn and Keogh (2002) or Snedcor and Cochran (1980).

Models with Covariates

When one or more discrete factors and one or more continuous covariates are also of interest, the general approach is to identify the form of the dependence of **Y** (shape) on the covariate, and then to remove the variance in **Y** explained by the covariate, assuming no interaction between the covariate and factor(s). Having removed the variance explained by the covariate, one can then proceed to test the significance of the factors, as discussed

TABLE 9.10 *F*-Ratios for a Three Factor, Univariate, Balanced Design in Which All Factors are Fixed or Just One is Random

Source	A, B, C Fixed	A,B Fixed, C Random
A	MS_A/MS_{error}	MS_A/MS_{AC}
B	MS_B/MS_{error}	MS_B/MS_{BC}
C	MS_C/MS_{error}	MS_C/MS_{error}
$A \times B$	MS_{AB}/MS_{error}	MS_{AB}/MS_{ABC}
$A \times C$	MS_{AC}/MS_{error}	MS_{AC}/MS_{error}
$B \times C$	MS_{BC}/MS_{error}	MS_{BC}/MS_{error}
$A \times B \times C$	MS_{ABC}/MS_{error}	MS_{ABC}/MS_{error}

earlier. In the univariate case, the term "common slope model" is used to describe a homogeneous response to the covariate across all levels of all factors. If the responses to the covariate are *not* homogeneous, meaning that there is an interaction between the covariate and one or more factors, the variance explained by the covariate cannot be removed by regression. In such cases, the focus of the analysis must be on the interaction between factor(s) and the covariate rather than on the main effects of the factors. In a univariate analysis (ANCOVA), the typical approach is to compute the regression slopes for each level of *A* and then to test the null hypothesis that the slopes do not differ by comparing values of the derived univariate slopes (this approach does not have a simple multivariate analog).

More Complex Designs

As well as the models discussed above, there are a wide range of experimental designs for ANOVA and MANOVA, including randomized blocks, repeated measures and partly-nested designs (see Quinn and Keogh, 2002; Snedecor and Cochran, 1980). All of these can be adapted for unbalanced designs, and extended to multivariate analyses permutation based MANOVAs. Consequently, the range of existing experimental designs can be adapted for use with shape data. We will not present an exhaustive treatment of the models; readers who need models more complex than the few presented in this text will need to consult other sources particularly Snedecor and Cohran (1980), Quinn and Keogh (2002) and Rencher and Schaalje (2008). The approach taken by Lorenzen and Anderson (1993) to calculating expected mean squares and appropriate *F*-tests for univariate balanced designs may prove especially helpful to readers needing designs not found in standard tables.

UNBALANCED DESIGNS AND SUMS OF SQUARES

As we noted at several points, the analysis of unbalanced designs can be complex and there is some controversy about it. Designs can be unbalanced either because there are no observations in one or more cells, or because there are observations in all cells but the sample sizes are unequal. We do not consider the first case here; this section concerns the case in which the sample sizes are not equal for all cells. The difficulty posed by unbalanced designs arises from the fact that the factors are not orthogonal even if there are no actual interactions among factors. That causes two problems. First, it is not possible to partition the variance cleanly into the main effects of the factors. The estimates of the main effects are ambiguous because we get different estimates depending on whether the means are weighted by the sample size. Second, the hypotheses tested using *F*-ratios become complex functions of the distribution of the sample sizes within the cells rather than being simple statements about the impact of the factors. As a result, there is some controversy over the meaning of the hypotheses underlying the *F*-tests as well as about the appropriate sums of squares to use in the tests. Minor departures from a balanced design have less severe consequences than more drastic departures, thus modest variation in sample sizes is probably not a concern, particularly when using permutation methods for testing the statistical significance of *F*.

There are at least six distinct approaches to calculating the sums of squares in unbalanced designs (for an overview see http://www.statsoft.com/textbook/general-linear-models). We will discuss only three of them because they are the ones applicable to cases in which every cell has at least one observation. These three types of sums of squares are routinely called Type I, Type II, and Type III Sums of Squares (following SAS usage). We discuss these three using a two factor model of the form:

$$Y = A + B + A \times B + \varepsilon \tag{9.25}$$

Type I Sums of Squares

Type I sums of squares are also called sequential or hierarchical sums of squares. The estimates for the sums of squares are obtained for each term by computing the sums of squares for two models, the model that lacks that term and the model that includes it. The sums of squares for the model lacking the term are subtracted from the sums of squares for the model including it. So, for the two factor model, we would compute the sums of squares for the model containing only *A*:

$$Y = A + \varepsilon \tag{9.26}$$

which are calculated as they would be for a balanced design. We would then compute the sum of squares for the model containing both *A* and *B*:

$$Y = A + B + \varepsilon \tag{9.27}$$

which we denote as SS_{A+B}. The Type 1 sum of squares for *B* is then $SS_{B|A} = SS_{A+B} - SS_A$, where B|A means the sum of squares due to *B* given the sum of squares due to *A*. Next we would calculate the sum of squares due to all terms: SS_{A+B+AB},

$$Y = A + B + A \times B + \varepsilon \tag{9.28}$$

from which we would calculate the sums of squares for the interaction term (AB), given the sums of squares for the two main factors, A and B, as:

$$SS_{AB|A,B} = SS_{A+B+AB} - SS_{A+B} \tag{9.29}$$

The error, or residual, sums of squares is then estimated as:

$$SS_{error} = SS_{Total} - SS_{A+B+AB} \tag{9.30}$$

When using Type I sums of squares, the order in which factors are entered into the analysis matters because the estimate for the sums of squares explained by a factor is conditional on the estimate for the sums of squares for the factor(s) already in the model. As a result, for factor A, there are two feasible estimates for its sums of squares: SS_A and $SS_{A|B}$. In a three factor model we would have even more options: SS_A, $\mathrm{SS_{A|B}}$ and $\mathrm{SS_{A|B,C}}$. Because the factors are confounded due to the unbalanced design, SS_A would typically contain some contribution from factor B. That contribution could be positive if A and B produce similar changes in Y, or it could be negative if A and B produce contrasting changes in Y. The term $SS_{B|A}$ is the contribution of factor B, given that we have removed the sums of squares due to A, so it is a contingent estimate of the effects of A. The interaction term may also be altered by the correlation between A and B.

Type 1 sums of squares are sometimes called the "improvement sums of squares" because they are determined by the improvement in fit of the model caused by adding each term to the model. A characteristic of Type I sums of squares is that the sums of squares for all the terms sum to the total sum of squares (SS_{total}) so Type I sums of squares yield an additive partitioning of variance, unlike the other two approaches. When computing the F-ratio, the Type I sums of squares are substituted for the sums of squares computed for a balanced design.

Type II Sums of Squares

This method for computing sums of squares involves computing the sums of squares for the model, including the factor of interest and all other factors of the same order (e.g. all other main effects, or all other pairwise interactions or all other three-way interaction terms), then subtracting the sums of squares for the model lacking the factor of interest from that sum. So, for this two-factor case, to find the sums of squares for factor A, we would first compute the sums of squares for the model containing both A and B (SS_{A+B}) as:

$$Y = A + B + \varepsilon \tag{9.31}$$

And then compute the sums of squares for the model containing only B (SS_B):

$$Y = B + \varepsilon \tag{9.32}$$

And we would then calculate the sums of squares for A ($\mathrm{SS_{A(II)}}$) as:

$$SS_{A(II)} = SS_{A+B} - SS_B, \tag{9.33}$$

The notation $SS_{A(II)}$ means a Type II estimate of the sum of square for *A* (SS_A refers to a SS for *A* calculated in the same manner as for a balanced design).

Similarly:

$$SS_{B(II)} = SS_{A+B} - SS_A. \tag{9.34}$$

The interaction term is computed as:

$$SS_{A\times B(II)} = SS_{A+B+A\times B} - SS_{A+B} \tag{9.35}$$

When computing the *F*-ratio, the Type II sums of squares are substituted for the sums of squares computed for a balanced design. Unlike the Type I sums of squares, Type II sums of squares do not depend on the order in which terms at the same or lower level are entered. In this context, "lowest" means the main effects of the factors, pairwise interaction terms are at a higher level than the main effects, and three-way interactions are at a higher level than pairwise interactions, etc. The contribution made by a term is assessed by comparing the model that contains that term to a model that lacks it. When comparing the model for a factor to a model that lacks it, the interaction terms are omitted from the model. Another important distinction between Type II and Type I sums of squares is that, in the case of Type II sums of squares, the sums of squares are not additive – the sum of all the sums of squares need not equal the total sum of squares.

Type III Sums of Squares

Type III sums of squares are also called "marginal sums of squares" because the method is based on the marginal means: the grand mean (i.e. the overall mean) is calculated from the means of the means. To see the distinction, look again at Table 9.1 where we tabulated the sample sizes for handedness by sex. We could compute the mean for right-handed individuals, summing the values for the 50 right-handed females and 70 males and dividing by 120, and doing the same for the left-handed individuals: summing the values for the 50 females and 30 males and dividing by 80. Alternatively, we could compute the mean value for right-handed females, and the mean for right-handed males and then compute the mean of those two means, doing the same for the left-handed means. That second approach is the one used in computing the Type III sums of squares.

When testing hypotheses, the method based on Type III sums of squares compares the sum of squares explained by the full model to the sum of squares explained by the model without the factor of interest (the reduced model) so, in this case, the Type III sums of squares for *A* ($SS_{A(III)}$) is given by:

$$SS_{A(III)} = SS_{A+B+A\times B} - SS_{B+A\times B} \tag{9.36}$$

Where $SS_{B+A\times B}$ is obtained by fitting

$$Y = B + A \times B + \varepsilon \tag{9.37}$$

Similarly, the Type III sums of squares for *B* ($SS_{B(III)}$) is given by:

$$SS_{B(III)} = SS_{A+B+A\times B} - SS_{A+A\times B} \tag{9.38}$$

and the Type III sums of squares for the interaction term AB ($SS_{AB(III)}$) is given by

$$SS_{AB(III)} = SS_{A+B+A\times B} - SS_{A+B} \tag{9.39}$$

When computing the F-ratio, the Type III sums of squares are substituted for the sums of squares computed for a balanced design. The Type III sums of squares do not depend on the order in which the factors are entered and do not yield an additive partitioning of variance.

Which Sums of Squares to Use?

There is considerable controversy among statisticians regarding the merits of these various types of sums of squares. Some authors contend that Type III sums of squares are the appropriate ones to use, as a general rule, because the hypotheses being tested are more straightforward, not being functions of the sample sizes for the cells (Quinn and Keogh, 2002). It is certainly true that we do not usually frame our hypotheses in terms of the number of observations per group, and hypotheses framed in those terms may indeed seem nearly nonsensical. But, on the other hand, we usually rely on the additivity of sums of squares when defining the main effects of the factor. That property of additivity is unique to Type I sums of squares. An important consideration favoring Type I sums of squares is that, if factors are confounded (or if interaction terms are statistically significant) we would not normally interpret the main effects as if they do not depend on the design. We will not tell you which sums of squares to use, but we suggest that, whenever possible, examine both Type I and III sums of squares and, when using Type I sums of squares, which are sensitive to the order in which factors are entered, enter the factors in different orders to develop your understanding of your data.

WORKING WITH MULTIVARIATE SUM OF SQUARES

The sums of squares for multivariate data are sum of squares and cross products (SSCP) matrices, rather than scalars. In this section, we first consider multivariate statistical approaches to testing hypotheses based on classical analytic multivariate methods and then permutation-based methods based on inter-specimen (Procrustes) distances.

Classical Analytic Approaches to Significance Testing of GLM Models

Given a model, we can calculate the SSCP of the hypothesis that we wish to test (SS_H), a denominator SSCP ($SS_{denominator}$), which may be the residuals or some other SSCP depending on the structure of the desired F-ratio test, as well as the total SSCP (SS_{Total}). There are a range of different analytic multivariate tests based on either the relationship between the SSCP of the hypothesis and the total, $SS_H(SS_{total})^{-1}$ or on the relationship between the SSCP of the hypothesis and the denominator, $SS_H(SS_{denominator})^{-1}$ (the negative exponent indicates a matrix inversion). The SSCP matrices are assumed to follow the multivariate distributions of partitioned Wishart matrices (Mardia et al., 1979). There are four major analytic test statistics: Wilk's Lambda, Pillai's trace, Hotelling–Lawley trace and Roy's largest root, all based on the properties of the eigenvalues of these matrices (Quinn and Keogh, 2002). Pillai's trace is thought to be the most robust (Johnson and Field, 1993), but all the tests should converge in the large sample limit. The analytic models used in the tests based on these statistics assume that the SSCP matrices follow the Wishart distribution, which means that the matrices are of the form $\mathbf{SS} = \mathbf{Y'Y}$ where $\mathbf{Y}$ is a centered matrix (i.e. the column means are equal to zero) with identically distributed normal distributions of all elements in $\mathbf{Y}$. The variances need not be equal across all the variables and they can also be correlated and the tests are thought to be reasonably robust to violations of the assumption of normality, although tests of normality are available. The methods also assume equality of the error variance–covariance matrices within each factor. Details of these tests may be found elsewhere (Mardia et al., 1979; Rencher, 1995; Searle, 1997, 2006; Quinn and Keogh, 2002; and references therein) and most software packages that carry out GLM or MANOVA or MANCOVA will report several, if not all, of these tests.

For the statistical analysis of shape data, the major issue is the need to estimate these matrices accurately, and to invert the sum of square and cross products matrices. A matrix cannot be inverted unless it is of full rank, which will not be the case when there are fewer degrees of freedom in the data than there are measured variables. Even when the degrees of freedom are relatively close to the number of variables, these tests can yield wildly inaccurate results when applied to relatively small data sets. One of us (HDS) has observed a substantial overestimate of the variance explained by factors that are estimated using SSCP matrices at a sample size roughly two to three times the degrees of freedom in the data set. One rule of thumb is that we need four times the number of specimens as landmarks (Bookstein, 1996). When the data contain a large number of semilandmarks, it will be difficult to invert these SSCP matrices and even if we have only landmarks we will still have more variables than degrees of freedom. Using semilandmarks exacerbates the problem because semilandmarks, in two dimensions have two coordinates but only one degree of freedom, and using enough semilandmarks to get good coverage of the morphology increases the number of specimens needed to estimates the SSCP. For these reasons, permutation methods based on Procrustes distances appear to be more useful approaches for shape data.

PERMUTATION APPROACHES TO GENERAL LINEAR MODELS

McArdle and Anderson (2001) note that the information contained in the sum of squares and cross products matrix, $SS_{Total} = \mathbf{Y'Y}$, of any centered matrix, which is also referred to as the "inner product" matrix, is also contained in the "outer product" matrix $\mathbf{YY'}$, obtained from the matrix of pairwise distances among the n specimens. One approach to hypothesis testing follows from this, which is to form *pseudo F statistics* (McArdle and Anderson, 2001). Remember that a single factor MANOVA based on the model

$$\mathbf{Y} = \mathbf{XB} + \varepsilon \tag{9.40}$$

has an F-test of the form:

$$F = [(\mathbf{SS_H})/(J-1)]/[(\mathbf{SS_{error}})/(n-J)] \tag{9.41}$$

McArdle and Anderson then discuss pseudo F-tests of the form:

$$F = \text{tr}(\mathbf{SS_H})/(J-1))/(\text{tr}(\mathbf{SS_{error}})/(n-J)) \tag{9.42}$$

where the trace is simply the sum along the diagonal of the matrix. The pseudo F-test thus does not require a matrix inversion. The calculation of the SSCP is based on the partitioning:

$$\text{tr}(\mathbf{SS_{Total}}) = \text{tr}(\mathbf{SS_H}) + \text{tr}(\mathbf{SS_{error}}) \tag{9.43}$$

where $\mathbf{SS_{total}} = \mathbf{Y'Y}$, $\varepsilon = \mathbf{Y(I-H)}$, where $\text{Y}_{pred} = \mathbf{XB} = \mathbf{HY}$ and $\mathbf{H}$ is $\mathbf{X(X'X)^{-1}X'}$
so that **SS** under the hypothesis is

$$\mathbf{SS_H} = \mathbf{Y'Y} - \varepsilon'\varepsilon \tag{9.44}$$

and the F-ratio is formed from the trace of the SSCP matrices of the J groups and of error sum of squares (for n specimens and J levels). In more complex factorial designs, the terms in the F-ratio would be the same as those of the univariate F, but would use the trace of the related SSCP instead of the univariate SS. We can therefore consider how to carry out the partitioning and formation of F-tests based on the outer product matrix $\mathbf{YY'}$, which may be derived from the matrix of all pairwise interspecimen distances. In studies of shape, that distance will typically be the Procrustes distance, although the methods are more widely applicable to any distance metric.

Expressing GLM Models in Terms of Distance Matrices A and B for Which We Can Compute Both AB and BA

$$\text{tr}(\mathbf{AB}) = \text{tr}(\mathbf{BA}) \tag{9.45}$$

then

$$\text{tr}(\mathbf{Y'Y}) = \text{tr}(\mathbf{YY'}) \tag{9.46}$$

and we can partition $\text{tr}(\mathbf{YY'})$ as

$$\text{tr}(\mathbf{YY'}) = \text{tr}(\mathbf{Y_{pred}Y_{pred'}}) + \text{tr}(\varepsilon/\varepsilon') \tag{9.47}$$

Noting that $\mathbf{Y_{pred}Y_{pred}}' = \mathbf{H(YY')H'}$ and $\varepsilon\,\varepsilon' = \mathbf{(I-H)YY'(I-H)'}$, we can partition $\mathbf{YY'}$ by simply using $\mathbf{YY'}$ and $\mathbf{H}$. We don't need to know $\mathbf{Y}$, in fact, we need only $\mathbf{YY'}$, which can be obtained from the matrix of all pairwise inter-specimen distances ($\mathbf{D}$).

Given $\mathbf{D}$ we can compute $\mathbf{A}$, whose elements are $\alpha_{ij} = -1/2d_{ij}^2$, and then we can calculate Gower's centered matrix $\mathbf{G}$ from $\mathbf{A}$

$$\mathbf{G} = (\mathbf{I} - (1/n)\mathbf{II'})\mathbf{A}(\mathbf{I} - (1/n)\mathbf{II'}) \tag{9.48}$$

and $\mathbf{YY'} = \mathbf{G}$. This means that if we have any matrix of interspecimen distances, we can write it in squared, centered form, and use it in the same manner as an SSCP matrix, i.e. as a way of describing the variance–covariance structure (the SSCP matrix is simply a multiple of the variance–covariance matrix). Using the distance matrix and the related pseudo-F-test(s) in permutation tests, allows us to test the same types of GLM and MANOVA models that are tested using the analytic methods.

In calculating these mean squares, we typically use the univariate degrees of freedom rather than multiplying the univariate degrees of freedom by the dimensionality of the data (e.g. $2K-4$ for K landmarks in two dimensions, $3K-7$ for K landmarks in three dimensions, etc.). It therefore seems reasonable to ask whether the multivariate degrees of freedom should also be used in calculating means squares. There are two reasons why they do not seem necessary. The first is that we chose to think of variance as a property of the configuration, not of each coordinate separately. Consequently, dividing the sums of squares by the univariate degrees of freedom gives a mean square that represents the contribution that an individual makes to the variance rather than the contribution made by a landmark coordinate. The second is that we are using F-ratios, and the numerator and denominator both are altered by the same multiplicative factor so it does not matter whether we use the multivariate or univariate degrees of freedom. Moreover, we are using permutations to test the statistical significance of F, and the degrees of freedom do not enter into those tests. However, the degrees of freedom are important if we use analytic tests. When using analytic multivariate tests, the degrees of freedom should be the multivariate degrees of freedom.

Permutation Tests Based on the Distance Matrix

When we want to test a particular null hypothesis, there is an implied statement about the *exchangeability* of specimens if the null hypothesis is true (Anderson, 2001a,b). If we want to test the significance of a particular factor (A), then the null hypothesis is that specimens belonging to different levels of this particular factor (A_i) could be exchanged with one another and, if the null hypothesis is true, this should not alter the sums of squares predicted by the model in any significant way. So, to test the null hypothesis, we might compute an F-ratio as described in the previous section, for the original labeling of specimens. We might then randomize the labeling of specimens that indicate their level on A_i, leaving labels for other factors intact, and then recompute the F-ratio, doing this repeatedly to estimate a distribution of F-ratios. We would then compare the observed F-ratio to that distribution to estimate a p-value for the observed F-ratio. One advantage of the use of the distance matrix is that since the data in the distance matrix are distances between specimens I and J, we can permute the rows and columns in the distance matrix without having to permute the individuals themselves. As a result, we do not have to recompute all distances between all specimens for each permutation. Instead, we can simply rearrange the existing distances within the matrix (see Anderson, 2001a,b, for a complete discussion of this approach).

An important feature of all permutation methods is that the units being permuted be exchangeable, which is equivalent to saying that the variables are independent and identically distributed, although they need not be normally distributed (Anderson and ter Braak, 2003). The assumption that error variances are equal among groups is also required, just as it was in the classical analytic approaches to multivariate analogs of the F-ratio. While the pseudo F-ratios do not explicitly address the variance–covariance matrix of the measured variables, the permutation methods do. That is because the individual measurements are not exchangeable, instead, it is the individual organism with all its measurements that is treated as a unit. It is the individual's landmark configuration that is exchangeable, not individual landmarks. So the estimated significance of an observed *F*-ratio is based on a permutation that preserves the variance–covariance pattern among the individual elements in **Y** (i.e. landmark positions). Consequently, the observed variance–covariance matrix is taken as a given when assessing the significance of the *F*-ratio.

Types of Permutations

There is a range of possible approaches to the permutation of data to determine the significance of a test statistic, such as the *F*-ratios discussed here. An *exact* statistical test is one in which the Type I error rate (*alpha*) is exactly equal to the *a priori* chosen alpha value. The approach to achieving an exact test for a one-way ANOVA (or MANOVA) seems to be straightforward and well understood (Good, 1994; Manly, 1997; Anderson and ter Braak, 2003). However, the approach is less straightforward when it is difficult to form exact tests, such as when there are multiple factors (Anderson and ter Braak, 2003), and is a subject of debate (ter Braak, 1992; Edginton, 1995; Manly, 1997). The simplest approach is to permute the raw data, meaning that specimens are randomly reassigned to groupings based on all factors at once. This approach is fine for a single factor, and can be used to form an exact permutation test. However, once there is more than one factor, permutation of residuals becomes an alternative approach.

One approach is to calculate the residuals based on the most complex model under consideration (the *full model*), and then base the estimates of *p*-values for all simpler models (meaning for models of individual factors) on the distributions of *F*-values obtained under this permutation of the residuals from the full model. In a two factor model, with factors **A**, **B** and an interaction $\mathbf{A} \times \mathbf{B}$, the residuals would be computed based on $\mathbf{A} + \mathbf{B} + \mathbf{A} \times \mathbf{B}$, so that the significance of the factor **A** would be based on the full residuals. In a permutation based on a *reduced model*, the factor **A** would be tested using residuals permuted with a fixed **B** value. The permutation is done with all the terms other than the one(s) being tested held fixed (Anderson and ter Braak, 2003), an approach also discussed by Edginton (1995). Anderson and ter Braak (2003) recommend using permutations under the reduced model on the grounds that the power should be greater than or equal to that of the exact test. Permutations of residuals always yield approximate tests, which should asymptotically approach the exact test.

MODELS WITH MULTIPLE FACTORS

To exemplify the analysis of models with multiple factors, we consider first the analysis of alpine chipmunk jaw shape with two fixed factors, sex and region (the Yosemite region and the southern Sierras). Rather than restricting the analysis to a balanced design, as we did in the last chapter, we use all the animals collected between 1911 and 1919 from these two regions. Because there are 117 animals, but 15 landmarks and 85 semilandmarks, we clearly cannot invert the variance–covariance matrix. However, we can reduce the dimensionality of the data using principal components analysis. To ensure that we have four times as many individuals as variables, we can use the first 30 principal components, which explain 95.4% of the variance. The results are shown in Table 9.11; both main effects are statistically significant (although only sex had a significant effect on jaw size, see Table 9.5). The interaction term is not statistically significant. As an alternative approach, we can use the distance-matrix permutation method; those results, based on the sequential (Type I) sums of squares, are shown in Table 9.12. Again, both main factors are statistically significant and the interaction term is not. Because we are using sequential sums of squares, we might also wish to look at the results after reordering the terms in the model. Table 9.13 shows the results, now with region entered first. Again, the two main effects are statistically significant and the interaction term is not, but the proportion of variation explained by the terms does change, albeit slightly. We might also wish to examine the marginal (Type III) sums of squares and, in this case, as shown in Table 9.14, there is one important difference – the interaction term *is* statistically significant. That is important because it means that we cannot generalize about the "main effects". The effects of each factor are not general because the impact of each one depends on the level of the other. This ambiguity is the consequence of the unbalanced design.

The second case that we will consider is the model for fluctuating asymmetry, a two-factor mixed model. The right and left side of each individual's jaw is measured twice; the random factor is individual and the fixed factor is side. The interaction term (individual × side) is the estimate of fluctuating asymmetry. This is usually tested by

TABLE 9.11 A Two-Factor Model for Alpine Chipmunk Jaw Shape

Source	df	Pillai Approx	F	df_1	df_2	P
Sex	1	0.39386	1.8194	30	84	0.0173
Region	1	0.80366	11.4607	30	84	<2e − 16*
Sex × Region	1	0.30464	1.2267	30	84	0.231
Residuals	113					

TABLE 9.12 A Two-Factor Model for Alpine Chipmunk Jaw Shape, Fitted to the Matrix of Pairwise Procrustes Distance, and Tested by Permutations Using Sequential (Type I) Sums of Squares

Source	SS	df	MS	F	R^2	P
Sex	0.002034	1	0.0020338	4.88	0.0369	0.001
Region	0.005370	1	0.0053697	12.88	0.0973	0.001
Sex × Region	0.000679	1	0.0006790	1.635	0.0123	0.068
Residuals	0.047114	113	0.0004169			
Total	0.055197	116	1			

Sex entered as the first term in the model.

TABLE 9.13 A Two-Factor Model for Alpine Chipmunk Jaw Shape, Fitted to the Matrix of Pairwise Procrustes Distance, and Tested by Permutations Using Sequential (Type I) Sums of Squares

Source	SS	df	MS	*F*	R^2	*P*
Region	0.0057994	1	0.0057995	13.91	0.106	0.001
Sex	0.001604	1	0.0016040	3.85	0.0291	0.001
Region × Sex	0.000679	1	0.0006790	1.65	0.0123	0.075
Residuals	0.047114	113	0.0004169			
Total	0.055197	116	1			

Region entered as the first term in the model.

TABLE 9.14 A Comparison Between the Results Using Sequential (Type I) and Marginal (Type III) Sums of Squares for the Impact of Sex and Region on Alpine Chipmunk Jaw Shape Fitted to the Matrix of Pairwise Procrustes Distance, and Tested by Permutations

		Sequential		
Source	**SS**	*F*	R^2	*P*
Sex	0.0020	4.88	0.0369	0.001
Region	0.0054	12.88	0.0973	0.001
Sex × Region	0.0007	1.635	0.0123	0.068
		Marginal		
Source	**SS**	**F**	R^2	**P**
Sex	0.0020	4.40	0.0368	0.0020
Region	0.0058	13.50	0.1051	0.0020
Sex × Region	0.0009	2.01	0.0171	0.0260

For the analysis using sequential sums of squares, sex is entered first.

what Klingenberg calls a "Procrustes Anova" (Klingenberg and McIntyre, 1998; Klingenberg and Zaklan, 2000). Table 9.15 shows the results and, in this case, it is the interaction term (which is tested again measurement error) that is of greatest interest. That interaction term is highly significant. Figure 9.1 shows the first two principal components of the symmetric (Figure 9.1A,B) and fluctuating symmetric (Figure 9.1 C,D) components of alpine chipmunk jaw shape.

MODELS WITH COVARIATES

If we examine a simple model with a single factor **A** and a single covariate **X**, it will have the general form

TABLE 9.15 A Two-Factor Mixed Model for Shape Using a Generalized Goodall's Test: Fluctuating Asymmetry of Chipmunk Jaw Shape

Source	SS	df	MS	F	R^2	P
Individuals	0.1479	9435	0.0013	6.71	0.83	<0.0001
Sides	0.00034	111	3.0843E-06	1.45	0.002	<0.0001
Individual × Sides	0.02204	9435	1.9847E-04	5.90	0.124	<0.0001
Measurement error	0.00755	19092	6.8018E-05		0.042	

Fluctuating asymmetry is estimated by the "Individual × Sides" term.

$$\mathbf{Y} = \mathbf{A} + \mathbf{M(A)X} + \varepsilon \tag{9.49}$$

where **Y** is the centered matrix of the dependent data, **X** is the centered matrix of covariate values, **A** is the factor, **M(A)** is a matrix of coefficients (or slopes), one per variable in **Y**. The coefficients in **M(A)** are assumed to depend on the levels in **A**, and ε is again the residual or error term. This would be the "full" model because it contains all the interactions of interest. We would like to compare it against a model in which **M** is independent of **A**

$$\mathbf{Y} = \mathbf{A} + \mathbf{M}_0\mathbf{X} + \varepsilon \tag{9.50}$$

which is a reduced model relative to the model with **M(A)**. We would probably also consider the model in which there is no dependence of **Y** on **X**

$$\mathbf{Y} = \mathbf{A} + \varepsilon \tag{9.51}$$

If the second model, which turns out to be significant, and the first is not, we might also consider the model with no dependence of **Y** on **A**.

$$\mathbf{Y} = \mathbf{M}_0\mathbf{X} + \varepsilon \tag{9.52}$$

As discussed above in the context of univariate analyses, the typical approach is to compute the regression slopes for each level of **A** and then to test the null hypothesis that the slopes do not differ by comparing values of the derived univariate slopes. A multivariate approach, presented by Rencher and Schaalje (2008), tests the hypothesis that the slopes are all equal using an *F*-test, in which they compute the variance explained by the full model (of independent slopes) beyond that explained by the reduced model (of homogeneous slopes) as the sums of squares in the numerator, and use the sums of squares of the error term of the full model (independent slopes). This produces an *F*-ratio of the form:

$$F_{independent\ slopes} = \frac{(SS_{independent\ slopes} - SS_{common\ slopes})/(J-1)}{(SS_{residuals/independent\ slope\ model})/(n-2J)} \tag{9.53}$$

noting that the numerator has $df = (J-1)$ because the independent slopes model has J slopes estimated for J levels in **A**, whereas the common slope model estimates only one slope. The residuals of the independent slope model have J estimated slopes, and J estimated means, leaving $n-2J$ degrees of freedom remaining for n specimens. This ratio can

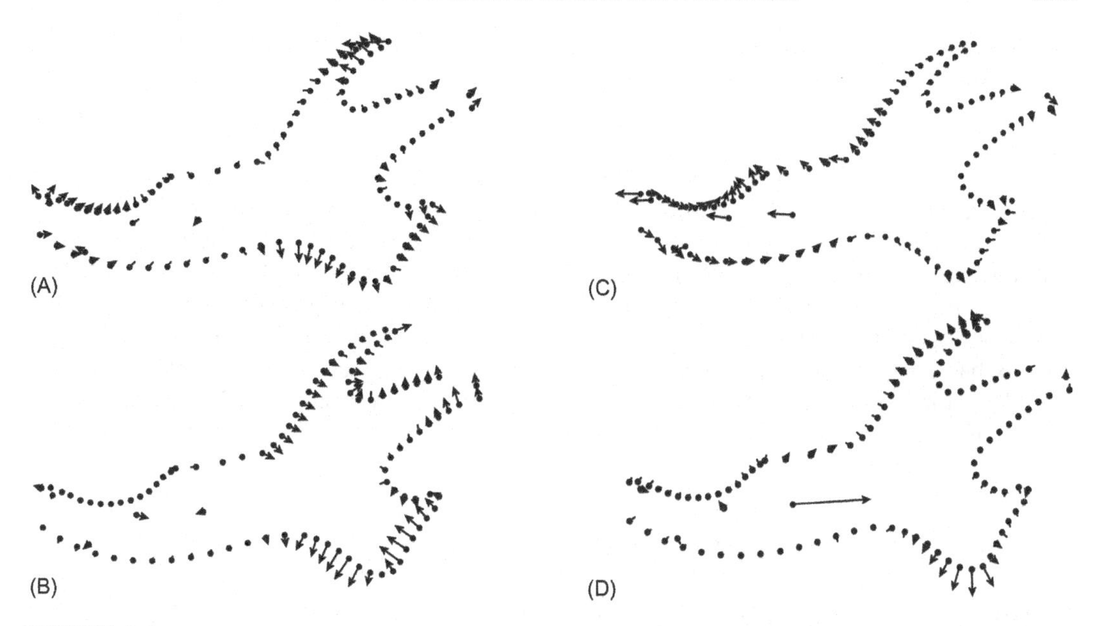

FIGURE 9.1 Principal components of symmetric and fluctuating asymmetric variation of alpine chipmunk mandible shape. (A) PC1 of the symmetric component of variation; (B) PC2 of the symmetric component of variation; (C) PC1 of the fluctuating asymmetric component of variation; (D) PC2 of the fluctuating asymmetric component of variation.

be tested using permutation methods as discussed earlier and, if the independent slopes model is significant, no further testing of the significance of the factor is warranted.

If the test for independent slopes is *not* significant, then the next step would be to test the common slope model against the reduced model with no covariate,

$$F_{common\ slopes} = \frac{(SS_{common\ slopes} - SS_{zero\ slopes})/(1)}{(SS_{residuals/independent\ slope\ model})/(n-J-1)} \tag{9.54}$$

If the common slope model is significant, then the variance due to the common slope is removed and the analysis of **A** proceeds as in the single-factor case discussed earlier. If neither the common nor independent slope models are significant, then one also proceeds to test **A**.

MODELS WITH MULTIPLE FACTORS AND A COVARIATE

In a situation where there are multiple factors **A** and **B**, as well as the covariate **X**, the full model would be

$$\mathbf{Y} = \mathbf{A} + \mathbf{B} + \mathbf{A} * \mathbf{B} + \mathbf{M}(\mathbf{A} \times \mathbf{B})\mathbf{X} + \varepsilon \tag{9.55}$$

with slopes dependent on the interactions of **A** and **B**. One might test this against both possible reduced models (as implied by the discussion in Rencher and Schaalje, 2008)

$$\mathbf{Y} = \mathbf{A} + \mathbf{B} + \mathbf{A} \times \mathbf{B} + \mathbf{M}(\mathbf{A})\mathbf{X} + \varepsilon \tag{9.56}$$

$$\mathbf{Y} = \mathbf{A} + \mathbf{B} + \mathbf{A} \times \mathbf{B} + \mathbf{M}(\mathbf{B})\mathbf{X} + \varepsilon \tag{9.57}$$

which have slopes dependent on **A** or **B**, but not both. These models might then, in turn, be compared to the common slope model:

$$\mathbf{Y} = \mathbf{A} + \mathbf{B} + \mathbf{A} \times \mathbf{B} + \mathbf{MX} + \varepsilon \quad (9.58)$$

using basically the same approach to forming *F*-tests.

To exemplify a model with multiple factors and a covariate, we add a covariate, size, to our model for alpine chipmunk jaw shape. As shown in Table 9.16, all three factors have a significant impact on jaw shape. These results are based on the Type 1 (sequential) sums of squares, with size entered first, then shape, and then region because we are primarily interested in the impact of region. The results can be read as saying that size has a significant impact on shape, that sex has a significant impact controlling for size, and that region has a significant impact controlling for size and sex. An interesting pattern can be seen in Figure 9.2: all three factors have a substantial effect on the angular process, differing in where and how they affect it.

ANALYZING MEASUREMENT ERROR

Measurement error contributes to the unexplained variation in the data and, by increasing the noise, measurement error makes it more difficult to pick out the influence of the factors of interest, especially when their effects are subtle. Measurement errors are generally thought of as being either *systematic* or *random* (see Arnqvist and Martensson, 1998). Systematic errors are consistent biases in a measurement meaning that all measures are incorrect to a consistent degree or extent. One of us (HDS) was thrilled to find a set of inexpensive plastic rulers which were roughly 3 to 5% shorter than claimed by the markings on them, perhaps because of shrinkage of the plastic, or poor mold-making. All measurements made with these rulers were uniformly short by a fixed factor, one specific to a particular ruler, which made them invaluable in an introductory physics lab on

TABLE 9.16 A Three-Factor Multivariate Analysis of Covariance: The Impact of Size, Sex and Region on Alpine Chipmunk Jaw Shape (Using Sequential Sums of Squares)

Source	SS	df	MS	*F*	R^2	*P*
Size	0.001497	1	0.001497	3.6151	0.02712	0.001
Sex	0.001587	1	0.0015865	3.8312	0.02874	0.001
Region	0.005194	1	0.0051943	12.5434	0.09411	0.001
Size × Sex	0.000362	1	0.0003625	0.8753	0.00657	0.549
Size × Region	0.000597	1	0.0005972	1.4423	0.01082	0.128
Sex × Region	0.000655	1	0.000655	1.5816	0.01187	0.081
Size × Sex × Region	0.000167	1	0.0001666	0.4023	0.00302	0.990
Residuals	0.045138	109	0.0004141	0.81776		
Total	0.055197	116	1			

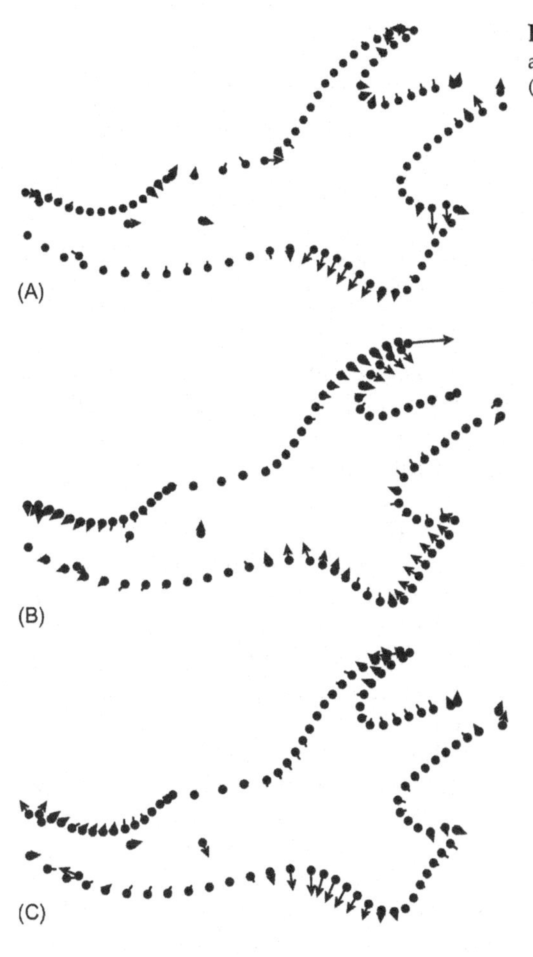

FIGURE 9.2 The effects of size, sex, and region on alpine chipmunk mandible shape: (A) size; (B) sex; (C) region.

measurement error. Systematic errors are most worrisome when they interact with a biological factor, such as parallax, which is most pronounced at the edge of the camera's field of view. It may therefore have a greater influence on larger specimens than on smaller ones if the field of view is held fixed. In contrast to systematic errors, random errors do not occur in a regular manner, but they are typically modeled by a statistical distribution, most commonly by a normal distribution. To reduce the impact of measurement error we can first decompose the error into its sources by a repeated-measures nested MANOVA, taking multiple measurements of the same specimens and treating each potential source of error as a factor. For example, we might have a choice between two imaging methods, e.g. CT scans and photography. We would then image each specimen several times using both methods, and digitize each image several times. We would then have the factors: (1) individual (the biological variation among members of the population); (2) imaging technology; and (3) measurement. All are random and "Imaging technology" is nested within "Individuals" and "Measurement" is nested within "Imaging technology". By decomposing the variance into these sources we can determine whether the technologies produce

TABLE 9.17 Mutlivariate Analysis of Variance of the Alpine Chipmunk Jaw Shape Measured Twice on the Right Side of the Jaw

Source	SS	df	MS	EMS
Ind	0.084781	93	0.00091162	$\sigma^2_{ME} + 2\sigma^2_{Ind}$
ME	0.003619	94	0.00003850	σ^2_{ME}
Total	0.088400	187		

The expected mean squares (EMS) are equated to the MS and used to compute the Individual (Ind) and measurement error (ME) variance components.

different results, and whether one is more prone to measurement error than another, and whether the difference is large enough to matter given the magnitude of the biological variation. This approach is especially valuable when you can choose your imaging and measurement methods. Even when you cannot, it is still useful to do repeated measures of the same specimens so that you can quantify measurement error.

Measurement error is often quantified as repeatability (R) using a ratio of two variance components, that for the among-individual to the sum of the among-individual and measurement error components. These components can be calculated from the MANOVA table by equating the mean squares (MS) to the expected mean squares (EMS). The EMS for the Individual term is $\sigma^2_{ME} + k\sigma^2 \mathrm{I}_{Ind}$, where k is the number of replicate measurements. To calculate the repeatability of alpine chipmunk jaw shape data, we would use Table 9.17, which gives the MS and EMS for the right sides of the jaw, each measured twice. In this case, $k = 2$. So, to compute the value for Individual variance component, we subtract the measurement error MS from the Individual MS and divide by the number of replicates: $(0.000912 - 0.0000385)/2 = 0.0004365$. We then compute the ratio between that component and the total, with the total being $\sigma^2_{ME} + \sigma^2_{Ind}$ $(0.0000385 + 0.0004365 = 0.00047525)$. We would then take the ratio between σ^2_{Ind} and $\sigma^2_{ME} + \sigma^2_{Ind}$: $0.0004365/0.00047525 = 0.919$. So the repeatability of shape is 0.92. We can reduce measurement error by averaging the repeated measures of the same specimens.

IMPLEMENTING GLM

Several software packages can be used to analyze GLM for shape. For permutation MANOVAs or permutation tests of general linear models, one usually needs to specify for each hypothesis:

How sums of squares and cross products should be calculated (i.e. Type I, II, III)
What the numerator and denominator of the *F*-tests should be
What factors (or labels) are exchangeable under the null hypothesis.

Some programs, such as adonis in the vegan package (Oksanen et al., 2011) in R (R_Development_Core _Team, 2011), are highly automated so all that you do is to specify the model. Others, such as DISTLM (Anderson, 2004) require you to input the design matrices for the factor(s) of interest, as well as the design matrices for other terms (including covariates) in the model, and for the denominator term, as well as to designate the exchangeable units. Not surprisingly, the highly automated programs are often either limited in the types of experimental designs that they can handle, or in the flexibility that they offer. For example, adonis (as currently implemented) cannot analyze a mixed model. More flexible programs, such as DISTLM, can analyze any model (using sequential sums of squares) but require far more work from the user. More details on the implementation of GLM using multivariate test criteria (applied either to the coordinates or principal components), as well as on the Procrustes Anova and the use of permutation-based tests of distance matrices are discussed in the workbook.

References

Anderson, M. J. (2001a). A new method for non-parametric multivariate analysis of variance. *Austral Ecology, 26*, 32–46.

Anderson, M. J. (2001b). Permutation tests for univariate or multivariate analysis of variance and regression. *Canadian Journal of Fisheries and Aquatic Sciences, 58*, 626–639.

Anderson, M. J. (2003). *XMATRIX: A FORTRAN computer program for calculating design matrices for terms in ANOVA designs in a linear model*. Department of Statistics, University of Auckland.

Anderson, M. J. (2004). *DISTLM: Distance-based multivariate analysis for a linear model*. Department of Statistics, University of Auckland.

Anderson, M. J., & Robinson, J. (2001). Permutation tests for linear models. *Australian & New Zealand Journal of Statistics, 43*, 75–88.

Anderson, M. J., & ter Braak, C. J. F. (2003). Permutation tests for multi-factorial analysis of variance. *Journal of Statistical Computation and Simulation, 73*, 85–113.

Arnqvist, G., & Martensson, T. (1998). Measurement error in geometric morphometrics: Empirical strategies to assess and reduce its impact on measures of shape. *Acta Zoologica Academiae Scientiarum Hungaricae, 44*, 73–96.

Bookstein, F. L. (1996). Combining the tools of geometric morphometrics. In L. F. Marcus, M. Corti, A. Loy, G. J. P. Naylor, & D. E. Slice (Eds.), *Advances in morphometrics* (pp. 131–151). New York: Plenum.

Edginton, E. S. (1995). *Randomization tests*. New York: Marcel Dekker.

Good, P. (1994). *Permutation tests: A practical guide to resampling methods for testing hypotheses*. New York: Springer-Verlag.

Johnson, C. R., & Field, C. A. (1993). Using fixed-effects model multivariate analysis of variance in marine biology and ecology. *Oceanography and Marine Biology Annual Review, 31*, 177–221.

Klingenberg, C. P., & McIntyre, G. S. (1998). Geometric morphometrics of developmental instability: Analyzing patterns of fluctuating asymmetry with Procrustes methods. *Evolution, 52*, 1363–1375.

Klingenberg, C. P., & Zaklan, S. D. (2000). Morphological integration between developmental compartments in the Drosophila wing. *Evolution, 54*, 1273–1285.

Lorenzen, T. J., & Anderson, M. J. (1993). *Design of experiments: A no-name approach*. New York: Marcel Dekker.

Manly, B. F. (1997). *Randomization, bootstrap and Monte Carlo methods in biology*. London: Chapman & Hall.

Mardia, K. V., Kent, J. T., & Bibby, J. M. (1979). *Multivariate analysis*. San Diego: Academic Press.

McArdle, B. H., & Anderson, M. J. (2001). Fitting multivariate models to community data: A comment on distance-based redundancy analysis. *Ecology, 82*, 290–297.

Myers, P., Lundrigan, B. L., Gillespie, B. W., & Zelditch, M. L. (1996). Phenotypic plasticity in skull and dental morphology in the prairie deer mouse (*Peromyscus maniculatus bairdii*). *Journal of Morphology, 229*, 229–237.

Oksanen, J., Blanchet, F. G., & Kindt, R., et al. (2001). vegan: Community Ecology Package.

Quinn, G. P., & Keogh, M. J. (2002). *Experimental design and data analysis for biologists*. Cambridge: Cambridge University Press.

R_Development_Core _Team (2011). *R: A language and environment for statistical computing*. Vienna, Austria: R Foundation for Statistical Computing.

Rencher, A. C. (1995). *Methods of multivariate analysis*. New York: John Wiley & Sons.

Rencher, A. C., & Schaalje, G. B. (2008). *Linear models in statistics*. New York: John Wiley & Sons.

Searle, S. R. (1997). *Linear models*. New York: John Wiley & Sons.

Searle, S. R. (2006). *Linear models for unbalanced data*. New York: Wiley.

Searle, S. R., Casella, G., & McCulloch, C. E. (1992). *Variance Components*. New York: Wiley.

Snedecor, G. W., & Cochran, W. G. (1980). *Statistical methods*. Iowa State University Press.

ter Braak, C. J. F. (1992). Permutation versus bootstrap significance test in multiple regression and ANOVA. In K. H. Jöckel, G. Rothe, & W. Sendler (Eds.), *Bootstrapping and related techniques* (pp. 79–86). Berlin: Springer-Verlag.

Underwood, A. J. (1997). *Experiments in ecology: Their logical design and interpretation using analysis of variance*. Cambridge: Cambridge University Press.

PART 3

APPLICATIONS

CHAPTER

10

Ecological and Evolutionary Morphology

In this chapter we discuss methods and approaches for studying evolutionary transformations of shape. This topic covers a wide variety of studies, so we focus on a few commonly investigated areas of interest: correlations between size and shape (evolutionary allometry), correlations between form and function (functional morphology, ecomorphology), comparisons of evolutionary trends (studies of parallelism and convergence), and quantitative descriptions of the morphospace occupation (evolution of morphological diversity). For each area, we examine the typical research questions and the principal methods of answering those questions.

Many of the studies of shape evolution have focused on identifying the factors that determine its direction. One question might be: Is there a correlation between size and shape? Another might ask: Do species that differ in diet also differ in shape and is that difference related to the strength or the speed of their bite? Having found such a relationship, the next step might be to ask if other lineages exhibit a similar trend, or what conditions would cause a change in the direction of that trend. Answers to these questions might lead to new questions about the generation of morphological diversity: for example, are some habits or habitats more conducive than others to increasing morphological diversity?

The vast majority of studies on evolution of shape involve comparing morphologies of multiple species, sometimes several dozen species. Many of these interspecific comparisons can be performed using the same analytic methods as would be applied to intraspecific comparisons of individuals (e.g. regressing size on shape or performing ANOVA to test for differences of means among diet classes). However, to perform these analyses correctly for interspecific comparisons, it is necessary to include the phylogenetic relationships of the taxa. Therefore, we begin this chapter with a discussion of the role of phylogeny in comparative studies.

INCORPORATING PHYLOGENY IN COMPARATIVE STUDIES

The problem phylogenetic relationships present for statistical inference have been discussed at length in numerous publications (Cheverud et al., 1985; Felsenstein, 1985;

Geometric Morphometrics for Biologists
DOI: http://dx.doi.org/10.1016/B978-0-12-386903-6.00010-1

Martins and Garland, 1991; Garland et al., 2005; Martins and Hansen, 1997; Rohlf, 2001, 2006). To understand the core of this problem, it is important to keep in mind that to perform any sort of statistical analysis, we need a model of a random distribution so we can determine whether our data deviate meaningfully from that model distribution. We need that model if we want to determine whether the mean of one trait differs between samples or if we want to determine whether traits are correlated in their distribution among samples. We must also be able to assume that our samples are randomly drawn from the population they represent; only then can we determine whether our samples are more or less similar than expected. The Brownian motion model of evolutionary change can be used to generate an expected distribution of values for a randomly evolving trait (as can other evolutionary models). The problem is that taxa representing an evolving lineage usually cannot be treated as equally independent samples of that distribution.

For a single, unbranching lineage, the Brownian motion model predicts that most changes will be small and no one direction of change is more likely than any other; and although a large change (either a single step or a run) is unlikely, longer branches provide more opportunities for one to occur. If one had a set of taxa with the same trait value at a given starting time, the mean of the set at any later time is expected to be the starting value, but the variance of that trait is expected to increase as a function of the elapsed time. More important, there is no expectation that one pair of taxa will be more similar than another pair of taxa. This unlikely scenario, often represented by a star phylogeny (Figure 10.1A), differs greatly from the usual situation in which taxa vary in degree of relatedness, that is, the amount of time that has passed since splitting from their most recent common ancestor (Figure 10.1B).

In a branching lineage, the Brownian motion model predicts that similarity will be a function of relatedness. If the lineage branched early in the clade's history (group X, see Figure 10.1B), it is likely the common ancestor (the branch point) had a small but non-zero deviation from the root, but its descendants are more likely to have deviations that are larger and in different directions. A lineage that branched more recently (group Y, see Figure 10.1B) is much more likely to have a common ancestor that diverged far from the root and descendants that differ little from their relatively younger common ancestor. Therein lies the crux of the statistical problem – similarity is predicted by common ancestry, so taxa cannot be treated as independent samples.

The non-independence of taxa is a problem for analyzing correlations as well as for analyzing differences between means. The structure of phylogenetic relationships predicts the same pattern of similarity for all traits. If two taxa are closely related, they will be similar in jaw length, and tooth shape, and every other trait. In Figure 10.2, one set of taxa is clustered around one pair of ancestral values; another set of taxa is clustered around a

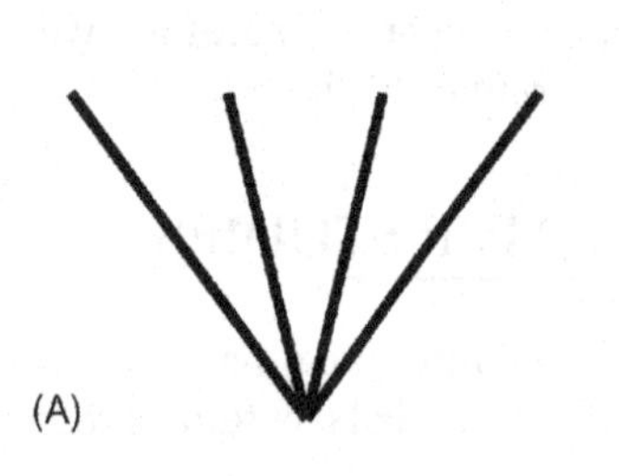

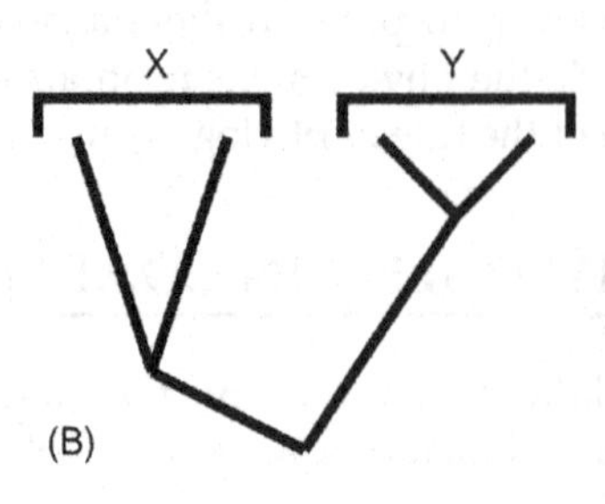

FIGURE 10.1 Two hypothetical phylogenies that differ in variation of relatedness of tip taxa. (A) A star phylogeny with four lineages diverging simultaneously from a single common ancestor. (B) The root gives rise to two lineages that branch at different times.

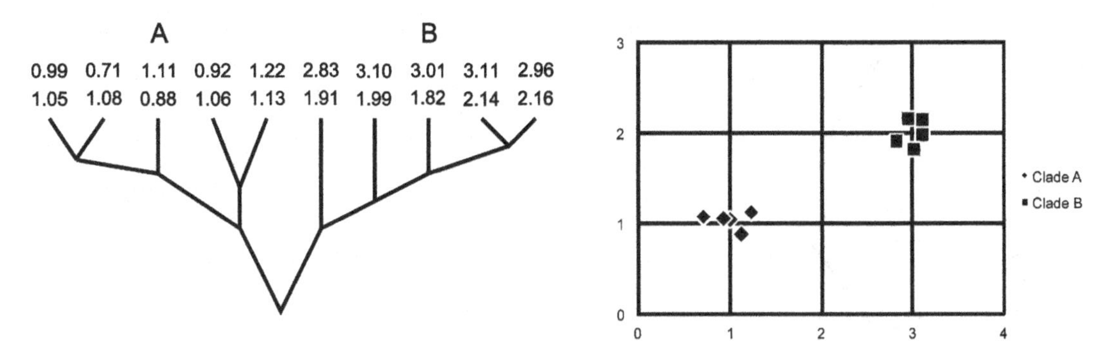

FIGURE 10.2 Influence of phylogeny on inference of a correlation. (A) Phylogeny with trait values for tip taxa. (B) Scatter-plot of trait values showing clustering by descendant lineage.

different pair of ancestral values. An analysis that fails to account for the phylogenetic relationships of this group is apt to infer a regression that really only characterizes the difference between one pair of ancestors: their respective most recent common ancestors. That ancestral divergence may not be representative of subsequent evolution in either group, which also may not be similar to each other. In more formal terms, degrees of freedom will be overestimated, confidence intervals and p-values underestimated, all leading to an elevated probability of incorrectly rejecting a true null hypothesis (type I error).

The most commonly used method of incorporating phylogenetic information in a comparative analysis is computation of Felsenstein's (1985) phylogenetically independent contrasts. A module for this technique is included in MorphoJ. Contrasts are net evolutionary differences between sister taxa (Figure 10.3). For two tip taxa, the contrast is simply the difference between the observed values (usually the means of the sampled individuals). For internal nodes, the contrast is the difference between the inferred values of the respective most recent common ancestors (MRCAs) of the descendants, which is the net change in each branch since the earlier MRCA. Under Brownian motion, evolution on one branch is independent of evolution on its sister branch, and divergence of daughters is independent of the evolution that produced their MRCA; consequently, contrasts computed in a non-recursive, top-down manner will be independent. Brownian motion also predicts a mean of zero and variance proportional to divergence time (branch length), so standardized contrasts (mean 0, variance 1) can be produced by dividing the contrasts by their branch lengths. Squared change parsimony optimization, weighted by independent estimates of branch lengths, produces node values that are consistent with this model and therefore can be used in the computation of contrasts (Maddison, 1991).

Standardized contrasts can be used in any conventional multivariate analysis, including PCA, regression and ANOVA. Note that the number of contrasts is one less than the number of taxa, so the total number of degrees of freedom in statistical tests is decremented by 1. Figure 10.4 compares regressions of jaw shape on size in North American tree squirrels using species means and contrast scores (Swiderski and Zelditch, 2010). The analysis of species means found a highly significant correlation ($p = 0.000011$) that accounted for 78% of the shape variation. Using contrasts, the regression accounts for only slightly less variation (74%) and the p-value was higher (0.000092) but still very significant. Also, the pattern

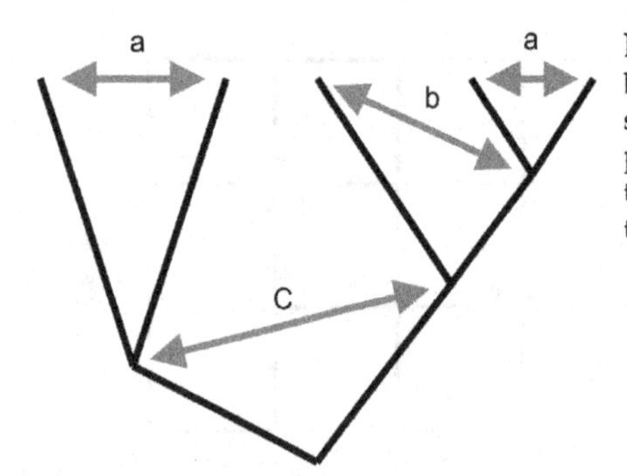

FIGURE 10.3 Independent contrasts. Contrasts are differences between sister taxa, as indicated by the gray arrow. For tips that share a common ancestor (a), they are the difference between sample means. For unpaired tips (b), they are the difference between that tip and the MRCA of its sister clade. For internal nodes (c), they are the difference between the respective MRCAs.

of shape change inferred from contrasts was not visibly different from that inferred from the species means.

You may have noticed that the regression line for contrasts passes through the origin (the y-intercept is zero). This is because contrasts are computed from the differences between taxa; so if there is no change in one trait, there should be no change in the other, whatever the slope happens to be ($\beta \times 0 = 0$). For the same reason, all values of the independent variable, size, are positive. There is no reason to choose between subtracting taxon A from taxon B or the reverse. What *is* important is computing the contrast in the same direction for both traits (e.g. A − B for independent and dependent variables) to preserve the positive or negative slope of the relationship between traits.

Unlike many previous studies, our contrast analysis found a somewhat steeper slope than the analysis based on trait values (0.0124 vs 0.0094), but as Rohlf (2006) points out there is no *a priori* reason to expect a difference in one direction rather than the other. Regression on untransformed species means and regression on contrasts are both unbiased estimators of the sample slope and correlation. This does not mean the two produce exactly the same result, only that underestimating and overestimating the true value are equally likely. The purpose of using contrasts is to judge correctly whether the slope can be considered different from zero or some other reference value, which is achieved by correcting the numbers of degrees of freedom and accounting for the expected covariance resulting from differences in relatedness. This may be clearer for the next method.

The most commonly used alternative to phylogenetic independent contrasts (PIC) is phylogenetic generalized least squares (PGLS; Martins and Hansen, 1997; see also Rohlf, 2001). PGLS also uses the expectation that change is proportionate to time under Brownian motion drift to weight observations, but the difference is that the transformation is applied directly to the data in PGLS, not to contrasts computed from the data as in PIC. Recall the regression formula, $y = \mathrm{b}\mathbf{X} + \varepsilon$, in which ε is the error term. In ordinary least squares regression, the elements of ε are expected to be independent and normally distributed with a mean of zero but, in comparative data, ε is expected to exhibit variances and covariances that are predicted by the phylogenetic variance–covariance matrix – that is, the heights of tips and internal nodes above the root. Factoring that covariance matrix out of the equation, thus incorporating it in the computation of b, yields a corrected error matrix, more accurate estimates of p, and better type I error rates.

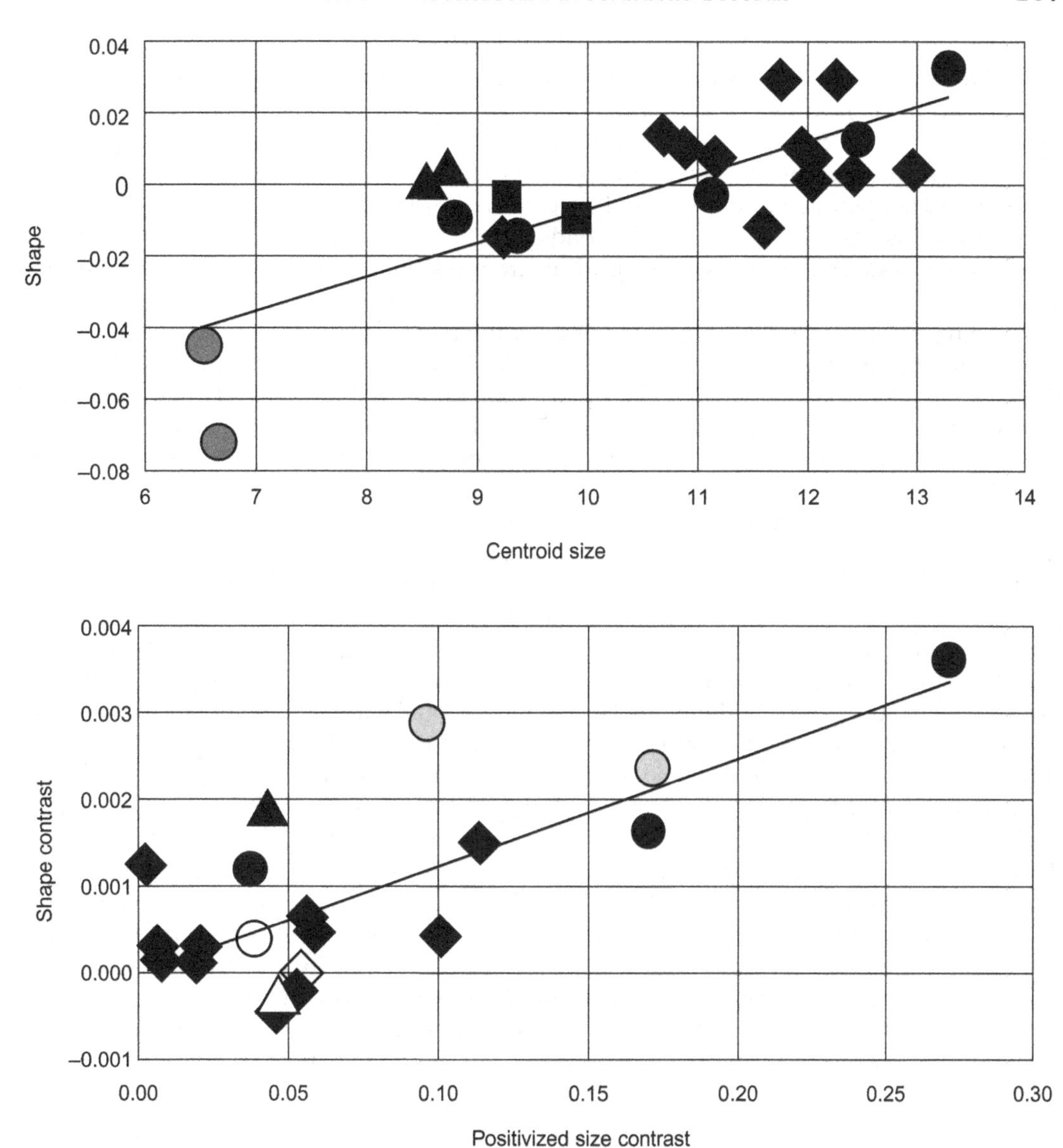

FIGURE 10.4 Comparison of regression on means to regression on contrasts. (A) Mean shape regressed on mean centroid size for jaws of 23 sciurine squirrel species. Triangles: Tamiasciurus: squares: Old World *Sciurus*; diamonds: North American *Sciurus*; circles: South American *Sciurus*; gray circles: South American dwarf species. (B) Regression on contrasts for the same species. Black symbols are within group contrasts for same taxa as in A; white symbols are contrasts between that group and its sister group; gray circles are contrasts between the dwarf species and their respective South America sister groups.

For both PGLS and PIC, errors in phylogenetic inference (both branching pattern and branch lengths) have the potential to produce misleading results. These phylogenetic errors increase the standard errors of the estimated statistical parameters, but the estimates are still unbiased (Rohlf, 2006). Numerous studies have demonstrated that these methods are robust to branch length errors (Martins and Garland, 1991; Diaz-Uriarte and Garland, 1998; Rohlf, 2006). Stone (2011) shows that changes in branch length and even small changes in topology usually have little effect on the phylogenetic covariance matrix and, thus, have little effect on results. Small topological changes only have large effects when they change the position of a particularly influential data point (analogous to the effect of an influential point on a regression). Large errors in topology will have more substantial effects (Martins and Garland, 1991; Symonds, 2002), but Martins and Garland argue that topological errors so large as to be misleading are unlikely for any reasonably well-studied taxon.

PGLS regression includes an estimate of the y-intercept. This is the same value as would be obtained by forcing PIC regression through the PGLS mean. The mean computed from PGLS is equal to the estimated value of the root under the squared change parsimony optimization. In fact, PGLS can be used to infer trait values at all nodes, and these estimates also are the same as the reconstructions using squared change parsimony optimization.

It is now well established that PIC and PGLS produce the same statistical results when the Brownian motion model is used to predict the expected covariances. The advantages of PGLS are that it is a more generalized method and more readily adapted to other evolutionary models (by making the appropriate modifications of the phylogenetic variance–covariance matrix; Rohlf, 2001). Another advantage, demonstrated by Revell (2009), is that it is possible to generate standardized values for taxa in the original measurement space, which then can be used for visualization or for analyses in other applications, (e.g. an ANOVA on size-standardized data). As Revell points out, it would still be advisable to use phylogenetic methods for the subsequent analysis.

A third approach to the problem of phylogenetic non-independence is to simulate evolution of the trait(s) on the tree using an appropriate evolutionary model, build a null distribution, and evaluate the position of the observed data relative to the population of simulations (Martins and Garland, 1991; Garland et al., 1993). One potential advantage of this approach is that it has the capacity to simulate evolution under various models, including ones for which the expected error structure is not easily derived. The interested reader is referred to Garland et al. (2005) and references therein for further guidance. For possible extensions of this approach, for example to test competing hypotheses about evolutionary processes, the reader is directed to on-going projects by Harmon and colleagues (e.g. Eastman et al., 2011).

A Note on Phylogenetic "Signal" or "Constraint"

Several writers have characterized the methods discussed here as correcting for phylogenetic signal, and some have taken the further step of inferring that this signal is evidence of constraints on the evolutionary process. It should be clear by this point that the signal to which they are referring is nothing more than congruence with phylogeny. Rohlf (2006) makes the point that the phylogenetic comparative methods do not correct the parameter estimate. They do not remove the proportion of the variance that is due to common ancestry, so they should not be regarded as analogous to size standardization by computing residuals from a regression. What the phylogenetic comparative methods correct is the type I error rate – the probability of incorrectly favoring the alternative hypothesis over a true null hypothesis. That correction insures that when we claim the traits we hypothesize to be integrated or otherwise constrained have evolved in concert, they actually exhibit greater covariation than expected by chance. Further assurance can be obtained by computing the K statistic proposed by Blomberg et al. (2003), or the permutation test presented by Klingenberg and Gidaszewski (2010), to test explicitly that the trait distribution is more congruent with the phylogeny than would be expected of a randomly evolving trait. The phylogenetic comparative methods may not be able to correct our inferences regarding the slope of that relationship, but they do provide firmer grounds for claiming that such relationships exist. In this very specific sense, they can and do give us a better basis for testing hypotheses of historical or phylogenetic constraint, but only if we keep in mind that there is much more to those hypotheses than the similarity of sister taxa. Congruence with phylogeny that is consistent with a model of random evolution is no more evidence of constraint than it is evidence of adaptation.

EVOLUTIONARY ALLOMETRY

Gould (1966) characterized allometry as "the study of size and its consequences". This description may seem rather extravagant, but does capture the importance of size for many aspects of biology. Many studies of allometry have investigated the influence of size on ecological role or functional performance, sometimes finding complex relationships that produce unexpected results. For example, a bigger snake may be able to eat absolutely bigger fish, but changes in jaw proportions that allow snakes to catch bigger and faster fish may also force them to choose relatively smaller ones. Conversely, a different jaw allometry that allows snakes to eat relatively larger prey might alter their strike mechanics and force them to switch to prey that are also relatively slower. Because geometric morphometrics is able to partition morphology into independent size and shape components, it is able to provide more direct answers to questions about relationships of size and shape than could be extracted from older methods.

Questions about the evolutionary role of morphological allometry concern many more topics than its influence on mechanical functions or ecological interactions. One of these questions is what proportion of shape variation is correlated with size? This may seem very specific and narrow, but the answer may determine the ability of the allometry to constrain evolutionary change. The long-term stability of allometric patterns may also be a factor in their role as constraints. A closely related question is whether allometric patterns provide an evolutionary line of least resistance, a notion predicated on their potential to predict or constrain responses to selection (Schluter, 1996; Marroig and Cheverud, 2005). The converse question is whether selection can change allometric patterns to direct evolution along a new axis. Dynamic evolution of allometric patterns could play an important role in the rapid morphological diversification of an adaptive radiation.

Gould's expansive representation of the importance of allometry should also serve as a reminder that not all allometric studies are concerned with the relationship of size and shape, not even all studies of morphological allometry. A study of the piscivorous snakes mentioned above could quite reasonably be focused more on the relative lengths of the various parts of the jaws than it would be on the shapes of those parts. Those ratios may be critical for determining function parameters like bite force or jaw closing speed, or the dimensions of the prey that can be accommodated (Vincent et al., 2009). However, an analysis of shape might still be useful for finding correlates of the changes in relative lengths. In the absence of an *a priori* mechanical model, or lacking a basis to choose between alternative models, analyzing the changes in shape associated with changes in size could be useful as an exploratory analysis.

Figure 10.5 shows positions of several lever arms on the squirrel jaw and regressions of their lengths on jaw centroid size. Most lines have similar slopes and deviations of individual contrasts from the line tend to be small. Large deviations are only found at the low end of the scale, where contrasts are small and the influence of measurement error is apt to be relatively large. None of these lines differs significantly from isometry (slope = 1), and we can expect their ratios to show similar stability. This differs sharply from the earlier result for shape (see Figure 10.4), which showed significant allometry (slope > 0.0). The pattern of shape change associated with a positive size contrast (size increase) may suggest an explanation for this incongruity. The deformation grids illustrating that shape change (Figure 10.5C) suggest that the angular relationships of the lever arms are changing even as the ratios of their lengths remain the same. The deformation also suggests the shape change may be more substantial in areas that are not well covered by the length measurements. Thus, shape analysis may suggest changes in performance or behavior that were not evident from an analysis of length measurements.

In our analysis of squirrel jaw allometry, we used centroid size of the jaw as our size measure. Centroid sizes were computed in the course of quantifying shape differences, so they were readily available. Jaw size was also deemed relevant to our study because jaw size is intimately associated with jaw function. Preliminary analyses also showed a strong correlation between jaw size and summer body weights of healthy animals, suggesting jaw size is a reasonable proxy for whole organism size. Had another measure of size been judged to be more relevant (e.g. linearized body mass or molar surface area), shape could have been regressed on it just as easily.

FORM AND FUNCTION

One of the recurrent themes in evolutionary morphology is the attempt to understand the relationship between differences in form and differences in function. These studies might focus on the relationship of morphology to the environments in which the organisms are found or their ecological roles in those environments (e.g. relating foot shape to substrate or jaw shape to diet). Other studies might focus more on the relationship of morphology to some measure of the organism's ability to perform an important function (sprint speed or bite force). For either type of study, the underlying question might be about the role of selection in producing the observed or hypothesized association between form and function.

Descriptive studies like those mentioned above are often a preliminary step in a study that is primarily aimed at investigating factors that might limit the ability of selection to

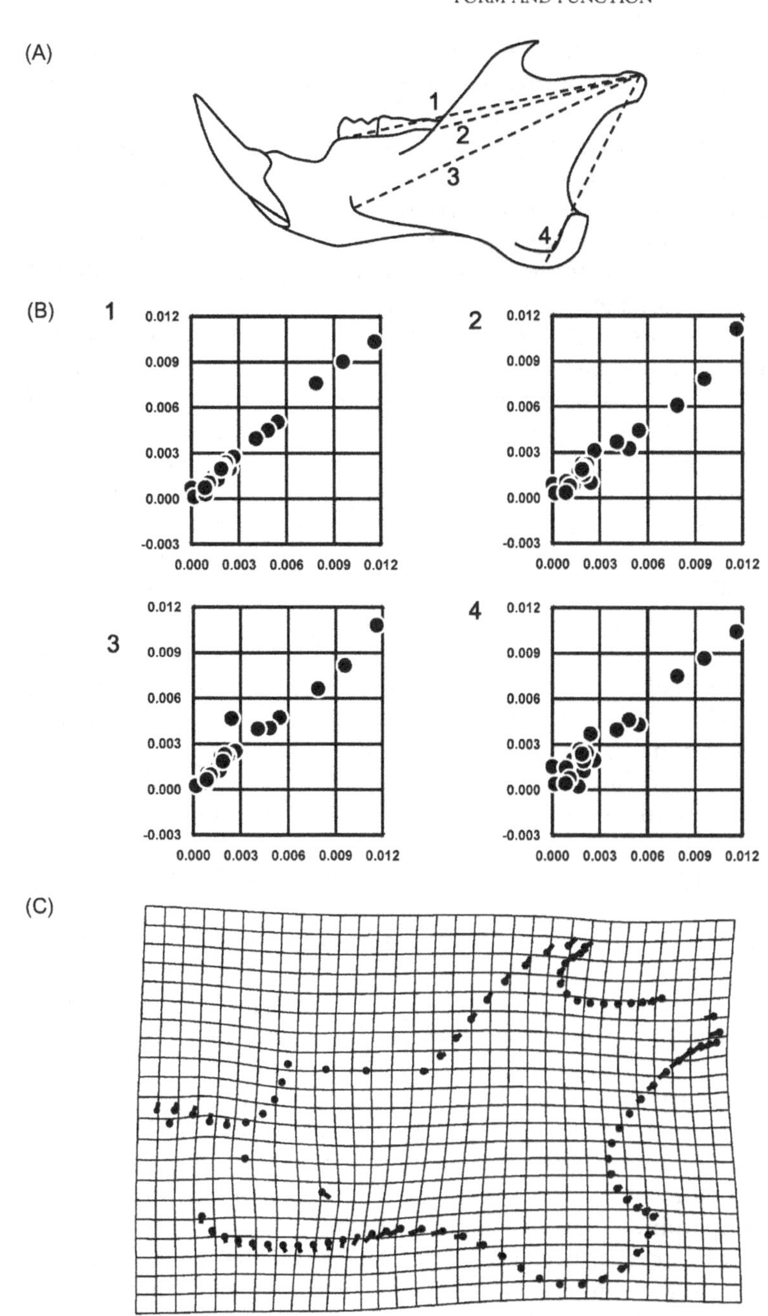

FIGURE 10.5 Allometry of lever arm lengths and shape. (A) *Sciurus* jaw with selected lever arm lengths shown as dashed lines. 1: anterior molar output arm; 2: anterior temporalis input arm; 3: anterior deep masseter input arm; 4: superficial masseter input arm. (B) Scatter-plots of contrasts for the four lever arms for 23 sciurine species. (C) Deformed grid showing shape change correlated with size increase inferred from contrasts (Figure 10.4).

produce evolutionary change. Numerous potential constraints on adaptive evolution have been proposed, from the limited mechanical properties of materials used to build organisms, to correlations of parts that are not congruent with the pattern of correlated changes favored by selection (Seilacher, 1979; Gould, 1984; Maynard Smith et al., 1985). Such constraints may prove to be short-lived, since theoretical models predict that phenotypic and genetic (co)variance structures evolve to match patterns of developmental and functional integration (e.g. Lande, 1980; Cheverud, 1982, 1984; Wagner, 1988; Wagner and Altenberg, 1996). This matching is expected to result from differential elimination of pleiotropic effects between components of different functional complexes, combined with the maintenance (or augmentation) of pleiotropic effects within a complex.

Whatever the ultimate goal might be, studies of form–function relationships share a common analytic task: testing whether differences in one set of variables are associated with differences in another set of variables measured on the same individuals or taxa. As we showed in previous chapters, tests of this type can be performed for shape variables using the same techniques as have been used for other morphometric variables. Suppose we want to know if there is a difference in jaw shape between tree squirrels in arid environments and those in more humid environments. We might expect a difference between the squirrels because different species of trees are found in those environments, with nuts and leaves and other edible bits that have different properties. We could collect squirrels from the different environments, digitize their jaws and perform an ANOVA to test the hypothesis that the shape of their jaws differs between environments. The test could be quite simple, with only two groups from habitats classified as humid and arid. A more complex, and perhaps, more realistic analysis might have many groups from many different habitats. If we have *a priori* grounds for expecting that humidity is the single controlling factor we could regress jaw shape on some measure of humidity evaluated in all the habitats from which we collected squirrels. If we regard the environmental influences as multidimensional, perhaps expecting that temperature, precipitation, elevation and other characteristics all have effects and may not be entirely independent, we might use MANOVA, or General Linear Modeling to identify the influential factors and their effects. And, as we discussed above, performing these analyses on interspecific comparisons requires that we include phylogenetic information to avoid inflated type I error rates.

For complex questions, or complex covariates of shape, it may be useful to apply techniques that look for patterns in both data sets simultaneously. One such approach is Partial Least Squares, which we discussed in more detail in Chapter 7. This method finds the axes of covariation within each data set that maximally covary between sets. Because PLS focuses attention on one dimension of variation in each data set, it is often useful to combine this analysis with a technique that provides a broad perspective, like matrix correlation (e.g. Monteiro and Nogueira, 2009; Zelditch et al., 2009). Matrix correlation compares distances computed from one data set to those computed from the other. For evolutionary morphology studies, one data set could be shape data, the other might be ecological traits, geographic distances, time or climate. Matrix correlation can be a powerful tool in the right context, but it should only be used if there is a valid distance metric for both data sets (Harmon and Glor, 2010). For shape data, the distance metric is the Procrustes distance, so this method can be used to compare distances in two sets of shape data (shapes of two parts, or shapes of the same part in juveniles and adults). Other data sets that might be sensibly compared to shape distances include physical geographic distances between localities and genetic distances between populations.

Returning to our squirrel jaw example, we may wish to consider other factors besides climate, which may seem to be only indirectly associated with jaw shape. We might instead prefer to characterize the differences in what is eaten (hard vs soft, tough vs brittle, thin shell vs thick). Given that most animals have varied diets, we might choose to classify diets according to preferred foods or the preponderance of foods eaten, or the ones that are critical to winter survival. Care must be taken in the construction of these variables to ensure that the variables and the scores on those variables are independent. For example, scoring diet components as percentages of the total (e.g. 10% vertebrates, 50% arthropods, 40% vegetation) could produce errors of inference from miscounting the number of independent variables (and thus, degrees of freedom) as well as incurring the other problems that arise from the use of ratios (Atchley et al., 1976).

Correlation of jaw shape with material properties of food may still be too indirect an explanation for variation in shape. We may still question why squirrels that eat a high proportion of hard foods have differently shaped jaws from those that eat a low proportion of hard foods. To answer this question, it may be necessary to measure jaw performance directly (e.g., the relationship between forces exerted by muscles and those applied to the food, or the deformations experienced by the bone when muscles contract). Correlations of these performance measures with shape may still not answer our ecomorphological question because shape may not directly translate into a relevant functional parameter like mechanical advantage – a ratio of two lever arm lengths. Indeed, if the answer to the question lies in an analysis of variables that are not shape and is sufficiently addressed by them, then a geometric analysis of shape differences may be uninformative or even misleading. On the other hand, if those non-shape parameters are only part of the answer, then a shape analysis may capture both the changes in those parameters and the broader morphological context that frames the transformation.

Use of shape analysis to answer a functional morphology question can be illustrated by examples from our study of squirrel jaws. There are several ways to change a jaw so the animal can produce a larger bite force and eat harder foods than the competition can. Many of the alternatives involve moving muscles or their lines of action farther from the joint, increasing the mechanical advantage of the jaw by reducing the length of the output arm relative to the input arm reducing its length, thereby increasing output force relative to input. Another possibility is to shorten the distal part of the jaw, bringing the teeth closer to the muscles (shortening the output arm). Even if these changes have the same effect on bite force, they may differ in net adaptive value because some of these changes might reduce gape more than others, which would be detrimental if the harder items in the diet are also the larger items. Yet another way to increase output force for eating harder foods would be to increase the thickness of the jaw and change the curvature of the incisor, allowing it to bear larger loads. This may be less efficient with respect to input–output ratio, but more productive in evolutionary terms because it allows feeding on harder foods within imposing a smaller gape that might limit choices.

For rigid structures like individual bones or fixed composites like most mammalian skulls, it is easy to conceive of the structure having a shape and to frame a question in terms of changing that shape in response to ecological or functional demands. In contrast, flexible or articulated structures (e.g. limbs) may not be seen to have "a" shape, but rather to exhibit different shapes under different conditions. (The elements each have a shape, and one could analyze their correlations, but it is not clear what the shape of the composite should be.) For such structures, it may seem more reasonable to compare proportions of articulated elements, and the question of shape analysis may not ever arise. However, such structures can also be fixed or photographed in standardized positions (Adams and Nistri, 2010). If the position is functionally relevant (jaws at maximum gape, for example), variation in the shape of the structure in that position might explain variation in performance of a particular function (suction velocity or prey capture success rate).

COMPARING TRAJECTORIES

Up to this point in this chapter, we have been focused on methods of testing for a single evolutionary pattern in the taxon of interest. In this section, we turn to comparing patterns in different lineages or in different parts of a lineage. We may be interested in such a comparison because a previous analysis failed to find a good fit to the entire data set. In the study of tree squirrel allometry, we found that the shape change associated with size reduction in two *Microsciurus* species differed from each other as well as from the general allometric pattern of the clade. A large Brazilian squirrel (*S. spadiceus*) also differed from the general pattern. Foraging behavior of *Microsciurus* already is known to be unusual and it is reasonable to suppose that *S. spadiceus* also has unusual behavior or dietary preferences. Were the South American clade better represented in our data, we could test for different allometric patterns.

The tree squirrel results also suggest some of the theoretical questions that often motivate a comparison of trends: for example we might ask if shifts in habitat or diet (which may represent entry to a new adaptive zone; Simpson, 1953) regularly produce a change in the direction of morphological evolution. A related question might be whether different lineages that undergo equivalent habitat shifts (deep water to shallow, dense forest to open grassland) also undergo equivalent morphological shifts. The shift into a new adaptive zone may have been facilitated by a key innovation, a morphological novelty that provided new evolutionary opportunities (Liem, 1973; Lauder, 1981). If we can confirm that novelty, we may want to test whether it led to multiple descendent lineages following the same new direction or a radiation of many new evolutionary directions. Many of the questions about similarity or persistence of trends concern the efficacy or persistence of constraints. As mentioned in the last section, shared constraints may explain similarity of evolutionary trajectory and the evolutionary transformation of those constraints may make it possible for later descendants to evolve in different directions from that taken by earlier ones.

In the simplest case, we want to know the angle between two vectors. The vectors could be defined by two ancestors and their respective descendants, or a common ancestor and two descendants. We can compute the angle between the two vectors, then, by resampling the samples at the endpoints, determine the uncertainty in the estimates, i.e. the confidence intervals around the vectors (Figure 10.6, Kim et al., 2002). By comparing the widths

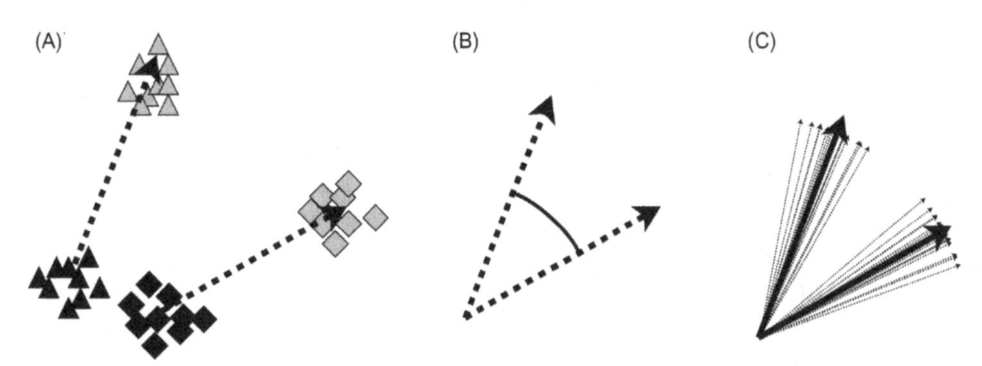

FIGURE 10.6 Vector comparison by resampling. (A) Two pairs of ancestor and descendant species, with vectors indicating directions of change. Triangles and squares represent two different lineages; black filled symbols are ancestors, gray are descendants. (B) Vectors translocated to show the angle between them (arc). (C) Same vectors with simulated resampling sets, so the span of resampling sets can be compared to angle between the original vectors.

of these confidence intervals to the size of angle between the vectors, we can judge whether the two directions are significantly different.

In a somewhat more complex case (Figure 10.7A), we might sample our evolving lineages at several successive points (stratigraphic levels or possibly locations along a cline). We could still compare the net difference (between end points) or the average directions (resampling all intermediate levels, as well as the endpoints) but, even in the latter case, we would be discarding much of the data on changes in direction. Adams and Collyer (2009) propose treating the trajectory as a shape. The trajectory shapes are obtained by Procrustes superimposition (centering, rescaling to centroid size = 1, rotating to minimize the summed squared distances between corresponding points; the only difference is that the shape trajectories are likely to have higher dimensionality than anatomical shapes). After superimposition (Figure 10.7B), comparison of populations of trajectories can be performed by MANOVA but, based on expectations of small sample sizes, Adams and Collyer suggest that a resampling strategy is more likely to give informative results.

This approach may be useful for comparing phyletic trends such as responses of herbivores to climate change, or invasion of soft substrate by diverse bivalves. It would be possible to test whether representatives of different clades responded differently (e.g. antelope vs deer), however, there are some important limitations to bear in mind. First, as with Procrustes superimposition of anatomical shapes, the trajectory shapes must have the same number of points. It is not imperative that the levels have the same range or spacing, but they are likely to be more informative if they are on similar scales. Comparing responses to climate change over millions of years in two geological epochs is apt to be more meaningful than comparing those historical trends to altitudinal or latitudinal gradients. Second, the method is not designed to compare branching or reticulate patterns. At first glance this may not seem like a big problem; however, differences in branching time relative to directional changes would pose serious challenges to the comparability of the trajectories.

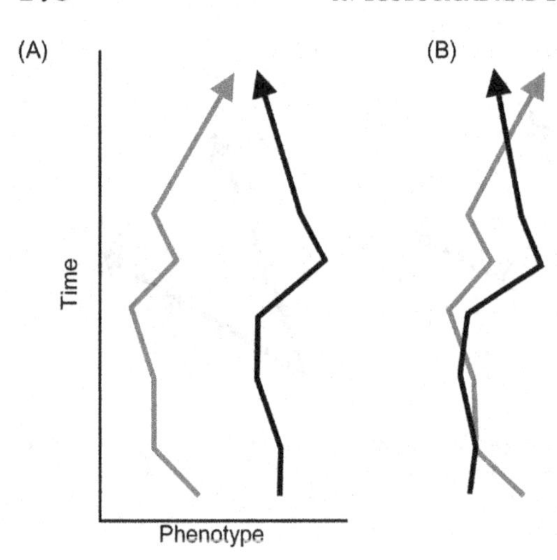

FIGURE 10.7 Comparing complex histories of phenotypic change by superimposition. (A) Two lineages exhibit multiple changes in the direction of phenotypic evolution. (B) Procrustes superimposition of the trajectories permits visualization and quantification of differences between the two trajectories.

MAGNITUDE AND STRUCTURE OF MORPHOLOGICAL DIVERSITY

Many of the questions we discussed in the previous sections are questions about the generation and dispersion of morphological variation. In this section, we focus on methods for quantifying the magnitude of morphological variation (also called disparity in interspecific analyses).

Disparity and variation are closely allied concepts – both refer to the general idea of "variety". Disparity usually signifies the variety of a group of species and is the outcome of evolutionary processes; variation, on the other hand, refers to the variety of individuals within a single (homogeneous) population and is the raw material necessary for evolution. Although there is a large and important theoretical distinction between disparity and variation, the metric (or formula) for measuring them is the same, and so we cover them both here. Still, to avoid confounding concepts that have little in common aside from a metric, we begin by reviewing their biological meanings, then turn to the issue of measurement.

Disparity

Disparity may be an unfamiliar term to many biologists, but it has emerged as a major theme in the paleobiological literature. The term was introduced to clarify the distinction between two notions of diversity that were often confounded: (1) phenotypic variety (often but not always morphological), and (2) taxonomic richness. Due largely to work by Foote (especially Foote, 1990, 1993a, 1993b), the distinction between them has been clarified – a major step towards increasing both conceptual clarity and methodological rigor. In the early literature, the number of taxa was often used as a measure of "disparity" but, as Foote showed (1993b), and as many other studies have confirmed, the number of taxa increases even as their morphological variety decreases.

To date, most studies of disparity have focused on its temporal dynamics over a geological time scale. The chief questions addressed by such studies are:

1. What is the temporal pattern of disparity?
2. What evolutionary processes explain those patterns?

Such studies are almost invariably based on fossils because they require sampling disparity at multiple times in the geological record. Some groups studied in this way include Cambrian marine arthropods (Foote and Gould, 1992; Wills et al., 1994), Paleozoic blastozoans (e.g. Foote, 1992), stenolaemate bryozoans (Anstey and Pachut, 1995), crinoids (e.g. Foote, 1994; Ciampaglio, 2002), gastropods (Wagner, 1995) and Ordovician trilobites (Miller and Foote, 1996). The growing empirical literature on disparity repeatedly documents a surprising historical pattern: disparity initially increases and then stabilizes or even decreases while the number of taxa increases.

Efforts to explain this pattern have focused on two classes of hypotheses: ecological and developmental. Ecological hypotheses postulate that ecological space is initially open and then becomes saturated; limits on disparity are thought to arise from the structure of the ecological space. In contrast, developmental hypotheses propose an intrinsic explanation for limits on disparity – the acquisition of developmental constraints that stabilize morphology (see Wagner, 1995 and Ciampaglio, 2002 for reviews of hypotheses and approaches to testing them). Whether any explanation is even needed has been questioned in a profound (if difficult) theoretical analysis (Gavrilets, 1999). At present, it is not clear what we ought to expect from disparity under plausible models; nor is it clear what role artifacts might play in the patterns detected by empirical analyses. It is also difficult to isolate causal factors that might explain the temporal dynamics of disparity because of the multiplicity of uncontrollable factors that can influence those dynamics, including rates of speciation and extinction, selectivity of extinction or speciation that is non-random with respect to morphology, the magnitude of change within a lineage, and factors potentially limiting that magnitude (such as developmental and selective constraints).

Of the various factors that can influence disparity, constraints may be the least understood – partly because they are rarely documented prior to analyzing disparity. Instead, constraints are inferred from the data, even though it is not clear how either developmental or selective constraints ought to influence disparity. Both sorts of constraints are thought to limit disparity, which may seem intuitively obvious; however, like many intuitions, it may be faulty. We know little about the impact of either sort of constraint on disparity, and determining their impacts will require studies that document constraints independently of such supposed effects. We cannot simply infer constraints from decreases in disparity when we do not know if they generally decrease disparity. Instead, we need to determine whether development is constrained or not, and then ask how those constraints affect disparity. In at least one case, developmental constraints are inferred to increase disparity (Zelditch et al., 2003).

Studies of disparity of living taxa are still relatively rare, but they have been used to address basic issues in evolutionary biology – such as whether decoupling of integrated parts increases disparity (Schaefer and Lauder, 1996), whether biomechanical and morphological disparity are related to each other (Hulsey and Wainwright, 2002), and whether developmental constraints might limit disparity (Zelditch et al., 2003). Studies relating ecological heterogeneity and morphological disparity have become more common (e.g. Ricklefs and Cox, 1977; Ricklefs and Travis, 1980; Viguier, 2002; Collar et al., 2010; Carlson and Wainwright, 2010), but much more work is needed.

Any biological explanation for an empirically documented pattern rests on the assumption that the pattern is real. Whether it is real or an artifact depends partly on how disparity is measured, and also on the sampling design. Both metrics and sampling designs have been foci of critical reviews. In particular, a number of critics have taken issue with the phenetic approach to disparity implicit in the use of a variance as its metric (e.g. Wills et al., 1994). Alternative metrics, which measure change along branches of a phylogeny, have been recommended, but they are difficult to apply when ancestors have not been sampled (or are unknown). They also pose an interpretative challenge because they redefine disparity, replacing the idea of variation around an average with that of directed change away from the ancestor (see Wills et al., 1994; Wagner, 1997; Smith and Lieberman, 1999). A second criticism is that measures of disparity typically do not consider the biological significance of the contributing variables. It is conceivable that large morphological changes could have few biological consequences, and some small changes affecting just a few morphological details could have profound consequences for function. In that light, weighted measures of disparity that take the biological significance of the changes into account might seem more justified than measures of disparity per se (see Wagner, 1995).

For reviews of the literature, including critical discussions of metrics and methods, and summaries of empirical studies, see Foote (1997), Ciampaglio et al. (2001) and Wills (2001).

Variation

Variation within populations is a major theme in evolutionary biology because it is so fundamental to evolution – phenotypic variation provides the opportunity for selection to act, and genetic variation enables selection to effect change. Variation is the raw material on which selection acts, and its structure can influence the outcome of selection. Because evolution can be constrained by limited or biased variance, the variance–covariance matrix is sometimes viewed as an intrinsic constraint on evolution; such limits or biases arising from developmental processes are developmental constraints (see Maynard Smith et al., 1985). Although that view of variation emphasizes its role as a potential constraint, the structure of (co)variation itself may be molded by selection. Theoretical models predict that phenotypic and genetic (co)variance structures evolve to match patterns of developmental and functional integration (e.g. Lande, 1980; Cheverud, 1982, 1984; Wagner, 1988; Wagner and Altenberg, 1996). This matching is expected to result from differential elimination of pleiotropic effects between members of different functional complexes, combined with the maintenance (or augmentation) of pleiotropic effects within a complex. There is much empirical evidence that phenotypic and/or genetic covariances reflect developmental and functional relationships among traits, a conclusion based on many exploratory studies (Olson and Miller, 1958; Berg, 1960; Van Valen, 1962, 1970; Gould and Garwood, 1969). In addition, many studies have deduced the structure of (co)variation among measurements from developmental and functional theories (e.g. Cheverud, 1982, 1995; Zelditch and Carmichael, 1989; Kingsolver and Wiernasz, 1991; Marroig and Cheverud,

2001). Most studies concentrate on a single developmental stage, but a few have examined the ontogenetic dynamics of variance (e.g. Foote, 1986; Zelditch, 1988; Zelditch and Carmichael, 1989; Zelditch et al., 1993).

The concept of variation is also central to systematic studies, both because systematists study evolutionary processes and also because the systematic value of a character is partly a function of its variability. In the systematics literature, the term "variation" is sometimes used very broadly, such as when talking about "ontogenetic variation". In that context, the "variation" results from the mixture of ages in the sample; because individuals differ in age, they differ in everything that changes with age. Ontogeny is thus the factor explaining the variation within the sample, but that is not the variance on which selection acts (unless we seriously entertain the idea that selection favors adults over juveniles, which is unlikely in the first place and would not have any evolutionary consequences in the second). To study the variance on which selection could act, we would first need to remove the variation resulting from the heterogeneity of the sample. Should removing that variation strike you as an improper manipulation of the data, ask yourself whether it is reasonable to imagine that selection acts on it.

A classic hypothesis linking variance to disparity is often called the "Kluge–Kerfoot" phenomenon: traits that vary the most (within populations) are also the ones that most differentiate populations (Kluge and Kerfoot, 1973). The original empirical support for the hypothesis was harshly criticized on methodological grounds (e.g. Sokal, 1976; Rohlf et al., 1983), but the hypothesis has re-emerged in the recent literature with more impressive empirical support; the dimension of greatest (genetic) variance is sometimes regarded as the evolutionary line of least resistance (e.g. Schluter, 1996).

Metrics for Disparity and Variance

As mentioned above, there is no universally accepted metric for disparity (there is for variation, so we will focus on disparity throughout this section). One major distinction among the available metrics is whether they measure the variety of forms in a sample or the diversification along branches of a cladogram. The first could be viewed as a static measure of disparity, the second as a dynamic measure of diversification. We will focus on the first approach for two reasons: the first is that we define disparity in terms of variety rather than in terms of magnitudes or rates of diversification; the second is that ancestral morphologies are rarely observed and known to be ancestral. Without direct observations of known ancestors, ancestral morphologies must be inferred, and the methods for inferring ancestral morphologies are still a matter of dispute.

Metrics for the variety of observed forms can be subdivided into two broad classes: (1) those applied to continuously valued variables (such as size and shape) and (2) those applied to ordinal or categorical data. The distinction (which is based on the type of data) is important, because continuously valued variables are measured on an unambiguous scale, which is not the case for ordinal or categorical data. For example, if we want to know how different two organisms are, and one is 10 mm while the other is 12 mm, we can say that their difference is 2 mm. Given a third, which is 14 mm, we would say that the difference between the first and third is 4 mm, and the difference between the second and third is 2 mm. Because 2 mm is equal to 2 mm, we can say that the difference between the first and second organisms is equal to that between the second and third. We might choose a scale that takes proportions into account, so that 2 mm counts for more when organisms are near 1 mm than when they are near 100 mm, but still the scale is unambiguous and measurements are mathematically commensurable. In contrast, if we classify morphologies into three types – "one", "two" and "three" – "one" and "two" are taken to be one unit apart, as are "two" and "three", but we cannot say that the difference between "one" and "two" is equal to the difference between "two" and "three". Perhaps the first two types differ by the presence or absence of a notochord, whereas the second two differ by the presence or absence of a tubercle on the tibia. The problem faced here does not arise when coding discrete classes for phylogenetic analyses because the characters may be equally informative in that context. However, weighting them equally in studies of disparity implies that they contribute equally to morphological variety. Fortunately, size and shape data are continuously valued variables, so we will concentrate on metrics of disparity suited to continuously valued variables.

The metrics for continuously valued variables can be either Euclidean or non-Euclidean distances, although most workers use Euclidean distances. We can also distinguish among metrics by whether the measures are of: (1) linear distances between forms (corresponding to a standard deviation); (2) squared distances between forms (corresponding to a variance); or (3) volumes. Measures of volume might seem most desirable because they could appear to capture the most information about the size of the occupied morphospace. Unfortunately, no satisfactory measure of volumes is available yet, because measuring them involves multiplication rather than addition. When distances along dimensions are multiplied, a trivial distance along one deflates the size of the space. For example, if we multiply distances along several dimensions, such as 0.4, 0.3 and 0.2, we get a volume of 0.024 and, if we multiply that product by 0.002, we get 0.000048 – therefore, adding information about that fourth dimension reduces the size of the space to nearly zero. Logically, we would expect that the additional information would only increase the size of the space. Another disturbing feature of this volume-based approach to disparity is that the volume of several slightly disparate variables can be far larger than the volume of three very disparate variables and one nearly invariant variable. In the example above we had three disparate variables and one that is nearly invariant. We might have another case in which there are also four variables, each with a disparity of 0.1; the product of (0.1)(0.1) (0.1) (0.1) = 0.0001, which is more than twice the volume of the first case (0.000048). In contrast, if we restrict our analysis to only the first three variables, the disparity would be (0.1)(0.1) (0.1) = 0.001 – substantially less than that of the first case (0.024).

If we had an objective and non-arbitrary method for ignoring some dimensions (so that their low levels of disparity do not deflate the space), we could circumvent these problems. However, all methods for deciding whether to exclude a variable depend on subjective arguments, and the decision about whether to exclude a variable can have an enormous impact on the results. For that reason, we prefer metrics based on standard deviations and variances. Both standard deviations and variances are equally useful metrics, and there is no reason to debate which of them is preferable because one is easily derived from the other. The major reason for using a variance is that variances are additive. Because of that property, we can calculate the overall disparity of a group, then partition it into the contribution made by each taxon (the partial disparity of that taxon; Foote, 1993a). The additivity of variances means that the sum of partial disparities equals the overall disparity. However, it is worth noting that the two measures weigh outliers differently, and consequently their results can differ. Standard deviations and variances are not linearly related, and a highly distinctive taxon has a much greater impact on a variance than on a standard deviation.

Measuring Disparity

To measure morphological disparity (MD) by a variance, we calculate:

$$MD = \frac{\sum_{j=1}^{N} D_j^2}{(N-1)} \tag{10.1}$$

where D_j is the distance of species j from the overall centroid (which is the grand mean calculated over the n species or other groups being analyzed). We can use Equation 10.1 to calculate both size and shape disparity. For size data, D_j is the difference between the centroid size of an individual species and the grand mean centroid size. For shape data, D_j is the Procrustes distance between the average shape of an individual species and the grand mean shape. We can compute shape disparity directly by estimating those Procrustes distances, or we can calculate the variances of coordinates obtained by a generalized least squares Procrustes superimposition (GLS) or variances of partial warp scores (including scores on the uniform component). All three approaches yield the same results because the sum of squared coordinates obtained by GLS equals the squared Procrustes distance to the mean, as does the sum of squared partial warp scores. In those analyses the grand mean shape is the consensus, so if we are using partial warps we can use the formula:

$$MD = \frac{\sum_{j=1}^{N} PW_j^2}{(N-1)} \tag{10.2}$$

where PW represents the partial warp scores for an individual, so the formula tells us to sum all the squared partial warp scores for each individual over all individuals. Because the grand mean shape is the consensus, its partial warp scores are all zeros, so Equation 10.2 is equivalent to Equation 10.1.

Both are also equivalent to:

$$MD = Tr\{S\} \tag{10.3}$$

where Tr is the trace of a matrix (the sum of its diagonal elements) and S is the variance–covariance matrix of the partial warp scores (including the uniform component, and computed using the grand mean as the consensus). The diagonal elements of a variance–covariance matrix are the variances, so this formula tells us to sum the variances of the variables, which takes us back to the squared distances from the consensus.

To exemplify the analysis of disparity, we will measure the disparity of adult body shape of nine species of piranhas sampled at the 16 landmarks shown in Figure 10.8.

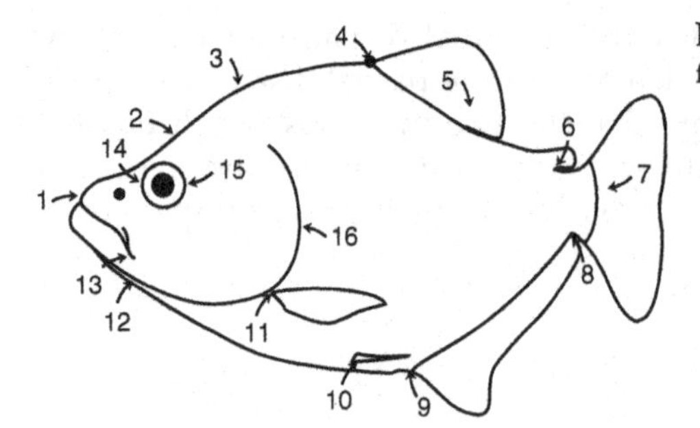

FIGURE 10.8 Landmarks sampled on the external body form of piranhas.

Before doing this analysis, we remove the shape variance within each species that is due to ontogeny, allowing us to estimate the shape of an average adult (this is done by standardizing each species to its maximum adult size). Each species is represented by a single data point, the mean shape for that species. There are nine species, so $N = 9$. The result of the analysis is that $MD = 0.00398$. Of course, we cannot yet interpret this number – we cannot say if that value is large or small, or how uncertain it is. Before we can go any farther, we need to deal with the issue of uncertainty.

To place confidence intervals on *MD*, we need first to consider the various parameters being estimated. In general, there is uncertainty in the estimate of the mean shape of each species, and in the estimate of the consensus. Both uncertainties must be taken into account when computing confidence intervals. Additionally, when the mean shape of each species is calculated by removing the variance due to ontogeny (or some other factor), we must also account for the uncertainty of the regression model used to standardize the shapes. We may also need to take a further source of uncertainty into account – the sampling of species, because unless we have measured them all we must consider the uncertainty of the grand mean that arises from our sampling of species. If we do not consider this particular source of uncertainty, we cannot generalize from our sample of species to the larger group that includes them, although we can make statements about our particular sample of species that takes the uncertainty of our sampling of them into account.

The confidence intervals might look odd because they frequently are not symmetric about the mean, even when the distribution of shapes around the GLS consensus *is* symmetric. That symmetric distribution of shapes implies that the uncertainty in the estimate of the mean is roughly equal in all directions (i.e. it is a hyperspherical solid). Turning to the estimates of disparity, we can see why the uncertainty in the distance of a species from grand mean is not symmetric about the mean distance even then. The hyperspherical distribution of uncertainty in the mean yields a non-symmetric distribution of distances – there are many more possible locations of a species' mean that increase the distance than there are that decrease it. As we can see in Figure 10.9, the line joining the grand mean to a species' mean is in a single direction in a high dimensional space; random variation in the position of the sample mean rarely lies along the line between the species' mean and grand mean. In Figure 10.9, *D* is the distance from the species' mean to the grand mean shape, and the circle around X represents the range of uncertainty about the species' mean. The region within the circle that is a distance *D* or less from the grand mean is shaded, and this region is clearly smaller than the unshaded region that is farther than *D* from the grand mean. This effect is even more pronounced in higher dimensions.

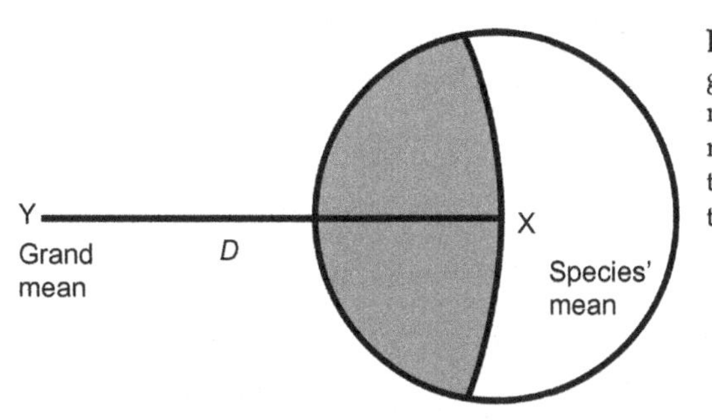

FIGURE 10.9 The line joining a species' mean to the grand mean; random variation in the position of the mean only rarely lies along the line within the shaded region. Changes in the position of shapes orthogonal to that line or within the unshaded region increase the distance to the mean.

We can construct confidence intervals and standard errors for *MD* by bootstrapping. When we need to take the uncertainty of the regression into account, we first fit a regression model to the data, determine the residuals, predict the shape expected for each size, bootstrap the residuals and randomly allocate them to each predicted shape, then refit the regression model to the data to generate a standardized data set for the bootstrap set. This is iterated N times (where N is the number of bootstrap sets). If we do not need to take the uncertainty of the regression into account, we simply resample (with replacement) from each of the samples. For each bootstrap set of standardized values, we calculate the disparity of that sample using the formula for *MD* above. In the case of the adult piranhas discussed above, the estimate of $MD = 0.00398$; the 95th percentile over the bootstrap sets gives us the two-tailed confidence interval on that estimate, 0.00377 to 0.00440.

We still do not know if that value is large or small because we have still not compared it to the disparity of anything else. We will thus continue the analysis, comparing the levels of adult disparity to that of juveniles, and comparing the disparities of several piranha clades (Figure 10.10). Table 10.1 gives the disparities (*MD*) of juvenile and adult shapes, as well as the standard errors (*SE*) for the estimates. We can use a *t*-test to determine whether derived traits like mean disparities are significantly different:

$$t = \frac{MD_1 - MD_2}{\sqrt{\left(\frac{(N_1-1)N_1SE_1^2 + (N_2-1)N_2SE_2^2}{N_1+N_2-2}\right)\left(\frac{N_1+N_2}{N_1N_2}\right)}} \qquad (10.4)$$

with $(N_1 + N_2 - 2)$ degrees of freedom. Because *MD* is computed from the mean shapes of species, N_1 and N_2 are the numbers of species in the respective clades. We can also use a bootstrap procedure like that used to test whether two Procrustes distances are different. We begin by computing the disparities of the two groups and the difference between those disparities, and then we resample each data set with replacement, repeating the calculations of the disparities and the difference between them. After a sufficient number of bootstraps, we can determine the 95% interval for the range of differences. If this range excludes zero, we can conclude that the observed difference is significant at the 95% level.

For the most inclusive piranha group (Clade 1), disparity decreases significantly over ontogeny, as it does in Clade 2. In Clade 3, disparity increases statistically significantly, but the change is slight – in contrast to the dramatic increase in Clade 4. In Clades 5 and 6, disparity is constant throughout ontogeny. A perhaps counterintuitive result is that adult disparities of Clades 3 and 4 are significantly greater than that of the group as a

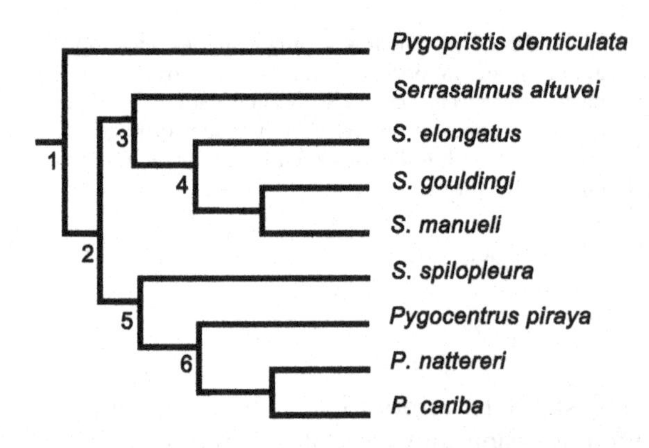

FIGURE 10.10 Cladogram of the piranhas analyzed in this chapter; nodes are numbered to designate clades.

TABLE 10.1 Disparities of Clades (Numbered as in Figure 10.10), Measured at Two Ontogenetic Stages

Taxon	Juvenile Disparity	Standard Error	Adult Disparity	Standard Error
Clade 1	0.00543	0.0003	0.00398	0.0002
Clade 2	0.00575	0.0003	0.00405	0.0002
Clade 3	0.00431	0.0004	0.00550	0.0003
Clade 4	0.00229	0.0002	0.00603	0.0004
Clade 5	0.00116	0.0002	0.00151	0.0001
Clade 6	0.00073	0.0002	0.00051	0.0002

Disparities of juveniles are measured at the transition from larval to juvenile growth; those of adults are measured at maximum body size attained by each species.

whole (Clade 1), which may seem impossible, but disparities measured this way are not additive. In these analyses, we are measuring the disparity of each clade relative to that clade's own mean – hence a low disparity indicates that few species differ by much from the mean of that clade. Consequently, a group comprising three or four species that differ a great deal from each other (and from the group mean) can have a much higher disparity than a larger group that includes those species. That is because the additional species in the larger group may all be much closer to the grand mean. Consequently, their values of D_j are small and contribute relatively less to $\sum D_1^2$ whereas the addition of each species increases $N-1$ by one. The net effect is that MD decreases. For that reason, a large group containing only a few species that are far from the grand mean can be less disparate than a small group with the same number of species far from the mean. That is one reason why morphological disparity can decrease while taxonomic diversity increases.

Partial Disparity

When we want to quantify the contribution that a particular taxon makes to the overall disparity of a larger group, we want a metric that allows us to partition disparity additively. Therefore, we need an alternative to the method discussed above. The alternative does allow us to estimate partial disparity (*PD*) of the species, and the partial disparities sum to the total disparity. We estimate partial disparities (*PD*), following the procedure outlined by Foote (1993a), in terms of the variance contributed by each individual species:

$$PD = \frac{D_i^2}{N-1} \tag{10.5}$$

where D_i is the distance of the *i*th species from the grand mean and *N* is the total number of species (or other groups). If we wish to calculate the partial disparity of several species (e.g. a subclade in a larger clade), we can sum their individual partial disparities, yielding the partial disparity of that group. The utility of this approach can be seen clearly in a comparison of adult partial disparities for clade 4 of the piranhas. The sum of partial disparities for all nine species is the same as the disparity of clade 1 (within rounding error). The sum of partial disparities for Clade 4 is 0.00203, which is 51.1% of the total. The partial disparity of a single species, *S. elongatus*, accounts for 36.3% of the total disparity of adults of these nine species (Table 10.2). Quantifying partial disparities is one method for estimating the phenotypic distinctness of a particular taxon, which may have a practical application in conservation biology (e.g. optimization of preserved biodiversity).

Measuring Variation

Studies of variation, like those of disparity, use a variance as a metric. The major computational difference between analyses of disparity and variance are that (1) studies of variance use the mean of a single homogeneous population as the grand mean, and (2) individuals (rather than mean shapes of species) are the data points in studies of variance. One quick method for estimating the variance in shape is to calculate the variance for

TABLE 10.2 Partial Disparities (*PD*) of Adults, and the Standard Errors of *PD*

Species	PD	% MD	Standard Error
P. denticulata	0.00039	9.82	0.00032
S. elongatus	0.00144	36.27	0.00029
S. gouldingi	0.00026	6.55	0.00031
S. manueli	0.00033	8.31	0.00032
S. altuvei	0.00014	3.53	0.00032
S. spilopleura	0.00023	5.79	0.00032
P. cariba	0.00036	9.07	0.00028
P. nattereri	0.00039	9.82	0.00027
P. piraya	0.00043	10.83	0.00031

all the coordinates obtained by a GLS superimposition and sum those variances over all landmarks (this is exactly the same as calculating the trace of the variance–covariance matrix, and can be done in any spreadsheet). This method, while quick and intuitive, will not provide confidence intervals. It can also be risky if it leads to thinking of variances as being *at* landmarks (recall that changes in relative landmark positions are distributed across landmarks, a topic discussed in context of superimposition methods). Just as change is not located *at* a landmark, neither is variance.

ANALYZING THE STRUCTURE OF DISPARITY

To this point we have talked solely about the magnitudes of disparity and variance; but in many studies we want to know if shapes are randomly distributed throughout the morphospace. A closely related question (do samples occupy the same subspace?) will be addressed in Chapter 11. Here, we focus on two questions concerning the homogeneity of morphospace occupation. The first asks how widely shapes are dispersed, i.e. are they as close together as we would expect if they were randomly distributed. This question is answered using nearest-neighbor analysis. The second question asks whether there are clusters and gaps indicating hierarchical structure (which could be phylogenetic or ecological or both). This question is answered by combining cluster analysis to infer the hierarchical structure with the cophenetic correlation test to determine whether the inferred clustering accurately reflects the morphological distances between samples.

Nearest-Neighbor Analysis

Nearest-neighbor analysis, as the term implies, examines the smallest distances between shapes. From those distances, we can ask whether shapes are more (or less) similar than expected by chance. If they are closer than expected by chance, we would reject the null hypothesis in favor of one of clustering; conversely, if they are further apart than expected by chance, we would reject the null model in favor of a hypothesis of "over-dispersion" (or "repulsion"). Because the null model is the distribution expected by chance, it is important to consider what the reasonable null model might be. One reasonable null model is that the probability of being at any location in the morphospace is equal (uniform) over the entire space, and is independent of the shape of any other species. Another reasonable null model is that shapes follow a normal (Gaussian) distribution. The uniform model is a reasonable null for comparisons among species, whereas the Gaussian model is more reasonable when analyzing distributions of individuals around the mean of a homogeneous sample. Having two null models allows us to guard against accepting a hypothesis of a *particular* random distribution.

Nearest-neighbor analysis is another method pioneered by Foote (1990), so we begin by reviewing his approach, and then we extend it to geometric shape data. The first step in a nearest-neighbor analysis is to compute the nearest-neighbor distance D_i for each of the N species (or other groups) in the study. For the sake of brevity, we will refer to "species" as the units of analysis, but the analysis follows the same protocol even when the units are individual specimens. The next step is to construct a second data set using Monte Carlo simulations. That is done by estimating the mean and range of each variable; from the data, $N-1$ simulated specimens are generated with values randomly drawn from the observed range. Monte Carlo simulations are similar to bootstraps in that they simulate data based on a given null model and an observed set of data, but they differ in that bootstrapping is carried out using a non-parametric resampling procedure whereas Monte Carlo simulations are based on a distributional model. The distribution of the original data set is parameterized, and those parameters are used to generate a simulated dataset having the distribution of the observations. Given the simulated data, a second nearest-neighbor distance, R_i, is computed between each observed specimen and the one closest to it in the Monte Carlo set (note that R_i is not a nearest-neighbor distance between Monte Carlo specimens, but rather the distance between an *observed specimen* and the *nearest Monte Carlo simulated specimen*).

Foote provides a measure that allows us to compare the fit of the simulated distances to the observed ones, the proportional distance P_i for the ith specimen. This is a ratio whose numerator is the difference between the two distances (D_i, the observed nearest-neighbor distance, and R_i, the Monte Carlo nearest-neighbor distance) and whose denominator is the Monte Carlo nearest-neighbor difference:

$$P_i = \frac{D_i - R_i}{R_i} \tag{10.6}$$

If the random model fits the data, we would expect that, on average, D_i would equal R_i, and hence the mean Pi over all specimens (P_{mean}) is zero. When P_{mean} is less than zero the observed specimens are more clustered than expected by chance; conversely, if P_{mean} is greater than zero they are further apart than expected by chance. To determine whether zero lies within the confidence interval, we estimate the range of P_{mean} by running the Monte Carlo simulation many times.

To generate a Monte Carlo set under a multivariate normal (Gaussian) model, we must estimate the mean and standard deviation of each variable; to generate a Monte Carlo set under a uniform distribution model, we must estimate the upper and lower bounds of the range for each variable. It can be difficult to estimate the range accurately when sample sizes are small because, at small sample sizes, the observed minimum and maximum will underestimate the "true" range. Thus, rather than using the observed minimum and maximum values to estimate the range, Foote uses estimators developed by Strauss and Sadler (1989) for the "true" minimum (Y) and the "true" maximum (Z) of a distribution:

$$Y = \frac{NA - B}{N - 1} \tag{10.7}$$

$$Z = \frac{NB - A}{N - 1} \tag{10.8}$$

where A is the lowest observed value and B is the highest observed value in N specimens. Rather than use the *observed* minimum and maximum values, Foote determines the mean and the standard deviation of a normal distribution fitted to the data. He uses normal theory (citing Feller, 1968) to predict the mean and standard deviation:

$$X_{mean} = Y + \frac{(Z - Y)}{2} \tag{10.9}$$

$$SD_X = \left\{\frac{(Z-Y)^2}{12}\right\} \tag{10.10}$$

and he uses those to estimate the range parameters:

$$Y = X_{mean} - 3^{1/2} SD_X \tag{10.11}$$

$$Z = X_{mean} + 3^{1/2} SD_X \tag{10.12}$$

Extending nearest-neighbor analysis to geometric data is straightforward. Distances D_i and R_i are measured by Procrustes distance; estimates of means, standard deviations or ranges used in the Monte Carlo simulation are obtained by calculating the statistics from the coordinates of each landmark. The rest is straightforward: a Monte Carlo data set is generated and R_i is calculated for each specimen, and these are used to estimate P_{mean}. The simulation is reiterated numerous times, yielding the distribution of P_{mean} values over the Monte Carlo sets. It is then possible to carry out all the usual statistical tests using this distribution.

We illustrate nearest-neighbor analysis by testing two hypotheses about piranha disparity:

1. Piranha body shapes, both juvenile and adult, are further apart than expected.
2. Those shapes are more clumped than expected.

The reason for testing these hypotheses separately is that a conservative test of one is a liberal test of the other. For the hypothesis of over-dispersion, the conservative approach uses the Strauss and Sadler estimator of the range – the estimator enlarges the range so that large distances between points will not necessarily be further apart than expected. However, that expansion of the range can lead to a liberal test of clumping (under-dispersion) because, within that expanded range, observations may be closer than expected. To be conservative, we would test the hypothesis of over-dispersion using the enlarged range, but we would use parameters of the observed range to test a hypothesis of clustering. Each hypothesis will be tested using two null models, one uniform and the other Gaussian, because we have no good reason to view one as a more plausible random model.

Using the uniform model, the average P_{mean} of the juveniles is −0.2810 and the 95% range of P_{mean} is from −0.3551 to −0.1792, an interval that excludes zero. This result suggests a non-random distribution, with distances being smaller than expected under a random uniform model. Using the Gaussian model, the average $P_{mean} = -0.2758$ and its range is from −0.3450 to −0.1950, an interval that again excludes zero. Both results thus argue against the hypothesis of a random distribution and also against over-dispersion. Instead, they suggest clustering, the hypothesis we will explicitly test after we have tested the hypothesis of over-dispersion for adults.

Using the uniform null model, the average P_{mean} of the adults is −0.267 and the range is −0.3365 to −0.1689, an interval that excludes zero. This result also suggests a non-random distribution, with distances being smaller than expected under a random uniform model. Using the Gaussian model, the average $P_{mean} = -0.2636$ and the range is −0.3312 to −0.2036, an interval that also excludes zero. As we found for the juveniles, the data argue against the null hypothesis of a random distribution, and also against over-dispersion. Therefore, we now explicitly test the hypothesis of clustering.

We now test the hypothesis of clustering using the narrower estimate of the range. For the juveniles, based on the uniform model, the average $P_{mean} = -0.3172$ with a range of −0.3813 to −0.2247, an interval that excludes zero and supports the hypothesis of clustering. Analyzing the data under the null Gaussian model, the average $P_{mean} = -0.3006$ with a range from −0.3700 to −0.2372, an interval that again excludes zero. Taking these results altogether, they suggest that juvenile piranha body shapes are more tightly clustered than expected under either null model.

For the adults, the uniform null model yields an average $P_{mean} = -0.2537$ and a range of −0.3092 to −0.1788, an interval that excludes zero. These results again support the inference of clustering. Analyzing the data under the Gaussian null model, the average $P_{mean} = -0.2388$ with a range from −0.3091 to −0.1598, an interval that also excludes zero. Taking these results altogether, they suggest that adult piranha body shapes are more tightly clustered than expected under either null model.

Nearest-neighbor analysis can be used to examine patterns of variation as well as disparity. To exemplify this, we return to the ontogenetic variation in mouse skulls. Considering that each sample comprises individuals from a single homogeneous population, we would expect random variation to follow a Gaussian distribution. Results of analyses based on both range estimators (i.e. the parameter values estimated using the Strauss–Sadler estimate of the range (SS), and those estimated from the data (DP)) are given in Table 10.3. It is difficult to argue that the data suggest a departure from random variation. When the parameter estimates are based on an expanded range, the two youngest samples seem to be more clustered than expected under the null hypothesis of a Gaussian distribution. That expansion seems appropriate in light of the small sample sizes, but using it could be considered an overly liberal test of clustering. When estimates are

TABLE 10.3 Nearest-Neighbor Analysis of Skull Shape Variation in *M. m. domesticus*, Sampled at 5-Day Intervals (Average P_{mean} and the Range of P_{mean} Obtained from 100 Monte Carlo Simulations)

Age	SS (P_{mean})		DP (P_{mean})	
	Average	Range	Average	Range
10	−0.0929	(−0.1356)–(−0.0276)	−0.0028	(−0.0425)–(0.0377)
15	−0.0944	(−0.1503)–(−0.0334)	0.0153	(−0.0326)–(0.0598)
20	−0.0409	(−0.0963)–(−0.0178)	0.0126	(−0.0313)–(0.0658)
25	−0.0745	(−0.1343)–(−0.0051)	0.0122	(−0.0495)–(0.0654)

Parameter estimates are based either on the Strauss–Sadler estimators (SS) or on the parameters of the data (DP).

based on the observed values, the range of P_{mean} invariably includes zero and, for that reason, we cannot rule out the Gaussian null model.

Cluster Analysis

The test for clustering in the nearest-neighbor analysis tells us that species tend to be closer together than expected for a random distribution, but it does not tell us if there is one large cluster, or several smaller ones. In other words, it does not tell us about the homogeneity of the clustering. Several methods have been developed to identify groups or clusters based on the distances between species, and to depict that clustering in a dendrogram showing nested sets of species. Below, we present a few of these methods to illustrate their variety and discuss their limitations. Each method will be used to analyze the same data set: mean shapes of mandibles from 31 squirrel species, including tree squirrels, ground squirrels and flying squirrels (Figure 10.11A). Scores on the first two principal components of these data (Figure 10.11B), which account for 60% of the variation, suggest there may be some hierarchical structure in the data; but that structure appears to be more consistent with ecological similarity than with the phylogenetic relationships that have been inferred (Figure 10.11C).

The simplest clustering approach is called, appropriately enough, single linkage. The first link connects the two closest taxa. The second link connects the next two closest taxa, which may be a different pair, or it may connect a third taxon to a member of the first pair. It often occurs that the first pair forms a nucleus and other species are successively linked to this growing core; because they are on the periphery of the core, they tend to be closer to a species that is already in the group than to another species that is still on the outside. Consequently, this method tends to link taxa in long chains, successively linking singletons in a comb-like dendrogram that suggests little hierarchical structure (Figure 10.12A). In those cases where two groups of several species are linked, they are linked through the one species in each group that are closest to each other (like the tips of two ellipses aligned on the same axis). There is no implication of any general similarity among the other members of the group.

A somewhat more complicated method is the average linkage method, or unweighted pair-group method with arithmetic means (UPGMA). Again, the two closest taxa are linked first. At each successive step, the taxa that are joined are the ones that would have the smallest increase in average linkages. This is analogous to linking two ellipses that are side-by-side before linking ones that are tip-to-tip. The average of the linkages between the side-by-side pair will tend to be smaller than the average between the tip-to-tip pair unless the tips are very much closer than the sides. Similarly, this method will tend to avoid creating groups that are elliptical or making them more eccentric than they already are. This tends to create a more tree-like dendrogram (more hierarchical structure) than single-linkage does. In the UPGMA dendrogram (see Figure 10.12B), less than ¼ of the species were linked as singletons to existing clusters, in contrast to more than ⅔ linked as singletons in the single linkage dendrogram.

A still more complicated method is Ward's (1963) minimum variance. In this method, the taxa that are joined are the ones that create the smallest increase in variance in the new

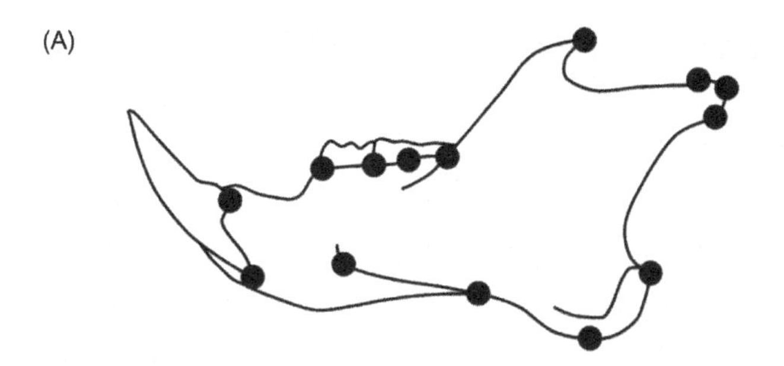

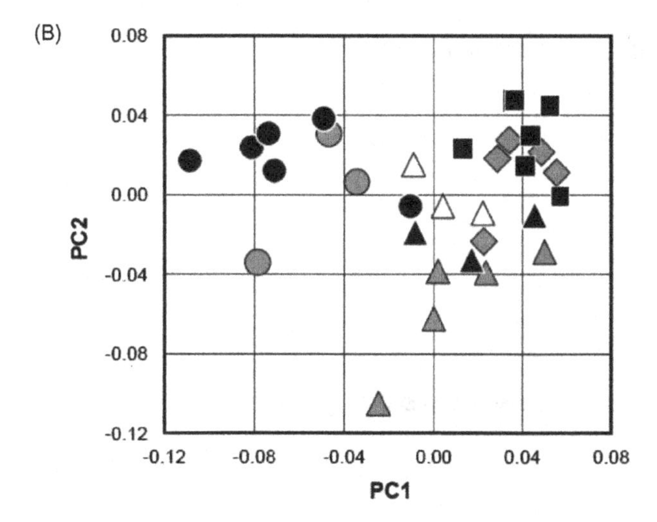

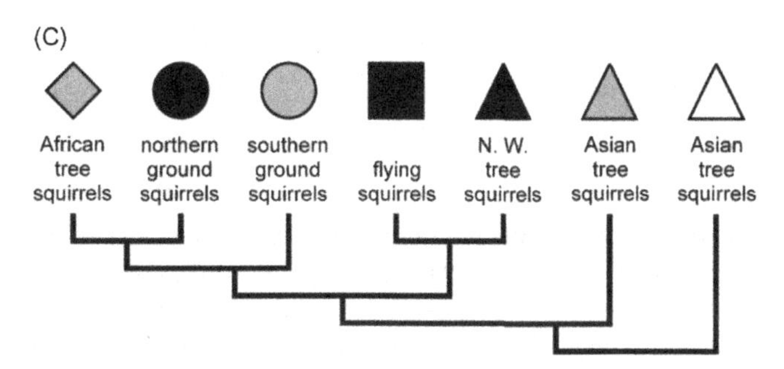

FIGURE 10.11 Cluster analyses of squirrel mandibles, part I. (A) Landmarks on representative *Sciurus* (North American tree squirrel). (B) Scores on first two PCs; symbols represent seven major monophyletic groups within the family, which are geographically and ecologically distinct. (C) Phylogenetic relationships of the clades, based on Mercer and Roth (2003).

group. It is broadly similar to UPGMA, in that some measure of within-group differences is minimally increased at each step. Since average distances are related to variances, UPGMA and Ward's method will tend to produce similar clusters, as shown in Figure 10.12C. In fact, in this example, the biggest difference between UPGMA and Ward's dendrograms is the greater variety of levels at which groups are linked; there are very few differences in group composition.

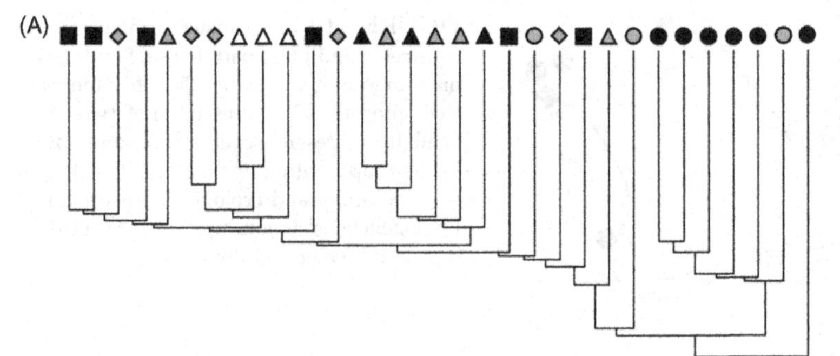

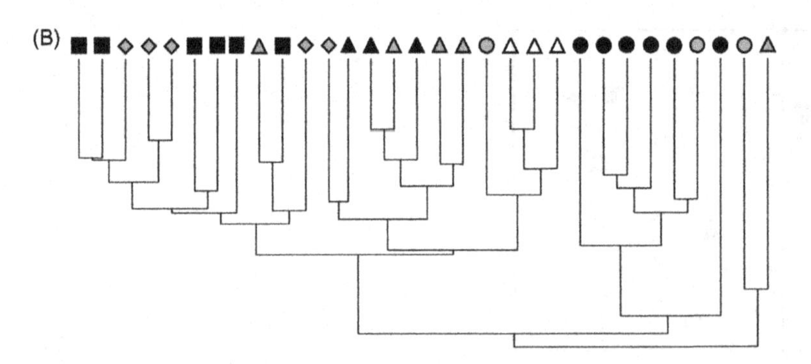

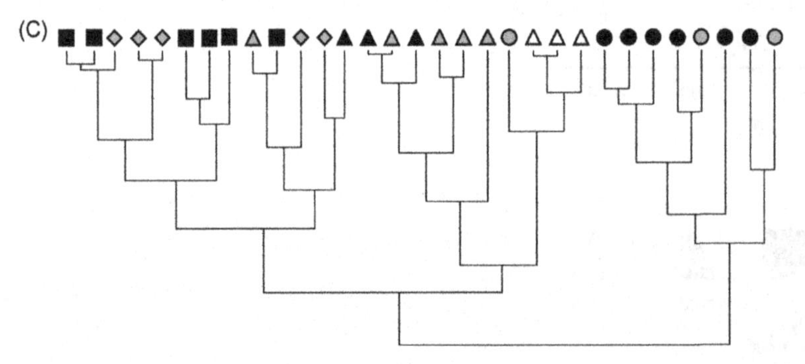

FIGURE 10.12 Cluster analyses of squirrel mandibles, part II. (A) Single linkage; (B) average linkage; (C) Ward's linkage.

In choosing between methods (or results), it is important to recognize that none of these clustering methods performs a statistical test to determine whether one linkage is significantly better than the alternatives. Consequently, it seems unwise to place much trust in short branches. But even for clusters separated by long branches from the next linkage, there is no explicit test to determine that the members of that cluster are distinct from the other taxa or other clusters. One could take the dendrogram as the starting point for a MANOVA hypothesis, but you should be wary of testing one *ad hoc* hypothesis after another. It would be better to start with an *a priori* hypothesis of group memberships, test those by MANOVA, and then use clustering and nearest-neighbor analyses to understand the distribution of similarities and differences among taxa.

Because all of these methods will generate a hierarchical dendrogram whatever the distribution of interspecific distances happens to be, it is important to test whether the structure of the diagram accurately reflects the hierarchical structure in the data. The cophenetic correlation coefficient (Sneath and Sokal, 1973) compares heights of tips above the node at which they are joined to the observed correlation matrix (transformed to a distance matrix by subtracting each correlation from 1.0) for the data used to generate the dendrogram. Values less than 0.85 are generally indicative of poor fit.

References

Adams, D. C., & Collyer, M. L. (2009). A general framework for the analysis of phenotypic trajectories in evolutionary studies. *Evolution, 63*, 1143–1154.

Adams, D. C., & Nistri, A. (2010). Ontogenetic convergence and evolution of foot morphology in European cave salamanders (Family: Plethodontidae). *BMC Evolutionary Biology, 10*.

Anstey, R. L., & Pachut, J. F. (1995). Phylogeny, diversity history, and speciation in Paleozoic Bryozoans. In D. H. Erwin, & R. L. Anstey (Eds.), *New approaches to speciation in the fossil record* (pp. 239–284). Columbia University Press.

Berg, R. L. (1960). The ecological significance of correlation pleiades. *Evolution, 14*, 171–180.

Blomberg, S. P., Garland, T., Jr., & Ives, A. R. (2003). Testing for phylogenetic signal in comparative data: Behavioral traits are more labile. *Evolution, 57*, 717–745.

Carlson, R. L., & Wainwright, P. C. (2010). The ecological morphology of darter fishes (Percidae: Etheostomatinae). *Biological Journal of the Linnean Society, 100*, 30–45.

Cheverud, J. M. (1982). Phenotypic, genetic and environmental integration in the cranium. *Evolution, 36*, 499–512.

Cheverud, J. M. (1984). Quantitative genetics and developmental constraints on evolution by selection. *Journal of Theoretical Biology, 110*, 155–172.

Cheverud, J. M. (1995). Morphological integration in the saddle-back tamarin (*Saguinus fuscicollis*) cranium. *American Naturalist, 145*, 63–89.

Cheverud, J. M., Dow, M. M., & Leutenegger, W. (1985). The quantitative assessment of phylogenetic constraints in comparative analyses: Sexual dimorphism in body weight among primates. *Evolution, 39*, 1335–1351.

Ciampaglio, C. N. (2002). Determining the role that ecological and developmental constraints play in controlling disparity: Examples from the crinoid and blastozoan fossil record. *Evolution & Development, 4*, 170–188.

Ciampaglio, C. N., Kemp, M., & McShea, D. W. (2001). Detecting changes in morphospace occupation patterns in the fossil record: Characterization and analysis of measures of disparity. *Paleobiology, 27*, 695–715.

Collar, D. C., Schulte, J. A., O'Meara, B. C., & Losos, J. B. (2010). Habitat use affects morphological diversification in dragon lizards. *Journal of Evolutionary Biology, 23*, 1033–1049.

Diaz-Uriarte, R., & Garland, T., Jr. (1998). Effects of branch length errors on the performance of phylogenetically independent contrasts. *Systematic Biology, 47*, 654–672.

Eastman, J. M., Alfaro, M. E., Joyce, P., Hipp, A. L., & Harmon, L. J. (2011). A novel comparative method for identifying shifts in the rate of character evolution on trees. *Evolution, 65*, 3578–3589.

Feller, W. (1968). *An introduction to probability theory and its applications*. John Wiley and Sons.

Felsenstein, J. (1985). Phylogenies and the comparative method. *American Naturalist, 125*, 1–15.

Foote, M. (1986). Developmental buffering as a mechanism for stasis. *Evolution, 42*, 396–399.

Foote, M. (1990). Nearest-neighbor analysis of trilobite morphospace. *Systematic Zoology, 39*, 371–382.

Foote, M. (1992). Paleozoic record of morphological diversity in blastozoan echinoderms. *Proceedings of the National Academy of Sciences of the United States of America, 89*, 7325–7329.

Foote, M. (1993). Contributions of individual taxa to overall morphological disparity. *Paleobiology, 19*, 403–419.

Foote, M. (1993). Discordance and concordance between morphological and taxonomic diversity. *Paleobiology, 19*, 185–204.

Foote, M. (1994). Morphological disparity in Ordovician-Devonian crinoids and the early saturation of morphological space. *Paleobiology, 20*, 320–344.

Foote, M. (1997). The evolution of morphological diversity. *Annual Review of Ecology and Systematics, 28*, 129–152.

Foote, M., & Gould, S. J. (1992). Cambrian and recent morphological disparity. *Science, 258*, 1816.

Garland, T., Jr., Dickerman, A. W., Janis, C. M., & Jones, J. A. (1993). Phylogenetic analysis of covariance by computer simulation. *Systematic Biology, 42*, 265–292.

Garland, T., Jr., Bennett, A. F., & Rezende, E. L. (2005). Phylogenetic approaches in comparative physiology. *Journal of Experimental Biology, 208*, 3015–3035.

Gavrilets, S. (1999). Dynamics of morphological diversification on the morphological hypercube. *Proceedings of the Royal Society of London, Series B, 266*, 817–824.

Gould, S. J. (1966). Allometry and size in ontogeny and phylogeny. *Biological Reviews, 41*, 587–640.

Gould, S. J. (1984). Morphological channeling by structural constraint: Convergence in styles of dwarfing and gigantism in *Cerion*, with a description of two new fossil species and a report on the discovery of the largest *Cerion*. *Paleobiology, 10*, 172–194.

Gould, S. J., & Garwood, R. A. (1969). Levels of integration in mammalian dentitions: An analysis of correlations in *Nesophantes micrus* (Insectivora) and *Oryzomys couesi* (Rodentia). *Evolution, 23*, 276–300.

Harmon, L. J., & Glor, R. E. (2010). Poor statistical performance of the Mantel test in phylogenetic comparative analyses. *Evolution, 64*, 2173–2178.

Hulsey, C. D., & Wainwright, P. C. (2002). Projecting mechanics into morphospace: Disparity in the feeding mechanics of labrid fishes. *Proceedings of the Royal Society of London, Series B, 269*, 317–326.

Kim, K., Sheets, H. D., Haney, R. A., & Mitchell, C. E. (2002). Morphometric analysis of ontogeny and allometry of the Middle Ordovician trilobite *Triathrus becki*. *Paleobiology, 28*, 364–377.

Kingsolver, J. G., & Wiernasz, D. C. (1991). Development, function, and the quantitative genetics of wing melanin pattern in *Pieris* butterflies. *Evolution, 45*, 1480–1492.

Klingenberg, C. P., & Gidaszewski, N. A. (2010). Testing and quantifying phylogenetic signals and homoplasy in morphometric data. *Systematic Biology, 59*, 245–261.

Kluge, A. G., & Kerfoot, C. (1973). The predictability and regularity of character divergence. *American Naturalist, 107*, 426–464.

Lande, R. (1980). The genetic covariance between characters maintained by pleiotropic mutations. *Genetics, 94*, 314–334.

Lauder, G. V. (1981). Form and function – structural analysis in evolutionary morphology. *Paleobiology, 7*, 430–442.

Liem, K. F. (1973). Evolutionary strategies and morphological innovations – cichlid pharyngeal jaws. *Systematic Zoology, 22*, 425–441.

Maddison, W. P. (1991). Squared-change parsimony reconstructions of ancestral states for continuous-valued characters on a phylogenetic tree. *Systematic Zoology, 40*, 304–314.

Marroig, G., & Cheverud, J. M. (2001). A comparison of phenotypic variation and covariation patterns and the role of phylogeny, ecology, and ontogeny during cranial evolution of New World monkeys. *Evolution, 55*, 2576–2600.

Marroig, G., & Cheverud, J. M. (2005). Size as a line of least evolutionary resistance: Diet and adaptive morphological radiation in New World monkeys. *Evolution, 59*, 1128–1142.

Martins, E. P., & Garland, T., Jr. (1991). Phylogenetic analyses of correlated evolution of continuous characters: A simulation study. *Evolution, 45*, 534–557.

Martins, E. P., & Hansen, T. F. (1997). Phylogenies and the comparative method: A general approach to incorporating phylogenetic information into the analysis of interspecific data. *American Naturalist, 149*, 646–667.

Maynard Smith, J., Burian, R., & Kauffman, S., et al. (1985). Developmental constraints and evolution: A perspective from the Mountain Lake conference on development and evolution. *Quarterly Review of Biology, 60*, 265–287.

Mercer, J. M., & Roth, V. L. (2003). The effects of Cenozoic global change on squirrel phylogeny. *Science, 299*, 1568–1572.

Miller, A. I., & Foote, M. (1996). Calibrating the Ordovician radiation of marine life: Implications for Phanerozoic diversity trends. *Paleobiology, 22*, 304–309.

Monteiro, L. R., & Nogueira, M. R. (2009). Adaptive radiations, ecological specialization, and the evolutionary integration of complex morphological structures. *Evolution, 64*, 724–744.

Olson, E. C., & Miller, R. L. (1958). *Morphological integration*. University of Chicago Press.

Revell, L. J. (2009). Size-correction and principal components for interspecific comparative studies. *Evolution, 63*, 3258–3268.
Ricklefs, R. E., & Cox, G. W. (1977). Morphological similarity and ecological overlap among passerine birds on St Kitts, British West Indies. *Oikos, 28*, 60–66.
Ricklefs, R. E., & Travis, J. (1980). A morphological approach to the study of avian community organization. *The Auk, 97*, 321–338.
Rohlf, F. J. (2001). Comparative methods for the analysis of continuous variables: Geometric interpretations. *Evolution, 55*, 2143–2160.
Rohlf, F. J. (2006). A comment on phylogenetic correction. *Evolution, 60*, 1509–1515.
Rohlf, F. J., Gilmartin, A. J., & Hart, G. (1983). The Kluge–Kerfoot phenomenon: A statistical artifact? *Evolution, 37*, 180–202.
Schaefer, S. A., & Lauder, G. V. (1996). Testing historical hypotheses of morphological change: Biomechanical decoupling in loricariod catfishes. *Evolution, 50*, 1661–1675.
Schluter, D. (1996). Adaptive radiation along genetic lines of least resistance. *Evolution, 50*, 1766–1774.
Seilacher, A. (1979). Constructional morphology of sand dollars. *Paleobiology, 5*, 191–221.
Simpson, G. G. (1953). *Major features of evolution*. New York: Columbia University Press.
Smith, L. H., & Lieberman, B. S. (1999). Disparity and constraint in olenelloid trilobites and the Cambrian radiation. *Paleobiology, 25*, 248–272.
Sneath, P. H. A., & Sokal, R. R. (1973). *Numerical taxonomy*. San Francisco: Freeman.
Sokal, R. R. (1976). The Kluge–Kerfoot phenomenon reexamined. *American Naturalist, 110*, 1077–1091.
Stone, E. A. (2011). Why the phylogenetic regression appears robust to tree misspecification. *Systematic Biology, 60*, 245–260.
Strauss, D., & Sadler, P. M. (1989). Classical confidence intervals and Bayesian probability estimates for ends of local taxon ranges. *Mathematical Geology, 21*, 411–427.
Swiderski, D. L., & Zelditch, M. L. (2010). Morphological diversity despite isometric scaling of lever arms. *Evolutionary Biology, 37*, 1–18.
Symonds, M. R. E. (2002). The effects of topological inaccuracy in evolutionary trees on the phylogenetic comparative method of independent contrasts. *Systematic Biology, 51*, 541–553.
Van Valen, L. (1962). Developmental gradients in the dentition of *Peromyscus*. *Evolution, 16*, 272–277.
Van Valen, L. (1970). An analysis of developmental fields. *Developmental Biology, 23*, 456–477.
Viguier, B. (2002). Is the morphological disparity of lemur skulls (Primates) controlled by phylogeny and/or environmental constraints? *Biological Journal of the Linnean Society, 76*, 577–590.
Vincent, S. E., Brandley, M. C., Herrel, A., & Alfaro, M. E. (2009). Convergence in trophic morphology and feeding performance among piscivorous natricine snakes. *Journal of Evolutionary Biology, 22*, 1203–1211.
Wagner, G. P. (1988). The influence of variation and of developmental constraints on the rate of multivariate phenotypic evolution. *Journal of Evolutionary Biology, 1*, 45–66.
Wagner, G. P., & Altenberg, L. (1996). Complex adaptations and the evolution of evolvability. *Evolution, 50*, 967–976.
Wagner, P. J. (1995). Testing evolutionary constraint hypotheses with early Paleozoic gastropods. *Paleobiology, 21*, 459–470.
Wagner, P. J. (1997). Patterns of morphologic diversification among the Rostroconchia. *Paleobiology, 23*, 115–150.
Ward, J. H. (1963). Hierarchical grouping to optimize an objective function. *Journal of the American Statistical Association, 58*, 236–244.
Wills, M. A. (2001). Morphological disparity: A primer. In J. M. Adrain, G. D. Edgecombe, & B. S. Lieberman (Eds.), *Fossils, phylogeny, and form: An analytical approach* (pp. 55–144). Kluwer Academic: Plenum Publishers.
Wills, M. A., Briggs, D. E. G., & Fortey, R. A. (1994). Disparity as an evolutionary index – a comparison of Cambrian and recent arthropods. *Paleobiology, 20*, 93–130.
Zelditch, M. L. (1988). Ontogenetic variation in patterns of phenotypic integration in the laboratory rat. *Evolution, 42*, 28–41.
Zelditch, M. L., & Carmichael, A. C. (1989). Ontogenetic variation in patterns of developmental and functional integration in skulls of *Sigmodon fulviventer*. *Evolution, 43*, 814–824.

Zelditch, M. L., Bookstein, F. L., & Lundrigan, B. L. (1993). The ontogenetic complexity of developmental constraints. *Journal of Evolutionary Biology, 6*, 121–141.

Zelditch, M. L., Sheets, H. D., & Fink, W. L. (2003). The ontogenetic dynamics of shape disparity. *Paleobiology, 29*, 139–156.

Zelditch, M. L., Wood, A. R., & Swiderski, D. L. (2009). Building developmental integration into functional systems: Function-induced integration of mandibular shape. *Evolutionary Biology, 36*, 71–87.

CHAPTER

11

Evolutionary Developmental Biology (1): The Evolution of Ontogeny

Studies of evolving ontogenies are grounded in two important insights. The first is that all evolutionary change arises from changes in ontogeny and therefore we need to understand how ontogenies evolve in order to understand the origins of morphological diversity, i.e. disparity (Zelditch et al., 2003b; Adams and Nistri, 2010; Drake, 2011; Frederich and Vandewalle, 2011; Gerber, 2011; Ivanovic et al., 2011; Piras et al., 2011). What modifications of ontogeny are responsible for the disparity of a group, and whether those modifications increase or decrease disparity can be answered by comparative analysis of ontogenies. As found in a study of damselfishes, disparity of both body shape and diet increase over ontogeny (Frederich and Vandewalle, 2011). But disparity of body shape decreases over ontogeny despite an increase in disparity of diet in piranhas (Zelditch et al., 2003b). In European cave salamanders, foot shape and interdigital webbing both decrease in disparity over ontogeny because some species maintain a webbed juvenile foot (and juvenile foot shape) as they grow whereas others increase webbing and change foot shape as they grow; the result is similar adult morphologies via different developmental processes (Adams and Nistri, 2010). By combining comparative studies of ontogeny with analyses of disparity at two or more developmental stages, it is possible to test hypotheses about the developmental origins of disparity.

The second insight is that organisms have time-extended phenotypes. An organism's phenotype is not static – it changes from age to age in both form and function. To comprehend that dynamic form and function relationship, as it evolves, we need to understand how developing organisms negotiate the ontogenetic transformations in form and function. Several studies have examined the relationship between ontogenetic transformations in shape and measures of performance such as bite-force or biomechanical parameters such as mechanical advantage of the masticatory apparatus (Birch, 1999; Abdala et al., 2001; Pfaller et al., 2010; Tanner et al., 2010b; La Croix et al., 2011a,b). For example, studies of carnivores found that feeding performance matures far later than skull size,

Geometric Morphometrics for Biologists
DOI: http://dx.doi.org/10.1016/B978-0-12-386903-6.00011-3

mandibular shape and feeding biomechanics, and that hyenas are far more protracted in their development than coyotes (Tanner et al., 2010a; La Croix et al., 2011a,b).

The concept central to all studies of evolving ontogenies is the "ontogenetic trajectory", introduced by Alberch and colleagues (1979) to signify the complete record of the physical appearance of the organism. It has been defined as much by an iconic diagram (Figure 11.1) as by words or formula. The diagram depicts a phenotype as a trajectory in a space of three dimensions: size (s), age (a), and shape (σ). Two of these (size and age) are one-dimensional, but shape obviously is not. When the diagram was initially drawn, shape was typically characterized by a single ratio but, as explicitly stated by Alberch and colleagues (p. 299), the picture remains the same no matter how many shape coordinates are required to specify the system. Now that we have methods for multivariate shape analysis, we can construct the ontogenetic trajectory for a multidimensional ontogeny of shape, although in the case we show here (Figure 11.2), the piranha *Serrasalmus gouldingi*, we do not have any data on age. We thus have only two axes – size and shape. We begin with this example, despite our lack of data on age, for two reasons. First, many studies do not have information about age so empirical studies of ontogenetic trajectories are often restricted to size and shape data, and second, the ontogeny of shape for this case is simple. By "simple" we mean that the direction of shape change is constant throughout ontogeny – it is not a function of size (or age). We can therefore represent the ontogeny of shape by a single vector, and score each individual for its position along it relative to size. These scores are obtained by projecting the shape data onto a line in the direction of the ontogenetic shape change, i.e. the vector of regression coefficients when shape is regressed on log-transformed centroid size (Drake and Klingenberg, 2008).

When we have age data, as we do for the cotton rat, *Sigmodon fulviventer*, we can include that in the diagram (Figure 11.3). This diagram is more difficult to read because it shows three-dimensions projected onto a two-dimensional plane. To see the relationship between shape and size, and between shape and age, we can show each pair of axes separately (Figure 11.4A, B, respectively). As evident in the plot for shape versus size, the relationship between shape and size is linear but that between shape and age is not. The other non-linearity is not evident in this plot because the regression vector is obtained by linear regression of shape on size or age. But the ontogeny of shape changes its direction from age to age (Figure 11.5). In this case, the ontogeny of shape cannot be represented by a line. Between two ages, such as birth and 10 days, it can be represented by a line, but that

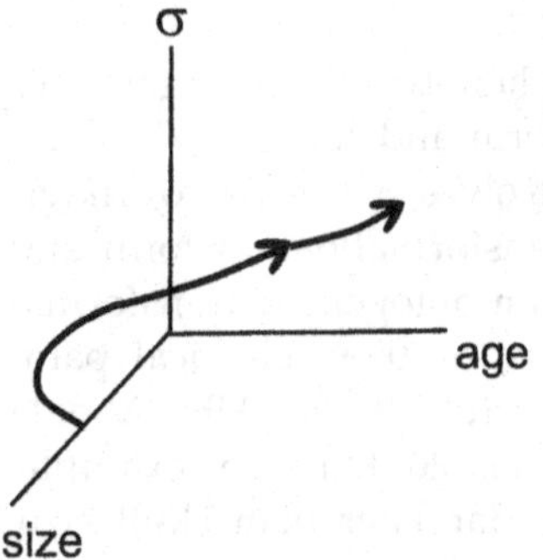

FIGURE 11.1 The ontogenetic trajectory, as depicted by Alberch et al., 1979.

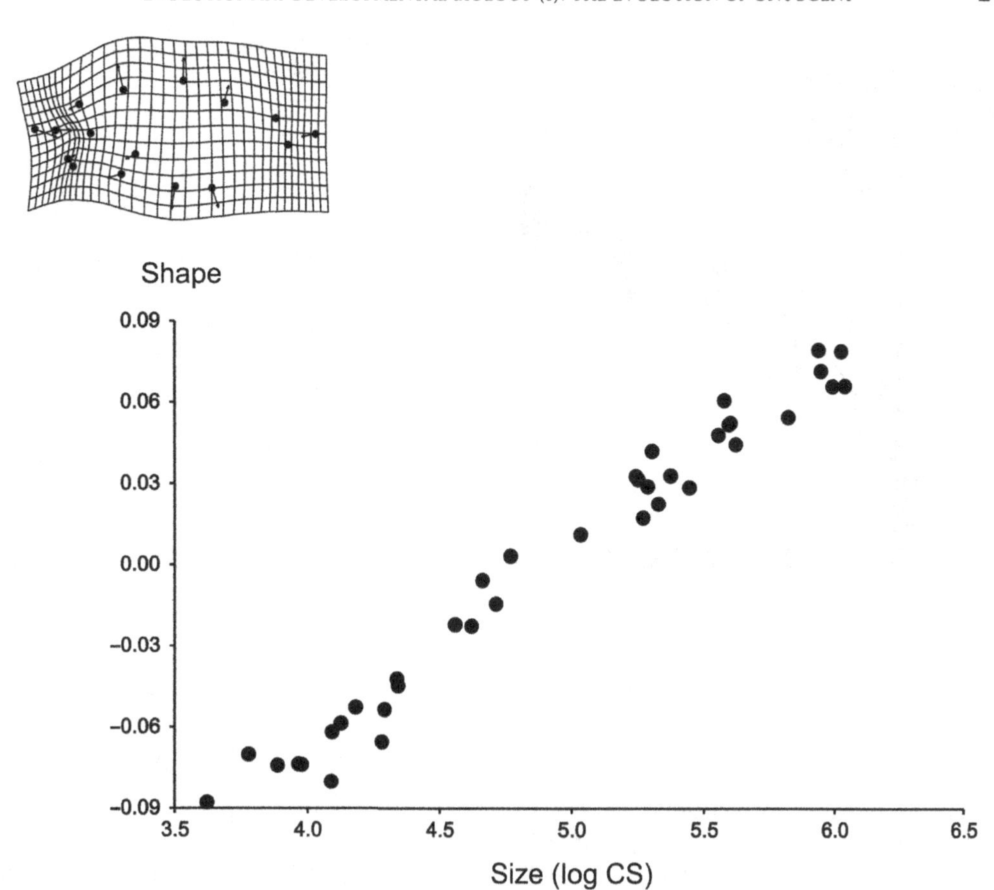

FIGURE 11.2 In the absence of any data on age, the ontogenetic trajectory of shape and size for a piranha, *Serrasalmus gouldingi*. The plot shows the Procrustes distance from the juvenile, indicating the magnitude of change; the deformation grid illustrates the change in shape.

line reorients and points in another direction between 10 and 20 days, and it reorients again between 20 and 30 days, and points in yet another direction from that age through sexual maturity. To represent the ontogeny of this species, we need a curve through a multidimensional shape space not a shape axis. The ontogenetic trajectory, as represented in the iconic diagram, will thus be an oversimplification of actual ontogenies, especially when extended to multivariate data. However, that does not compromise the value of the concept. Even when it takes more than three dimensions to draw it, the ontogenetic trajectory is nonetheless a function of age, size and shape.

The difficulty of drawing complex trajectories does have important implications for depicting comparative analyses, a subject we discuss in more detail below. The difficulty is that it is not easy to visualize even three dimensions. Even to show the relationships between shape, size and age we eased the task of visualization by flattening the picture, projecting it onto two pairs of dimensions. We did not try to show the whole ontogeny of

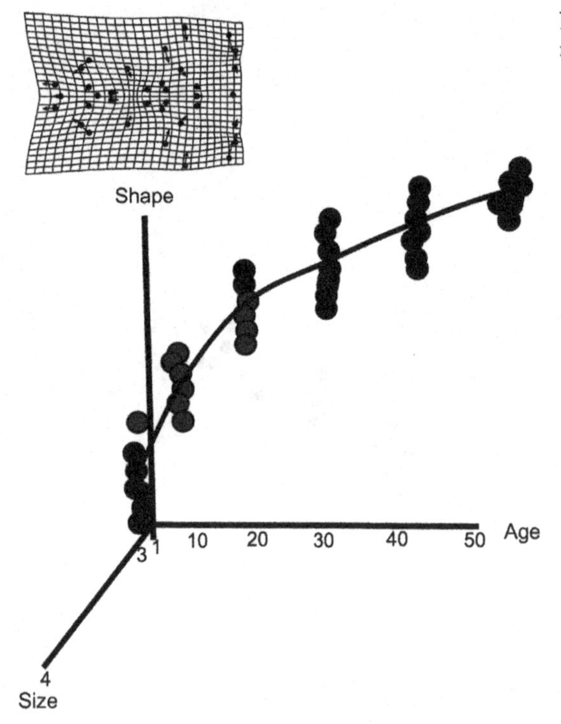

FIGURE 11.3 The ontogenetic trajectory of a cotton rat, *Sigmodon fulviventer*.

shape as it changed direction from age to age. If we try to compare two species, the diagrams become even more complicated because we would then have two (curving) trajectories and we would need to represent the geometrical relationships of these curves within and between the trajectories. Projections of two or more complex multidimensional ontogenetic trajectories can be misleading and distort those geometric relationships so we cannot rely on the pictures to see by how much, or in what, two or more trajectories differ. Fortunately, we do not use these projections to formulate or test hypotheses.

In this chapter, we discuss the range of hypotheses about the evolution of ontogenetic trajectories that can be tested and how to test them. Some hypotheses cannot be tested without information about chronological age. We emphasize "chronological" age because of the contrast some authors have made between "chronological" and "biological" or "developmental" age. This contrast has been central to some of the arguments about the necessity for age data; one argument for that necessity is that information about age is needed for process-oriented studies – it is growth relative to time that provides the information about process and therefore without age data, the analysis devolves into a merely pattern-oriented study (Blackstone, 1987). One counterargument is that the mathematical model for allometry is the solution of the differential equation for growth rates relative to time (Strauss, 1987). A second is that *chronological* time does not have any theoretical priority over other estimates of biological age, which includes size, for the description of comparison of growth patterns (Strauss, 1987). The argument is not that size is a *proxy* for *chronological* age but rather that size provides information about *biological*

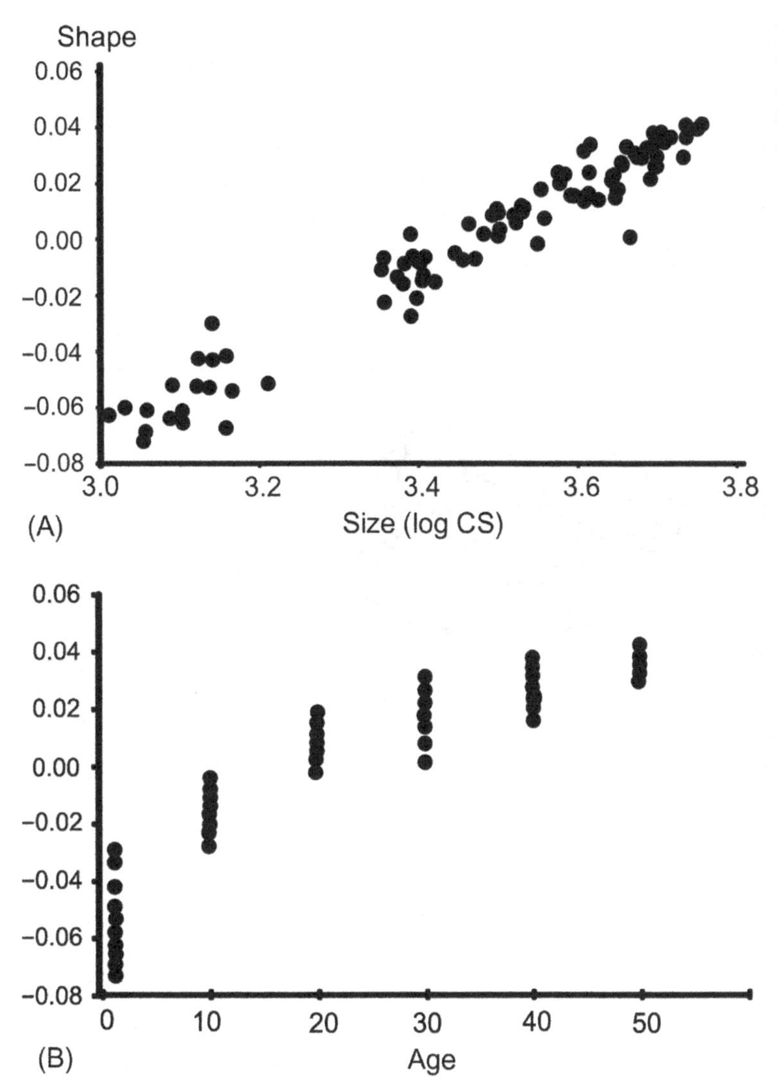

FIGURE 11.4 Projecting the three-dimensional trajectory onto a pair of planes, shape versus size, measured as log centroid size and shape versus age.

(developmental) age. Whether age or size provides the more relevant information is sometimes addressed empirically, such as by analyzing the dependence of the variable of interest on both size and age, as in the study of woody plant stem structure and function (Rosell and Olson, 2007). When size is a better predictor of the feature of interest, and when species are at different developmental stages at the same chronological age, size might be viewed as a better predictor of both structure and function.

Despite the arguments regarding the merits of analyses based on developmental age, or size, some questions cannot be answered without information about chronological age because they are not about either intrinsic biological age or the predictability of structure and/or function from size (or developmental age). Instead, they are specifically about the

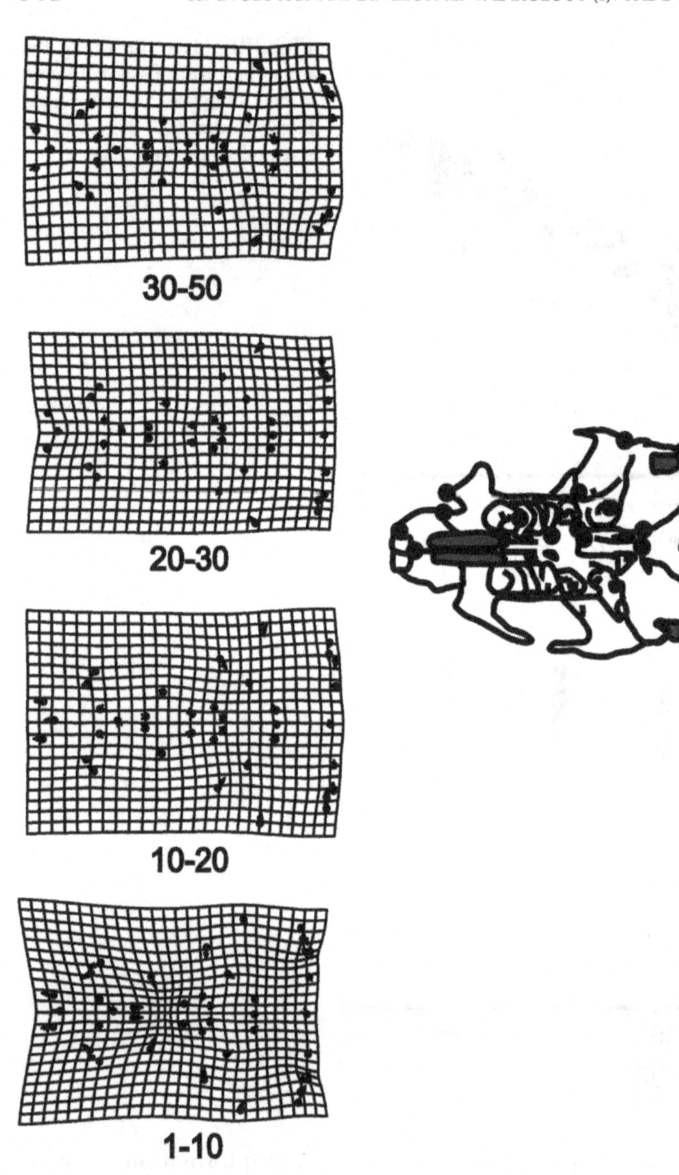

FIGURE 11.5 The directions of ontogenetic change of the cotton rat skull measured between successive ages (in days postnatal).

link between ontogeny and extrinsic time, as in studies that relate evolving ontogenies to ecology. That requirement for age data may be most obvious for the case of heterochrony, which concerns the changes in developmental rate and timing that produce the parallelism between ontogeny and phylogeny (Gould, 1977). Without age data, it is not possible to distinguish between various changes that produce the same morphological outcome, such as a later onset of development, a slower rate of development or a shorter duration of development. All produce a descendant that resembles the ancestral juvenile. In another case, we also need information about age because the questions are specifically about developmental rate and timing. The focus is on timing because the issue is the coordination between periodic biological phenomena (phenology) and the environment, both abiotic and biotic. This has become an important issue recently because one of the best documented responses to global warming is a shift in phenologies. One of the most intriguing (and worrying) patterns is the discordant shifts in phenologies across trophic levels, leading to a mismatch between organisms and their foods (Both and Visser, 2005; Post and Forchhammer, 2008; Post et al., 2008; Both et al., 2009; Miller-Rushing et al., 2010). To investigate such (mis)matches, we need to relate timing of development events (such as birth and weaning) to the environment, both thermal and biotic.

The link between ecology and developmental rate has motivated many studies of heterochrony as well (Gould, 1977; McKinney, 1986; Emerson et al., 1988; Schweitzer and Lohmann, 1990), but the theory of heterochrony does not supply a general enough framework for analyzing changes in phenology because the theory of heterochrony as developed by Gould (1977) and formalized by Alberch et al. (1979) makes specific predictions about morphology. Those predictions may be wrong even if developmental rate and timing do evolve because heterochrony predicts that *only* developmental rates or timings evolve. Developmental rate and timing can evolve and so can ontogenies of shape, hence the descendant shape might not be predictable by extrapolating the ancestral ontogeny. That the predictions are empirically refuted for some cases does not compromise the value of the formalism devised by Alberch et al. for the study of heterochrony, a formalism that we discuss in detail below, because the formalism was intended to apply *solely* to studies of heterochrony and it is applicable to all cases of heterochrony. The scheme is nonetheless limited in its applicability because it applies solely to studies of heterochrony, as Gould and Alberch et al. defined "heterochrony". We highlight that matter of definition because there are multiple definitions of heterochrony in the literature, which are often inconsistent and even mathematically incommensurate. That could be considered merely a matter of semantics (McKinney, 1999), but the predictions that follow from a theory, as well as the formal representation of a theory, depend on what the theoretical terms mean.

The long-standing fascination with heterochrony has made age data seem necessary for virtually all studies of evolving ontogenies, even when the questions are not about the link between ontogeny and ecology. But the relationship between shape and size is no less important than that between shape and time. Both in context of function and development, allometry is interesting in its own right. We thus begin with a discussion of allometry, and why it is interesting in its own right, and then how to analyze it. We first review the formalism for the analysis of allometry using traditional morphometric variables because some hypotheses make more sense when framed in terms of those (size) measurements. The interpretation of the results requires scaling coefficients. We then more briefly discuss the geometric analysis of ontogenetic allometry (which was presented in Chapter 8). Although the regression coefficients are not readily interpretable, geometric morphometrics has notable advantages for testing a range of hypotheses about the evolution of ontogeny, the final subject of this chapter.

This chapter begins with the review of the formalism for the analysis of allometry using traditional morphometric variables, then briefly recalls the analysis of allometry using geometric morphometric data, and then examines a series of hypotheses that can be tested about the evolution of ontogenetic trajectories. As well as comparing ontogenetic trajectories, we also incorporate analyses of disparity to dissect the developmental sources of disparity. These analyses are done without any information about age, so we next consider hypotheses that require age data to test, and how to test them when species differ in their ontogenetic trajectories. Finally, we discuss the disparity of ontogeny itself.

WHY ALLOMETRY IS INTERESTING IN ITS OWN RIGHT

On purely biomechanical grounds, we often expect organisms to change shape when they change size in either ontogeny or evolution. Were organisms to grow (or evolve to a larger size) without changing shape, they would likely decrease their ability to perform such vital functions as respiration, locomotion and feeding. That is because, under geometric scaling, a length of x scales to an area of x^2 and a volume of x^3. To see why that might impair performance, consider the cross-sectional area of a weight-bearing limb bone. If the length, area and mass of a small organism arc in the proportions of 2:4:8, those for a somewhat larger organism would be 4:16:64, and for a much larger organism they would be 10:100:1000. Thus, length has increased fivefold whereas area increased 25-fold and volume increased 125-fold. Geometric scaling could cause limbs to buckle under the more rapidly increasing mass, and also cause bones loaded by force-generating muscles to bend. For that reason, changes in size are expected to lead to changes in shape to maintain functional equivalence. This reasoning does not predict changes in proportions of length measurements (because they scale to the same power). They are functionally equivalent at constant proportions, i.e. two jaws with the same ratio between input and output lever arms are equal in their mechanical advantage.

Allometric scaling maintains functional equivalence over a range of sizes for certain basic physical properties (such as surface area:volume relationships). These might be expected to scale predictably over an entire ontogenetic series even though young animals are not just small they are also young and often ecologically different than older (larger) members of their own species. For that reason, they do not face the same functional demands but they might for length:surface area or for surface area:volume relationships. Otherwise we might expect scaling relationships that alter proportions by more (or less) than predicted from the scaling of lengths to areas to volumes. That by itself is interesting because it means that, over an individual's life-time, it is increasing its size, changing its shape, and experiencing transitions in functional demands and that, at every age, the organism must be competent to perform whatever functions it currently has while it is continually changing both form and function. How these transformations in size, shape and function are interrelated is a central question in studies of ontogenetic allometry.

One question is whether the ontogenetic trajectory is directed towards the optimal *adult* shape or instead towards an optimum weighted in favor of the most vulnerable age. The trajectory may be oriented towards the adult morphology because that morphology is stable for longest, at least in organisms that have determinate morphogenesis. It may therefore be more consequential to fitness than the morphologies that precede it. Yet the adult shape will never be reached at all if organisms do not survive vulnerable pre-adult phases. However, the direction of the trajectory may represent a compromise direction across a sequence of age-specific optima because it could change direction from age to age, as in the case of the cotton rat discussed above. The direction of the trajectory, as well as the rate at which shape develops, may be related, in part, to age-specific functional demands or mortality rates. Curving trajectories for shape have been documented for several mammalian species (Zelditch et al., 2003a; Tanner et al., 2010b; La Croix et al., 2011a), although linear trajectories are detected or assumed in others (e.g. Monteiro et al., 1999; Ponce de Leon and Zollikofer, 2001; Strand Vioarsdottir et al., 2002; Bastir and Rosas, 2004). In organisms that undergo metamorphosis, transitions in direction may be abrupt, decoupling larval from juvenile morphogenesis (Strauss and Altig, 1992; Ivanovic et al., 2007, 2011). Whether trajectories are linear, smoothly curving or abruptly reorienting can have important implications for the evolution of morphology because the decoupling between phases could allow for their independent evolution. For example, one study of scapular ontogeny concludes that the shape of the infant constrains that of the adult (Young, 2008) – the shape of the infant is hypothesized to drive the pattern of postnatal growth. Another view of constraints limiting the evolutionary flexibility of ontogenetic allometries is based on analyses of craniofacial form in muroid rodents; according to this hypothesis, functional constraints of later development, more specifically those due to the biomechanics of mastication, lead to conserved postweaning ontogenies. Early ontogenies are more flexible, being less constrained by function.

In studies that emphasize the biomechanical or other functional interpretations of scaling coefficients, the ratios between linear, areal and volumetric measurements are biologically meaningful. There are theories that predict the expected values, so it is those values that need to be measured. Allometric coefficients are also interpretable developmentally, and many studies of allometry are concerned with the developmental interpretation of the coefficients. To lay the foundation for that interpretation we introduce the conventional formalisms for studies of ontogenetic allometry.

FORMALISMS FOR THE ANALYSIS OF ONTOGENETIC ALLOMETRY: TRADITIONAL MORPHOMETRIC DATA

The traditional formalism for the study of allometry relates the increase in size of one part (Y) to that of another (X). Often, X is intended to represent the size of the whole organism. To make our discussion of allometry as concrete as possible, and to ease the transition from geometric to traditional morphometric data, we will focus on the case of the piranha, *Serrasalmus gouldingi*, one of the examples we have used throughout this text. To analyze its ontogenetic allometry using traditional morphometric data we measure a variety of lengths and depths (Figure 11.6). For our measure of body size we will use the measurement extending from landmark 1 to landmark 7, which is termed "standard length" (SL) and is frequently used as the measurement of body size in studies of teleosts, so, in our example, $X = SL$. The other 29 measurements are the measures represented by the vector $\{Y_1, Y_2, Y_3, \ldots Y_{29}\}$. We first discuss the mathematical analysis of allometry, then follow this with an interpretation of the coefficients obtained by the analysis, and then consider their developmental significance.

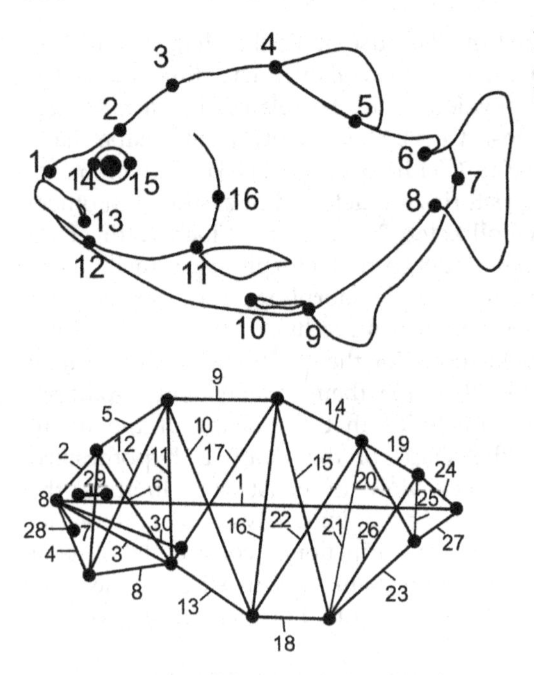

FIGURE 11.6 Landmarks sampled on *S. gouldingi*, and the traditional morphometric measurement scheme based on those landmarks.

The relationship between X and Y often fits a model, the power law (Huxley, 1932):

$$Y = bX^k \tag{11.1}$$

where k is the growth rate of part Y relative to X, and b is the size of Y when X is at unit size. To ease fitting the model to data, it is often rewritten in a linear form:

$$\log(Y) = \log(b) + k\log(X) \tag{11.2}$$

Expressed in this form, we can use linear regression to estimate the parameters b and k; they are the intercept (because $\log(1) = 0$) and slope, respectively, of a linear regression of $\log(Y)$ on $\log(X)$. Table 11.1 gives the regression coefficients, b and k, of the variables shown in Figure 11.6 regressed on *SL*. We should note that the literature is inconsistent on the symbols used for these two coefficients, but because b and k are widely used in the literature on allometry, we follow that convention.

Usually, the coefficients are estimated by simple bivariate regression, but multivariate regression yields the same estimates as obtained from bivariate analysis. We can therefore treat the bivariate estimates of k as components of the vector $\{k_1, k_2, k_3, \ldots k_P\}$, where P is the number of measurements. The estimates of $\log(b)$ are then components of the vector $\{\log(b_1), \log(b_2), \log(b_3), \ldots \log(b_P)\}$. In studies of traditional size measurements, allometric coefficients are often estimated by principal components analysis (PCA), following Jolicoeur (1963) who first proposed that PC1 is a multivariate allometry vector when PC1 is extracted from a variance–covariance matrix of log-transformed measurements. Conceptually, multivariate regression and PCA differ in that PCA does not single out one variable as the independent size variable. Instead, size is a linear combination of the

TABLE 11.1 Allometric Coefficients for the 30 Measurements of *Serrasalmus gouldingi* Shown in Figure 11.6

Variable	b	k
v2	−0.939	0.806
v3	−0.613	0.885
v4	−1.163	0.85
v5	−2.643	1.225
v6	−1.396	1.042
v7	−1.512	1.002
v8	−1.383	0.928
v9	−1.974	1.104
v10	−1.761	1.22
v11	−1.931	1.198
v12	−1.595	1.136
v13	−2.378	1.186
v14	−2.228	1.116
v15	−1.681	1.21
v16	−1.781	1.225
v17	−1.551	1.171
v18	−2.572	1.17
v19	−1.938	1.085
v20	−1.685	1.104
v21	−1.991	1.225
v22	−1.834	1.217
v23	−1.473	1.072
v24	−1.674	0.939
v25	−2.686	1.12
v26	−1.558	1.129
v27	−1.815	0.898
v28	−1.231	0.811
v29	−1.160	0.751
v30	−0.635	0.903

variables that explains the correlations among them. When the variables are very highly correlated, as they usually are in ontogenetic studies of size measurements, least squares regression and PCA tend to give very similar results. For example, Table 11.2 shows the estimates of the slope for the measurements of *S. gouldingi* obtained by regression and PCA. The numbers may appear to be quite different, but the differences disappear when the coefficients are rescaled to make each one the ratio between k for a dependent variable and k for SL, the independent variable in the regression. Because SL is the independent variable, $k_{SL} = 1$. Rescaling the coefficients by dividing each by the one for SL gives the values shown in Table 11.3. The estimates obtained by multivariate regression and PCA are identical. The important distinction between regression and PCA is that PCA does not provide estimates of b.

Interpreting Allometric Coefficients

The interpretation of k is straightforward – it is the growth rate of the measured part relative to that of a standard (X), such as overall body size. When k is 1.0, the growth of the measured part keeps pace with X. Their proportions are constant throughout growth. Such measurements are termed *"isometric"*. When k is greater than 1.0, the measured part grows more rapidly than X so the relative size of that part increases; these measurements are termed *"positively allometric"*. When k is less than 1.0, the measured part grows more slowly than X, so its relative size decreases even though its absolute size increases; these measurements are termed *"negatively allometric"*. The problem with classifying coefficients solely in terms of ratios between the part-specific growth rate (k) and the growth rate of X is that *all* ks are relative growth rates. When several part-specific ks equal each other, their proportions (relative to each other) are constant. Several measurements have nearly equal allometric coefficients, including the four measurements of body depth (measured from landmarks 4 and 5, which are at the anterior and posterior bases of the dorsal fin; v15, v16, v21 and v22). All four are positively allometric relative to body length, so the body (in that region) deepens relative to its length. Among the negatively allometric measurements are the most anterior lengths (v2, v3, v4, v8, v28, v29, v30) and the two most posterior ones (v24, v27). This means that measurements in the anterior head and caudal regions shorten relative to the whole body (of course they do not actually shorten – they lengthen in an absolute sense, it is just that they shorten relative to the length of the body). Consequently, the head and caudal region form a relatively smaller fraction of body length in adults than in juveniles.

To look at a second example, this one of mammalian craniofacial growth, we consider the ontogenetic allometries of the two rodents mentioned above, the cotton rat (*Sigmodon fulviventer*) and house mouse (*Mus musculus domesticus*). These two species differ strikingly in life-history; cotton rats are precocial, meaning that they are relatively mature at birth. Cotton rats open their eyes within a day of birth, and can also hear and even walk at that time. In contrast, house mice (like most muroid and sciurid rodents) are altricial; their eyes do not open for 10 days, when their ears also open and they begin to walk. Despite that striking difference in maturity at birth, both species wean at approximately 21 days. The ontogenetic allometries for 12 craniofacial measurements, relative to skull length, v1

TABLE 11.2 Allometric Coefficients (*k*) Computed by Multivariate Regression (R) and PCA

Variable	R	PCA
v1	1	0.694
v2	0.806	0.559
v3	0.885	0.614
v4	0.85	0.589
v5	1.225	0.851
v6	1.042	0.723
v7	1.002	0.696
v8	0.928	0.644
v9	1.104	0.766
v10	1.22	0.847
v11	1.198	0.832
v12	1.136	0.789
v13	1.186	0.823
v14	1.116	0.775
v15	1.21	0.84
v16	1.225	0.85
v17	1.171	0.813
v18	1.17	0.812
v19	1.085	0.753
v20	1.104	0.766
v21	1.225	0.85
v22	1.217	0.845
v23	1.072	0.744
v24	0.939	0.651
v25	1.12	0.777
v26	1.129	0.783
v27	0.898	0.623
v28	0.811	0.563
v29	0.751	0.522
v30	0.903	0.627

v1 is (standard length) the independent variable in the multivariate regression so its value for *k* must be included to make these vectors comparable.

TABLE 11.3 Ratios Between Allometric Coefficients (*k*) of Each Variable and Standard Length (v1) for the Coefficients Computed by Multivariate Regression (R) and PCA

Variable	R	PCA
v2	0.81	0.81
v3	0.89	0.88
v4	0.85	0.85
v5	1.23	1.23
v6	1.04	1.04
v7	1	1
v8	0.93	0.93
v9	1.1	1.1
v10	1.22	1.22
v11	1.2	1.2
v12	1.14	1.14
v13	1.19	1.19
v14	1.12	1.12
v15	1.21	1.21
v16	1.23	1.22
v17	1.17	1.17
v18	1.17	1.17
v19	1.09	1.09
v20	1.1	1.1
v21	1.23	1.22
v22	1.22	1.22
v23	1.07	1.07
v24	0.94	0.94
v25	1.12	1.12
v26	1.13	1.13
v27	0.9	0.9
v28	0.81	0.81
v29	0.75	0.75
v30	0.9	0.9

(Figure 11.7) are shown in Table 11.4. The several measurements that are strikingly negatively allometric in both species (v4–v7) are all measurements of width, especially the widths of the cranial base. The most striking positive allometries are, for the house mouse, v9–v11, which are all measurements of facial length except that v10, which extends from

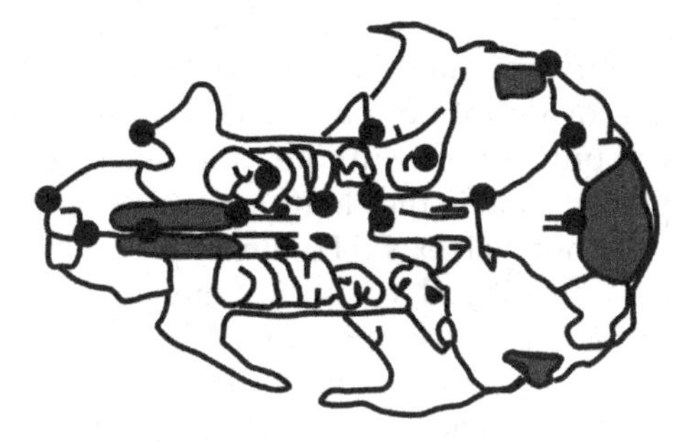

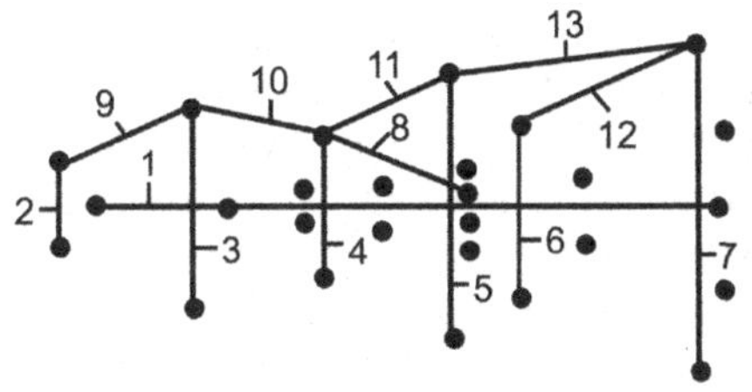

FIGURE 11.7 Landmarks sampled on ventral skull of the cotton rat, *Sigmodon fulviventer*, and house mouse, *Mus musculus domesticus*, and the traditional morphometric measurement scheme based on those landmarks.

TABLE 11.4 Ontogenetic Allometric Coefficients (k) for Postnatal Growth of the House Mouse (*Mus musculus domesticus*) and Cotton Rat (*Sigmodon fulviventer*) for the Craniofacial Measurements Shown in Figure 11.7

Variable	House Mouse	Cotton Rat
v2	1.10	1.39
v3	0.95	0.75
v4	0.61	0.57
v5	0.45	0.63
v6	0.37	0.65
v7	0.57	0.62
v8	0.96	0.93
v9	1.34	1.10
v10	1.45	1.68
v11	1.27	1.02
v12	1.13	0.94
v13	1.07	1.09

the tip of the zygomatic arch to the first molar, and thus measures the lengthening of the zygomatic spine as well as facial lengthening. For the cotton rat, the most pronounced positive allometries are for v2 and v10, the most anterior premaxilla and that distance between zygomatic spine and molar. Thus, these coefficients convey what we know to be general features of mammalian development – the skull generally lengthens relative to its width, the face lengthens even more so, and widens relative to the width of the braincase.

The interpretation of b (or log b) is less straightforward, and there has been some controversy about its biological meaning. One reason for doubting that b has any general biological significance is that its value depends on the units of measurement; unlike k, b is not a dimensionless quantity. However, a more important one is that $\log(b)$ is the value of $\log(Y)$ when $\log(X)$ is zero, a size at which Y might not yet exist. For example, when the body is 1 mm long, the dorsal fin might not have developed yet so it cannot have a meaningful size. Additionally, $\log(b)$ is estimated under the assumption that k is constant from $\log(X) = 0$, not just that it is constant over the range of values actually sampled.

Under one condition, b does have a simple interpretation: when species do not differ in k. In that case, differences in b will persist throughout the entire ontogeny. Although we might reasonably hesitate to infer a value for $\log(Y)$ when $\log(X) = 0$, the difference between species *at any point* in ontogeny will be invariant over ontogeny. Although we might also be hesitant to infer that species have diverged either when $\log(X) = 0$ or before, we could conclude that the difference arose prior to the stage when we first observe them and persists throughout the rest of ontogeny. That difference in b says how those populations will differ *at any given value of* X. Under other conditions, b can be viewed as just a parameter needed to predict Y at a given value of X. To determine whether species differ at a particular age of interest (such as birth, or the transition from larval to juvenile growth, or at weaning), we can use the regression equation to determine the predicted values for the dependent variables at the relevant value for X.

The Developmental Meaning of b and k

Most of the literature on ontogenetic allometry has focused on the developmental meaning of k because b is static – it is not a descriptor of development but rather of where the regression line intersects the Y-axis. At the heart of the literature is the view of growth as a multiplicative process. This was the rationale given by Huxley (1932) for the power law, and it is the basis for cellular models of allometric growth (Katz, 1980). Within that context, the meaning of k has been viewed from both spatial and temporal perspectives. Huxley (1932) emphasized the spatial interpretation of k, proposing that differences in k over the organism indicate spatially organized "growth intensities". He noted that values of k tend to be spatially coherent, rising and falling in organized patterns across the body. To help visualize spatial patterns in k, we can first put the coefficients on the organism rather than in a table (Figure 11.8). We can see that they increase from the head to the middle of the body, then fall towards the tail, although not to a level as low as found in the head. This is (approximately) an inverted U-shaped gradient, which is interesting because it is the inverse of the gradient found in several teleost larvae (Fuiman, 1983). This suggests that the allometry of juvenile growth compensates for that of larval

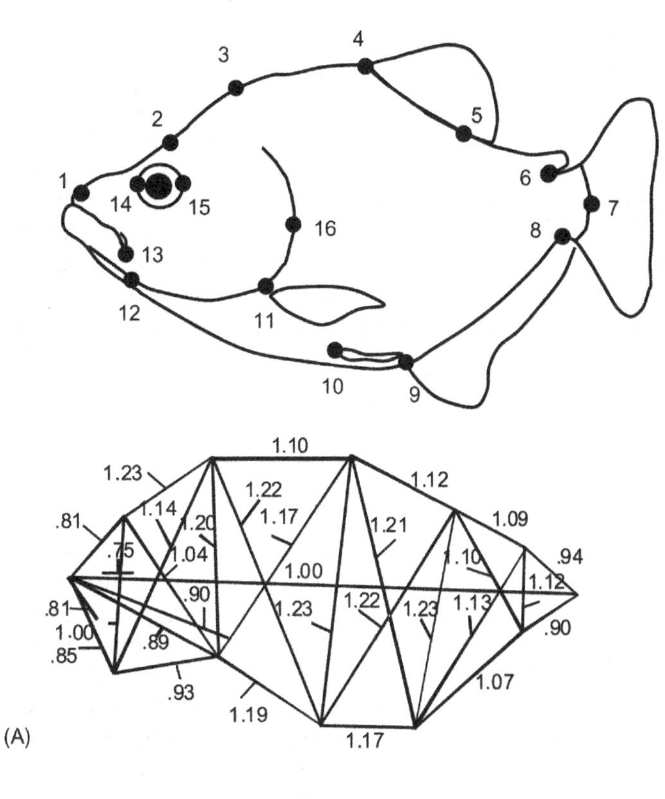

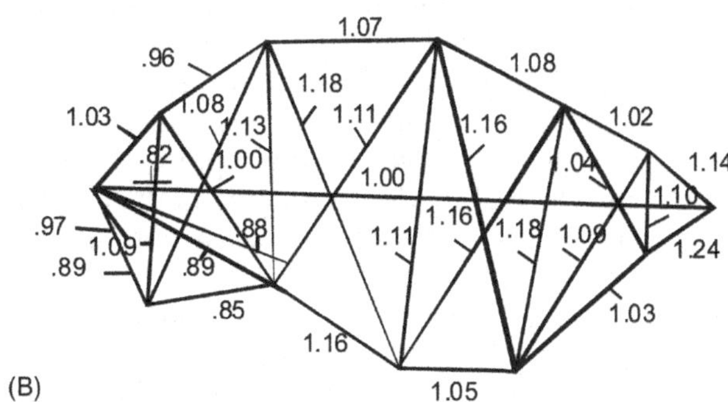

FIGURE 11.8 Allometric coefficients of *Serrasalmus gouldingi*. Coefficients higher than 1.0 indicate positive allometry; coefficients lower than 1.0 indicate negative allometry; and coefficients near 1.0 indicate isometry.

growth: the head and caudal body initially grow very rapidly and, later, during juvenile growth, the middle of the body catches up. The spatial distribution of growth rates can also suggest an anteroposterior gradient, meaning that the rates fall off linearly from the head to the tail. To analyze these patterns more rigorously Huxley constructed "growth profiles", which are plots of allometric coefficients as a function of their position along body axes.

Laird and colleagues have stressed the temporal significance of allometric coefficients (Laird, 1965; Laird et al., 1968). Even though time is not explicitly incorporated in studies of allometry, it is nonetheless implicit. This becomes evident when considering why the power law holds in the first place. As mentioned before, the primary biological explanation for allometry is that growth is a multiplicative process. When analyzing the relationship between size and time, the best-fitting models are usually not linear but rather are sigmoidal in form. An important feature of these models is that growth rates decay over time. Similarities in decay rates are interpreted by Laird (1965) as the explanation for the linear relationship among log-transformed measurements. In effect, all measurements follow the same growth curve; their differing values of k tell us how they are displaced relative to each other in time – different parts of the body reach the same point on their growth curves at different times. Laird et al. (1968) elaborated on this theory, stating the relationship between k and lag time (ΔT) as:

$$\Delta T = -\frac{1}{\alpha}\ln(k) \tag{11.3}$$

where α is the decay rate and k is an allometric coefficient.

To measure decay rates we need information about age, but we can use Equation 11.3 to understand the temporal relationships among growth curves even without known age samples so long as we are willing to assume that all measurements whose logs are linearly related have the same decay rates. Because growth rates decay over time, we would intuit that a more negatively allometric part has decayed over a longer time, and that it has decayed for longer because it began growing earlier. The increment of time by which we need to shift one curve to match another that starts growing later is ΔT. Based on this interpretation of allometric coefficients, we would conclude that those for piranha body growth mean that the head and caudal peduncle develop before the midbody, that the eye is the first structure to develop, and that the body elongates before it deepens. In the case of the mammalian skull, we would interpret the allometric coefficients to mean that the broad bulbous cranium of the neonate, resulting from the rapid prenatal growth of the brain, becomes relatively narrowed over postnatal growth by the rapid elongation of the skull as a whole and especially of the face, which also widens relative to the cranial base, as part of the overall increase in facial dimensions relative to braincase and cranial base.

The spatial and temporal perspectives on allometric coefficients are not antagonistic. The spatial coherence noted by Huxley, interpreted within the temporal framework of Laird, suggests that growth is spatiotemporally organized. There is no reason to think that either space or time is primary. We do not need to adopt one view over the other – they are mutually consistent, and help explain each other. With increasing information about the spatial determination of development, in conjunction with that on its temporal organization, we can relate allometric coefficients to the underlying developmental processes that explain them.

To interpret these coefficients in terms of both growth and function, we can apply theories about scaling relationships to growth. Applied to ontogenetic series, such theories may explain ontogenetic allometry in terms of the ontogeny of function. For example, in many larval teleosts the head and caudal region are highly positively allometric, which is due to the early demands imposed by swimming, feeding and respiration (see, for example, van Snik et al., 1997). The converse allometric pattern is seen later, in juvenile growth, as exemplified by the coefficients of *S. gouldingi*. These patterns are hardly surprising, which is reassuring if our aim is to make sense of ontogenetic allometry in functional and ecological terms. A striking example of a study of allometric scaling is one that tests the hypothesis that scaling maintains functional equivalence of the mandibular symphysis (where the right and left sides come together) and resistance to "wish-boning" (lateral transverse bending). Previous study of adults showed that positive evolutionary allometry of symphyseal measures (particularly its width) maintains similar load resistance capabilities as increasing symphyseal curvature, and consequent stress concentrations increase with size (Hylander, 1985). As a result, adults of species that differ in size maintain a similar capability to resist loads. Studying the ontogeny of symphyseal curvature and width, as well as relative stress, in two species of macaques, Vinyard and Ravosa (Vinyard and Ravosa, 1998) found no difference between the two ontogenies in relative stresses, and also that stress does not change significantly throughout ontogeny in either species. Thus, ontogenetic allometry maintains the functional equivalence in stress and strain levels during postnatal growth.

Because theories about developmental controls over the spatiotemporal organization of relative growth, as well as theories about the functional significance of scaling relationship, are most easily expressed in terms of traditional morphometric measurements, studies of allometry using traditional morphometric measurements will remain an important part of evolutionary developmental biology.

Revisiting Geometric Morphometric Analyses of Allometry

In Chapter 8, we introduced multivariate regression, the method we use to analyze the relationship between shape and size. To review that, the model for allometry is:

$$\{Y_1, Y_2, Y_3, \ldots Y_P\} = \{m_1, m_2, m_3, \ldots m_P\}X + \{b_1, b_2, b_3, \ldots b_P\} + \{\varepsilon_1, \varepsilon_2, \varepsilon_3, \ldots \varepsilon_P\} \tag{11.4}$$

where $\{Y_1, Y_2, Y_3 \ldots Y_P\}$ is the vector of shape variables, X is centroid size and $\{m_1, m_2, m_3, \ldots m_P\}$, $\{b_1, b_2, b_3, \ldots b_P\}$ and $\{\varepsilon_1, \varepsilon_2, \varepsilon_3, \ldots \varepsilon_P\}$ are vectors of slope coefficients, intercepts and residuals, respectively. We would not use principal components analysis for geometric analyses of allometry because PC1 of geometric shape data need not be aligned with the ontogenetic trajectory whenever there are factors other than age in the data (e.g. sexual dimorphism). Of course it could be aligned with the ontogenetic trajectory, but there is no reason to use PCA when multivariate regression is guaranteed to give the optimal description of the dependence of shape on size. Also, given that we have a size metric (centroid size), there is no need to estimate "size" by a linear combination of the measured (shape) variables.

As is always the case, it does not matter which shape variables we use because we obtain the same trajectory regardless of whether we use coordinates obtained by a generalized Procrustes analysis, by partial warps plus uniform components or the full set of PC scores. But even though the complete description of the ontogenetic change does not depend on the choice of variables, the coefficients obviously do. Thus, in striking contrast to the coefficients obtained from traditional morphometric data, those obtained from geometric shape data are not interpretable one by one. Another notable difference is the ability of geometric analyses to synthesize all the data (including all interlandmark distances that were not measured). We can best appreciate that difference by comparing two pictures, one showing the allometric coefficients drawn on the skulls and the other showing the deformed grids depicting the ontogenetic change in skull shape (Figure 11.9). What we can see more readily from the grids are the striking changes in palatal proportions medially versus more laterally.

Another notable difference is that geometric data yields more interpretable results in comparative studies. We first present a series of hypotheses about the evolution of ontogeny, and then discuss how to test them.

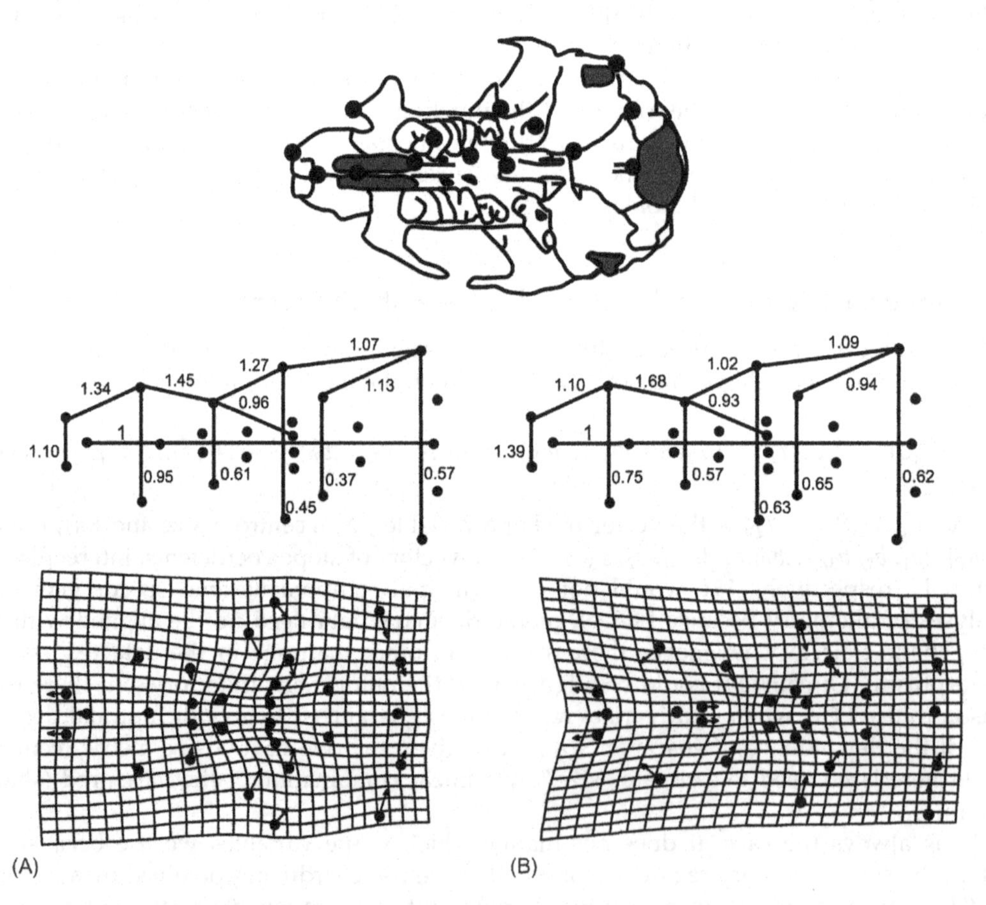

FIGURE 11.9 Allometric coefficients of the two rodent skulls; (A) cotton rat; (B) house mouse. Coefficients higher than 1.0 indicate positive allometry; coefficients lower than 1.0 indicate negative allometry; and coefficients near 1.0 indicate isometry.

HYPOTHESES ABOUT THE EVOLUTION OF ONTOGENETIC TRAJECTORIES

Until recently, most studies of evolving ontogenies focused on three possibilities: (1) ontogenetic scaling; (2) heterochrony; and (3) "transpositional" allometry. None of these three alter the ontogenetic trajectory. These are shown as vectors in a two-dimensional shape space (Figure 11.10A, B, C), which makes the contrasts among them easiest to see but which precludes depicting the relationships between shape and size or age. The first two alter the relationship between shape and either size or age, and the first is actually a special case of the second. In the case of ontogenetic scaling (Figure 11.10A), the descendant's shape can be predicted from the ancestral regression equation – the descendant has the shape that the ancestor would at the descendant size. So, if the descendant grows to a larger size, the descendant juvenile has the shape of the ancestral adult. Conversely, if the descendant grows to a smaller size, the descendant adult has the shape of the ancestral juvenile. The evolutionary change in shape thus results from the difference in rate or timing of growth as increases or decreases in growth rate, or longer or shorter durations of growth, alter the descendant's adult size. In the second case, it is the relationship between shape and age that is altered (Figure 11.10B). The descendant has the shape predicted for a younger or older ancestor. Ontogenetic scaling is a special case of this because ontogenetic scaling preserves the relationship between shape and size; in the more general case, evolutionary changes in size and shape can be dissociated from each other. In the case of heterochrony, it is the rate or timing of shape change relative to age that is modified. The shape of the descendant appears in ancestral trajectory or can be obtained by extrapolation of that ontogeny, but the regressions of shape on age or size are not identical. Heterochrony, and the special case of ontogenetic scaling, are encompassed by the classic definition of heterochrony (Gould, 1977; Alberch et al., 1979).

The third case is called "transpositional" allometry because the two trajectories are identical; it is just that one is translated (transposed) along the y-axis relative to the other (see Figure 11.10C). In this case, the trajectories point in the same direction, but the ontogenies diverged before the youngest observed age. As a result, the trajectories are parallel rather than coincident. At no point subsequent to divergence (and therefore, at no point in the observed phase of development) does the shape of the descendant appear in the ontogeny of the ancestor. The two are consistently different – the same features that distinguish the two at birth distinguish them as adults. It is possible that the ancestor and descendant

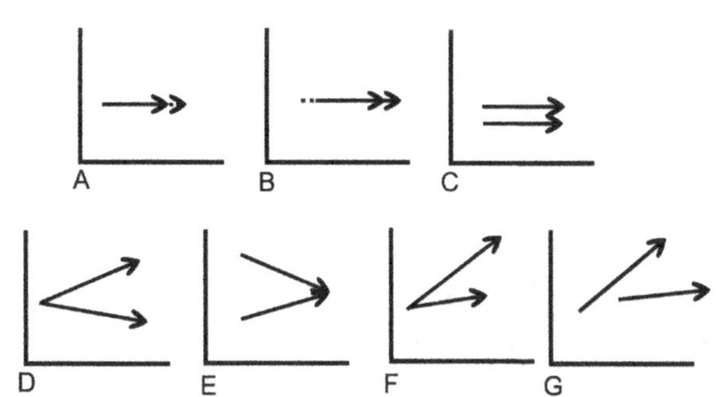

FIGURE 11.10 Hypotheses about the evolution of ontogenetic trajectories drawn as two vectors in a two-dimensional shape space. Three hypotheses predict no change in the ontogenetic trajectory of shape (A, B, C). The remaining three predict change in the ontogenetic trajectory of shape and early morphogenesis (D, E) and/or lengths of the ontogenetic trajectories (F, G).

also differ in rates or timings of growth and/or morphogenesis, so there are more than just the three possibilities shown in Figure 11.10.

There are also at least four cases in which the ontogenies of shape are modified (see Figure 11.10D, E, F, G). In the first case, the two trajectories differ solely in direction – the species are identical at the outset of development but progressively diverge over time (see Figure 11.10D). There is no general term for this case, although one suggestion is "allometric repatterning" (Webster and Zelditch, 2005). In the second case, the two trajectories differ in the starting point as well as direction; they differ in shape at the outset of the observed development but their ontogenetic trajectories point towards the same adult shape so they converge on the same adult form (see Figure 11.10E). This has been termed "ontogenetic convergence" (Adams and Nistri, 2010). Of course, the divergent trajectories could also lead to divergent adult morphologies; the species could diverge further over the course of development, increasing the distance between the ancestral and descendant shapes. In the third case, the two trajectories differ in length plus direction (see Figure 11.10F); the two species develop at different rates, in different directions. To our knowledge, there is no term for this possibility. In the fourth case, the two species differ in all three attributes: shape at the outset of the measured phase, direction and length (see Figure 11.10G). This is another case for which, to our knowledge, there is no term. Species that have complex, curving ontogenetic trajectories add to these possibilities.

At present, we do not know which of these possibilities occurs most frequently, nor do we know which contributes most to disparity. Until fairly recently, most studies focused on ontogenetic scaling, in particular, or heterochrony, more generally. As a result, these two topics dominate the literature. But that does not mean that either is especially common or that either makes a large contribution to disparity. One reason for focusing on ontogenetic scaling was to find the traits that do *not* evolve by extending or truncating the ontogenetic trajectory because such extensions or truncations were expected when body size evolves; consequently, the traits that do not exhibit such extensions or truncations were thought to require a specific, adaptive explanation. In that sense, ontogenetic scaling simply served as a "criterion for subtraction" (Huxley, 1932; Gould, 1966). Heterochrony, however, was seen as especially interesting, one that challenged conventional evolutionary theory. Whether frequent or not, heterochrony was seen as worthy of special attention. The reasons why Gould thought that heterochrony is especially interesting are important for understanding his analytic scheme as well as its reformulation by Alberch and colleagues. Gould's arguments about the theoretical meaning of heterochrony are grounded in intuition rather than formal theory, but those intuitions motivated the fascination with heterochrony. As Gould construed "heterochrony", it referred to *the changes in developmental rate and/or timing that produce the parallelism between ontogeny and phylogeny.* Because ontogenetic scaling is a special case, we include it in our discussion of heterochrony.

Heterochrony

Gould (1977) devoted his entire book on ontogeny and phylogeny to heterochrony because he regarded it as especially interesting and as challenging to traditional evolutionary theory. The first reason why he regarded heterochrony as especially interesting is that he thought that it could yield relatively large morphological changes by simple modifications of development, yielding large bursts of change in a short amount of time (Gould, 1977, 1982; Maderson et al., 1982). Second, heterochrony was thought to occur by selection on size and/or life-history parameters, leading to predictable changes in morphology as a by-product (Gould, 1977, 1982, 1988; Maderson et al., 1982). In regarding morphological evolution as a by-product of selection on size or life-history, Gould followed Huxley's (1932) reasoning that shape evolves by selection on size. The morphological changes might have no selective value; they are simple correlates of selection on size and/or life-history (Maderson et al., 1982; Gould, 1988). This connection between heterochrony and evolutionary allometry is seemingly obvious, but Gould was the first to recognize and emphasize it. The idea that morphology evolves in a specific direction due to selection on size or life-history is why Gould thought that heterochrony provides the best empirical data for the study of developmental constraints (Gould, 1988). The third reason why Gould found heterochrony to be so interesting is that it linked ecology to morphology; he argued, for instance, that progenesis (truncated development, see below) would be causally linked to unstable (r-selected) environments that promote rapid maturation whereas neoteny (slow development) would be linked to stable environments (Gould, 1977, 1988). Under contrasting ecological conditions, heterochrony would yield the same morphological outcomes – descendant adults that resemble ancestral juveniles, under contrasting ecological conditions.

As well as regarding heterochrony as interesting, and even as especially informative about developmental constraints, Gould also aimed to rehabilitate the concept of recapitulation. He denied that recapitulation is a general rule, but he, nevertheless, argued that the idea had been unfairly dismissed and for reasons unrelated to the failure of the theory. Certainly, he was not alone in attempting to rescue the idea of recapitulation; others, especially Cope (1887) also tried to do so, but by applying the concept to individual parts (or measurements). Unlike them, Gould took an organismal, multivariate view of parallelism. He strongly opposed the trait-by-trait approach to morphology, whereby each individual organ (or measurement) is accorded its own explanation. Instead of that approach, which he called "atomistic", he favored viewing organisms as integrated entities, bound together by developmental correlations. Formal models predicting the evolutionary response to selection on size (or any other trait) for the bivariate (Lande, 1979) and more general multivariate case (Lande and Arnold, 1983) confirm Gould's intuition that selection on body size or life-history trait can indirectly affect morphology, just not necessarily in the direction of ontogenetic scaling or heterochrony.

The fundamental idea underlying all these implications of heterochrony is that growth, morphogenesis and maturation can be dissociated from each other. Growth refers to an increase in size (with size being equated to geometric scale), morphogenesis refers to the process that alters shape, and maturation to the attainment of sexual maturity. Although this separation, especially between size and shape, has sometimes been viewed as justified solely on operational grounds, i.e. as necessary for the construction of Gould's clock model, Gould justified separating them on the grounds that their dissociability is necessary for heterochrony. Of course, they are correlated within any ontogeny, but they are potentially dissociable in their evolution. When growth and morphogenesis are dissociated from age, the descendant has the ancestral shape and size at a different developmental stage. For example, the descendant adult could have the ancestral juvenile shape and size. If growth is dissociated from morphogenesis and maturation, then the descendant adult would have the ancestral adult shape at a different size. It might, for example, have the ancestral adult shape at the ancestral juvenile size. Finally, if morphogenesis is dissociated from growth and maturation, then the descendant adult would have the shape found in the ancestral ontogeny at a different age and size.

The purpose of the clock model (Figure 11.11) was to reveal the modifications of growth, morphogenesis and maturation because the same morphological outcome could arise from different modifications of life-history. Thus, to understand the links between ecology and heterochrony, we need to distinguish between the modifications of development that produce the same morphological outcomes. Although the clock model is rarely used now, understanding it is important because it supplied the context for the scheme that replaced it. The face of the clock contains two arcs and one bar. One arc is a shape axis, the other is a size axis and the bar is the age axis. The values of the ancestral shape are plotted along the arc, with the values for the youngest age on the left. The ancestral sizes are plotted on the size axis, lining up the ages at which the ancestor has that size, and also lining up the shape at that size. The entire ontogeny of the ancestral shape is represented on the clock, but the descendant is analyzed at one single stage. Not surprisingly, the need to single out one stage for comparison prompted much discussion about what that stage should be and whether it should be chosen according to chronological age, developmental age, or even size. Whatever standard is used, the objective is to find the matching ancestral size and shape at that point. When found, the hands of the clock are arranged to point to it; if the matching shape occurs at an earlier stage in the ancestor, the "shape hand" of the clock will point to the left. Similarly, if the matching size occurs at an earlier stage in the ancestor, the "size hand" also points to the left. Differences between ancestor and descendant in chronological age at the developmentally comparable stages are indicated by the filled portion on the age bar.

The clock, and the terms defined by it, proved confusing and the scheme was soon replaced by the one devised by Alberch and colleagues, whose intent was to clarify the terminology of heterochrony. To that end, Alberch et al. (1979) redesigned Gould's formalism, using a more conventional representation of a three-dimensional space: three mutually orthogonal axes (Figure 11.12). They also replaced Gould's static comparative framework by a dynamic one; the descendant ontogeny (not just one point along it) is analyzed in conjunction with the ancestral ontogeny. Each ontogeny is represented as a vector in the three-dimensional space defined by the ancestral values of size and shape. The comparisons are made with respect to four parameters: (1) α, the age at the onset of development; (2) β, the age at offset of development; (3) k_{σ}, the rate of development

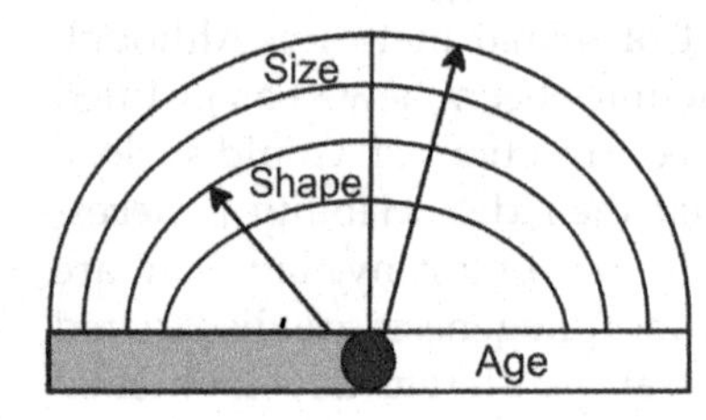

FIGURE 11.11 Gould's (1977) clock model. The descendant's size, shape and age at one developmental stage are compared to the ancestor's ontogeny of shape, size and age. The hands of the clock show the change from ancestral to descendant values pointing from the descendant's age-specific shape and size to the corresponding ancestral values (no hand is shown if there is no change). The shape hand points to the left, so the descendant adult has the morphology of a younger stage in the ancestral ontogeny.

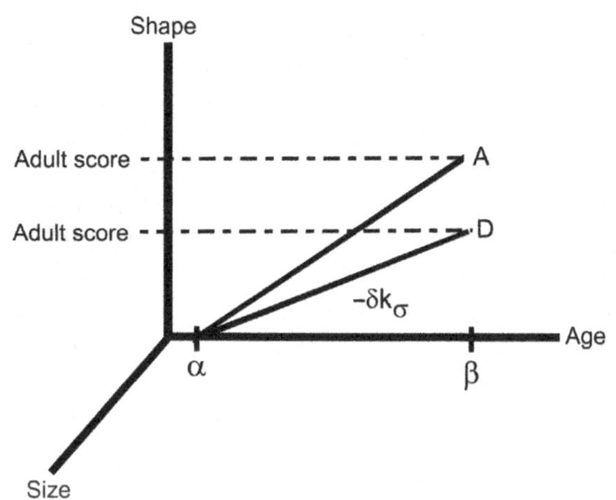

FIGURE 11.12 Alberch et al. (1979) formalism for the study of heterochrony. The clock is redrawn by representing the ancestral shape, size and age as three mutually orthogonal axes. Species are compared with respect to the age at onset of development (α), rate of development (k_σ), rate of growth (k_s), and age at termination of development (β). Shown is an example of neoteny, a decrease in developmental rate ($-\delta k_\sigma$). The score for the descendant's adult shape is lower than that of the ancestor's, so the descendant adult has the morphology of a younger stage in the ancestral ontogeny. See Table 11.5 for the names of the heterochronic perturbations defined by changes in these three parameters.

TABLE 11.5 Definitions of the Eight Pure Heterochronic Perturbations and their Morphological Expression, as Defined by Alberch et al. (1979)

Control Parameter	Incremental Change	Process	Morphological Expression
α	$-\delta\alpha$	Predisplacement	Peramorphosis
	$+\delta\alpha$	Postdisplacement	Pedomorphosis
β	$-\delta\beta$	Progenesis	Pedomorphosis
	$+\delta\beta$	Hypermorphosis	Peramorphosis
$K\sigma$	$-\delta k\sigma$	Neoteny	Pedomorphosis
	$+\delta k\sigma$	Acceleration	Peramorphosis
ks	$-\delta ks$	Proportional giantism	
	$-\delta ks$	Proportional dwarfism	

(i.e. the rate of change in shape); and (4) k_s the rate of growth (i.e. the rate of change in size). Each parameter can differ in two directions (and increase or decrease) yielding the eight pure heterochronic perturbations (Table 11.5). Each produces either a pedomorphic (=childlike) descendant or the converse, a peramorphic descendant, i.e. one who goes "beyond" the endpoint of the ancestral trajectory. There are also two perturbations of growth rate yielding proportional giantism and dwarfism, which are not usually considered to be heterochronic perturbations because they do not yield either pedomorphic or peramorphic descendants. Instead, they yield giant or miniature replicas of the ancestral shape. Nevertheless, they are usually included for the sake of completeness.

Of course, combinations of these pure cases are also possible. Naming the combinations is less straightforward and if we found a combination of $+\delta k_s$ and $-\delta k_\sigma$, for example, we would not likely want to construct a compound name from the labels for each one. For that example, the compound term would be some form of "proportional giantism plus neoteny" but, by definition, proportional giantism produces a giant replica of the ancestral morphology whereas neoteny produces an adult that resembles the ancestral juvenile. The combination of the two is self-contradictory. But the fact that there are no terms for modifications in size plus shape is problematic, although it is an interesting feature of the scheme that only two perturbations predictably affect growth or size. The others affect only development and shape. As a result, one cannot use this scheme to predict how heterochrony will affect size. Even though ontogenetic scaling is a special case of heterochrony, there are no terms for perturbations that lead to extensions or truncations of both size and shape.

Shea (1983a) introduced terms for perturbations that affect size and shape, including rate hypermorphosis or hypomorphosis, and time hypermophosis or hypmorphosis. These are sometimes used in the anthropological literature, but for readers unaccustomed to them they can be confusing because they combine the feature that is modified (growth rate or time) with a term for a morphological outcome of a change in developmental timing ($+\delta\beta$). The distinction being made is between an increase or decrease in rate (with no change in the duration of growth) and an increase or decrease in duration (with no change in the rate of growth). Rather than using a term that means a delayed offset in development for an increased rate of growth, it seems clear enough, if more verbose, to say that durations of growth are extended or truncated, or that rates of growth and development are increased or decreased.

As should be evident by this point, both the clock model and the Alberch et al. formalism can only be applied when the ancestral and descendant ontogenetic trajectories shape are the same. The trajectories can differ only by extension or truncation. That is not a limitation of either scheme because, by definition, heterochrony and scaling refer to extensions or truncations of conserved ontogenetic trajectories of shape. The two schemes are thus intended to be used solely for the cases in which the ancestral and descendant ontogenies differ only by extension or truncation. If the two ontogenies differ otherwise, the hypotheses of heterochrony and scaling are (or should be) rejected. If that hypothesis of a conserved ontogenetic trajectory of shape is not rejected, the two schemes can be used to diagnose the heterochronic perturbation. If the hypothesis is rejected, the question is whether the trajectories are parallel or not.

Parallel Trajectories (Transpositional Allometry)

Parallel trajectories (see Figure 11.10C) are at least as intriguing as heterochrony and ontogenetic scaling because they mean that early development is less conserved than later. Consequently, adults differ in precisely the same direction that larvae or infants did. This is surprising in light of the conventional view that early development is more conservative than later, although the larval stage is still quite late in development. The conservative developmental phase is the "phylotypic period" – the stage at which embryos of all members of a phylum look the same (Seidl, 1960; Sander, 1983; Slack et al., 1993). That period begins at onset of neurulation and ends with somitogenesis (see Kimmel et al., 1995). Whether it actually is so conservative is controversial; several studies have challenged the idea that embryos initially diverge and then converge, an idea that has been termed the "hour-glass" model (e.g. Bininda-Emonds et al., 2003) on the grounds that the phylotypic period is not as conservative as generally thought. But even if distantly related species do resemble each other through somitogenesis, it would not be surprising to find that larvae or neonates differ substantially in shape. Even fairly modest differences in development are likely to have large impacts on morphology when rates of development are high. What is surprising is not that divergence occurs during larval or fetal development but rather than divergence occurs only then – over the entire observed phase of development, the ontogenetic trajectories are parallel.

One hypothesis that could explain parallel trajectories is that later stages of ontogeny are subject to more severe functional constraints. This has been postulated for the case of sigmodontine rodents; the reason why post-weaning growth allometries might be conservative is the biomechanical constraints of masticatory function; that earlier developmental stages are less conserved could result from the absence of such constraints in the fetus or suckling pup (Voss and Marcus, 1992). Another explanation is serial correlations between developmental stages; just as modifications early in development will have cascading effects on later stages, selection on adult morphology will lead to divergence at earlier stages. This hypothesis presupposes that individuals who deviate from the mean as adults also do so (and in the same direction) much earlier in ontogeny. Although there is some evidence for that correlation between developmental stages, one recent study of human craniofacial shape found a weak correlation between newborns and adults, although correlations are high between three year olds and adults (Bulygina et al., 2006). An alternative hypothesis, which also assumes a high correlation between early and late developmental stages, is that selection for changes of early ontogeny leads to modifications of adult morphology; the shape of the adult may even be constrained by the shape of the infant (Young, 2008).

Divergent Ontogenies of Shape

The remaining hypotheses predict that ontogenies diverge in ontogenetic trajectories for shape. What differentiates the hypotheses is what else is expected to differ and whether morphologies are expected to diverge or converge over the course of development. The reason for expecting one of these patterns, as opposed to heterochrony or parallel trajectories, is that there is no good *a priori* reason to expect that the spatiotemporal patterning of morphogenesis, or any other aspect of morphogenesis, growth or maturation is conserved. Rather, ontogenies can evolve by modifications of allometry, by accelerating or retarding growth, or by increasing or decreasing the durations of growth or by increasing or decreasing either the rates or durations of morphogenesis. Thus, in the absence of any reason for expecting that ontogenies are constrained in their evolutionary possibilities, to determine which modifications occur we need to consider all the possibilities shown in Figure 11.10 as well as all variants upon them.

TESTING HYPOTHESES ABOUT THE EVOLUTION OF ONTOGENY

Ontogenetic scaling and heterochrony are the most frequently studied and reported kinds of modifications of ontogeny (e.g. Shea, 1983b, 1992; McKinney, 1986; Wayne, 1986; McKinney and McNamara, 1991; German et al., 1994; Neige et al., 1997; Cronier and Courville, 2003; Lieberman et al., 2007; Galatius et al., 2011; Gerber, 2011; Piras et al., 2011). That is partly because the definition of "heterochrony" was broadened to the point that any and all modifications of ontogeny qualify as "heterochrony" (see, especially McKinney and McNamara, 1991). But even if we discount studies that make heterochrony synonymous with the evolution of ontogeny, we are still left with an enormous number of studies that support the hypothesis of either heterochrony or ontogenetic scaling. Many of these use traditional morphometric data, so we first consider some of the methodological issues posed by these data and whether methodology might explain why heterochrony and scaling appear to be the predominant modes of evolutionary change in ontogenies. However, we would also point out many studies use ontogenetic scaling primarily as a criterion of subtraction. The idea is to identify the traits that do not evince scaling rather than to support a hypothesis of scaling.

Numerous studies do reject a hypothesis of heterochrony or scaling in favor of parallel ontogenies. For example, parallel trajectories have been inferred for postnatal facial ontogeny of several hominins (e.g. Ponce de Leon and Zollikofer, 2001; Ackermann and Krovitz, 2002; Zollikofer and Ponce de Leon, 2010), for postnatal mandibular ontogeny of common and pygmy chimpanzees (Boughner and Dean, 2008), for postnatal scapular development of anthropoids (Young, 2008) as well as for postnatal development of the postcranial skeleton for callitrichines (Falsetti and Cole, 1992) and anthropoids (Jungers and Cole, 1992). Additionally, this pattern has been detected in post-weaning craniofacial ontogeny of seven pairs of congeneric sigmodontine rodents (Voss and Marcus, 1992) and for the evolution of giant damselfishes (Frederich and Sheets, 2010). Some of these cases, however, have been vigorously challenged, most notably that of postnatal hominin facial ontogenies (e.g. O'Higgins et al., 2001; Strand Vioarsdottir et al., 2002; Cobb and O'Higgins, 2004; Bastir et al., 2007) on the grounds that hypothesis of a conserved ontogenetic trajectory can be rejected when tested statistically.

The question raised by the case of postnatal hominin facial ontogenies is whether methods commonly used to infer shared ontogenies fail to detect divergent ones even when trajectories do in fact diverge. Because many studies have used traditional morphometric data, and these data remain valuable for the reasons discussed above, we first consider the methodological issues raised by them.

Framing Hypotheses About the Ontogeny of Shape in Terms of Size Variables

Both the clock model and the Alberch et al. scheme framed the hypothesis of heterochrony (and scaling) in terms of shape. Thus, to test the hypotheses using traditional morphometric data, the hypotheses must be translated into expectations for size variables. Sometimes this step is skipped, even intentionally, on the grounds that shape is merely a derivative of size, making analysis of shape change just a comparison of relative sizes

(McKinney, 1988; McKinney and McNamara, 1991). Skipping that translation step redefines all the terms. For example, "neoteny" no longer means a decrease in developmental rate. Instead it means a decrease in growth rate of any measurement regressed on size (McKinney, 1986, 1988; McKinney and McNamara, 1991). To be consistent with Gould's and Alberch et al.'s definition of the term (as well as with much of the preceding literature), the hypothesis of neoteny must be translated as predicting that positively allometric coefficients will decrease but negatively allometric coefficients will increase. If this seems counterintuitive, consider what it means to be pedomorphic (i.e. "childlike"). A pedomorphic descendant resembles the ancestral juvenile. The extreme case would be an adult that does not depart *at all* from the juvenile shape as it grows. For that to be the case, growth must be isometric (shape does not change over ontogeny). In less extreme cases, the descendant's ontogeny is more nearly isometric than the ancestor's. Based on that reasoning, positively allometric coefficients will *decrease* in slope, in the direction of isometry, and negatively allometric coefficients will *increase* in slope, also in the direction of isometry. Positively and negatively allometric coefficients approach isometry from opposing directions. Of course, the coefficients must all change by the appropriate amount, not just in the appropriate direction.

Considering that pedomorphosis results from truncating the ancestral ontogeny, and peramorphosis from extending it, we would anticipate that the vectors of allometric coefficients would point in the same direction – the two vectors differ only by an extension or truncation, and thus in length, not in direction. From geometry, it should be obvious that when the regression vectors actually are the same line, they point in the same direction and therefore the angle between them is 0°. Because the correlation between the two vectors is the cosine of the angle, many studies have measured the correlation between the vectors to determine if they differ by much. They often are very highly correlated, for example, in the case of sigmodontine rodents, the correlation between the ontogenetic trajectories of post-weaning trajectories are, on average 0.981 in comparisons between congeneric species and 0.962 in comparisons between genera, corresponding to angles of 11.2° and 15.84° (Voss and Marcus, 1992). Similarly, comparisons between pygmy chimpanzee (*Pan paniscus*), common chimpanzee (*Pan troglodytes*) and the gorilla (*Gorilla gorilla*) yield correlations that range from 0.964 to 0.977, corresponding to angles from 12.31° to 15.42°. When angles are that small, testing them for a significant deviation from 0° may not seem necessary, and they typically were not tested. However, it is worth examining even such obviously interpretable results with a jaundiced eye because explicit statistical testing can yield surprising outcomes.

Calculating the Angle Between Two Vectors

The angle between any two vectors **A** and **B**, each with P components, may be computed by taking the dot product (also called the "inner product") of the two vectors. The dot product is calculated by multiplying the corresponding components of the two vectors together, then summing those products. For example, if we have two vectors, **A** and **B**, with $\mathbf{A} = [A_1, A_2, A_3, \ldots A_P]$ and $\mathbf{B} = [B_1, B_2, B_3, \ldots B_P]$, the dot product is:

$$A \cdot B = A_1B_1 + A_2B_2 + A_3B_3 + \ldots A_PB_P \tag{11.5}$$

To use the dot product to calculate the angle between two vectors of size variables, we would first estimate the regression coefficients for each one, then normalize the vectors to unit length (meaning that the square root of their summed squared coefficients equals one). To calculate the angle for shape variables, we would similarly first estimate the regression coefficients for each such component, such as regression coefficients for coordinates obtained by GPA or partial warp scores. We then calculate the dot product by multiplying the allometric coefficient of one species by the allometric coefficient of that same variable in the other, then multiply the coefficient of the next variable in one species by the allometric coefficient of that same variable in another, and so for all coefficients. Finally, we would sum all those products. This gives the correlation between the vectors (R_v). Because a correlation is a cosine of an angle, we can also write the equation for the dot product as:

$$A \cdot B = |A||B| \cos \theta \tag{11.6}$$

where $|A|$ is the magnitude (length) of **A**, which is calculated by $(A_1^2 + A_2^2 + \ldots A_P^2)^{1/2}$ and similarly, $|B|$ is the length of **B**, calculated by $(B_1^2 + B_2^2 + \ldots B_P^2)^{1/2}$, and θ is the angle between them.

If **A** and **B** are unit vectors, the two lengths $|A|$ and $|B|$ are both one, so, to find the angle between the two vectors we solve for θ by:

$$\theta = \arccos(A \cdot B)/((|A||B|)) \tag{11.7}$$

When two vectors are parallel, the angle between them is 0° and the vector correlation between them is 1.0; in contrast, when two vectors point in exactly the opposite direction (which is termed being anti-parallel), the angle between them is 180° and the vector correlation between them is −1.0. The angle between perpendicular (orthogonal) vectors is 90°, and the correlation between them is 0.0.

Testing the Statistical Significance of the Angle

Once we have computed an angle between two regression vectors, we are left with the question of whether it is statistically significant. Rather than attempt to find an analytic test of significance, we can rely on a bootstrap or permutation procedure (see Chapter 8 for an overview of resampling methods and bootstrapping, and Chapter 9 for a more detailed discussion of permutation tests). Using bootstrapping, we can determine a confidence interval for the range of angles between regression vectors that can be produced by random variation within each group. At issue is whether the uncertainty of our estimate of each vector (due to sampling) is so large that we cannot reject the null hypothesis of no difference.

To estimate the range of angles within each species, we estimate the residuals from the regression of shape on the independent variable. Each individual gives a multidimensional set of residuals that describe the deviation of that individual from its expected shape. We then form a pair of bootstrap sets for each group that will be used to calculate the angle between the vectors. These pairs are constructed by resampling the residuals (with replacement) and randomly assigning them to expected values of shape (derived from the original regression model) at the values of size observed in the original data. This procedure preserves the covariance structure among variables and is a multivariate extension of the standard approach to estimation of uncertainties of regression slopes by resampling.

From the paired samples, we calculate the angles between the vectors, reiterating this procedure to generate a distribution of within-group angles. Because sample sizes can differ for different groups, the two bootstrap sets formed from the group with the larger sample size match the sample sizes of the two groups (i.e. one of the bootstrap sets will have a sample size equal to that of the group with more observations and one will have the sample size equal to that of the group with fewer observations). Both bootstrap sets formed from the data of the group with the smaller sample size have that group's smaller sample size because we ought not form bootstrap sets larger than the original data set. We then determine the statistical significance of the inter-group angle by comparing it to the 95th percentile of the range of both within-group angles. Should it be larger, the inter-group difference is judged to be statistically significant at a 5% level.

We can also use a permutation test to determine whether the difference between the trajectories is greater than expected by chance (Adams and Collyer, 2009). One approach is to use the residuals from the reduced model, which includes two factors "species" and "size" (or "age") but not the interaction term "species × size" (or "species × age"). The reason for using the residuals from the reduced model is to hold constant the relationship between shape and size (age) within each species. The residuals are then randomly assigned to the species and the randomized residuals are added to the predicted values. The full model is then used to calculate the predicted values from the random data. Repeating that procedure numerous times yields the empirical null distribution. The p-value is determined by the proportion of values that are as extreme or more extreme than the observed value.

We might also want to know if two species are no more similar than expected by chance. To test this null hypothesis, we can randomly permute the coefficients of the allometric vectors and compute the mean correlation between random vectors and the 95% and 97.5% confidence intervals. If the observed correlation is higher than 97.5% of the correlations obtained by randomly permuting the coefficients, then the observed correlation is higher than expected by chance.

A Traditional Approach to Estimating the Contribution of Scaling Makes to Morphological Variation

If the null hypothesis of no difference between trajectories is rejected, we might still want to know whether heterochrony or scaling is the dominant cause of disparity – the two trajectories might not be identical but the difference between them might be slight and have very little impact on the evolution of morphology. We would also like a method for assessing the degree to which scaling accounts for the variation in morphology. One widely used method is to conduct a PCA of the pooled ontogenies series, i.e. a multigroup PCA (Shea, 1985). PC1 is expected to show ontogenetic scaling, or the shape variation resulting from extension or truncation of shared allometry and PC2 and subsequent components show differences between groups in allometric coefficients as well as differences due to transpositions. The contribution that ontogenetic scaling (and the differential truncation/extension of it) makes to the overall variation is then quantified by the eigenvalue of the first component relative to subsequent ones (Shea, 1985).

TABLE 11.6 Two-way Multivariate Analysis of Variance (MANOVA) of Piranha Body Shape Between Species ("Taxa") and Developmental Stage ("Stage")

Effect	df	Pillai's Trace	Approx F	Num df	Den df	*P*
Taxa	1	0.99751	1642	30	123	<0.0001
Size	1	1	1230192	30	123	<0.0001
taxa:level	1	0.9971	1411	30	123	<0.0001
Residuals	152					

Examples: Applying These Methods to Data

We first compare the ontogenetic trajectories of the two piranha species (*S. gouldingi* and *S. manueli*) and then those of the two rodent species, the cotton rat (*S. fulviventer*) and house mouse (*M. musculus domesticus*) using traditional morphometric data. Next we conduct both comparisons using geometric morphometric data.

The hypothesis of ontogenetic scaling predicts that species differ only in adult body size, not in either juvenile shape (measured at the intercept, or at a comparable developmental stage), or the direction of ontogenetic shape change. Given this hypothesis, we can use MANCOVA (see Chapter 9) to test it. But MANCOVA presents a problem because the substantive biological hypothesis (ontogenetic scaling) is equivalent to the statistical null. Normally, the null is the hypothesis that we would like to reject, and we use various strategies to ensure that we do not reject it too readily. But the hypothesis of scaling is the one we wish to *accept*, so we are put in an odd position. Procedures that prevent rejecting a false null hypothesis, such as using a very conservative test, and factors that can reduce our ability to reject it, such as small sample size, can lead us to accept a false hypothesis of scaling. Thus, the inference of ontogenetic scaling will not be convincing if the test is conservative or the sample size is small. Presuming that the sample size *is* large enough to reject a false null hypothesis, finding that only the covariate is a significant term supports a hypothesis of ontogenetic scaling. In the analyses of both the two piranhas and the two rodent species, we use MANOVA rather than MANCOVA because we are comparing the mean shapes for each species at two developmental stages rather than using a continuous factor (size) as a covariate. We find that both factors, plus the interaction term, are highly significant statistically in the comparison of the two piranhas (Table 11.6). Nevertheless, the angle between allometric vectors appears to be very small, 6.27°, which corresponds to a very high correlation of 99.4. Nonetheless, using the bootstrapping procedure described above, the within-species angles are just 1.5° and 2.0° so 6.27° *is* significantly greater than 0.0° ($P < 0.05$). A permutation test similarly determines that the two species differ significantly (at $P < 0.001$). Both factors, plus the interaction term, are also highly significant statistically in the case of the two rodents (Table 11.7) but, again, the correlation between the vectors is very high: 0.984, corresponding to an angle of just 10.4°. Again, using the bootstrapping procedure to test the statistical significance of the angle between vectors, we find that within-species angles are just 4.2° and 3.4° so the observed angle of 10.4° is significantly greater than expected by chance. Thus, in both cases, the

TABLE 11.7 Two-way Multivariate Analysis of Variance (MANOVA) of Rodent Craniofacial Shape Between Species ("Taxa") and Ages ("Age")

Effect	df	Pillai's Trace	Approx F	Num df	Den df	*P*
Taxa	1	0.97848	748.46	13	214	<0.0001
Age	8	2.56727	8.03	104	1768	<0.0001
taxa:level	4	1.06715	6.07	52	868	<0.0001
Residuals	226					

pairs of species differ in their ontogenies, but based on the tiny angles, the differences appear very small.

To determine whether scaling might be the dominant cause of morphological variation we conduct a PCA of the covariance matrix of the log-transformed measurements and compare the eigenvalues for the first two axes. In the case of the two piranhas, the eigenvalue of PC1 is 2.61; this axis accounts for 99.9%. For PC2, the eigenvalue is 0.010 and this axis accounts for just 0.4% of the variance. Thus, ontogenetic scaling (and the differential extension/truncation of ontogenetic allometry) appears to account for nearly all the variation in the data. This dominance of scaling is evident in the plot of the PC scores for the two species (Figure 11.13). We obtain a similar result for the two rodents. The eigenvalue of PC1 is 0.915; this axis accounts for 95.5% of the variation in the data. For PC2, the eigenvalue is 0.017 and this axis accounts for merely 1.8% of the variance. Thus, PC1 accounts for an overwhelming proportion of the variance. As evident from the plot, the older house mice overlap the neonatal cotton rats along PC1, due to the enormous size difference between the species (Figure 11.14) and the two species are separated along PC2. The ontogenetic trajectories are slightly oblique to PC1, suggesting that the two species differ in allometries and not just by a consistent difference in shape (a consistent difference would be perpendicular to PC1). Thus, for both species, we could conclude that the data are explained well by a hypothesis of ontogenetic scaling even though the ontogenetic trajectories are not strictly the same.

Repeating these same analyses using geometric data, we obtain strikingly different results. In the comparison of ontogenetic trajectories between the two piranhas we obtain an angle of 34.9°, corresponding to a correlation of 0.819. The angle is large relative to those obtained by resampling (11.0°, 16.6°). A permutation test similarly determines that the species differ significantly ($P < 0.001$). The statistical result agrees with what we obtained from the traditional morphometric data, but an angle of 35.0° is not small. Analysis of the two rodent ontogenies yields an angle of 42.7° between the two species, corresponding to a correlation of 0.74, which again are larger than those obtained by resampling within-species (13.2°, 10.4°). A permutation test similarly determines that the species are more different than expected by chance ($P < 0.001$). Again, the statistical result agrees with what we obtained from the traditional morphometric data, but an angle of 42.7° is not small.

We can obtain further insight into the meaning of the angles by testing the other null hypothesis, i.e. that the vectors are no more similar than expected by chance. For the comparison between the piranhas, the angles between randomly permuted vectors of

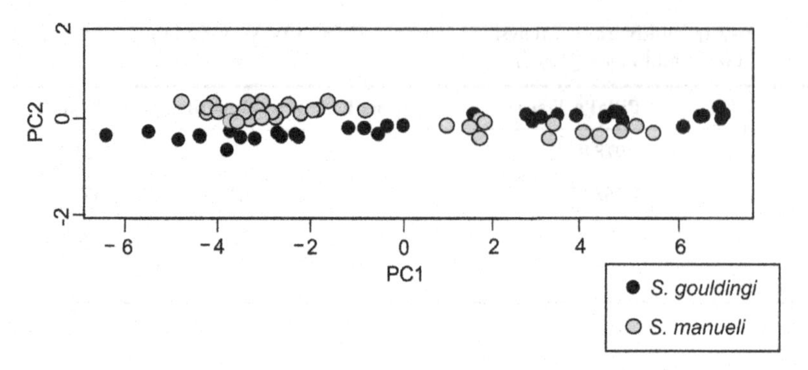

FIGURE 11.13 Principal components analysis of the traditional measurements, pooling the ontogenetic series of *S. gouldingi* and *S. manueli*.

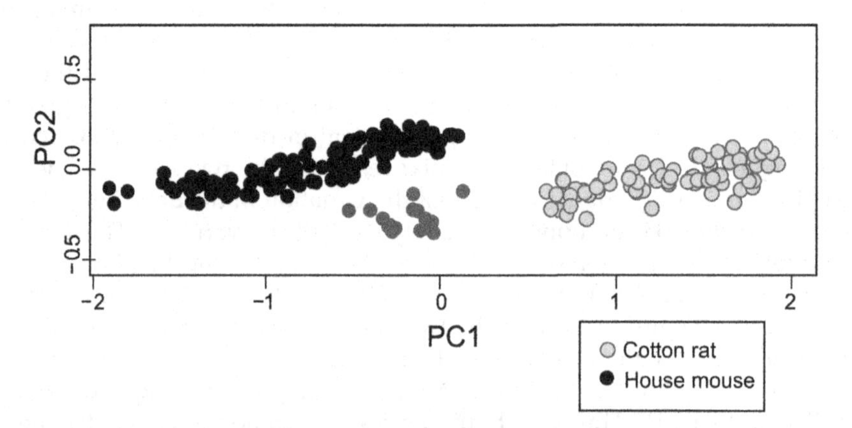

FIGURE 11.14 Principal components analysis of the traditional measurements, pooling the ontogenetic series of cotton rat and house mouse.

allometric coefficients of traditional data average 0.981, corresponding to an angle of 11.13°. Thus, only correlations higher than 0.987 or angles smaller than 9.10° indicate any greater similarity than expected by chance. The correlation obtained from the data is indeed higher than 0.987, but that value of 0.994 no longer seems remarkably high. For the two rodent species, the average correlation obtained by randomly permuting the allometric coefficients of traditional data is 0.89, corresponding to an angle of 27°. So, in this case, only correlations higher than 0.96, or angles smaller than 16.3°, indicate any greater similarity than expected by chance. The observed correlation of 0.984 is higher than 0.96, but again, 0.984 no longer appears to be impressive evidence of a conserved ontogeny. In striking contrast, in analyses based on geometric data, the average correlation between randomly permuted vectors is −0.002, corresponding to an angle of 89.9°; the value that we would expect for randomly related vectors.

To understand why angles between random vectors of traditional data are so far from 90°, recall how allometric coefficients (*k*) are calculated as well as how the angles between vectors are calculated. The coefficients are the power to which body size is raised in the power law (Equation 11.1). As long as structures grow rather than shrink over ontogeny, *k* is invariably a positive number. Moreover, the coefficients rarely differ by much – the most extreme values for *S. gouldingi* are 0.75 (for eye diameter) and 1.23 (for mid-body depth and posterodorsal head length). That difference may seem very large because one is highly negatively allometric whereas the other is highly positively allometric, but the difference is still numerically very small (in this case, it is less than 0.5). The angles are computed by taking the dot product between the two vectors (after they are normalized to unit length), which means that we multiply k_1 of one species by k_1 of the other, and add that to the product of k_2 in one species by k_2 of the other, and so forth. We are therefore summing products of corresponding allometric coefficients. Because all the elements of both vectors are positive numbers, the sum of their products cannot be zero – much less negative. To produce an angle of 90°, the sum would have to be zero (because the cosine of 90° is zero). The angle is necessarily smaller than that – often very much smaller (as in the two comparisons, above). In striking contrast, allometric coefficients obtained from geometric data can be negative as well as positive, the angle between random vectors is near 90.0°. Thus, results from studies based on geometric data yield angles that are more easily interpreted in light of our expectation that the angle between random vectors should be 90°.

Re-evaluating the Traditional Method for Estimating the Variation Explained by Scaling

Given that ontogenetic scaling clearly does not explain either data set well, we need to re-evaluate the inference that we drew from the PCA. When analyzing the traditional morphometric data, we found that PC1 explained an overwhelming proportion of the variation in piranhas and the rodents alike. Conducting a PCA of the geometric data for the two piranhas (Figure 11.15), we again find that PC1 accounts for far more variance than PC2 does: 64.81% versus 9.45%. The eigenvalue of PC1 is 0.0023 and that for PC2 is 0.00034 so PC1 clearly dominates. But the picture no longer suggests either ontogenetic scaling or parallel trajectories. It even looks as if the ontogenetic trajectories of both species bend. What we know from computing the two trajectories that each is linear, and we also know from computing the angle between that the two vectors are at 34.9° to each other. The picture shows neither the linearity nor the degree of divergence well. The multigroup PCA (Figure 11.16) for the two rodents suggests that the two ontogenies are parallel except that the ontogenetic trajectories are oblique to PC1 which, in this data set, is not the size or age axis. PC1, which accounts for 53.61%, separates the two species at all ages. PC2, which accounts for 27.88% of the variance, is aligned with the averaged ontogeny. What the plot suggests is that the difference between the two species is not constant throughout ontogeny, implying a divergence in growth allometries.

The most notable difference between the PCs derived from traditional and geometric data is predictable – in analyses of geometric data, size no longer dominates the analysis. Changes in shape related to size are preserved in data, but geometric scale is not and

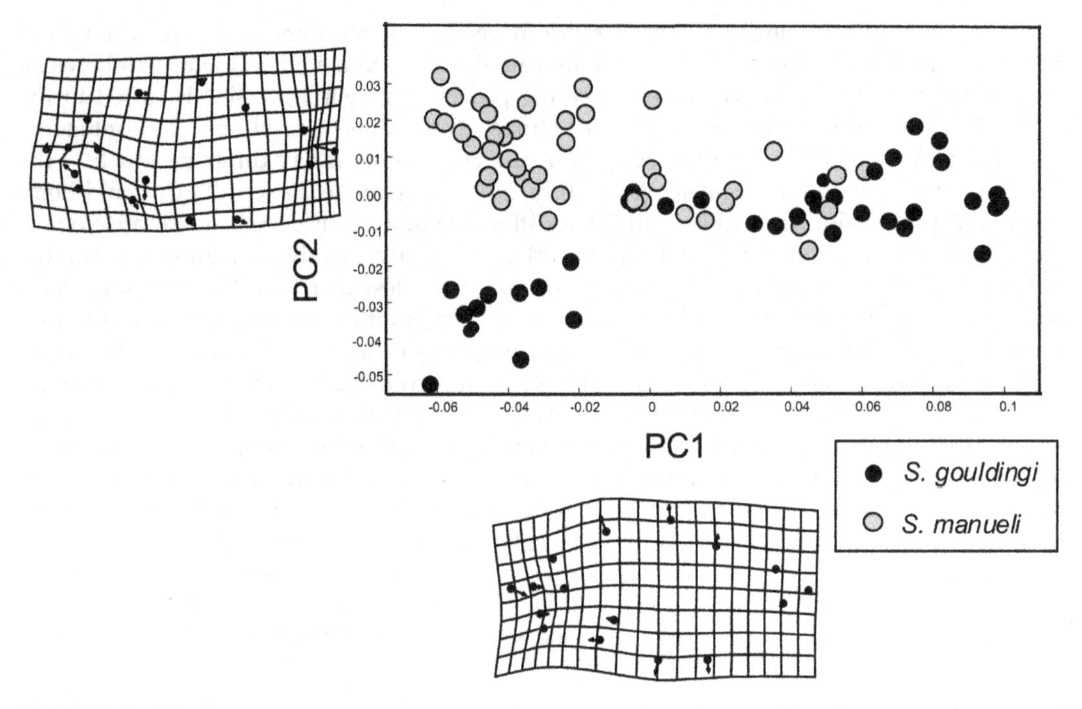

FIGURE 11.15 Principal components analysis of geometric shape data, pooling the ontogenetic series of *S. gouldingi* and *S. manueli.*

therefore the large ontogenetic increase in scale (and the differences between species in scale) no longer contributes to PC1. The consequence is self-evident in the plots for the two rodents – whereas PC1 is a size axis in the analysis based on traditional (size) data, PC1 separates the two ontogenies in the analysis based on geometric data. In that geometric analysis, the averaged ontogeny is more nearly aligned with PC2, although oblique to it. The eigenvalues of these axes cannot tell us which hypothesis is best supported nor quantify the proportion of the disparity that results from each modification of ontogeny. After all, PCA is an ordination method, not a statistical test of a hypothesis. In these plots, we cannot even detect that the two trajectories are at 42.7° to each other. Principal components analysis, whether of traditional or geometric data, is a low-dimensional projection of complex data. The eigenvalues of the PCs tell us how much of the total variation projects onto each axis, not how much each modification of ontogeny contributes to disparity.

DISSECTING THE DEVELOPMENTAL BASIS OF DISPARITY

Dissecting the developmental basis of disparity is obviously a complex task when there are many modifications of ontogeny and many species. Not only do we need to identify what differs among the ontogenies (and between which species they differ), but also we need to determine the impact of those modifications on disparity. Numerous studies, using

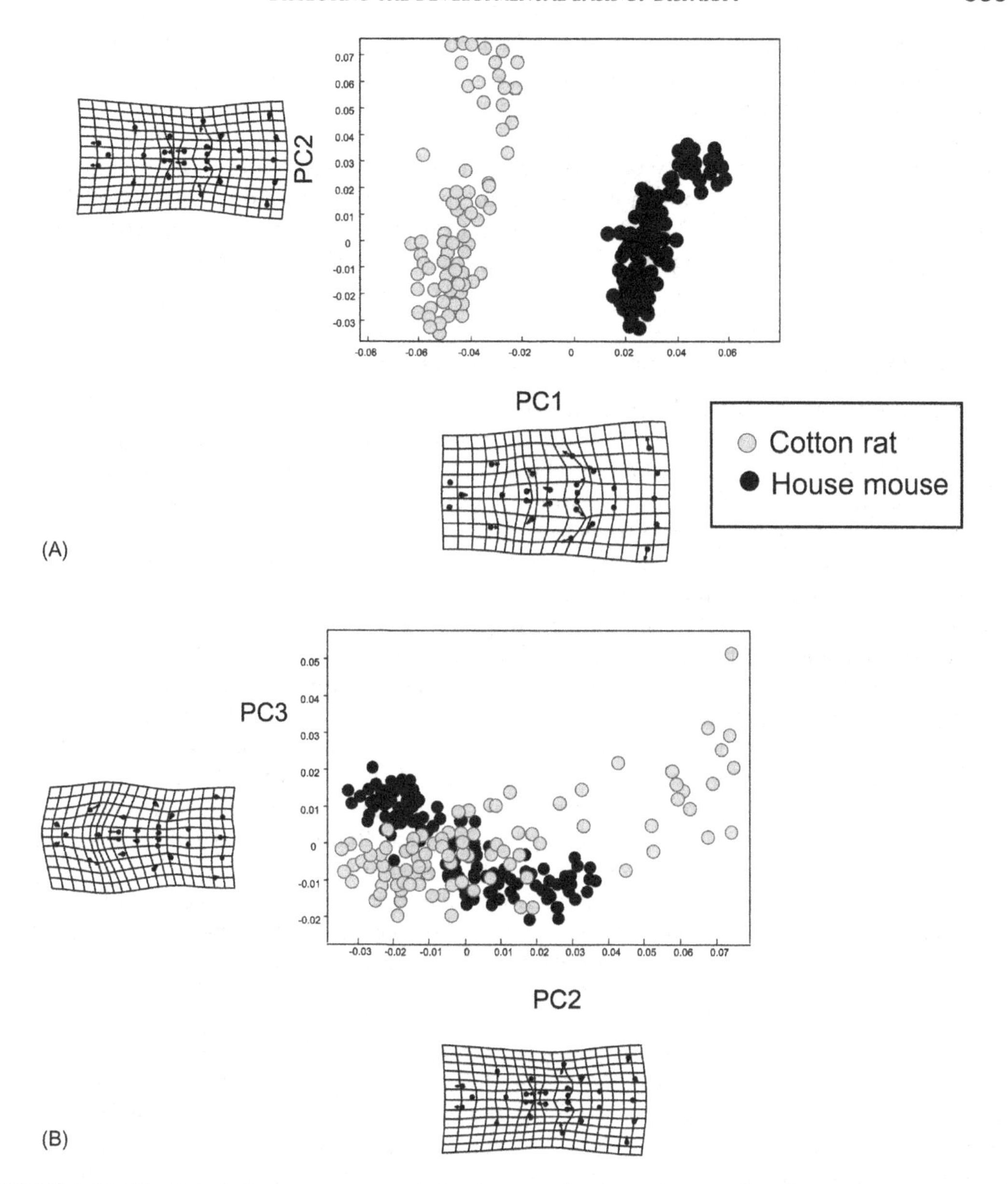

FIGURE 11.16 Principal components analysis of geometric shape data, pooling the ontogenetic series of cotton rat and house mouse.

traditional as well as geometric data have shown that species differ in shape at the outset of development, the trajectories diverge and they may also differ in length (e.g. Strauss and Fuiman, 1985; McKinney, 1986; Klingenberg and Froese, 1991; Zelditch et al., 2000, 2003a; Cardini and Tongiorgi, 2003; Bastir and Rosas, 2004; Cobb and O'Higgins, 2004; Mitteroecker et al., 2004a, 2004b; Larson, 2005; Cardini and Thorington, 2006; Bastir et al.,

2007; Webster, 2007, 2009; Frederich et al., 2008; Sanfelice and De Freitas, 2008; Piras et al., 2010; Drake, 2011; Frederich and Vandewalle, 2011). Relatively few, however, explicitly compare levels of disparity at different ontogenetic stages. In one case, multiple modifications of ontogeny increase disparity (Frederich et al., 2008). In other cases, combinations of modifications decrease disparity because disparate young develop along divergent trajectories towards similar adult morphologies (Zelditch et al., 2003b; Adams and Nistri, 2010; Piras et al., 2010; Ivanovic et al., 2011). To dissect the developmental basis of disparity, we need comparisons of the ontogenetic trajectories plus measures of disparity.

Testing Hypotheses About the Evolution of Ontogeny

To determine what differs between ontogenetic trajectories we need to conduct a series of tests. How we progress through these tests depends on the results of each one.

What ontogenetic scaling would look like is shown for a hypothetical case in Figure 11.17. In this case, the two species have the same shape at the outset of development, follow the same ontogeny of shape but one grows to a larger size, with size and shape maintaining the same relationship with each other that they had in the ancestral species. Thus, we see that the coordinates of the juveniles completely overlap (Figure 11.17A), the two ontogenies of shape are the same (Figure 11.17B), and the trajectories differ in length (Figure 11.17C). As a result, the coordinates of the descendant's adult morphology lie at a subadult position on the ancestral ontogeny (Figure 11.17D). However, ontogenetic scaling is not the only hypothesis consistent with these figures; with the exception of Figure 11.17A, the diagrams are also consistent with another hypothesis – heterochrony more generally. That is because the two hypotheses differ in only one respect – the association between size and shape. The hypothesis of ontogenetic scaling predicts that size and shape are associated in their evolutionary changes whereas the hypothesis of heterochrony predicts that they need not be. As a result, we might not find that the two species are identical in shape at any given *size*. The ancestral shape could be identical to that of the descendant at a different size. We would still find that the trajectories point in the same direction, and the coordinates for one species would be found at a subadult position for the other. But, in the case of ontogenetic scaling, the two species have the same regression equation whereas in the case of heterochrony they do not. We thus need to distinguish between identical regressions versus overlapping trajectories.

Mitteroecker and colleagues (2005) suggest two tests to make the distinction between those two cases. The first involves computing the multivariate regression of shape on size for each species (separately) and randomly assigning the summed squared residuals from the two regressions to species, recomputing the regression numerous times. If the two trajectories are identical, the test statistic for the observed case should not be an outlier in the distribution of summed squared residuals for the permuted data. So, for N permutations, the hypothesis of identical trajectories is rejected if $(C+1)/(N+1)<\alpha$, where C is the number of cases that produce a smaller test statistic than found for the data. For the case of overlapping trajectories, the test involves using the residuals normal to the regression, ignoring deviations *along* the trajectory (because the expectation is that the same shapes will be at different points along the trajectories). So this test, which is otherwise the same as the first, uses the summed squared distances from each shape to its nearest point on the regression curve rather than the summed

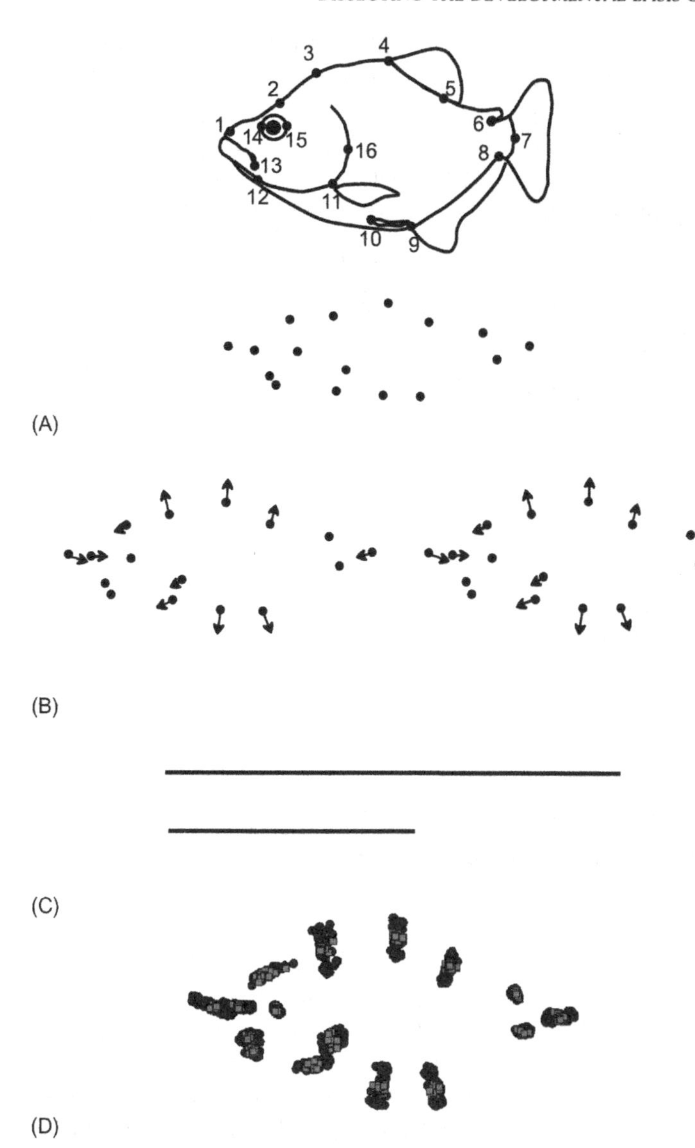

FIGURE 11.17 Ontogenetic scaling and heterochrony. (A) Superimposed coordinates of juvenile shapes; (B) ontogenies of shape; (C) lengths of ontogenetic vectors of shape. The two species have the same shape at the outset of the measured phase, follow the same ontogeny of shape, but differ in the length of their ontogenetic vectors; the descendant has a truncated version of the ancestral ontogeny. (D) Superimposed coordinates for showing the ontogenetic transformation of ancestral shape (black circles) and the descendant adult shape (gray squares). The descendant adult shape is at an intermediate position along the ancestral ontogeny.

squared residuals. Piras and colleagues (Piras et al., 2011) offer a modified version of these tests, using the mean squared error rather than the sums of squares.

A third hypothesis, parallel trajectories, also predicts that the two trajectories point in the same direction but, in this case, the two species never resemble each other. We would therefore expect that they have different shapes at the youngest comparable stage (Figure 11.18A), but subsequently follow the same ontogeny of shape (Figure 11.18B), perhaps to the same extent (Figure 11.18C). To test the hypothesis that only early development is labile, we can show that there is a significant difference in shape at the outset of the measured phase, but the ontogenies of shape do not differ. For the hypothetical species

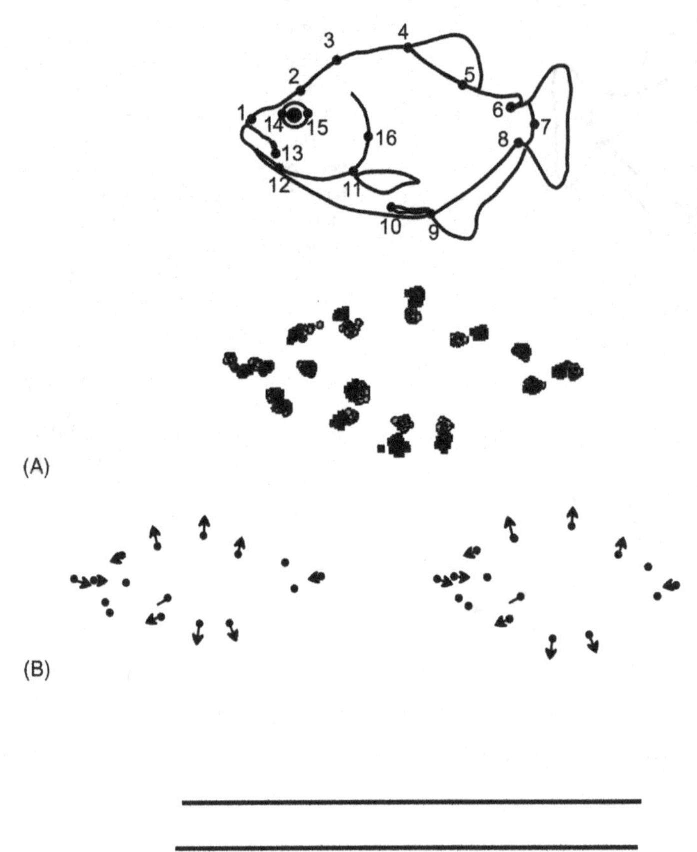

FIGURE 11.18 Change confined to early morphogenesis. (A) Superimposed coordinates of juvenile shapes; (B) ontogenies of shape; (C) lengths of ontogenetic vectors of shape. The two species differ in shape at the outset of the measured phase, but subsequently follow the same ontogeny of shape and do not differ in the length of their ontogenetic vectors.

shown in Figure 11.18, the difference between their shapes at the transition from larval to juvenile phases is highly significant ($P < 0.0001$) and the Procrustes distance between their means is large: 0.1247. The contrast between the shapes is particularly striking in the superimposed coordinates because we find little or no overlap between species in several of them (Figure 11.18A). But, as anticipated, there is no significant difference in their ontogenies of form; the angle between the two vectors is a tiny 1.9° (compared to the within-species angles of 4.0° and 3.7°). And the lengths of the ontogenetic vectors are statistically indistinguishable; the Procrustes distance between the youngest and oldest for one species is 0.1999 and for the other it is 0.2040. Thus, all that differs between the two trajectories is the shape at the outset of development. We can thus test for a difference between shapes at the youngest comparable stage, a difference between the ontogenies of shape and a difference in length of the ontogenies. Piras and colleagues (Piras et al., 2011) suggest testing the hypothesis of no transposition of the allometries, which is equivalent to a test of no difference in elevation of the parallel trajectories, using the distance between the predicted intercept shapes, comparing that distance to a distribution of distances obtained by

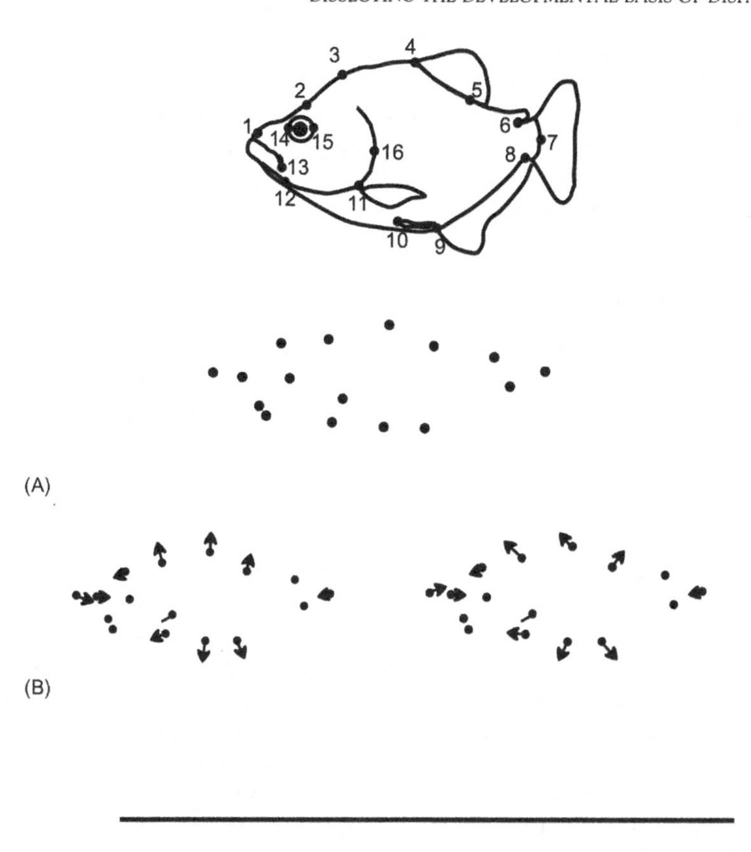

FIGURE 11.19 Change in the spatiotemporal pattern of development confined to late development. (A) Superimposed coordinates of juvenile shapes; (B) ontogenies of shape; (C) lengths of ontogenetic vectors of shape. The two species have the same shape at the outset of the measured phase, but subsequently follow different ontogenies of shape; they do not differ in the length of their ontogenetic vectors.

randomly permuting species affiliations of the data. Should the observed distance lie outside the 95th percentile for the random distribution of distances, the null hypothesis of no difference in elevation is rejected.

The remaining hypotheses predict that ontogenetic trajectories diverge; what differs among them is what *else* differs. Should all change be confined to late morphogenesis, we would expect to see no difference between the shapes of the two species at the youngest comparable developmental stage (Figure 11.19A), a difference in their ontogenies of shape (Figure 11.19B) and no difference in length of the trajectories (Figure 11.19C). The remaining hypotheses differ from this one in that they predict differences in either shape at the outset of the comparable phase and/or length of the trajectories. They also differ in whether the multiple changes lead to greater divergence in the adult shape(s) compared to that between the youngest comparable age. Testing these hypotheses requires comparing shapes at the youngest comparable age, comparing the directions of the ontogenetic trajectories of shape, comparing the lengths of the trajectories and comparing the distances between the youngest and oldest comparable stages. Comparisons between shapes at the youngest and oldest ages are most easily done by calculating the predicted shape for that stage (from the regression equation) then adding the residuals around the regression line to the predicted shape. The lengths of the trajectories can be calculated from the distance between the predicted shapes at the youngest and oldest stages, and the comparison between distances can be done by computing the distance between the shapes at the two stages, testing the null hypothesis that the distances do not differ.

Example: Ontogeny of Shape and Disparity

In this example, we combine a comparative analysis of ontogeny with an analysis of shape disparity, measured at two developmental stages – at the transition from larval to juvenile growth and at maximum adult body size. We use the approach presented by Adams and Collyer (2009) for the comparison of phenotypic trajectories. We begin with a Multivariate Analysis of Variance (MANOVA, see Chapter 9). Just as we did in the analysis of traditional morphometric data, in this one we use MANOVA rather than MANCOVA because we are comparing the mean shapes for each species at two developmental stages rather than using a continuous factor (size) as a covariate. We find that both factors, plus the interaction term, are highly significant statistically (Table 11.8). The species all differ statistically significantly in shape at the transition from larval to juvenile growth; some of the distances between species are large (Table 11.9). All but two species differ in their ontogenetic trajectories of shape (Table 11.10) and several (but not all) also differ in lengths of the trajectories (Table 11.11). The trajectories for all nine species are shown in the space of the first two principal components for body shape (Figure 11.20); this plot obviously cannot do justice to the complexity of these data. Nevertheless, it does show that three species on the left side of the plot (*S. manueli*, *S. gouldingi* and *S. elongatus*) have distinctive juvenile shapes and two of them, but not the third (*S. elongatus*), develop in the direction of the other *Serrasalmus*.

To determine what impact that combination of modifications has on disparity, we will compute the disparity of body shape at the transition from larval to juvenile development and again at maximum adult size. To do this, we will estimate the predicted shape at the two developmental stages and compute disparity of the means for each stage. Disparity is calculated as the square root of the average of the squared distances from each species to the mean of the distribution (see Chapter 10 for further discussion of measuring disparity). To determine if the disparities of juveniles and adults differ significantly, we can repeatedly resample individuals within species and repeat the calculation of the predicted values

TABLE 11.8 Two-way Multivariate Analysis of Variance (MANOVA) of Piranha Body Shape Across Species ("Taxa") and Developmental Stage ("Stage")

Effect	Df	Pillai's Trace	Approx F	Num Df	Den Df	P
Taxa	8	5.461	49.60	256	5904	<2.2e−16
Stage	1	0.99	2343.08	32	731	<2.2e−16
Taxa:Stage	8	5.04	39.26	256	5904	<2.2e−16
Residuals	762					

TABLE 11.9 Comparing Angles Between Ontogenetic Trajectories

	dent	alt	el	gould	man	spilo	pir	nat	car
dent	0	44.89	38.59	32.99	46.04	37.15	58.99	70.26	48.68
alt	0.001	0	39.11	31.97	30.80	35.22	63.74	65.67	41.83
el	0.001	0.001	0	38.39	45.73	26.31	48.19	54.57	33.06
gould	0.001	0.001	0.001	0	34.80	38.07	64.89	67.98	47.78
man	0.001	0.001	0.001	0.001	0	44.97	73.41	76.31	55.59
spilo	0.001	0.001	0.001	0.001	0.001	0	51.44	56.80	34.10
pir	0.001	0.001	0.001	0.001	0.001	0.001	0	22.70	40.30
nat	0.001	0.001	0.001	0.001	0.001	0.001	0.001	0	39.30
car	0.001	0.001	0.001	0.001	0.001	0.001	0.001	0.001	0

Angles between trajectories are given above the diagonal; *p*-values for the null hypothesis that the angle is no greater than 0° obtained by permutation of residuals are given below the diagonal. dent = *Pygopristis denticulate*; alt = *Serrasalmus altuvei*; el = *S. elongatus*; gould = *S. gouldingi*; man = *S. manueli*; spilo = *S. spilopleura*; pir = *Pygocentrus piraya*; nat = *P. nattereri*; car = *P. cariba*.

TABLE 11.10 Procrustes Distances Between Juvenile Shapes

	dent	alt	el	gould	man	spilo	pir	nat	car
dent	0	0.079	0.012	0.138	0.134	0.065	0.069	0.064	0.052
alt		0	0.107	0.126	0.107	0.064	0.094	0.088	0.085
el			0	0.060	0.071	0.120	0.152	0.143	0.127
gould				0	0.08	0.138	0.165	0.160	0.145
man					0	0.122	0.156	0.151	0.131
spilo						0	0.060	0.058	0.053
pir							0	0.033	0.038
nat								0	0.044
car									0

All differences between shapes are statistically significant, adjusting for the multiple comparisons. dent = *Pygopristis denticulate*; alt = *Serrasalmus altuvei*; el = *S. elongatus*; gould = *S. gouldingi*; man = *S. manueli*; spilo = *S. spilopleura*; pir = *Pygocentrus piraya*; nat = *P. nattereri*; car = *P. cariba*.

and disparity and compute the difference between the juvenile and adult disparities, repeating this calculation numerous times to generate the distribution of the difference in disparities between the two stages. If zero does not lie within the confidence interval of the difference in disparities, the two disparities differ significantly from each other. The disparity for juveniles is 0.00543 and for adults it is 0.00398 and the confidence intervals for these values do not even overlap (Zelditch et al., 2003b).

The comparison of the disparities at the two developmental stages gives an estimate of the overall level of disparity but it does not tell us how much each modification of

TABLE 11.11 Differences in Lengths of Ontogenetic Trajectories Above the Diagonal

	dent	alt	el	gould	man	spilo	car	nat	pir
dent	0.000	0.020	0.010	0.108	0.085	0.014	0.026	0.001	0.013
alt	0.007	0.000	0.030	0.088	0.065	0.034	0.046	0.021	0.033
el	0.132	0.001	0.000	0.118	0.095	0.004	0.016	0.009	0.003
gould	0.001	0.001	0.001	0.000	0.023	0.122	0.134	0.109	0.121
man	0.001	0.001	0.001	0.001	0.000	0.099	0.111	0.086	0.098
spilo	0.037	0.001	0.580	0.001	0.001	0.000	0.012	0.013	0.001
car	0.001	0.001	0.030	0.001	0.001	0.081	0.000	0.025	0.013
nat	0.836	0.001	0.096	0.001	0.001	0.012	0.001	0.000	0.012
pir	0.071	0.001	0.647	0.001	0.001	0.911	0.093	0.058	0.000

p-values for the null hypothesis that the lengths do not differ obtained by permutation of the residuals below the diagonal.
dent = *Pygopristis denticulate*; alt = *Serrasalmus altuvei*; el = *S. elongatus*; gould = *S. gouldingi*; man = *S. manueli*; spilo = *S. spilopleura*; pir = *Pygocentrus piraya*; nat = *P. nattereri*; car = *P. cariba*.

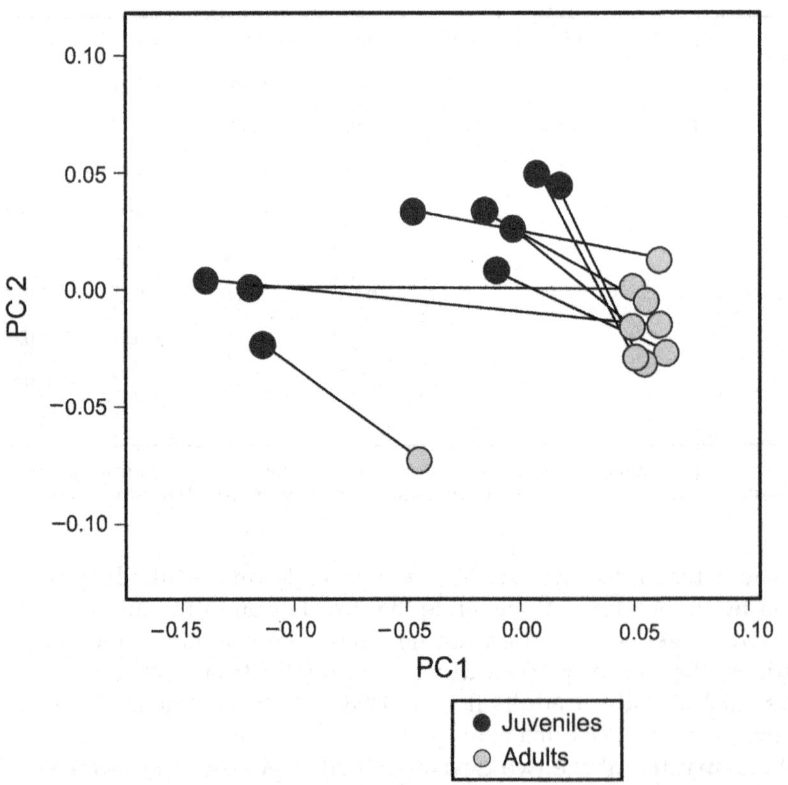

FIGURE 11.20 The ontogenetic trajectories of shape for the nine species of piranhas, in the space of the first two principal components of shape. The vectors connect each species' juvenile shape to its adult shape.

ontogeny contributes to disparity. What we need is a quantity similar to a partial disparity, but one that is the partial disparity for each modification rather than for each taxon. But the partial disparities for taxa are additive – the contribution that each one makes to the total disparity sum to the total disparity. In contrast, the contribution that each modification of ontogeny makes to the total need not sum to the total. Two or more modifications, taken separately, can produce more disparity than the two do taken together.

To quantify the disparity due to each modification, taken individually, we can either fix all but that one parameter of the trajectory, or we can fix only that one and allow the others to vary. To fix a parameter, we assign the same value to both species; the one that is free to vary has the values observed in the data. We first look at a simple case – the comparison between *S. gouldingi* and *S. manueli*. Figure 11.21A shows the disparity of adults, first for the data, then that produced by fixing all but one parameter. We can therefore compare the disparity produced by variation in juvenile shape, length and direction to that produced by varying only juvenile shape or length or direction. Regardless of which one varies, the disparity of adults is far higher than seen in the data, especially if length or direction varies. Figure 11.21B shows the disparity of adults when two parameters vary. Again, fixing any one parameter increases disparity over that observed in the data. The greatest increase is found when the juvenile shape is fixed and length and direction vary.

The analysis of all nine species is more complex because we could fix the parameters for any combination of species and fix them to any of a variety of values. In this case, we will

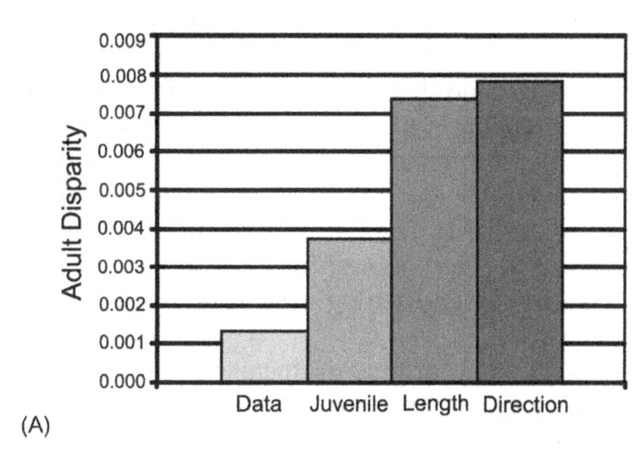

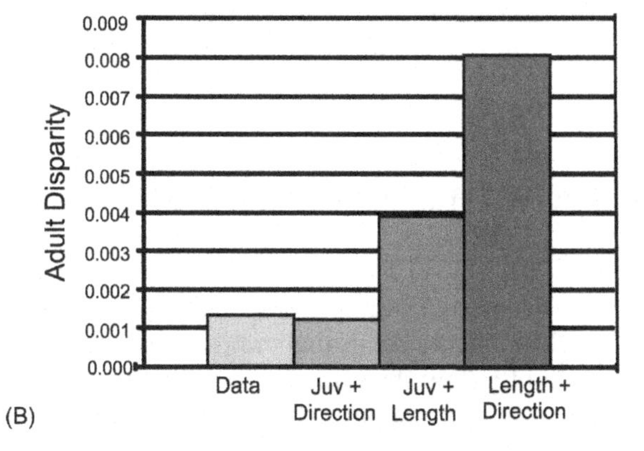

FIGURE 11.21 The impact of variation in ontogenetic trajectories on the difference in shape of adult *S. gouldingi* and *S. manueli*. (A) The disparity observed in the data ("Data") is compared to that obtained by variation in only juvenile shape ("Juvenile") or length of the ontogenetic trajectory ("Length") or direction of the ontogenetic trajectory ("Direction"); (B) the disparity observed in the data is compared to that obtained by variation in both juvenile shape and the direction of the trajectory ("Juv + Direction"), juvenile shape plus the length of the trajectory ("Juv + Length") or length and direction of the trajectory ("Length + Direction").

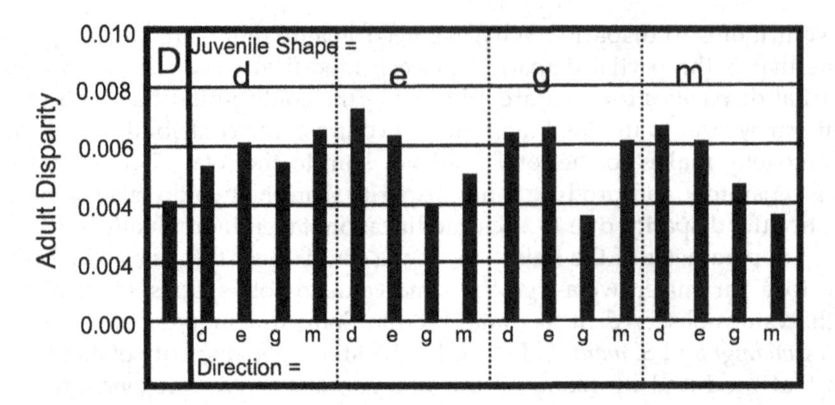

FIGURE 11.22 The impact of variation in attributes of ontogenetic trajectories on the disparity of the nine species of piranhas. Levels of disparity are indicated by the height of the bar. The level observed in the data (D) is compared to disparity produced by fixing juvenile shape and the direction of the ontogenetic trajectory of shape two attributes to be equal across the three species with shallow-bodied juveniles (*S. gouldingi, S. manueli* and *S. elongatus*). Thus only the length of the ontogenetic trajectory varies. The values for juvenile shape and the direction of the ontogenetic trajectory of shape are fixed to four values, values observed for *P. denticulata* (d), *S. elongatus* (e), *S. gouldingi* (g), or *S. manueli* (m).

focus on the three with the distinctive juvenile body shapes and fix their juvenile shapes, trajectory lengths or directions either to that seen in the outgroup species (*Pygopristis denticulata*) or to the value seen in the data for one of these three species. We will let length be the sole parameter that varies, but the result of that variation clearly depends heavily on the particular value to which the juvenile shape or direction is fixed (Figure 11.22). The lowest values are found when fixing both juvenile shape and direction in the three shallow-bodied species to the values observed in either *S. gouldingi* or *S. manueli* (both of which have relatively long ontogenetic trajectories). The greatest disparity results from fixing the juvenile shape of the three shallow-bodied juvenile species to the one seen in *P. denticulata,* fixing trajectory length to the large value seen in *S. gouldingi,* and letting each species follow the direction of its own trajectory (Zelditch et al., 2003b).

Clearly, even when we can dissect the developmental origins of disparity and quantify the variation in morphology that arises from each modification of ontogeny, it is not straightforward to answer the question: How much of the disparity arises from variation in any one parameter? The interactions between them complicate answering that question because the disparity produced by variation in any one parameter may be countered by that produced by variation in another.

AGE-BASED COMPARISONS OF GROWTH AND DEVELOPMENTAL RATES AND TIMINGS

Even after ruling out the hypothesis of heterochrony by finding that species follow different ontogenetic trajectories of shape we may still be interested in their rates and timings of growth and development. Because the species do not follow the same ontogenetic trajectories of shape, we cannot use the Alberch et al. formalism to compare rates and timings – the formalism requires that the same shapes appear in the ontogenies of all species being compared, just at different ages/sizes. So we need an alternative formalism that can be used when species follow species-specific ontogenetic trajectories. Gould (1977; pp. 385–388) suggested one such alternative, which is to compare the rates and timings at which species depart from their *own* juvenile forms, an approach similar to that used by Hingst-Zaher and colleagues (Hingst-Zaher et al., 2000), who measured the amount of shape change from age to age by the Procrustes distance between successive ages. They showed that the distance declines with age, meaning that the rate of development (like that of growth) decreases over time. To linearize the relationship between size and age, they regressed size on log(age +1) and the same transformation can be used to examine the relationship between development and age (Figure 11.23). By linearizing the relationship between shape and age, we can compare the developmental rates, which are 0.061 for the house mouse and 0.044 for the cotton rat.

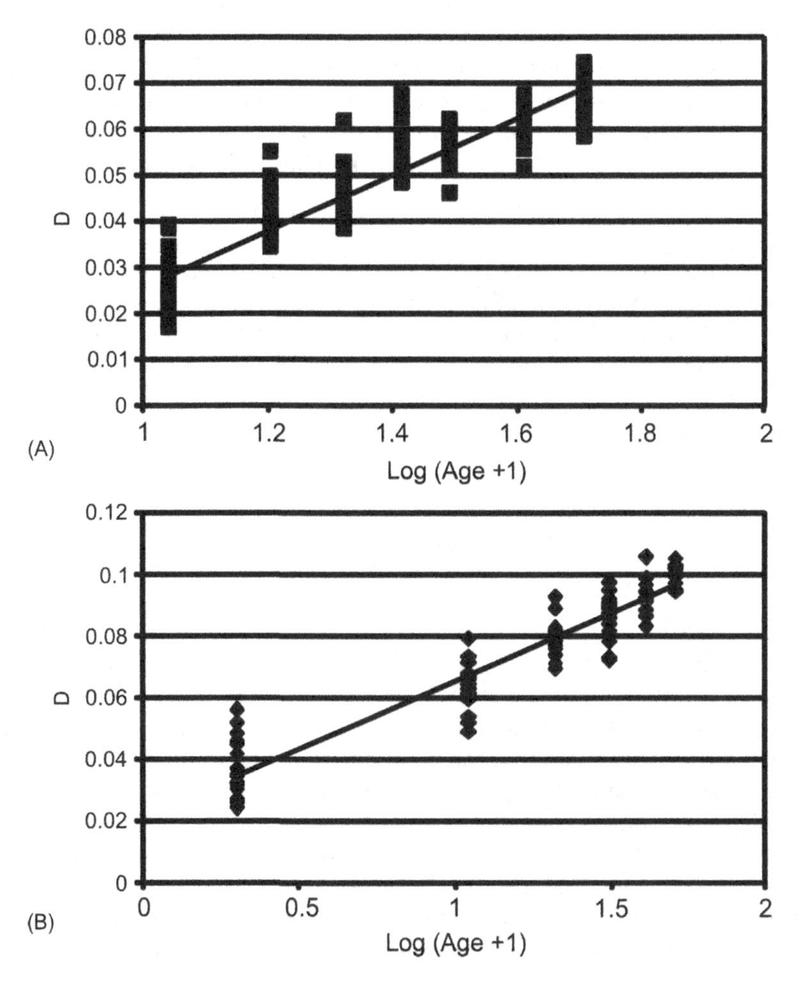

FIGURE 11.23 A linear regression of the Procrustes distance between each specimen and the mean of the youngest age class on log (age +1). (A) House mice; (B) cotton rats.

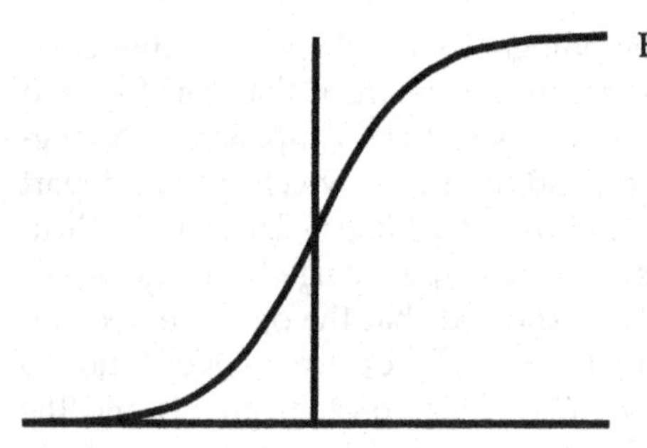

FIGURE 11.24 A sigmodial (logistic) growth curve.

An alternative method for analyzing rates of growth and development is to use non-linear models, which have been widely applied to studies of growth. One class of models is called "sigmoidal" because the curve is S-shaped (Figure 11.24). There are several growth models, including the monomolecular (Gaillard et al., 1997), von Bertalanffy and Gompertz ((Zullinger et al., 1884) and Fiorello and German (Fiorello and German, 1997). Relative fit of competing models can be assessed by the Aikake Information Criterion (AIC) which is a function of the log-likelihood of the parameters given the data and the number of parameters in the model (Akaike, 1974). The AIC is calculated as:

$$\mathrm{AIC} = 2k - 2\ln(\mathrm{likelihood})) \tag{11.8}$$

where k is the number of parameters in the model. The number of parameters is important because simple models will generally have lower likelihoods than more complex ones. The AIC effectively balances likelihood with model complexity, although the derivation of the AIC is based on information measures rather than a concept of balancing. To compare models, we can compute the difference in their AIC (ΔAIC) or estimate the AIC weight. AIC weight is an estimate of the relative probability that a particular model is true, and is computed as:

$$\mathrm{AIC\ Weight} = \exp(-0.5 * \Delta\mathrm{AIC}) \tag{11.9}$$

So if we have two models, one with an AIC of −4 and another with an AIC of −2, ΔAIC for the second model is $-2 - (-4) = 2$, and the AIC weight of that model is exp(−0.5 * 2) or 0.368, meaning that it is 0.368 as likely as the first. But whether the best fitting model fits well is also important because we would not want to choose the best of several poorly fitting models. Additionally, we would not pick a model that yields significant serial autocorrelations between residuals from the model even if it has the highest AIC weight. Although the data from growth series are serially autocorrelated, the residuals from the model should not be. Correlated residuals mean that the data violate the assumptions of the model – there is a mismatch between data and model. Thus, to decide if any model fits well, we can first look at the variance explained, then at the serial autocorrelation of residuals, and then, looking only at the models that have not been ruled out due to significant autocorrelation of residuals, we can choose the one that has the lowest AIC (or, equivalently, highest AIC weight).

Having chosen the model, and estimated its parameters, we can then predict Procrustes distance at any developmental age. For example, the monomolecular model is:

$$D(t) = A\{1 - e^{k(t_0 - t)}\} \tag{11.10}$$

where A is the asymptotic value for D, K is the rate of approach to the asymptotic value and t_0 is the age at the onset of development. We can therefore predict the value for D at any age and also predict the degree of developmental maturity, measured as the proportion of the asymptotic adult value attained at that age. For example, we can predict the degree of maturity attained by the house mouse and cotton rat at birth, eye-opening, weaning and sexual maturity. House mice are altricial, being blind, deaf, hairless and immobile at birth; eye-opening occurs at around 10 days. In contrast, the cotton rat is precocial, opening its eyes the day it is born, its ears shortly thereafter, being furred and mobile at birth. The two species differ in gestation length by 12 days (Zelditch et al., 2003a), which could explain their difference in degree of maturity at birth. But the cotton rat is not consistently 12 days more advanced than the house mouse. The two species wean and become sexually mature at nearly the same ages. As we can infer from the rates of shape maturation obtained from the linear model (above), the house mouse develops more rapidly. But non-linear growth models may give a better estimate of developmental rate.

Fitting the multiple growth models to the data shows that several models fit well (Table 11.12). The logistic model corresponds to the one that we fit above, by regressing D on log(age + 1); but this yields significant autocorrelated residuals for the house mouse so

TABLE 11.12 Evaluating Relative Fit of Growth Models to the Data for Developmental Maturity, Measured as the Procrustes Distance Between Each Specimen and the Mean for the Youngest Age

Species	Model	%Var	AC	AIC Weight
House mouse	Chapman-Richards	0.88	ns	0.1171
	Monomolecular	0.88	ns	0.3077
	von Bertalanffy	0.87	*	–
	Gompertz	0.86	*	–
	German Gompertz	0.87	ns	0.2976
	Logistic	0.87	*	–
	Quadratic	0.86	ns	0.2776
	Linear	0.78	*	–
Cotton rat	Chapman-Richards	0.90	ns	0.0615
	Monomolecular	0.90	ns	0.1654
	von Bertalanffy	0.90	ns	0.1628
	Gompertz	0.88	ns	0.1379
	German Gompertz	0.90	ns	0.1611
	Logistic	0.89	ns	0.1554
	Quadratic	0.89	ns	0.1559
	Linear	0.83	*	–

% Var = variance explained; AC = serial autocorrelation of residuals (statistical significance indicated by *).

we would not further consider it. The one that fits best is the monomolecular (Equation 11.10 above). So we can use the estimated values for K, A and t_0 to predict D at any age. We can also determine relative maturity by dividing the value of D at each age by A (which is the value of D at 100% of maturity). For the comparison between house mouse and cotton rat, we can then assess their degree of maturity at several life-history milestones (Figure 11.25).

DISPARITY OF ONTOGENY

As well as analyzing the disparity of shape, we can also analyze the disparity of the ontogenetic trajectories. We do this by computing the vector of allometric coefficients for each species; each vector is then entered as a row in the data matrix, i.e. the first column of the data matrix is the allometric coefficient for the first shape variable, e.g. the x-coordinate of landmark 1, or the x-component of partial warp 1 and the second column is the allometric coefficient for the next variable, etc. The second row contains the allometric vector for the next species and so forth. Recall that correlation between the vectors is the sum of the products of the corresponding coefficients; the patterns of variation among

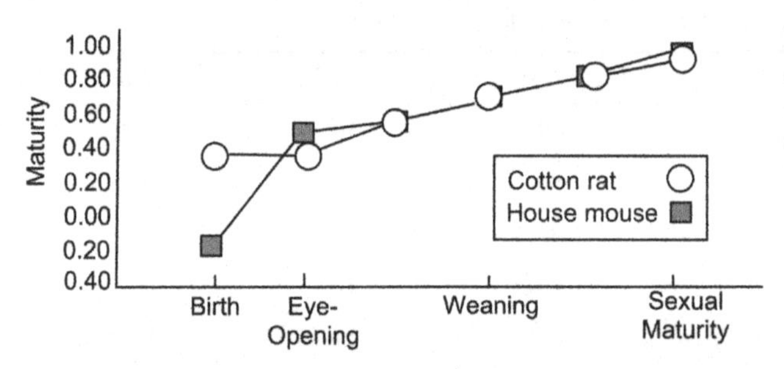

FIGURE 11.25 The degree of maturity shape (measured as the proportion of the asymptotic adult value attained at each age) against life-history milestones for the house mouse and cotton rat.

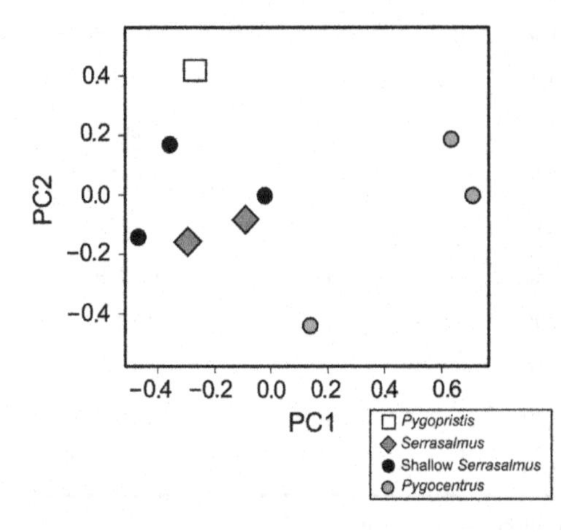

FIGURE 11.26 Allometric disparity: principal components of variation of ontogenetic trajectories.

allometric vectors can be analyzed by the principal components of the covariance matrix of these vectors, giving the orthogonal components of variation among allometric patterns (Klingenberg and Froese, 1991). This space has been termed an "allometric space" (Gerber et al., 2008, 2011; Wilson and Sanchez-Villagra, 2010) or "developmental morphospace" (Eble, 2003; Gerber et al., 2008, 2011; Gerber, 2011) and the analysis of disparity within this space has been termed "allometric disparity" (Gerber et al., 2008, 2011; Wilson and Sanchez-Villagra, 2010). Not surprisingly, in the analysis of allometric disparity of the nine species of piranhas (Figure 11.26) PC1, which explains 52.5% of the variation, shows that two *Pygocentrus* are distinctive in their allometries. These are the species whose ontogenetic trajectories are typically at 60° or more to the others.

References

Abdala, F., Flores, D. A., & Giannini, N. P. (2001). Postweaning ontogeny of the skull of *Didelphis albiventris*. *Journal of Mammalogy*, *82*, 190–200.

Ackermann, R. R., & Krovitz, G. E. (2002). Common patterns of facial ontogeny in the hominid lineage. *Anatomical Record*, *269*, 142–147.

Adams, D. C., & Collyer, M. L. (2009). A general framework for the analysis of phenotypic trajectories in evolutionary studies. *Evolution*, *63*, 1143–1154.

Adams, D. C., & Nistri, A. (2010). Ontogenetic convergence and evolution of foot morphology in European cave salamanders (Family: Plethodontidae). *Bmc Evolutionary Biology*, *10*. doi:10.1186/1471-2148-10-216.

Akaike, H. (1974). A new look at the statistical model identification. *IEEE Transactions on Automatic Control*, *19*, 716–723.

Alberch, P., Gould, S. J., Oster, G. F., & Wake, D. B. (1979). Size and shape in ontogeny and phylogeny. *Paleobiology*, *5*, 296–317.

Bastir, M., O'Higgins, P., & Rosas, A. (2007). Facial ontogeny in Neanderthals and modern humans. *Proceedings of the Royal Society B-Biological Sciences*, *274*, 1125–1132.

Bastir, M., & Rosas, A. (2004). Comparative ontogeny in humans and chimpanzees: similarities, differences and paradoxes in postnatal growth and development of the skull. *Annals of Anatomy-Anatomischer Anzeiger*, *186*, 503–509.

Bininda-Emonds, O. R. P., Jeffrey, J. E., & Richardson, M. K. (2003). Inverting the hourglass: quantitative evidence against the phylotypic stage in vertebrate development. *Proceedings of the Royal Society of London, Series B*, *270*, 341–346.

Birch, J. M. (1999). Skull allometry in the marine toad, *Bufo marinus*. *Journal of Morphology*, *241*, 115–126.

Blackstone, N. W. (1987). Allometry and relative growth: pattern and process in evolutionary studies. *Systematic Zoology*, *36*, 76–78.

Both, C., van Asch, M., Bijlsma, R. G., van den Burg, A. B., & Visser, M. E. (2009). Climate change and unequal phenological changes across four trophic levels: constraints or adaptations? *Journal of Animal Ecology*, *78*, 73–83.

Both, C., & Visser, M. E. (2005). The effect of climate change on the correlation between avian life-history traits. *Global Change Biology*, *11*, 1606–1613.

Boughner, J. C., & Dean, M. C. (2008). Mandibular shape, ontogeny and dental development in bonobos (*Pan paniscus*) and chimpanzees (*Pan troglodytes*). *Evolutionary Biology*, *35*, 296–308.

Bulygina, E., Mitteroecker, P., & Aiello, L. (2006). Ontogeny of facial dimorphism and patterns of individual development within one human population. *American Journal of Physical Anthropology*, *131*, 432–443.

Cardini, A., & Thorington, R. W. (2006). Postnatal ontogeny of marmot (Rodentia, Sciuridae) crania: allometric trajectories and species divergence. *Journal of Mammalogy*, *87*, 201–215.

Cardini, A., & Tongiorgi, P. (2003). Yellow-bellied marmots (*Marmota flaviventris*) "in the shape space" (Rodentia, Sciuridae): sexual dimorphism, growth and allometry of the mandible. *Zoomorphology*, *122*, 11–23.

Cobb, S., & O'Higgins, P. (2004). Hominins do not share a common postnatal facial ontogenetic shape trajectory. *Journal of Experimental Zoology (Mol Dev Evol)*, *302B*, 302–321.

Cope, E. D. (1887). *The origin of the fittest*. New York: McMillan.

Cronier, C., & Courville, P. (2003). Developmental variations in neodevonian phacopide trilobites. *Comptes Rendus Palevol, 2*, 577–585.

Drake, A. G. (2011). Dispelling dog dogma: an investigation of heterochrony in dogs using 3D geometric morphometric analysis of skull shape. *Evolution & Development, 13*, 204–213.

Drake, A. G., & Klingenberg, C. P. (2008). The pace of morphological change: historical transformation of skull shape in St Bernard dogs. *Proceedings of the Royal Society B-Biological Sciences, 275*, 71–76.

Eble, G. J. (2003). Developmental morphospaces and evolution. In J. P. Crutchfield, & P. Schuster (Eds.), *Evolutionary dynamics* (pp. 35–65). Oxford: Oxford University Press.

Emerson, S. B., Travis, J., & Blouin, M. (1988). Evaluating a hypothesis about heterochrony: larval life-history traits and juvenile hindlimb morphology in *Hyla crucifer*. *Evolution, 42*, 68–78.

Falsetti, A. B., & Cole, T. M. (1992). Relative growth of the postcranial skeleton in callitrichines. *Journal of Human Evolution, 23*, 79–92.

Fiorello, C. V., & German, R. Z. (1997). Heterochrony within species: craniofacial growth in giant, standard and dwarf rabbits. *Evolution, 51*, 250–261.

Frederich, B., Adriaens, D., & Vandewalle, P. (2008). Ontogenetic shape changes in Pomacentridae (Teleostei, Perciformes) and their relationships with feeding strategies: a geometric morphometric approach. *Biological Journal of the Linnean Society, 95*, 92–105.

Frederich, B., & Sheets, H. D. (2010). Evolution of ontogenetic allometry shaping giant species: a case study from the damselfish genus Dascyllus (Pomacentridae). *Biological Journal of the Linnean Society, 99*, 99–117.

Frederich, B., & Vandewalle, P. (2011). Bipartite life cycle of coral reef fishes promotes increasing shape disparity of the head skeleton during ontogeny: an example from damselfishes (Pomacentridae). *Bmc Evolutionary Biology, 11*. doi:10.1186/1471-2148-11-82.

Fuiman, L. A. (1983). Growth gradients in fish larvae. *Journal of Fish Biology, 23*, 117–123.

Gaillard, J.-M., Pontier, D., Allaine, D., Loison, A., Herve, J.-C., & Heizmann, A. (1997). Variation in growth form and precocity at birth in eutherian mammals. *Proceedings of the Royal Society B-Biological Sciences, 264*, 859–868.

Galatius, A., Berta, A., Frandsen, M. S., & Goodall, R. N. P. (2011). Interspecific variation of ontogeny and skull shape among porpoises (Phocoenidae). *Journal of Morphology, 272*, 136–148.

Gerber, S. (2011). Comparing the differential filling of morphospace and allometric space through time: the morphological and developmental dynamics of early Jurassic ammonoids. *Paleobiology, 37*, 369–382.

Gerber, S., Eble, G. J., & Neige, P. (2008). Allometric space and allometric disparity: a developmental perspective in the macroevolutionary analysis of morphological disparity. *Evolution, 62*, 1450–1457.

Gerber, S., Eble, G. J., & Neige, P. (2011). Developmental aspects of morphological disparity dynamics: a simple analytical exploration. *Paleobiology, 37*, 237–251.

German, R. Z., Hertweck, D. W., Sirianni, J. E., & Swindler, D. R. (1994). Heterochrony and sexual dimorphism in the pigtailed macaque (*Macaca nemestrina*). *American Journal of Physical Anthropology, 93*, 373–380.

Gould, S. J. (1966). Allometry and size in ontogeny and phylogeny. *Biological Reviews of the Cambridge Philosophical Society, 41*, 587–640.

Gould, S. J. (1982). Change in developmental timing as a mechanism of macroevolution. In J. T. Bonner (Ed.), *Evolution and development* (pp. 333–346). Berlin: Springer-Verlag.

Gould, S. J. (1977). *Ontogeny and phylogeny*. Cambridge, MA: Harvard University Press.

Gould, S. J. (1988). The uses of heterochrony. In M. L. McKinney (Ed.), *Heterochrony in evolution: a multidisciplinary approach* (pp. 1–13). New York: Plenum Press.

Hingst-Zaher, E., Marcus, L. F., & Cerqueira, R. (2000). Application of geometric morphometrics to the study of postnatal size and shape changes in the skull of *Calomys expulsus*. *Hystrix, The Italian Journal of Mammalogy, 11*, 99–113.

Huxley, J. S. (1932). *Problems of relative growth*. New York: MacVeagh. The Dial Press.

Hylander, W. L. (1985). Mandibular function and biomechanical stress and scaling. *American Zoologist, 25*, 315–330.

Ivanovic, A., Cvijanovic, M., & Kalezic, M. L. (2011). Ontogeny of body form and metamorphosis: insights from the crested newts. *Journal of Zoology, 283*, 153–161.

Ivanovic, A., Vukov, T. D., Dzukic, G., Tomasevic, N., & Kalezic, M. L. (2007). Ontogeny of skull size and shape changes within a framework of biphasic lifestyle: a case study in six Triturus species (Amphibia, Salamandridae). *Zoomorphology, 126*, 173–183.

Jolicoeur, P. (1963). The multivariate generalization of the allometry equation. *Biometrics, 19*, 497–499.

Jungers, W. L., & Cole, M. S. (1992). Relative growth and shape of the locomotor skeleton in lesser apes. *Journal of Human Evolution, 23*, 93–105.

Katz, J. M. (1980). Allometry formula: a cellular model. *Growth, 44*, 89–96.

Kimmel, C. B., Ballard, W. W., Kimmel, S. R., Ullmann, B., & Schilling, T. F. (1995). Stages of embryonic development of the zebrafish. *Developmental Dynamics, 203*, 253–310.

Klingenberg, C. P., & Froese, R. (1991). A multivariate comparison of allometric growth patterns. *Systematic Zoology, 40*, 410–419.

La Croix, S., Holekamp, K. E., Shivik, J. A., Lundrigan, B. L., & Zelditch, M. L. (2011a). Ontogenetic relationships between cranium and mandible in coyotes and hyenas. *Journal of Morphology, 272*, 662–674.

La Croix, S., Zelditch, M. L., Shivik, J. A., Lundrigan, B. L., & Holekamp, K. E. (2011b). Ontogeny of feeding performance and biomechanics in coyotes. *Journal of Zoology, 285*, 301–315.

Laird, A. K. (1965). Dynamics of relative growth. *Growth, 29*, 249–263.

Laird, A. K., Barton, A. D., & Tyler, S. A. (1968). Growth and time: an interpretation of allometry. *Growth, 32*, 347–354.

Lande, R. (1979). Quantitative genetic analysis of multivariate evolution, applied to brain: body size allometry. *Evolution, 33*, 402–416.

Lande, R., & Arnold, S. J. (1983). The measurement of selection on correlated characters. *Evolution, 37*, 1210–1226.

Larson, P. M. (2005). Ontogeny, phylogeny, and morphology in anuran larvae: morphometric analysis of cranial development and evolution in Rana tadpoles (Anura : Ranidae). *Journal of Morphology, 264*, 34–52.

Lieberman, D. E., Carlo, J., de Leon, M. P., & Zollikofer, C. P. E. (2007). A geometric morphometric analysis of heterochrony in the cranium of chimpanzees and bonobos. *Journal of Human Evolution, 52*, 647–662.

Maderson, P. F. A., Alberch, P., & Goodwin, B. C., et al. (1982). The role of development in macroevolutionary change. In J. T. Bonner (Ed.), *Evolution and development* (pp. 279–312). Berlin: Springer-Verlag.

McKinney, M. L. (1988). Classifying heterochrony: allometry, size and time. In M. L. McKinney (Ed.), *Heterochrony in evolution: A multidisciplinary approach* (pp. 17–34). New York: Plenum Press.

McKinney, M. L. (1986). Ecological causation of heterochrony: a test and implications for evolutionary theory. *Paleobiology, 12*, 282–289.

McKinney, M. L. (1999). Heterochrony: Beyond words. *Paleobiology, 25*, 149–153.

McKinney, M. L., & McNamara, K. J. (1991). *Heterochrony: the evolution of ontogeny*. New York: Plenum Press.

Miller-Rushing, A. J., Hoye, T. T., Inouye, D. W., & Post, E. (2010). The effects of phenological mismatches on demography. *Philosophical Transactions of the Royal Society B-Biological Sciences, 365*, 3177–3186.

Mitteroecker, P., Gunz, P., Bernhard, M., Schaefer, K., & Bookstein, F. L. (2004a). Comparison of cranial ontogenetic trajectories among great apes and humans. *Journal of Human Evolution, 46*, 679–697.

Mitteroecker, P., Gunz, P., Weber, G. W., & Bookstein, F. L. (2004b). Regional dissociated heterochrony in multivariate analysis. *Annals of Anatomy-Anatomischer Anzeiger, 186*, 463–470.

Mitteroecker, P., Gunz, P., & Bookstein, F. L. (2005). Heterochrony and geometric morphometrics: a comparison of cranial growth in *Pan paniscus* versus *Pan troglodytes*. *Evolution & Development, 7*, 244–258.

Monteiro, L. R., Lessa, L. G., & Abe, A. S. (1999). Ontogenetic variation in skull shape of *Thrichomys apereoides* (Rodentia : Echimyidae). *Journal of Mammalogy, 80*, 102–111.

Neige, P., Marchand, D., & Laurin, B. (1997). Heterochronic differentiation of sexual dimorphs among Jurassic ammonite species. *Lethaia, 30*, 145–155.

O'Higgins, P., Chadfield, P., & Jones, N. (2001). Facial growth and the ontogeny of morphological variation within and between the primates *Cebus apella* and *Cercocebus torquatus*. *Journal of Zoology, 254*, 337–357.

Pfaller, J. B., Herrera, N. D., Gignac, P. M., & Erickson, G. M. (2010). Ontogenetic scaling of cranial morphology and bite-force generation in the loggerhead musk turtle. *Journal of Zoology, 280*, 280–289.

Piras, P., Colangelo, P., & Adams, D. C., et al. (2010). The Gavialis–Tomistoma debate: the contribution of skull ontogenetic allometry and growth trajectories to the study of crocodylian relationships. *Evolution & Development, 12*, 568–579.

Piras, P., Salvi, D., & Ferrar, S., et al. (2011). The role of post-natal ontogeny in the evolution of phenotypic diversity in Podarcis lizards. *Journal of Evolutionary Biology, 24*, 2705–2720.

Ponce de Leon, M. S., & Zollikofer, C. P. E. (2001). Neanderthal cranial ontogeny and its implications for late hominid diversity. *Nature, 412*, 534–538.

Post, E., & Forchhammer, M. C. (2008). Climate change reduces reproductive success of an Arctic herbivore through trophic mismatch. *Philosophical Transactions of the Royal Society B-Biological Sciences, 363*, 2369–2375.

Post, E., Pedersen, C., Wilmers, C. C., & Forchhammer, M. C. (2008). Warming, plant phenology and the spatial dimension of trophic mismatch for large herbivores. *Proceedings of the Royal Society B-Biological Sciences, 275*, 2005–2013.

Rosell, J. A., & Olson, M. E. (2007). Testing implicit assumptions regarding the age vs. size dependence of stem biomechanics using Pittocaulon (similar to Senecio) praecox (Asteraceae). *American Journal of Botany, 94*, 161–172.

Sander, K. (1983). The evolution of patterning mechanisms: gleanings from insect embryogenesis and spermatogenesis. In B. C. Goodwin, N. Holder, & C. C. Wylie (Eds.), *Development and evolution* (pp. 137–159). Cambridge: Cambridge University Press.

Sanfelice, D., & De Freitas, T. R. O. (2008). A comparative description of dimorphism in skull ontogeny of *Arctocephalus australis, Callorhinus ursinus*, and *Otaria byronia* (Carnivora : Otariidae). *Journal of Mammalogy, 89*, 336–346.

Schweitzer, P. N., & Lohmann, G. P. (1990). Life-history and the evolution of ontogeny in the ostracode genus. Cyprideis. Paleobiology, *16*, 107–125.

Seidl, F. (1960). Körpergrundgestalt und Keimstruktur. Eine Erörterung über die Gundlagen der vergleichenden und experimentellen Embryologie un deren Gültigkeit bei phylogeneticschen Überlegungen. *Zoologische Anzeiger, 164*, 245–305.

Shea, B. T. (1983a). Allometry and heterochrony in the African apes. *American Journal of Physical Anthropology, 62*, 275–289.

Shea, B. T. (1983b). Paedomorphosis and neoteny in the pygmy chimpanzee. *Science, 222*, 521–522.

Shea, B. T. (1985). Bivariate and multivariate growth allometry: statistical and biological considerations. *Journal of Zoological Society of London (A), 206*, 367–390.

Shea, B. T. (1992). Ontogenetic scaling of skeletal proportions in the talapoin monkey. *Journal of Human Evolution, 23*, 283–307.

Slack, J. M., Holland, P. W., & Graham, C. F. (1993). The zootype and the phylotypic stage. *Nature, 361*, 490–492.

Strand Vioarsdottir, U., O'Higgins, P., & Stringer, C. (2002). A geometric morphometric study of regional differences in the ontogeny of the modern human facial skeleton. *Journal of Anatomy, 201*, 211–229.

Strauss, R. E. (1987). On allometry and relative growth in evolutionary studies. *Systematic Zoology, 36*, 72–75.

Strauss, R. E., & Altig, R. (1992). Ontogenetic body form changes in 3 ecological morphotypes of anuran tadpoles. *Growth Development and Aging, 56*, 3–16.

Strauss, R. E., & Fuiman, L. A. (1985). Quantitative comparisons of body form and allometry in larval and adult Pacific sculpins (Teleostei: Cottidae). *Canadian Journal of Zoology, 63*, 1582–1589.

Tanner, J. B., Zelditch, M. L., Lundrigan, B. L., & Holekamp, K. E. (2010). Ontogenetic change in skull morphology and mechanical advantage in the spotted hyena (*Crocuta crocuta*). *Journal of Morphology, 271*, 353–365.

van Snik, G. M. J., van den Boogaart, J. G. M., & Osse, W. M. (1997). Larval growth patterns in *Cyprinus carpio* and *Clarias gariepinus* with attention to the finfold. *Journal of Fish Biology, 50*, 1339–1352.

Vinyard, C. J., & Ravosa, M. (1998). Ontogeny, function, and scaling of the mandibular symphysis in papionin primates. *Journal of Morphology, 235*, 157–175.

Voss, R. S., & Marcus, L. F. (1992). Morphological evolution in muroid rodents II. Craniometric factor divergence in seven neotropical general, with experimental results from *Zygodontomys*. *Evolution, 46*, 1918–1934.

Wayne, R. K. (1986). Cranial morphology of domestic and wild canids: the influence of development on morphological change. *Evolution, 40*, 243–261.

Webster, M. (2007). Ontogeny and evolution of the early Cambrian trilobite genus Nephrolenellus (Olenelloidea). *Journal of Paleontology, 81*, 1168–1193.

Webster, M. (2009). Ontogeny, systematics, and evolution of the effaced early Cambrian trilobites peachella Walcott, 1910 and Eopeachella new genus (Olenelloidea). *Journal of Paleontology, 83*, 197–218.

Webster, M., & Zelditch, M. L. (2005). Evolutionary modifications of ontogeny: heterochrony and beyond. *Paleobiology, 31*, 354–372.

Wilson, L. A. B., & Sanchez-Villagra, M. R. (2010). Diversity trends and their ontogenetic basis: an exploration of allometric disparity in rodents. *Proceedings of the Royal Society B-Biological Sciences, 277*, 1227–1234.

Young, N. M. (2008). A comparison of the ontogeny of shape variation in the anthropoid scapula: functional and phylogenetic signal. *American Journal of Physical Anthropology, 136*, 247–264.

Zelditch, M. L., Lundrigan, B. L., Sheets, H. D., & Garland, T. (2003a). Do precocial mammals develop at a faster rate? A comparison of rates of skull development in *Sigmodon fulviventer* and *Mus musculus domesticus*. *Journal of Evolutionary Biology, 16*, 708–720.

Zelditch, M. L., Sheets, H. D., & Fink, W. L. (2003b). The ontogenetic dynamics of shape disparity. *Paleobiology, 29*, 139–156.

Zelditch, M. L., Sheets, H. D., & Fink, W. L. (2000). Spatiotemporal reorganization of growth rates in the evolution of ontogeny. *Evolution, 54*, 1363–1371.

Zollikofer, C. P. E., & Ponce de Leon, M. S. (2010). The evolution of hominin ontogenies. *Seminars in Cell & Developmental Biology, 21*, 441–452.

Zullinger, E. M., Ricklefs, R. R., Redford, K. H., & Mace, G. M. (1884). Fitting sigmoidal equations to mammalian growth curves. *Journal of Mammalogy, 65*, 607–636.

CHAPTER

12

Evolutionary Developmental Biology (2): Variational Properties

This chapter discusses methods for analyzing variational properties, including phenotypic plasticity, canalization, developmental stability, morphological integration and the related property of modularity. What all of them have in common is that they determine by how much, and in what combinations of traits, phenotypes *can* vary. We emphasize *can* vary rather than *do* vary, for two reasons. First, phenotypes can vary in more than they do – each phenotype is a single instance of what a genotype can produce and each population is a single case of the phenotypes that could be produced by that population's genotypes. Second, the variation within a population depends not only on the (intrinsic) variational properties of the genotypes – it also depends on the processes that determine the mix of genotypes within a population as well as on the environments they encounter during development. A population could be nearly invariant phenotypically for a variety of reasons, including that it is genetically homogeneous, that the environment is uniform, or that deviants die young. None of these are attributes of the genotypes within the population – none will travel with the individuals or be transmitted to their offspring. But a population could be nearly invariant because of developmental mechanisms that suppress the expression of variation. These mechanisms *can* travel with the individual and *can* be transmitted to its offspring. The distinction between the observed, realized variation within a population and the variation that could be produced motivates the distinction between "variation", i.e. the realized, observable variation in a population; and "variability", i.e. the propensity to vary (Wagner and Altenberg, 1996). Although we cannot directly observe propensities, the distinction between variation and variability is nonetheless useful because it highlights the difference between population-genetic processes that determine the mix of genotypes within a population and the intrinsic-developmental mechanisms that regulate the expression of (co)variation.

The variational properties are all conceptually related to each other. All of them can be regarded as attributes of the genotype–phenotype map, which provides a unifying theme

Geometric Morphometrics for Biologists
DOI: http://dx.doi.org/10.1016/B978-0-12-386903-6.00012-5

linking them all (Wagner and Altenberg, 1996). But how they are related biologically has been a matter of considerable controversy, especially when it comes to the relationships between the various forms of canalization (i.e. macroenvironmental, microenvironmental, and genetic canalization and developmental stability, which could be considered canalization against developmental noise). Less often debated, perhaps because less often discussed, is the relationship between properties that regulate the expression of variation and those that structure covariation. The processes regulating variation include all those listed above as various forms of canalization and those that structure covariation include morphological integration and modularity. They are obviously linked in the sense that covariation requires processes that induce variation, but they are now often linked methodologically because studies that examine the relationship between canalization and developmental stability now often compare *co*variance matrices, one representing the (co) variation among individuals, the other representing the (co)variation due to developmental noise. Thus, patterns of covariation are analyzed for insight into the processes that suppress variance and responses to noise. Additionally, the role that developmental modularity plays in structuring morphological integration is now often analyzed by comparing those same two covariance matrices.

There are, however, two important distinctions between studying the processes regulating variation (plasticity, canalization and developmental stability) on the one hand and those structuring covariation (integration and modularity) on the other. The two distinctions are related to each other because one concerns what needs to be measured, the other concerns how it is measured. Studies of plasticity, canalization and developmental stability assess the impact of various factors (either environmental or genetic) on the expression of phenotype variation. We know how to measure and decompose variation, and the techniques traditionally used to analyze plasticity, canalization and developmental stability are readily adapted to geometric morphometrics. Conducting the analyses is far from easy because the logistics of the experiments can be daunting, and the statistical models can be remarkably complex. But the analyses are all based on sums of squares. Morphological integration and modularity present more severe technical challenges because the concept that has been central to them, that of a "trait", has no obvious analog in geometric morphometrics. An additional problem is that the Procrustes superimposition itself imposes a pattern of covariances on the data (Rice, 1989; Rohlf and Slice, 1990; Rohlf, 2003; Adams et al., 2004). For both those reasons, adapting geometric methods to the study of integration and modularity did not prove straightforward. Recent progress has led to several new methods that replace the idea of "a trait" with "a subset of landmarks" (Klingenberg et al., 2003, 2004; Monteiro et al., 2005; Marquez, 2008; Klingenberg, 2009). But methods for studying integration and modularity are not as mature and well understood as those for studying plasticity, canalization and developmental plasticity.

This chapter is organized primarily by subject matter rather than by methods. We first discuss methods for analyzing plasticity, then canalization and then developmental stability, and conclude the first section of this chapter by discussing methods for testing hypotheses about the relationship among these three properties. We then present three methods for analyzing morphological integration and modularity, all of which are well grounded in geometric morphometric theory and implemented in freely available software.

PHENOTYPIC PLASTICITY: QUANTIFYING NORMS OF REACTION

The central concept in studies of phenotypic plasticity is the norm of reaction (Wolterek, 1909; Schmalhausen, 1949), which is the array of phenotypes produced by a given genotype reared in different environments. To determine the norm of reaction of a genotype we might, for example, split the offspring of highly inbred rats into two groups, feeding one the standard laboratory diet and the other a protein deficient diet. After letting these two groups grow to adulthood, we could then weigh them and measure their skulls to determine if body weight and/or skull size is responsive to protein deficiency. The norm of reactions for weight and skull size could then be represented by plotting the independent variable (diet in this case) on the x-axis and the phenotypic variable on the y-axis (Figure 12.1). Figure 12.1A shows a case in which the treatment has no effect – the phenotype is invariant across the two environments. This is what we would see if protein deficiency had no impact on the weight or skeletal size. Figure 12.1B shows the alternative – the phenotype does differ across the two environments. This is what we would see if protein deficiency has an impact on weight or skeletal size. In the statistical analysis of plasticity, the null hypothesis is that the treatment has no effect, a hypothesis tested by Analysis of Variance (ANOVA, see Chapter 8, and for more complex designs, see General Linear Models, Chapter 9).

Extending the analysis to two or more genotypes is reasonably straightforward because we simply add a factor (genotype) to the statistical model and test the interaction term, genotype × environment. Several of the possible norms of reaction are shown in Figure 12.2. One shows that the phenotypes of the two genotypes differ, but both are invariant across the range of environments (Figure 12.2A). In this case, only genotype has a significant effect on phenotype. The second possibility is that both genotypes exhibit plasticity, to the same degree and in the same direction, making their norms of reaction parallel (Figure 12.2B). In this case, the statistical analysis would reveal that both genotype and environment have significant effects but the interaction term would not be significant because the impact of the environment does not depend on genotype. A third possibility is that one genotype exhibits plasticity but the other one does not (Figure 12.2C). In this

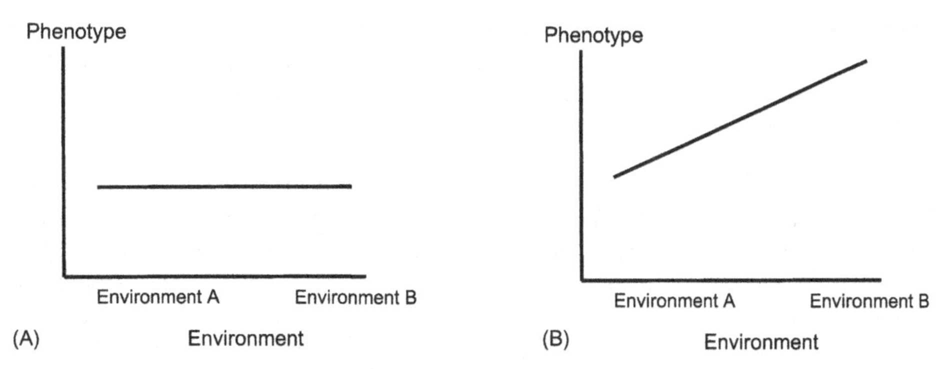

FIGURE 12.1 Norms of reaction for a single genotype reared in two environments. The environmental factor is plotted on the x-axis, the phenotype on the y-axis. (A) The phenotype does not differ across the two environments. (B) The phenotype differs across the two environments.

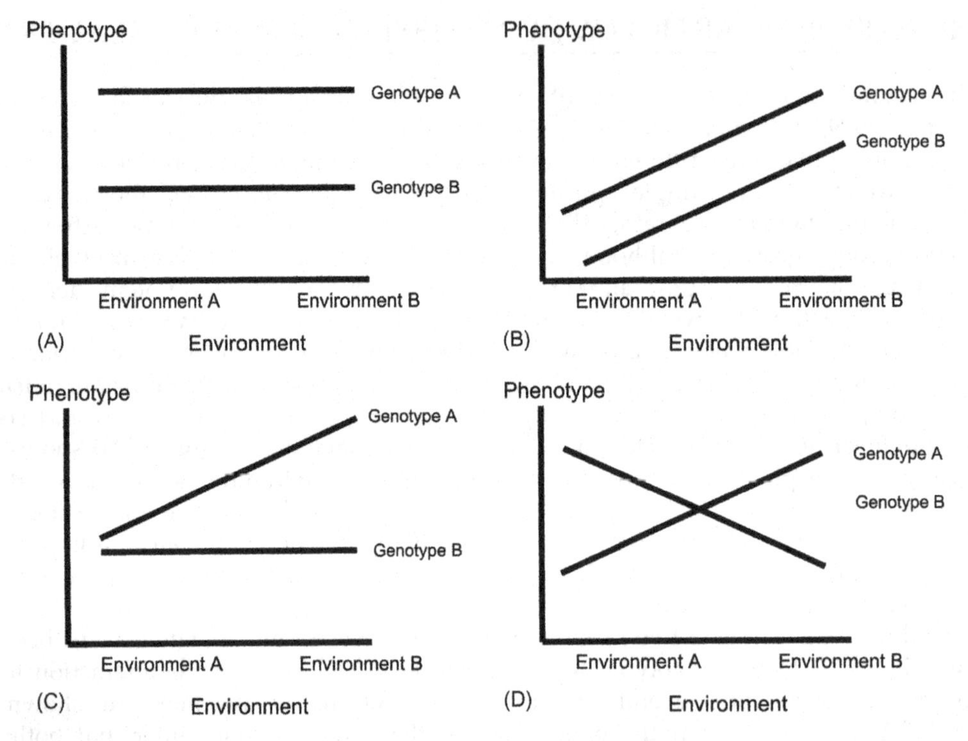

FIGURE 12.2 Norms of reaction for two genotypes reared in two environments. The environmental factor is plotted on the *x*-axis, the phenotype on the *y*-axis. (A) The phenotypes of the two genotypes reared in the same environment differ; both are invariant across the range of environments. (B) The phenotypes of the two genotypes reared in the same environment differ; both respond to the environment to the same degree and in the same direction; the norms of reaction are parallel. (C) The phenotypes of the two genotypes reared in the same environment differ and one responds to the environment of rearing whereas the other does not. (D) The phenotypes of the two genotypes reared in the same environment differ; both respond to the environment but in contrasting directions so that their norms of reaction cross.

case, both factors and the interaction term would be statistically significant as well because the impact of the environment depends on the genotype. We would also see a significant interaction term if both genotypes exhibit plasticity but to varying degrees or in different directions. A fourth possibility is that both genotypes exhibit plasticity but in contrasting directions – their norms of reaction cross each other (Figure 12.2D). In this instance, the statistical analysis would reveal no impact of either genotype or environment because the lines here cross at the mean, but the interaction term would be significant. When there are more than two genotypes or environments, the analysis of all groups is usually supplemented by pairwise comparisons to determine which means differ from which (and by how much).

Norms of reaction are more difficult to depict for multivariate data because those norms of reaction no longer describe a change in a single dimension. When the phenotype has only one dimension, its value on that dimension can only increase or decrease so every phenotype can be plotted on the same phenotype axis. But when the phenotype is

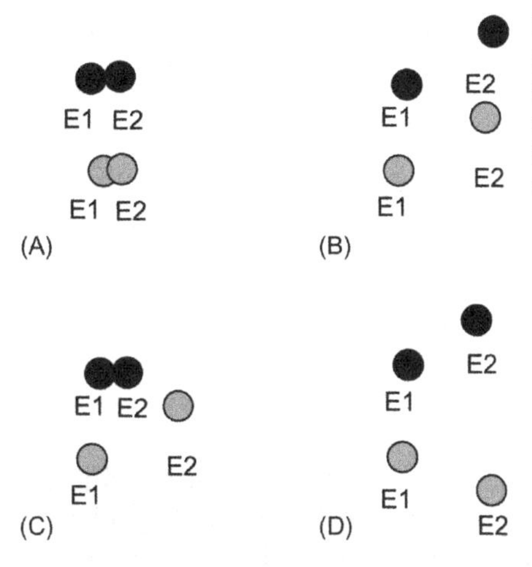

FIGURE 12.3 Two genotypes reared in two environments. (A) Genotypes differ in their phenotypes but neither responds to the environmental factor; (B) Both genotypes respond to the environmental factor, in parallel directions; (C) One genotype responds to the environmental factor, the other does not; (D) Both genotypes respond but in different directions.

multidimensional, we might not be able to depict the array of phenotypes produced by two genotypes in two environments. Those four shapes might not lie within a plane. It can therefore be difficult to depict norms of reaction (a subject we return to below), but for now we will project them onto a plane and show the endpoint of each vector as a shape in a two-dimensional space (Figure 12.3). Figure 12.3 shows two genotypes in two environments. In the first case, the two genotypes differ in their phenotypes but neither alters its phenotype in response to the environmental factor (Figure 12.3A). In the second case, both genotypes alter their phenotypes in response to the environmental factor, in parallel directions (Figure 12.3B). In the third case, one genotype exhibits a large response and the other does not (Figure 12.3C). In the fourth case, both genotypes exhibit a large response but in different directions (Figure 12.3D). What complicates the statistical analysis, which might otherwise be straightforward, is that a significant interaction term could mean that the genotypes differ solely in the magnitude of their response or that they differ solely in the direction of that response or that they differ in both. We therefore need to compare the phenotypic trajectories of plasticity or, as Adams and Collyer (2009) term them, phenotypic change vectors. Methods for comparing lengths and directions of ontogenetic trajectories were discussed at length in Chapter 11 and these are the same methods that we use to compare phenotypic trajectories, in general.

Studies of plasticity often use complex designs, measuring multiple replicates of each sampling as well as several such units responding to several environmental factors. These complex, multilevel designs make it possible to measure the overall plasticity of a genotype, or of a population or of even higher-level units (e.g. ecotypes, species) over an array of environmental factors. Overall plasticity would be measured by the variation of a given genotype over the array of environmental factors. In Figure 12.4, Genotype A varies little in response to all the environmental variables; we can see that all four of its four environment-specific phenotypes are tightly clustered, occupying a small region of the phenotype space. Genotype B responds to all the factors more than Genotype A, therefore the four environment-specific phenotypes occupy a larger area of the phenotype space. Genotype C responds to one environmental factor but not to the others.

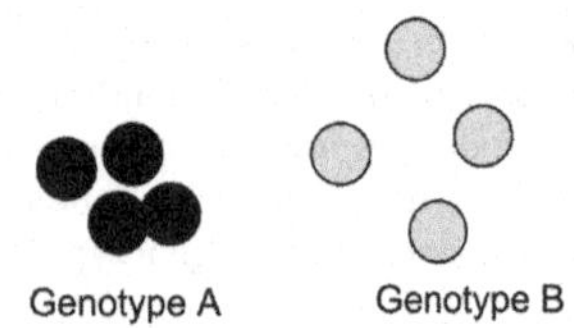

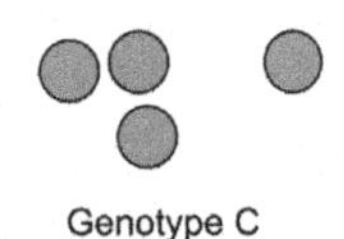

FIGURE 12.4 Norms of reaction for shape of three genotypes reared in four environments. Genotype A varies little over the four environments. Genotype B varies over the four environments, occupying a larger region of the phenotype space. Genotype C responds to one environmental factor but not to the others.

An exemplary recent analysis illustrates the methods used to study plasticity, including the structure of the comparisons that can be made and the methods for testing hypotheses. In this experiment, Hollander and colleagues examined two ecomorphs of one species of snail, *Littorina saxatilis* and another species, *L. littorea* (Hollander et al., 2006). *L. saxatilis* is polymorphic in shell traits; its morphs differ in shell size, thickness, and aperture area. An ecologically important distinction between the two species is that *L. saxatilis* releases miniature snails on the shore and therefore the juveniles experience the same habitat as the (sedentary) adults, whereas *L. littorea* releases drifting larvae which therefore experience a different environment from the adults. Because the environment of *L. saxatilis* does not vary across life-history stages, they are predicted to be less plastic than *L. littorea* even though the environment of the adults of both species is highly heterogeneous; adults of both species are found in wave-exposed rocky shores that are free of crabs (a predator) as well as in sheltered areas that are rich in the predator. The drifting larvae of *L. littorea* encounter the rocky-shore habitats largely at random and their environments need not resemble that of their parents, whereas the larvae of *L. saxatilis* remain in their parents' environment. The prediction was that *L. littorea* would be more plastic.

To test this hypothesis, the juveniles of both morphs of *L. saxatilis* and juveniles of *L. littorea*, were collected from various locations on the west coast of Sweden. The juveniles were pooled so that the samples encompassed the variation among local sites. The juveniles were exposed to simulated wave action, or a predator cue (water-borne effluent from a predator or water-borne effluent from crushed conspecifics) or to no treatment. Each group contained at least 50 individuals and each was replicated in three aquaria. Analysis of the variation among ecomorphs and species showed that the two ecomorphs are distinct as are the two species, thus the null model for the statistical analyses included these expected differences. Analyses were done to determine whether ecomorphs of *L. saxatilis* and *L. littorea* exhibit plasticity, whether all three groups exhibit equal levels of plasticity within and across treatments, and to identify the largest contrasts between treatments for each group. Multivariate analysis of variance (MANOVA) showed the treatments had a statistically significant impact on mean shape of each group.

To quantify the overall magnitude of plasticity, Hollander and colleagues measured disparity within each group and compared it across groups. Although the measure of disparity was introduced in Chapter 10, we repeat it here. Disparity is the variance across means, which measures the dispersion of treatment means from the grand mean for that group:

$$D_A = \frac{\sum_{i=1}^{a} (\overline{X}_i - \overline{X})^T (\overline{X}_i - \overline{X})}{\sum_{i=1}^{a} (N_i - 1)} \tag{12.1}$$

where there is a total of a treatments, $\overline{X}_i$ is mean for the ith treatment, X is the grand mean over all treatments, and N is the sample size for the group (e.g. for an ecomorph of *L. saxatilis*).

The statistical tests of *Littorina* disparity were done by randomly assigning snails to treatments within their own group, then computing the treatment means from those randomized data, then computing the disparity from those randomized data, iterating the procedure 9999 times. Adding the observed value to those 9999 gives 10 000 values. The number of times that a value as large as or larger than the observed one could be obtained by chance can then be calculated by counting the number of values equal to or exceeding the observed one.

To determine which particular treatments have a large effect relative to other treatments, and also to determine which treatments have a large effect in one group compared to that same treatment's effect in another group, distances between treatments were compared within and between groups. These are the Procrustes distances between means. For four treatments, there are six possible vectors extending between means: (1) control mean to predator effluent-treatment mean; (2) control to conspecific effluent-treatment mean; (3) control to simulated waves-treatment mean; (4) predator effluent-treatment mean to conspecific effluent-treatment mean; (5) predator effluent-treatment mean to simulated waves-treatment mean; and (6) conspecific effluent-treatment mean to simulated waves-treatment mean. As well as analyzing the lengths of these vectors, they also analyzed their directions by measuring the angles/correlations between them (the comparison of directions by the analysis of angles or correlations between them is discussed in detail in Chapter 11).

These comparisons of vector lengths and directions yielded 45 tests of vector lengths within ecotypes, 18 tests of vector lengths between ecotypes, 45 tests of vector correlations within ecotypes, and 18 tests of vector correlations between ecotypes. In light of the large number of tests, a sequential Bonferroni test (Rice, 1989) was used. A Bonferroni test divides the critical value, α, by the number of tests. So if there are 10 tests, and the desired table-wide value of α is 0.05, 0.05 is divided by 10, yielding 0.005. A sequential Bonferroni test is less conservative. Using this test, α is initially divided by the total number of tests and the smallest p-value is judged against that, then α is divided by the remaining number of tests and the next smallest p-value is judged against that and so forth until reaching one that is not significant.

Considering the large number of tests, especially the 45 tests of vector correlations within ecotypes, it is remarkable that any differences were statistically significant. With rare exceptions, the differences were very large (and statistically significant) or very small (and therefore not statistically significant). An interesting result is that, in general, both species exhibited similar magnitudes and directions of responses, contrary to the expectation that *L. littorea* would be more plastic. Perhaps even more surprising, the two ecomorphs of *L. saxitilis* exhibit less within-treatment disparity (i.e. exhibit greater micro-environmental canalization) when exposed to a treatment that mimics the environment of the *other* ecomorph.

Visualizing Norms of Reaction

Norms of reaction for shape can be difficult to depict unless groups differ solely in the magnitude of plasticity. If that is the sole difference, the norms of reaction can be depicted just as they are for a one-dimensional trait; groups then differ solely in their position along a single phenotype axis. The same vector would extend between any pair of means because all are arrayed along that one axis. But, in more complex cases, when the directions of the responses also differ, we cannot represent a norm of reaction by a single phenotype axis. Several recent studies have contrasted the outcomes of varied treatments on shape, depicting the norm of reaction by the contrasts between the resultant phenotypes. One approach is to project the array of phenotypes onto a low-dimensional space; for example, in the analysis detailed above of the impact of predator cues and wave-action on shell morphology of *Littorina saxatilis* and *L. littorea*, Hollander and colleagues (2006) showed the 12 treatment means projected onto the space of the first two principal components of shape variation. Another approach is to present a series of contrasts that depict the impact of the environmental factor on a single group. For example, a study of the impact of dietary consistency on marine, benthic and limnetic threespine sticklebacks displayed the effects by a series of deformed grids, each showing the impact of a benthic or a limnetic diet on marine, benthic or limnetic head shapes (Wund et al., 2008). Similarly, another experiment on the impact of diet on shape examined two attributes of diet, its hardness and calcium content; analyzing pharyngeal jaw shape (and size) of laboratory stocks of the cichlid *Amphilophus citrinellus* subjected to three diet treatments: (1) intact snails with shell, (2) peeled snails without shell, and (3) finely ground snails that were frozen, with the fish feeding on the soft thawed outer layer (Muschick et al., 2011). Pairs of treatments were compared by a Discriminant Function Analysis (see Chapter 6) and the shapes at the extremes of each function were depicted as interpolated outlines. Similarly, diet-induced plasticity of body shape of arctic char, *Salvelinus alpinus*, was analyzed by feeding young of each ecomorph a diet that mimics the natural diet of either a benthic or limnetic population of three ecotypes (Parsons et al., 2011). Both ontogenetic trajectories and age-specific phenotypes were compared between ecomorphs reared on the same diet in different lakes and within ecomorphs fed benthic versus limnetic diets. Discriminant function analysis was used to visualize the impact of diet on body shape for each ecomorph at two ontogenetic stages, depicting the contrasts by regressing shape on the discriminant function scores.

CANALIZATION: QUANTIFYING VARIATION

Canalization refers to the ability to produce the same phenotype despite variation in genotype and environments of rearing. A long-standing (but still controversial) hypothesis

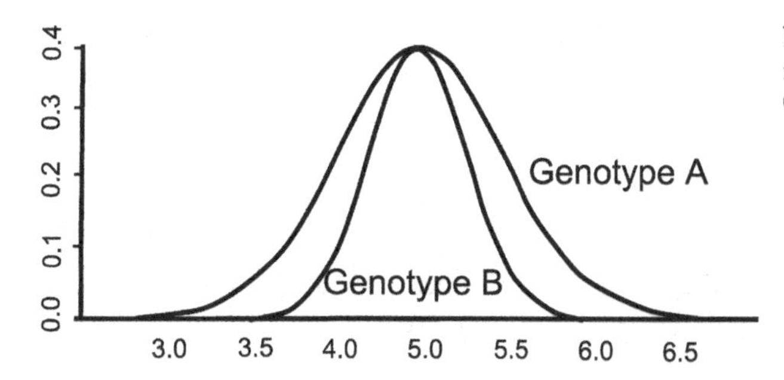

FIGURE 12.5 Canalization shown as a reduction in variance from that exhibited by Genotype A to that exhibited by Genotype B.

is that canalization evolves by stabilizing selection, which not only removes deviants from the population but also favors mechanisms that suppress the expression of variation (Waddington, 1942; Schmalhausen, 1949). As Schmalhausen (1949) put it, stabilizing selection produces a stable form by creating a regulating apparatus. Canalization, according to this hypothesis, is the long-term effect of stabilizing selection. If we look at the distribution of a phenotype, canalization is evident by a reduction in variance; compared to Genotype A, Genotype B is more canalized (Figure 12.5).

Canalization is now often subdivided into genetic and environmental canalization according to the source of the variation that is buffered. Canalization of genetic variation, including novel mutations, is termed "genetic canalization" (Kawecki, 2000; Elena and Lenski, 2001; de Visser et al., 2003), whereas canalization of environmental variation is termed "environmental canalization" (for recent reviews of these concepts, see Debat and David, 2001; Hallgrímsson et al., 2002; Willmore et al., 2007). That genetic and environmental canalization are distinguished in this classification scheme does not imply that the two forms of canalization are physiologically distinct. A still open question is whether they are. Answering that question is complicated by the several distinct forms of environmental canalization, which is now subdivided into the canalization of variation *across* environments ("macroenvironmental canalization"), and canalization within environments ("microenvironmental canalization"). Macroenvironmental canalization is the converse of phenotypic plasticity. One rationale for distinguishing macro- from microenvironmental canalization is that plasticity need not reduce the ability to buffer random variation within an environment. Even a steep norm of reaction might be well canalized.

Because macroenvironmental canalization is the converse of plasticity, macroenvironmental canalization is studied using the methods introduced in the previous section. Microenvironmental canalization is usually studied by comparing variances. In studies of shape, a variance can be calculated by measuring the Procrustes distance of each individual from the mean:

$$Var = \frac{\sum_{j=1}^{N} D_j^2}{(N-1)} \tag{12.2}$$

which is equivalent to measuring the variance of each coordinate, summed over all the coordinates:

$$Var = \frac{\sum_{j=1}^{n} (X_{ij} - \overline{X}_i)^T (X_{ij} - \overline{X}_i)}{N_i - 1} \tag{12.3}$$

If we have multiple groups, such as multiple treatments, and wish to test the hypothesis that one genotype (or population) is canalized over multiple treatments, we would calculate a pooled within-group variance by:

$$Var_{pooled} = \frac{\sum_{j=1}^{a} \sum_{j=1}^{n} (X_{ij} - \overline{X}_i)^T (X_{ij} - \overline{X}_i)}{\sum_{i=1}^{a} N_i - 1} \tag{12.4}$$

In the case of Var_{pooled}, the summed squared distances of j individuals from a treatment mean are summed over all treatments. Confidence intervals can be placed on these measures of variance just as they are on measures of disparity, a topic discussed in depth in Chapter 10.

One approach for testing the null hypothesis that two samples do not differ in their variance, using permutations of residuals, was discussed above in context of tests for plasticity. Another approach uses a *t*-test, computing the variance for each group at each iteration of a bootstrap or permutation procedure, then subtracting one group's variance from that of the other, and iterating the calculation of the variances and the difference between them at each iteration to generate the distribution of the difference between the two variances. If the confidence interval for that difference includes zero, we could not reject the null hypothesis that the two populations do not differ in their variances.

Example: Ontogenetic Decrease in Variance of Skull Shape

We exemplify an analysis of canalization by testing the hypothesis that variance diminishes over ontogeny in the absence of any selective deaths in the population. To that end, we compare the variance of skull shape across four ages, 10-, 15-, 20-, and 25-days postnatal, of the randombred Hsd/ICR strain of the house mouse (*Mus musculus domesticus*). The superimposed landmarks for each sample are shown in Figure 12.6; the estimates for the variance in shape at each age, and standard errors of the estimate, are given in Table 12.1. To compare the levels of variance between successive ages, we use the *t*-test described above to evaluate the difference between variances relative to the pooled standard errors of those variances. Over the initial 5-day interval, variance is halved; the difference in the variances for the two samples is 0.000279. The 95% upper bound on the confidence interval for the difference between the two variances is 0.00015639, and the magnitude of the observed difference was exceeded by none of the 200 permutations. After that point, variance is stable – no statistically significant differences are found between successive age classes later. The initial reduction of variance, in the absence of

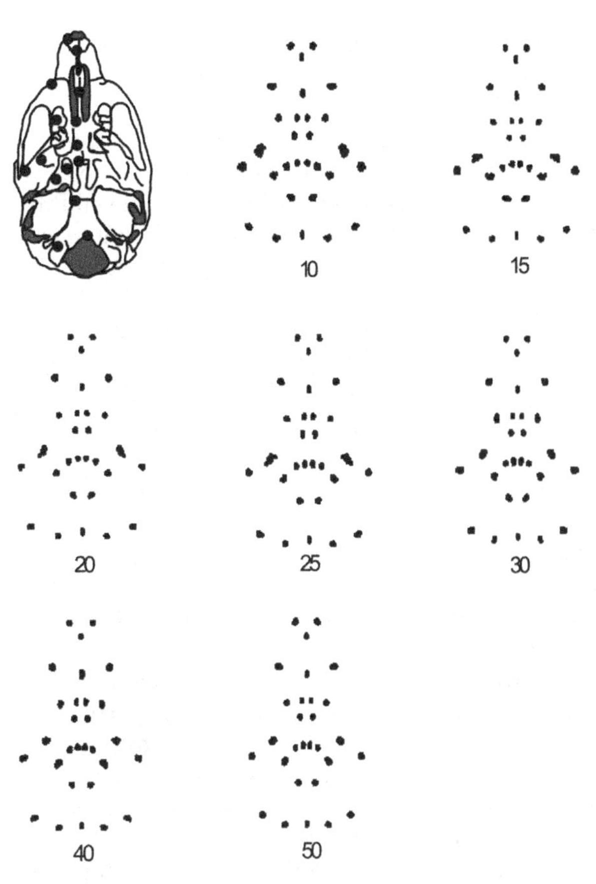

FIGURE 12.6 Ontogeny of variance for skull shape of the house mouse. The superimposed landmarks for skull shape of the house mouse, *Mus musculus domesticus.* Ages, in days after birth, are shown beneath the data. Variance and its standard error for four ages are given in Table 12.1.

TABLE 12.1 Skull Shape Variance of *Mus musculus domesticus* Sampled at Four Ages (Given in Days After Birth), and the Standard Errors of the Variance

Age	Variance	Standard Error
10	0.000628	0.0001
15	0.000349	0.00005
20	0.000316	0.0001
25	0.000410	0.0001

The superimposed landmarks are shown in Figure 12.6.

any selective deaths in the colony, indicates that variation is developmentally regulated and the later stability of the variance also suggests canalization because we would expect to see a continued production of variation in light of the ongoing process of skeletal development.

DEVELOPMENTAL STABILITY: QUANTIFYING DEVELOPMENTAL NOISE

"Developmental stability" refers to the ability of a genotype to produce the same phenotype under the *same* environmental conditions by buffering developmental noise (Reeve, 1960a; Zakharov, 1992; Clarke, 1998; Van Dongen and Lens, 2000). Developmental stability, or rather its converse, developmental instability, is usually measured by fluctuating asymmetry (FA), the random deviations from bilateral symmetry. An advantage of measuring developmental stability by FA is that we actually know the expected value for the trait barring developmental perturbation: both sides of a bilaterally symmetric individual have the same genotype and develop within the same environment so they should be identical, barring developmental perturbations (Reeve, 1960b; Palmer and Strobeck, 1986). Developmental stability has been intensively studied recently for at least three major reasons. First, decreased developmental stability may provide a sensitive indicator of environmentally or genetically stressed populations (Clarke, 1993; Graham et al., 1993; Estes et al., 2006). If that is generally the case, elevated FA could serve as a useful biomarker for stressed populations and aid conservation efforts. However, the causal connection between stress (environmental or genetic) and FA remains a contentious issue because many studies fail to find elevated FA in stressed populations (see Hoffmann and Woods, 2001; Hoffmann et al., 2005; Leamy and Klingenberg, 2005).

A second stimulus for studies of FA comes from theoretical models for the evolution of variability. One interesting hypothesis is that selection for one variational property indirectly selects for others; for example, selection for environmental canalization might indirectly select for genetic canalization (e.g. Wagner et al., 1997). This has been termed "plastogenetic congruence" by Ancel and Fontana (2000). Congruence between the classes of canalization (genetic, and macro- and microenvironmental) and developmental stability has thus become an important issue in evolutionary theory, stimulating several empirical investigations of the correspondence between them (e.g. Debat et al., 2000; Hallgrímsson et al., 2002; Dworkin, 2005b; Santos et al., 2005; Willmore et al., 2005; Breuker et al., 2006; Breno et al., 2011; Klingenberg et al., 2012). This is likely to remain an area of active investigation given the disparity and complexity of the results.

The Statistical Analysis of FA

FA is now usually studied by a two-way mixed model Analysis of Variance (ANOVA), whose two main factors are Individuals (a random factor) and Sides (a fixed factor); FA is quantified by the interaction term Individuals × Sides (Leamy, 1984; Palmer and Strobeck, 1986). This approach was extended to shape data by Klingenberg and colleagues (Klingenberg and McIntyre, 1998; Klingenberg et al., 2002) and has now been extended to symmetries more complex than bilateral (Savriama and Klingenberg, 2011). Here we restrict the discussion to bilateral symmetry and follow Klingenberg's exposition of the method.

In the case of bilateral symmetry, two kinds of symmetry can be distinguished (Mardia et al., 2000). One is "matching symmetry", which refers to the case in which there are two

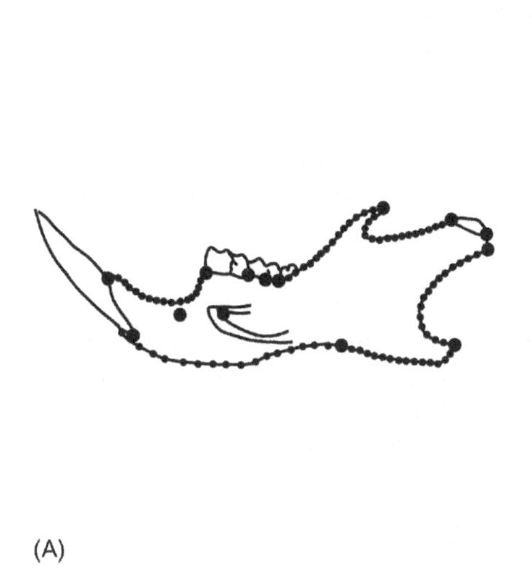

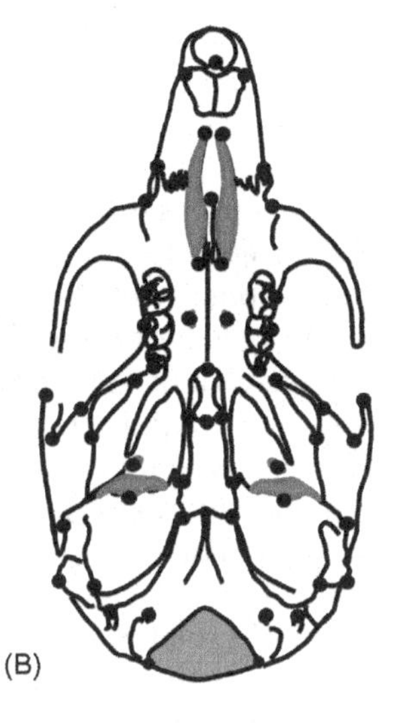

FIGURE 12.7 The two forms of symmetry for bilaterally symmetric forms. (A) Matching symmetry; (B) object symmetry

structures, one on the right side and the other on the left. In this kind of asymmetry, typified by the two halves of the mammalian mandible, all landmarks are present on both sides (Figure 12.7A). To analyze matching symmetry, we can compare the right to the left side, using one configuration of landmarks for each side. The alternative is "object symmetry", typified by the mammalian cranium, a single structure with an axis of symmetry along its midline (Figure 12.7B). Instead of analyzing two configurations, we analyze a single one, and there are midline landmarks in addition to the bilaterally paired landmarks. The ANOVA design and the analysis of the paired landmarks are similar for both kinds of symmetry. But the analyses differ according to the type of symmetry because matching symmetry is analyzed using two configurations per individual whereas object symmetry is analyzed using just one. One important consequence of having all the bilateral landmarks plus the midline landmarks within a single configuration is that we can analyze the relationship between the two halves when we have object asymmetry but not if we have matching asymmetry (Klingenberg et al., 2002).

The analysis of matching asymmetry is more straightforward so we begin with it. We will assume that every individual is photographed twice on each side. The first step in the analysis is to reflect all the configurations from one side, e.g. the left side, onto the other so that we can compare the two sides in the same orientation. If the photographs were not reflected before they were digitized, the reflection is done by changing the sign of the x-coordinate for every landmark on one side. Following that reflection, the configurations for both sides (for all replicates for all individuals) are superimposed using a standard least squares Procrustes superimposition. Then the right and left sides of all replicates for each individual are averaged, giving the estimate of the mean shape for that individual. Variation among the mean shapes of the individuals is the variation explained by the main factor "Individual". To calculate the variation explained by "Side", we compute the difference between the two sides from the difference in average shapes of each side. That is, for each individual we calculate its average right-side shape (over the replicate photographs) and its average left-side shape, and we then calculate the average for each side over all individuals in the population. The difference between those two averages is the variation explained "Side". Measurement error is quantified by the variation over replicates for each individual and side. This error is a combination of the photographic error plus digitizing error. The statistical significance of "Side", the fixed factor, is tested against the interaction term and the statistical significance of FA is tested against measurement error. The degrees of freedom for each term are the univariate degrees of freedom (e.g. 1 for side) multiplied by the dimensionality of the data.

The analysis of object asymmetry is more complex because we have only one configuration per replicate. As a result, we cannot superimpose the right side onto a reflected left side. Instead, we superimpose each configuration onto a reflected copy of itself. We thus start by making that reflected copy, which involves first reversing the sign of the x-coordinate of each landmark and then renumbering those landmarks to correspond with the number it had in the original copy. For example, if landmark 2 on the left side is bilaterally homologous with landmark 4 on the right side, we would relabel the reflected landmark 4 as landmark 2 and similarly relabel landmark 2 as landmark 4. Then the original and reflected configurations are superimposed (using the standard least squares Procrustes superimposition). Following superimposition, each individual is symmetric and the midline landmarks line up as do the midpoints between the paired landmarks. Because the right and left sides of the paired landmarks are now redundant, the whole configuration is described by the landmarks on just one side.

The variation of the paired landmarks is unconstrained – they can vary in all directions. But the variation of the midline landmarks is constrained – they can only vary in one direction, which is along the midline. Any variation away from the midline would mean that the midline is not the midline. Thus, when calculating the dimensionality of shape (and the degrees of freedom for each effect) the constraint on the variation of midline landmarks must be taken into account. For the symmetric component of variation, there are $2K + P - 2$ dimensions, where K is the number of paired landmarks and P is the number of midline landmarks (this notation differs from that of Klingenberg et al. (2002) because they use L for the number of midline landmarks but we have used L for the number of semilandmarks). So, if there are 20 paired landmarks and 5 unpaired landmarks, there are $40 + 5 - 2$ dimensions. For the symmetric component of a two-dimensional configuration, only one dimension is used up by translation and none are used up by rotation (hence -2 rather than -4).

In the analysis of object symmetry, the "Sides" component of the variation is calculated from the difference between the original and reflected configurations. In the asymmetric component of variation, just like in the symmetric component, the asymmetry of the paired landmarks can be in any direction. This means that for K pairs of two-dimensional landmarks there are $2K$ dimensions of variation. But asymmetry of the midline landmarks is possible only in the direction *perpendicular* to the midline (or median plane) – if the variation was along the midline it would not be asymmetric. One consequence of this restricted variation is the reduced dimensionality of the asymmetric component; the midline landmarks add P dimensions, but their sum must be zero, eliminating a degree of freedom (from both two- and three-dimensional data). The rotation step of the Procrustes superimposition removes one additional degree of freedom from two-dimensional data and two from three-dimensional data but none are used up by scaling or orientation of paired landmarks because variation in size or orientation of the asymmetric component is informative about shape asymmetry. Following the superimposition, there are thus $2K + P - 2$ in two-dimensional data, and $3K + P - 3$ in three-dimensional data. The sum of the dimensions for the symmetric and asymmetric component equal the total, i.e. $2(2K + P) - 4 = 4K + 2P - 4$ for two-dimensional data, and $3(2K + P) - 7 = 6K + 3P - 7$ for three-dimensional data.

What makes the remainder of the analysis yet more complicated is that the symmetric and asymmetric components are orthogonal to each other – they occupy orthogonal subspaces (see Mardia et al., 2000; Klingenberg et al., 2002). This means that the symmetric individual component is in a different subspace than the asymmetric Side component, and the symmetric and asymmetric components of measurement error are also in different subspaces. That poses no problem for testing directional asymmetry (i.e. Sides) because Sides is tested against FA and both are in the asymmetric component. But FA is tested against measurement error and the symmetric and asymmetric components of measurement error are in different subspaces. Klingenberg and colleagues (2002) suggest using the asymmetric component of the measurement error for the test, which means that measurement error is computed from the difference between the original and reflected configurations for each replicate. The average of those differences is the asymmetric component of the measurement error.

The statistical analysis can be done two ways. One is to use the sums of squares of the coordinates (see the discussion of Goodall's *F*-test, Chapter 8 and the discussion of the permutational Manova, Chapter 9). Klingenberg and colleagues call this a "Procrustes Anova". This test assumes the equality and independence of variance at each landmark. The alternative is a multivariate test that uses the whole covariance matrix, which requires inverting the variance–covariance matrix. Because there are more dimensions than degrees of freedom for shape, the covariance matrix cannot be inverted. One solution is to use a generalized inverse, or if the determinant of the Sums of Squares and Cross-Products matrix must be calculated (as in the case of Wilk's Λ), the product of the non-zero eigenvalues can be used instead.

Example: Fluctuating Asymmetry of Prairie Deer Mouse Mandibular and Cranial Shape

We exemplify the analysis of matching asymmetry by a study of prairie deer mouse (*Peromyscus maniculatus bairdii*) mandibular shape FA and the analysis of object asymmetry by a study of prairie deer mouse cranial shape FA. In the case of matching asymmetry, FA of mandibular shape is highly significant based on a Procrustes ANOVA, with all terms tested by permutations (Table 12.2). We can visualize the spatial structure of FA by its first principal component of variation (Figure 12.8A). FA is especially pronounced in the regions of the angular and coronoid processes, and there is also an interesting pattern of variation in the position and height of the molar alveolus. In the case of object asymmetry, FA is statistically significant as judged by a Procrustes ANOVA (Table 12.3) and also by MANOVA that takes the covariance matrix into account. FA is especially pronounced most anteriorly and in the lateral braincase, as evident from its first principal component (Figure 12.8B).

TABLE 12.2 Two-Way Mixed Model Procrustes ANOVA of Fluctuating Asymmetry of the Prairie Deer Mouse Mandible

Source	SS	DF	MS	*F*	p
Individuals	0.21283	17088	1.2455e-005	9.62	<0.001
Sides	0.00082	192	4.2432e-006	3.28	<0.001
Individuals × Sides	0.02211	17088	1.2941e-006	5.12	<0.001
Measurement error	0.00874	34560	2.5298e-007		

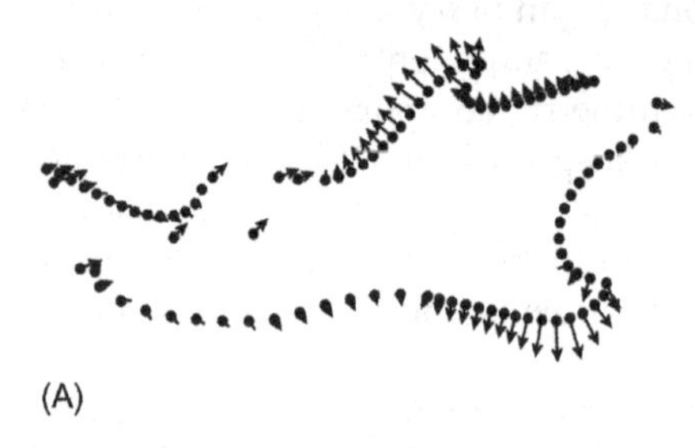

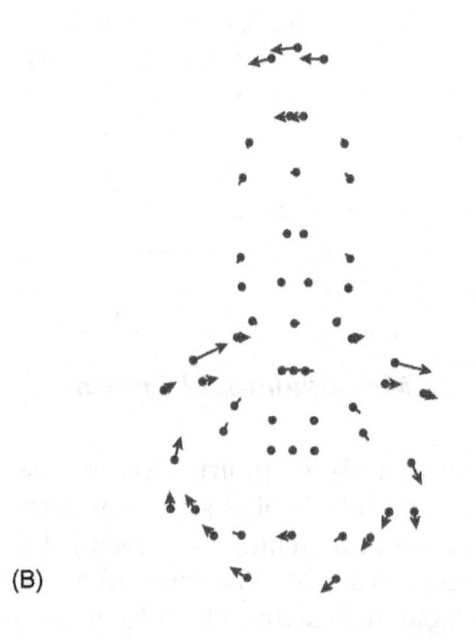

FIGURE 12.8 First principal component of variation of fluctuating asymmetry of the prairie deer mouse (*Peromyscus maniculatus bairdii*). (A) Mandible; (B) cranium.

Measuring the Overall Magnitude of FA

To compare levels of FA across populations we need a measure of the overall magnitude of FA. In the literature of traditional morphometric data, there is a large, even bewildering, array of such metrics. Palmer and Strobeck (2003) list 18 of them, some with two variants. Fortunately, only one of those is suitable for landmark data, and only two more have been added to the list. Of these, two are in units of Procrustes distance and the other

TABLE 12.3 Two-Way Mixed Model Procrustes ANOVA of Fluctuating Asymmetry of the Prairie Deer Mouse Cranium

Source	SS	D	MS	*F*	p
Individuals	0.15978	5980	0.0000267188	7.16	<0.001
Sides	0.0064622	52	0.000124272	33.31	<0.001
Individuals × Sides	0.0223116	5980	0.000003731	1.28	<0.001
Measurement error	0.0689126	23608	0.0000029190	–	

is in units of Mahalanobis distance (which takes the covariance structure into account). All three can be calculated by hand (or rather, in a spreadsheet). Of course, the calculation does not have to be done by hand – there are programs that will do it for you, but it is easy to understand the procedure if you can implement it yourself.

The first measure is a conventional Procrustes distance between each individual's right–left distance and the bilaterally symmetric mean shape. To calculate this in a spreadsheet, open the file that contains the superimposed right and left sides. Subtract the coordinates of one side from those of the other. This subtraction gives the difference between the right and left sides for that individual. Then square the differences for each coordinate, summing those squares over all the coordinates for that individual. The square root of that sum is the measure of overall FA for each individual. The second metric based on the Procrustes distance differs from the first only in that the average directional asymmetry is the standard instead of the bilaterally symmetric mean. This measure of FA is calculated just like the first except the population's average right–left difference is subtracted from each individual's right–left difference. No such subtraction was necessary in the first case because the average right–left difference for a bilaterally symmetric form is zero. After subtraction, the differences are squared and summed and the square root is taken of the sum.

The third metric is a Mahalanobis distance between the two sides (Klingenberg and Monteiro, 2005). This calculation, which is more involved than the other two, can also be done by hand. The first step is to compute the right–left differences for each individual. But these differences are not squared or summed over the coordinates. Rather, after calculating the right–left differences for each coordinates for all individuals, the data are subjected to a Principal Components Analysis (PCA) of the covariance matrix. Then the scores for the principal components (PCs) are standardized to unit variance (which is done by squaring them, summing the squares, taking the square root and dividing each score by that value). The number of PCs to use in this calculation depends on the dimensionality of the data. If the data consist solely of landmarks, and the PCA was done on the partial warps, all the PCs should be used when computing FA. If the analysis is instead done using the superimposed coordinates (or Procrustes residuals), the last four PCs should be excluded from the analysis because there are $2K-4$ dimensions but $2K$ PCs. When the data include semilandmarks, the number of PCs to use is $2K+L-4$ (for two-dimensional data). The same reasoning extends to three-dimensional data – the number of PCs used in the calculation should equal the dimensionality of the data. Fortunately, this procedure does not actually need to be done by hand.

Using any one of the measures of FA, it is possible to test hypotheses that predict elevated or reduced FA. The hypothesis that two or more populations differ in average FA can be tested just like the hypothesis that two or more populations differ in variance. One test that is particularly insensitive to deviations from normality is Levene's test (Levene, 1960), often used to test for differences in variance. What Levene's test compares is the average deviation of points from the mean of the sample. To carry out this test, which can be done in Excel, calculate the level of overall FA for each individual and subtract that from the mean (or median) value, and use the absolute value of that deviation in the test. If FA is high, the mean value of that deviation will be large. Then use a *t*-test or ANOVA to compare the mean values.

ANALYZING THE RELATIONSHIP BETWEEN PLASTICITY, CANALIZATION AND DEVELOPMENTAL STABILITY

Numerous studies have examined the relationship between plasticity, canalization and developmental stability (e.g. Scheiner et al., 1991; Debat et al., 2000; Hoffmann and Woods, 2001; Dworkin, 2005a,b; Santos et al., 2005; Willmore et al., 2005; Breuker et al., 2006; Hollander et al., 2006; Breno et al., 2011; Klingenberg et al., 2012). One goal of many studies is to test the hypothesis that there is a "general buffering capacity". What "general" means can differ between studies, but the usual aim is to test the hypothesis that the same mechanism(s) buffer variation arising from different sources. One possibility is that mechanisms that buffer phenotypes against the perturbations also canalize them against environmental perturbations, whether the environmental perturbations are macro- or microenvironmental or even developmental noise. Sometimes the question is framed more narrowly, such as whether mechanisms that enable phenotypes to respond to macroenvironmental variation make them more sensitive to developmental noise (e.g. Scheiner et al., 1991).

The hypothesis that the same mechanisms buffer more than one sort of perturbation can be tested in two ways. The first is to estimate the correlation between a measure of variance (genetic vs environmental or macro- vs microenvironmental) and/or a measure of FA. The second is by comparing covariance matrices. This is done by comparing the covariance matrix for one component of variation (e.g. "Individual") to another (e.g. "FA"). Using the first approach, the question is whether individuals who most deviate from the mean also most deviate from bilateral symmetry, i.e. whether an individual's deviation from the mean predicts its deviation from bilateral symmetry. In the second case, the question is whether the buffering mechanisms have the same morphological effects on variation. Both methods for testing the hypothesis are widely used, and both are often used in the same study.

The first approach is straightforward to apply. The first step is to calculate each individual's deviation from the relevant mean (e.g. the mean shape within an environment). The second is to calculate its deviation from the other relevant mean, or from bilateral symmetry. Given these two measures of each individual's deviations, the hypothesis is tested by measuring the correlation between them. If that correlation is statistically significant it would support the hypothesis of a common buffering mechanism. The second approach is less straightforward because it requires comparing covariance structures and methods for comparing covariance structures are a matter of some contention. However, the most commonly used method is to estimate the correlation between two matrices and to test its statistical significance by a Mantel test (Mantel, 1967). The matrix correlation is a standard Pearson product-moment correlation calculated over the corresponding entries in the two covariance matrices. Corresponding entries would be, for example, the covariance between the x-coordinate of the first two landmarks in the two matrices – that covariance in one matrix corresponds to that same covariance in the other matrix. The matrix correlation can be calculated by hand if the matrices are small; that involves arranging the two matrices in two column vectors (omitting the redundant elements and, if desired, the variances along the diagonal). The two matrices should then be matched up, with corresponding entries on each row. Once the matrices are turned into column vectors, the correlation is calculated between the two columns.

The Mantel test is the most common test of the null hypothesis that the two matrices are no more similar than expected by chance. To determine whether the correlation is significant, the elements in one matrix are randomly permuted and the correlation is measured between the permuted matrix and the other one, at each iteration, repeating this procedure many times. The correlations obtained from these permutations provide the distribution of the correlations between randomly related matrices. Given this distribution, the number of correlations that equals or exceeds the observed one can be counted. Usually, the observed correlation is included in the count, so if you do 100 permutations and obtain a p-value of 0.01, one value obtained by 99 random permutations plus the observed correlation are equal to or greater than the observed one. The test needs some modification to be used for geometric data because the standard version of the test would permute x-coordinates independently of y-coordinates. Adapted for geometric data, the Mantel test permutes landmarks as units (Klingenberg and McIntyre, 1998). The test has also been modified to allow for comparisons between the covariance matrix of object FA and individual variation, which, if you recall from the previous section on FA, are in different subspaces. The test is done by omitting the midline landmarks (and one whole side, which is redundant), thereby limiting the analysis to the paired landmarks on a single side (Klingenberg et al., 2002).

Examples: Comparing Phenotypic Variation to FA for Prairie Deer Mouse Mandibular and Cranial Shape

We first test the hypothesis that an individual's deviation from bilateral symmetry is correlated with its deviation from the mean shape, using the Procrustes distance from the bilaterally symmetric shape as our measure of FA. Similarly, we use that individual's Procrustes distance from the mean shape as our measure of its deviation from the mean. In the case of the mandible, the correlation is a very weak 0.09, which is not statistically significant ($p = 0.38$). For the cranial data, the correlation between these two measures is a weak 0.232, which is nonetheless statistically significant ($p = 0.012$). We then take the second approach, calculating the correlation between the covariance matrices for the FA and symmetric (among-individual) components of variation. The resemblance between these two covariance matrices can be examined visually by looking at each one's dominant dimensions of variation. The first two principal components for the mandible variation and FA are shown in Figure 12.9. When the diagonal is included, so that the variances as well as covariances are included in the analysis, the correlation is fairly high, 0.694, dropping very slightly to 0.643 when the diagonal is excluded. Not surprisingly, both correlations are statistically significant, $p < 0.001$. In the case of the cranium, whose first two PCs are shown in Figure 12.10, the correlation between the two matrices is also fairly high, 0.586, and statistically significant ($p < 0.001$) when the diagonal is included, but the correlation drops to a very low 0.038, which is not statistically significant ($p = 0.61$) when the diagonal is excluded.

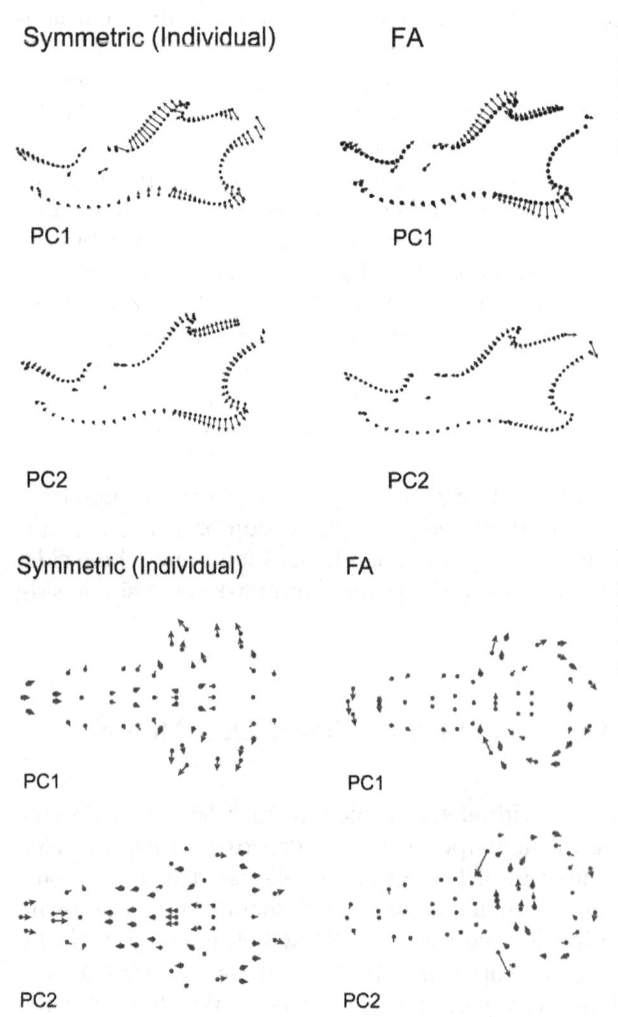

FIGURE 12.9 Comparing patterns of symmetric and fluctuating asymmetric covariation. First two principal components of symmetric variation (among individuals) and fluctuating asymmetry of the prairie deer mouse cranium.

FIGURE 12.10 Comparing patterns of symmetric and fluctuating asymmetric covariation. First two principal components of symmetric variation (among individuals) and fluctuating asymmetry of the prairie deer mandible.

The comparisons that we just did were intended to be merely exemplary of the methods used to test a hypothesis of a generalized buffering mechanism. These were not carefully controlled experiments that could isolate purely microenvironmental, macroenvironmental or genetic variation. There are, however, several carefully controlled studies that assess the relationship between FA and variation, both genetic and environmental. For example, Breuker and colleagues (2006) compared levels of FA and variation of *Drosophila melanogaster* wing shape across 115 genotypes. To quantify FA, they used both the Procrustes distance of each individual from the bilaterally symmetric shape and the Mahalanobis distance. Using the Procrustes distance, they found a correlation of 0.49 between FA and variation among individuals, which rose to 0.67 based on the Mahalanobis distance. They also examined the pairwise correlations between covariance matrices of variation and FA for the 115 genotypes. Including the variances in the analysis produced correlations ranging from 0.54 to 0.91; excluding the diagonals led to correlations ranging from 0.31 to 0.79. Other studies of insect wings have similarly found moderate to high correlations between covariance matrices of FA and variation (Klingenberg and McIntyre, 1998; Klingenberg and Zaklan, 2000; Klingenberg et al., 2001). But variance and FA do not always show a strong relationship to each other. In an exceptionally extensive analysis, Dworkin (2005b) found that genetic and environmental perturbations can have a profound impact on variance, but genetic and environmental canalization appear to be independent of each other. In contrast to studies of insect wings, those of the mammalian skull usually find very low or even non-significant matrix correlations between phenotypic variance and FA (Debat et al., 2000; Willmore et al., 2005; Breno et al., 2011). For example, using a quantitative-genetic analysis, Willmore and colleagues estimated the correlation between FA and the environmental component of variation at just −0.07; the highest correlation was between FA and phenotypic variation, which was merely 0.049.

Studies that compare levels of FA to variance, both genetic and environmental, address one form of the hypothesis of a general buffering mechanism. But the hypothesis could also be framed in terms of the developmental pathways being buffered. Even if there is no general relationship between macro- and microenvironmental, and/or between genetic canalization and developmental stability, there may still be general buffering mechanisms that indiscriminately buffer a developmental pathway against all sources of perturbations. But the results obtained by comparing covariance matrices of FA and variance may actually tell us more about the mechanisms of morphological integration than buffering.

PREDICTING THE STRUCTURE OF COVARIATION: MORPHOLOGICAL INTEGRATION AND MODULARITY

Morphological integration and the related property of modularity have long been the focus of quantitative evolutionary developmental biology. Recent developments in evolutionary theory have stimulated a resurgence of interest in both because of growing interest in "evolvability", i.e. the ability to evolve. The fact that organisms are able to evolve had not prompted much theoretical attention until recently, perhaps because that ability has been taken for granted (Wagner and Altenberg, 1996; Hansen, 2003). After all, it is obvious that organisms can evolve because they do. What raised questions about the ability to evolve was the miserable failure to produce functioning computer programs by the processes of mutation and selection of code, which randomized their behavior rather than improved them (for an overview of this work and its relationship to evolutionary biology, see Wagner and Altenberg, 1996). Not surprisingly, research in evolutionary computing turned to the question of what could enable programs to evolve. This question, in turn, prompted questions about what enables organisms to generate selectively useful variation?

Modularity is now regarded as one of the key attributes of evolvable systems because it makes it possible to improve one part (of both computer code and morphology) without interfering with already optimized parts. The integration of adaptively interdependent traits within modules, and the (quasi)-autonomy of individual modules, enables one functional complex to evolve when others are under stabilizing selection. This theory is the basis for one definition of "modularity", one which incorporates the idea of "selectively useful" variation into the definition of modularity itself. According to this definition, modules comprise traits that collectively serve a primary function, with different modules serving different primary functions (Wagner, 1996). Each complex is internally integrated due to the same genes affecting multiple traits within the complex and the complexes are genetically independent, or nearly so. This definition of modularity is represented by a classic diagram (Figure 12.11; after Wagner, 1996; Wagner and Altenberg, 1996) which shows two functions, Function 1 and Function 2, each served by multiple traits (T1–T7), with each trait being affected by many genes (G1–G6), most of which affect more than one trait. According to this diagram (and to the theory it represents), a gene typically affects two or more traits within a single module, with few genes affecting traits within different modules. This diagram presents a sharp contrast to one long-standing view of genetic architecture – universal pleiotropy. The idea of universal pleiotropy raised questions about the causes of uncorrelated traits; the explanation for the lack of a correlation is that positive and negative pleiotropic effects cancel out, i.e. "antagonistic pleiotropy", and it is one contrast to the theory portrayed by the diagram: independent traits are independent because pleiotropic effects are restricted to subsets of traits. As emphasized by Mezey

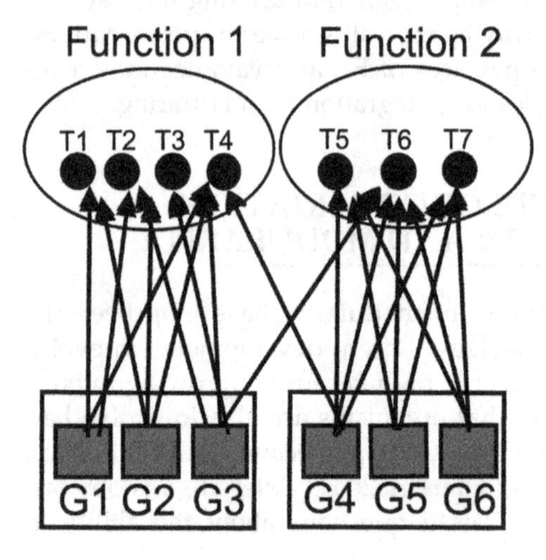

FIGURE 12.11 The classic depiction of modularity (after Wagner, 1996; Wagner and Altenberg, 1996) as the restriction of pleiotropic effects to complexes of traits serving the same primary function. Shown are two functional complexes, Function 1 served by traits T1–T4, and Function 2, served by traits T5–T7. The traits serving Function 1 are affected by genes G1–G3, all of which affect multiple traits within the complex and, with the exception of G3 that affects T6 belonging to the complex serving Function 2, all the affects of G1–G3 are restricted to the traits serving Function 1. Similarly, the traits serving Function 2 are affected by genes G4–G6, all of which affect multiple traits within the complex and, with the exception of G4 that affects T4 belonging to the complex serving Function 1, all the affects of G4–G6 are restricted to the traits serving Function 2.

and Houle (2003), when the organization of the genotype–phenotype map is modular, the effects of a group of genes are limited to distinct aspects of the phenotype.

The idea that effects of groups of genes are restricted to functionally coupled traits is not always incorporated in the definition of modularity. One alternative definition emphasizes the structural property of modularity, i.e. that modules are highly integrated internally and (quasi)-autonomous with respect to other modules. By this definition, it is not necessary for modules to contain functionally related traits, but it is necessary that they be conditionally independent of each other. The requirement of conditional independence also comes from the theoretical analysis of evolvability; Hansen and colleagues (Hansen, 2003; Hansen et al., 2003) define the "relevant evolvability" of a trait as its ability to respond to directional selection when the other traits are under stabilizing selection. Thus, it is the ability of one trait (or complex) to evolve when others are held constant that matters to evolvability. Two traits might be correlated and therefore each seems to lack the independence required to evolve individually when the other is held constant, but if their correlation is due to the mutual dependence on some other trait, the two may be independent when holding that third trait constant. If so, they are conditionally independent. A strong and purely structural definition of modularity is that modules comprise traits that are all mutually informative (conditionally dependent) and conditionally independent of the traits within other modules (Magwene, 2001). This structural definition is depicted in Figure 12.12. This figure shows a graph, in which each trait is a node; the edges between nodes connect conditionally dependent traits. In the absence of an edge, the traits are conditionally independent. Within each module there is an edge between every pair of traits, but between modules there are no edges. A weaker structural definition of modularity relaxes the requirement that *all* the traits within a module be mutually informative.

A third definition of modularity reframes the concept by highlighting its developmental origins (Figure 12.13; after Klingenberg, 2008). One notable distinction between this diagram and the one shown in Figure 12.11 is that this one replaces functional with developmental modules (M1, M2). A second distinction is that this diagram shows genes affecting developmental pathways rather than traits – the dotted lines show the genetic effects, the solid lines show the architecture of the pathways. Adding the pathways to the diagram rather than leaving them implicit makes it possible to depict two ways in which genetic correlations arise developmentally. One is by the same gene being expressed at two (or more) times or places, which Klingenberg terms "parallel variation". The other is by direct

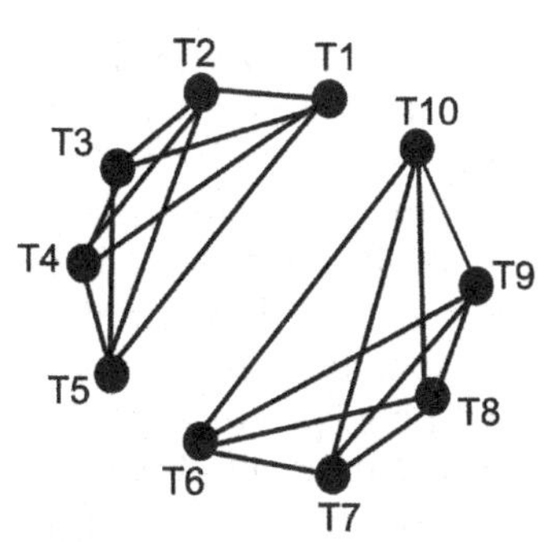

FIGURE 12.12 The structural concept of modularity. The nodes of the graph represent traits (T1–T10); those that are independent of each other, controlling for all others, are connected by an edge. Modules comprise traits that are all directly connected to each other. No edges connect the two modules; they are conditionally independent.

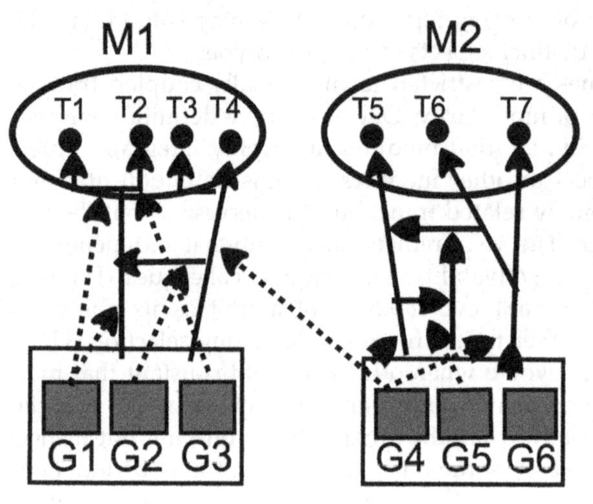

FIGURE 12.13 A developmental reframing of the classic concept of modularity, giving a developmentally explicit version of the genotype–phenotype map (after Klingenberg, 2008). There are two developmental modules (M1, M2) comprising two groups of traits (T1–T4, T5–T7). Genes affect developmental pathways rather than traits; the impact of the gene on a pathway is shown as a dotted line intersecting a pathway, shown as a solid line. A direct interaction between pathways is shown as an intersection between solid lines.

interactions within and between pathways, such as signaling interactions, partitioning of tissue, or other mechanisms that can transmit variation along and between pathways. The distinction is important because, in the case of parallel variation, the source of the variation is also the cause of the correlation. That is not the case for direct interactions. In the case of direct interactions, the cause of the correlation is the mechanism that regularly associates the development of the two traits. If variation arises upstream of a branch in the pathway, the variation will be transmitted downstream along both branches. Even if the variation is due to a random developmental perturbation, the variation will be transmitted downstream or from one pathway to another via a signaling interaction. Modules are thus highly integrated internally due to many, often strong, direct interactions within them, exceeding those that take place between modules (Klingenberg, 2005, 2008). This developmental view of modules not only reframes the hypothesis of modularity, it also yields a novel technique for testing hypotheses about modularity. Because direct interactions within developmental modules can regularly associate traits even when the source of the variation is a random developmental perturbation, fluctuating asymmetry becomes a useful tool for analyzing the structure of developmental modules. When FA and individual variation are highly similar in structure, direct interactions within developmental modules play a large role in integrating the phenotype (Klingenberg et al., 2001; Klingenberg, 2005, 2008). Whether the modular organization of development is an intrinsic feature of developmental systems and a potential constraint on the adaptive evolution of integration (Klingenberg, 2004, 2005, 2008) are open and provocative questions.

To this point, we have not tried to define morphological integration except in terms of modularity. But the ideas of integration and modularity can be partially separated. They are not separate when variation actually does have a modular structure, but variation might not actually be modular. It may be that genetic effects are spatially restricted but continuous rather than tightly clustered and even partially overlapping, producing what Roseman and colleagues (Roseman et al., 2009) termed "integration without modularity". Hallgrimsson has offered a developmental explanation for a non-modular organization of variation; he emphasizes the dynamics and complexity of development, and the fact that the structure of variation is an outcome of multiple variance-generating processes (Hallgrímsson et al., 2007a,b, 2009). Those processes may partially overlap spatially, amplifying each other in some regions and canceling out in others. Even if each individual process is modular, their net effect on variation need not be. We might also anticipate a non-modular organization of variation when functional complexes are not modular (Zelditch et al., 2008, 2009). Due to the dynamics and complexity of development, and/or the functional organization of morphology, we might find integration but not modularity.

The methods that we describe below test hypotheses of modularity. Although there are methods for exploring the data to find the best-fitting model (one of these is discussed below), the methods are used primarily to test hypotheses derived from developmental biology, functional morphology or any other source of theory that predicts the structure of variation. An important methodological consideration when it comes to choosing a method is the array of hypotheses against which your *a priori* hypothesis is tested. Regardless of the method you choose, it is important to remember that the best-fitting hypothesis is the best in a specific context – the alternatives that you entertained.

A Brief Overview of Methods for Analyzing Modularity

All three of the methods that we describe below require stating a hypothesis that predicts the modular structure of the data. In all three cases that is done by subdividing a configuration of landmarks into two or more subsets. For example, we can subdivide the mandible into two subsets of landmarks (plus semilandmarks) according to the hypothesis of mandibular modularity favored by quantitative genetic studies. The hypothesis posits two modules, one the tooth-bearing region, the other the muscle-bearing region (Cheverud et al., 1997, 2004; Mezey et al., 2000; Ehrich et al., 2003; Cheverud, 2004; Klingenberg et al., 2004). The division is usually made at the point where the molar alveolus separates from the coronoid process, and where the angular process can be distinguished from the horizontal ramus (Figure 12.14). (For brevity, we refer to this as the Front/Back model.)

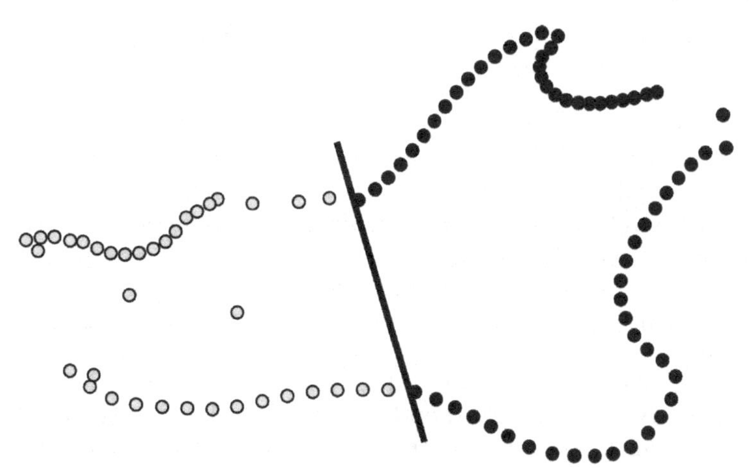

FIGURE 12.14 A hypothesis of mandibular modularity. The Front/Back hypothesis predicts that there are two modules, one comprising the tooth-bearing region of the jaw, the other comprising the muscle-bearing region of the jaw.

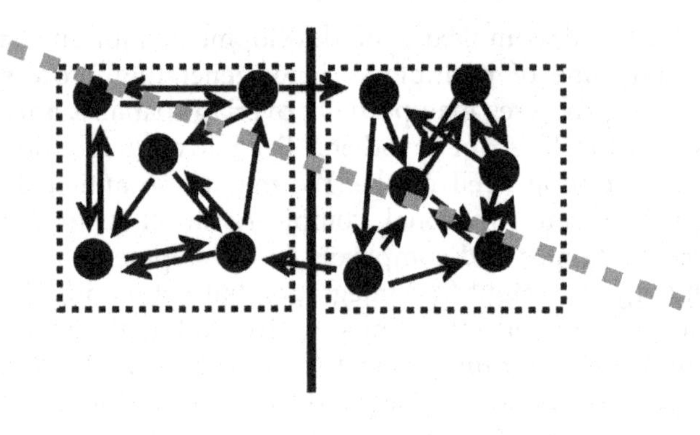

FIGURE 12.15 Logic of testing for low *RV*. Two developmental modules are shown, each enclosed in a rectangle, with the boundary between them represented by a black line. There are extensive interactions within each module, represented by the arrows connecting the landmarks. Few interactions occur between modules. The covariance between the two modules will be lowest if the two are separated along the black line, due to the few interactions between landmarks across that line. In contrast, the covariance between the two modules will be higher if the two are separated along the dotted gray line, due to the many interactions between the landmarks across that line.

One method for testing the hypothesis of modularity is to assess the covariance between the two modules relative to the covariance within them (Klingenberg, 2009). If the hypothesized boundary between the modules is correctly positioned, the covariance between the two blocks should be lower than the covariance obtained by any alternative partitioning, subject to the constraint that the alternative also contains two blocks having the same number of landmarks as in the proposed modules. The logic of the test is depicted in Figure 12.15 (after Klingenberg, 2008). The black line separates the two actual modules, one containing five landmarks, the other six. The covariances within each of the two subsets are high, as evident from the many arrows depicting interactions between them. Interactions between the two modules are few, so the covariance between them is low. The gray line separates the two subsets of landmarks that are hypothesized to be modules, one containing five landmarks, the other six. But the line separates landmarks belonging to the same actual modules. As a result, the covariances between the hypothesized modules will be high, far higher than the covariance that would be obtained by partitioning the landmarks along the black line. Because the covariance between modules should be lower than the covariance between randomly partitioned subsets of landmarks, the testing procedure determines whether the covariance between the hypothesized modules is significantly *lower* than expected by chance. This method, like the other two, has no name, so we will call it the "Minimum intermodular covariance method".

The second method tests a hypothesis of modularity by producing the covariance matrix predicted by the hypothesis and assessing the goodness of fit (Marquez, 2008). The covariance matrix predicted by the model is estimated by making the modules statistically independent of each other – they are placed into orthogonal subspaces. These intermodular covariances are fixed but the within-module covariances are estimated from the data. Having produced the covariance matrix predicted by the model, it can then be compared to the observed covariance matrix. The null hypothesis is that the difference between the observed and expected covariance matrices is no greater than expected by chance. This is tested by comparing the difference between observed and expected matrices to the range of values for the difference that could be obtained when the null hypothesis is true. The method allows many models to be fit to the data, with the best-fitting model being the one that deviates least from the data, taking into account the number of parameters fixed by the hypothesis. The number that is fixed is important because the free parameters (the intramodular covariances) are estimated from the data; models that have relatively few fixed parameters will always fit well. We will call this method the "Minimum deviance method".

The third method differs from the other two in two major respects. First, it produces a correlation matrix rather than a covariance matrix and second, it works with distance matrices rather than coordinates (Monteiro et al., 2005; Monteiro and Nogueira, 2009). The correlations are obtained by subdividing the data into modules, then calculating the pairwise Procrustes distances between all individuals for each module. Then the correlation between the distance matrices is estimated. For example, given the hypothesized two modules of the Front/Back model, we would divide the coordinates into those two subsets and calculate the pairwise Procrustes distances between all individuals for each subset of landmarks and then compute the matrix correlation between those two distance matrices. In this simple case of only two modules, we could test the hypothesis that they are independent of each other by the Mantel test. For more complex hypotheses, we could use any of the methods conventionally used in studies of morphological integration to evaluate the fit of the hypotheses to the correlation (or inverse) correlation matrix. We could also use exploratory methods to select the best-fitting model. We will call this method the "Distance-matrix method".

After discussing each method in more detail, we will apply all three of them to evaluate four hypotheses of mandibular modularity.

Minimum Intermodular Covariance Method

The test for modularity based on estimating the covariances between modules relative to those within modules was devised by Klingenberg and implemented in his software, MorphoJ (Klingenberg, 2011). The test statistic is Escoufier's *RV* coefficient (Escoufier, 1973). This was introduced in Chapter 7 (Partial Least Squares), but repeated here for convenience: the *RV* is a multivariate extension of the ordinary univariate squared correlation:

$$RV = \frac{trace(R_{12}R_{12}^t)}{\sqrt{trace(R_1R_1^t)trace(R_2R_2^t)}} \tag{12.5}$$

The numerator is the summed squared covariances between the two sets of variables and the denominator is the square root of the product of the summed squared variances within each block. *RV* ranges from 0 (no covariance) to 1 (complete covariance). When we introduced the *RV* in Chapter 7, we used it to test the hypothesis that the two blocks of landmarks covary so we tested the null hypothesis that the observed *RV* is no higher than expected by chance. But, as discussed above, in the context of a study of modularity, the expectation is that the *RV* will be *lower* than expected by chance.

The details of the test depend on two decisions that you make. The first is whether hypothesized modules should be treated as separate shapes or as parts of a whole. The distinction lies in how the test will use information about the connections between the subsets (e.g. the relative sizes of the partitions and their position relative to each other). In terms of the details of the method, the distinction lies in whether each module is superimposed separately or the entire configuration is superimposed only once. When each module is superimposed separately, information about the relative sizes and positioning of the modules is removed from the data. What matters is the covariance between shapes. When the variation in one shape is associated with the variation in another, that will generate covariance between the modules. In contrast, when the subsets are regarded as parts of a whole, any variation in the relationship between that generates covariances in relative sizes and positions is retained in the data. As Klingenberg points out, there is no right or wrong decision about this. In the case of the mandible, we might base the decision on considerations of function. Because the mandible is functionally a lever, it makes sense to treat these two parts of the lever as two parts of a whole. Alternatively, we could decide that function is immaterial and disregard the relationship between the two parts, testing the hypothesis that dividing the mandible into front and back produces a lower covariance between the two shapes than any alternative division of the mandible into two parts, having the same number of landmarks within them as our hypothesized modules.

The second decision is whether modules must be spatially continuous. What spatially continuous means, in this context, is that a module comprises all the landmarks that are adjacent to each other except for those on the boundaries between modules. More precisely, the definition of spatial contiguity for partitions of landmarks uses the graph theoretic concepts of node and edge introduced above in the context of a structural concept of modularity. When defining contiguity of landmarks, the nodes on the graph represent the landmarks and the edges connect them; a set of landmarks is spatially contiguous if every landmark within the set is connected to every other, either directly (by an edge between that pair of landmarks) or indirectly through the other landmarks in the set. It is thus possible to reach every landmark within the set by moving along the edges. This decision about continuity also has no right or wrong answer. It is reasonable to anticipate that developmental interactions act over spatially continuous regions, but those regions may not remain continuous through the whole course of development – they might be interrupted by morphogenetic movements, outgrowth or cell death. Also, in cases like the mandible, there is a landmark within the tooth-bearing region (the one on the masseteric fossa) that is a muscle insertion site so it might plausibly be regarded as part of the muscle-bearing region. The reason why this decision about continuity matters to the method is that the *RV* for the hypothesis will be compared only to the random, continuous (also called "contiguous") modules if continuity is a requirement for modules. For purposes of the example, we will restrict the analysis to contiguous modules.

When comparing the *RV* to all (or a subset) of possible alternatives, the comparisons are restricted to partitions that have the same number of landmarks as in the hypothesized modules. For example, if a hypothesis proposes that there are two modules, one having 10 landmarks and the other 15, all the alternative hypotheses will also comprise two modules, one having 10 landmarks and the other 15. If there are few landmarks in the configuration, it is possible to compare the *RV* for the data to all the possible alternatives. In the case of two partitions, one having p landmarks, the other $m-p$, the number of possible partitions is

$$\binom{m}{p} = \frac{m!}{(m-p)!p!} \tag{12.6}$$

With few landmarks, exhaustive enumeration is feasible, but if there are more than a few (e.g. >20) the number of alternatives becomes enormous. The requirement that modules be contiguous will result in far fewer than the total number of alternatives, but the number could still be very large. Rather than use exhaustive enumeration, the alternatives can be randomly sampled; Klingenberg recommends at least 10 000 permutations because we are interested in the left tail of the distribution.

The permutation procedure is straightforward when the subsets are separately superimposed but it becomes more complex when the subsets are simultaneously superimposed. When they are separately superimposed, the hypothesis of independence between the subsets can be tested by randomly permuting the observations in the two sets of landmarks. At each iteration of the procedure, the observations in one subset are randomly permuted and the *RV* is calculated; its statistical significance is assessed by the proportion of the cases in which the observed *RV* is equal to or higher than the observed one. The procedure is more complex when the subsets are simultaneously superimposed because the test must take into account the interdependence between partitions produced by the superimposition procedure. The procedure is thus modified to include a new Procrustes superimposition at each iteration, so the observations are randomly permuted in one of the two subsets, then they are combined into a single configuration. It is not likely that they are still optimally superimposed, so the superimposition is redone and the *RV* of this re-superimposed configuration is compared to the observed one. This whole procedure – random permutations of one of subset followed by a re-superimposition of the data, is done at each iteration.

To this point, we have talked about the analysis of just two modules, but the analysis is not limited to a two-module case even though the *RV* measures the covariance between two blocks of data. Klingenberg (2009) extended it to a multiblock case, introducing the multiset RV_M coefficient, which is the average of all the pairwise *RV* coefficients.

$$RV_M = \frac{2}{k(k-1)}\sum_{i=1}^{k-1}\sum_{j=i+1}^{k} RV(i,j) \tag{12.7}$$

where k is the number of subsets of landmarks and $RV(i,j)$ is the *RV* coefficient for the subsets i,j. Just like the pairwise *RV*, the multiset RV_M coefficient can be tested against the null hypothesis that the modules are independent, providing a test of overall integration. The test is done by computing RV_M for the hypothesized modules after which the landmarks of all but one subset are randomly permuted (and re-superimposed if the analysis is done using simultaneously superimposed landmarks).

Minimum Deviance Method

The second method assesses the goodness-of-fit of a model to the observed covariance matrix. This method, introduced by Marquez (2008) and subsequently modified to improve the assessment of relative fit (Parsons et al., 2012), is implemented in Mint (Marquez, 2012). As outlined briefly above, the expected covariance matrix is modeled by placing modules into orthogonal subspaces, one per module. This is done by making as many copies of the data as there are modules, and assigning a value of zero to the coordinates that do not belong to the hypothesized module. For example, given the hypothesis that the front and back are two modules, with landmarks 1, 2, 3, 4, 5, 6, 7 and 13 in the front and landmarks 8, 9, 10, 11, 12, 14 and 15 in the back, we would make two copies of the data and arrange them in the extended matrix of the Front/Back Model:

$$\text{Front/Back Model} = \begin{bmatrix} 1 & 2 & 3 & 4 & 5 & 6 & 7 & 13 & 0 & 0 & 0 & 0 & 0 & 0 & 0 \\ 0 & 0 & 0 & 0 & 0 & 0 & 0 & 0 & 8 & 9 & 10 & 11 & 12 & 14 & 15 \end{bmatrix}$$

Each of the numbers represent the x and y (and, if present, z) coordinates for that landmark. The values for the coordinates that belong to a module are taken from the data. The values for the coordinates that do not belong to the model are fixed to 0, 0. Before assessing the fit of the model, the matrix predicted under the model (e.g. Front/Back Model) is superimposed. The procedure is somewhat more complicated when the hypothesis predicts that some landmarks belong to two or more modules; the modules then partially overlap each other hence the subspaces are not orthogonal to each other and the variances of the overlapping landmarks must be allocated to multiple modules without altering the overall value of the variance. As currently implemented in Mint, the variance of the landmark's coordinates is equally partitioned across all the modules that contain that landmark.

The fit of the model to the data can be assessed by several metrics (and Mint offers three). We describe only one of them, the one that is fully standardized to allow for comparing results across data sets and also for assessing the relative fit of models that differ in the number of modules. This goodness-of-fit statistic is γ:

$$\gamma = \text{trace}((S - S_0)(S - S_0)^T) \tag{12.8}$$

where S and S_0 are the observed and expected (modeled) covariance matrices, respectively (Richtsmeier et al., 2005). For example, in the case of the Front/Back model, S_0 is the covariance matrix of the Front/Back Model. To make it comparable across data sets, γ is scaled by its maximum value, γ_{max}, by dividing γ for each model by γ_{max}, which is computed by comparing the data to the null model of "no integration". That null model is a diagonal matrix that has the variances of the coordinates along the diagonal and zeros for all the off-diagonal elements, which are the covariances. The second scaling step removes the dependence of γ on the number of fixed parameters (the landmarks whose coordinates are fixed to zero by model). This makes γ comparable across models that differ in the number of fixed parameters. The scaling is done by regressing the value of γ on the number of zeros within each model because γ is linearly related to the number of zeros. The residual from that regression, γ^*, can then be compared across models. When the model predicts that the covariances are zero and the observed values are indeed low, $\gamma^* < 0$; conversely, when the model predicts that the covariances are zero but they are actually high, $\gamma^* > 0$. Because this scaling step uses a regression, the method benefits from fitting *many* models to the data.

When testing model(s), the null hypothesis is that the difference between the observed and expected covariance matrices is no greater than expected by chance. A low probability means that the model does *not* fit the data. The test is done by comparing the observed value of γ^* to the range of values that could be obtained when the null hypothesis is true. To obtain the distribution of γ under the null hypothesis, the modeled covariance matrix and sample size are used to parameterize a Wishart distribution, which is the distribution of covariance matrices of a multivariate normal population (Wishart, 1928). The value of γ^* is calculated between each randomly drawn matrix and the expected matrix. The probability that the model differs from the data by no more than expected by chance is calculated from the proportion of cases in which the value of γ^* (computed by comparing the model to randomly drawn matrices) is larger than the observed one. Thus, a p-value of 0.9 means that in 90% of the cases in which the model is compared to random matrices, γ^* is larger than it is when the model is compared to the data.

When comparing multiple models, the best-fitting model is the one with the lowest γ^*. Multiple models, however, might be nearly equal in γ^* and all might fit well. All the expected covariance matrices might deviate little from the observed one. Then, the problem is to decide which model fits best. In principle, this can be decided by which has the lowest γ^*, but we may not be entirely confident either in γ^* or in its rank because both depend on the sampling of the observed covariance matrix. Before deciding that one of the models fits best, we want to be confident in the ranks of the models. To that end, we can resample the covariance matrices and rerun the analyses for each sample, calculating the number of runs in which the ranking is the same as we obtained for the observed covariance matrix. As implemented in Mint, the resampling is done by jackknifing the data, leaving out a proportion of the sample, refitting the model to the data, recalculating γ^* and the ranks of the models at each iteration.

Distance-Matrix Method

This method for analyzing modularity using correlations between pairwise Procrustes distance matrices was introduced by Monteiro and colleagues (Monteiro et al., 2005, Monteiro and Nogueira, 2009). As briefly outlined above, this method produces a correlation matrix from the matrix correlations between pairwise Procrustes distance matrices. Those Procrustes distance matrices preserve the information about the structure of variation within each module. If the analysis is done using Procrustes distances calculated separately for each module, the only information retained is the correlations between the shapes; any information about the relationships between the relative sizes and positions of the modules within the whole is disregarded. That is the procedure implemented in Coriandis (Marquez and Knowles, 2007). But it is possible to retain the information about the relative sizes and positions of the modules by not superimposing the modules separately after subdividing the subsets of landmarks. Rather than computing Procrustes distances, we can instead calculate Euclidean distances between each pair of individuals for each module (implemented in an R script in the workbook accompanying this text). Given those distance matrices, we can then calculate the correlation between each pair, which tells us whether variation in shape shows the same pattern for both modules. For example, if individuals who are most different from each other in the shape of one module are also most different in the shape of the other module, and those who are most similar to each other in the shape of one module are also most similar to each other in the shape of the other module, the correlation between the modules will be high. Should individuals who are most similar to each other in the shape of one module be the *least* similar to each other in shape of the other, the correlation will be high but negative. The correlation will be near zero when similarities among individuals in shape of one module do not predict similarities in shape of another module.

An important difference between this method and the other two that were introduced above is the treatment of correlations *within* modules. Using the present method, those intramodular correlations are not assessed – only the correlations between modules enter into the analysis. To overcome that limitation, each putative module can be subdivided into two or more parts, and the correlations between the parts of a module can then be assessed relative to the correlations between modules (Zelditch et al., 2008, 2009). However, if those correlations are included in the analysis, the strength of the intramodular correlations is not taken into account when analyzing the correlations between modules. In effect, the hypothesized intramodular correlations are treated no differently than the hypothesized intermodular correlations.

Once we have the correlation matrix we can analyze it by any of the methods conventionally used for testing hypotheses of morphological integration and modularity (Cheverud, 1982; Cowley & Atchley, 1990; Cheverud, 1995; Herrera et al., 2002; Young & Hallgrímsson, 2005). One method for assessing a hypothesis of modularity is to predict that the correlations between modules are zero, allowing the intramodular correlations to be estimated from the data. The question is whether this model fits the data. We can compare this to a model that fits the data perfectly and contains no fixed values. This model is usually termed the "saturated model", which can be represented as a graph in which the nodes are the subsets of landmarks and the edges between them are the correlations between the subsets and all the nodes are connected to all others (Figure 12.16). Figure 12.16 shows the saturated model for 12 subsets because the hypothesized modules were divided into two parts, whenever possible. Our objective is to reproduce the observed correlation matrix using as few edges as possible. To assess the fit of the model that includes only some of the edges relative to this saturated model, we use a measure of the deviance (D) between the models (Box, 1949; McCullagh & Nelder, 1989). D is -2 times the log-likelihood ratio of the model being tested compared to the saturated model:

$$D(y) = -2(\log(p(y|\hat{\theta}_o))) - \log(p(y|\hat{\theta}_s)) \tag{12.9}$$

where $\hat{\theta}_o$ are the fitted parameters of the model being tested and $\hat{\theta}_s$ are the parameters of the "saturated" model, the one shown in Figure 12.16. D is approximately distributed as a chi-square with degrees of freedom equal to the difference in the number of parameters in

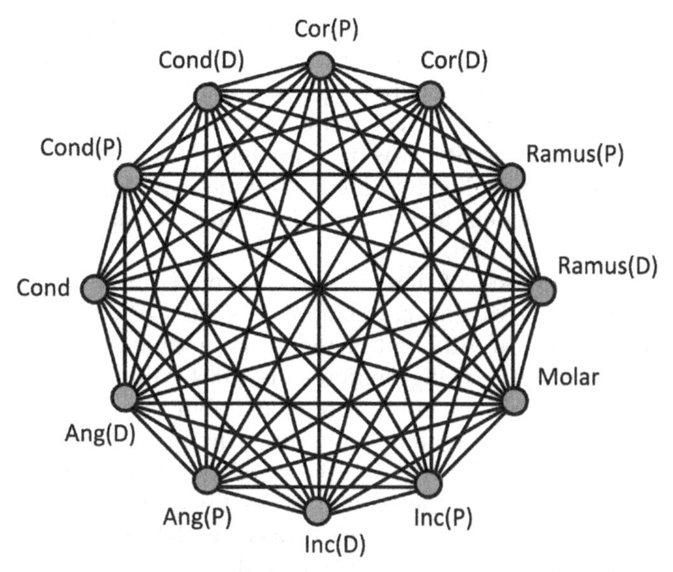

FIGURE 12.16 The saturated model. This model fits the data perfectly because all the edges are free parameters, estimated from the data. Its deviance is therefore zero and it has zero degrees of freedom. This is the null model; the objective of the analysis is to reproduce the observed covariance with as few edges as possible.

the two models. The saturated model has zero degrees of freedom, so the degrees of freedom for the chi-square are the number of fixed parameters in the model.

Models that contain few fixed parameters are likely to fit well, so we are looking not only for a model that fits well but one that fits well using as few edges as possible. When models are nested, meaning that one model is included within the other, we can compare the two by the chi-square difference test, subtracting the chi-square of the more complex model from that of the simpler model, and also subtracting the degrees of freedom of the more complex model from those of the simpler model. The resultant Δchi-square is distributed as a chi-square with the degrees of freedom given by the difference in degrees of freedom of the two models. If statistically significant, the more complex model improves upon the simpler one. When models are not nested, we need an alternative approach for judging the relative fit of two models. One is to use the Akaike Information Criterion (AIC), ranking models by their AIC. The AIC was introduced in Chapter 11, but we summarize it here for purposes of convenience. AIC is a function of the log-likelihood of the parameters given the data and the number of parameters in the model (Akaike, 1974), calculated as:

$$\text{AIC} = 2k - 2\ln(\text{likelihood}) \tag{12.10}$$

where k is the number of parameters in the model. To compare models, we can compute the difference in their AIC (ΔAIC).

We can also use exploratory methods to find the model that reproduces the observed correlation matrix using as few edges as possible. Alternatively, we can search for a model that reproduces the inverse correlation matrix, which gives the pairwise correlations holding all other variables constant. When the analysis is done using the inverse correlation matrix, the models are usually termed "Gaussian graphical models" and the search for the best-fitting model is called "concentration model selection". When the models are instead fitted to the correlation matrix, the models are termed "covariance graphical models" and the search for models that reproduce the correlation matrix is termed "covariance model selection". Using the inverse correlation matrix has the advantage that it allows us to test for the conditional evolvability of modules; the use of covariance graphical models for studies of morphological integration and modularity was recommended by Magwene (2001). But, for purposes of comparison with the other two methods, when we apply this method to our data, below, we will fit the models to the correlation matrix and do an exploratory analysis using covariance model selection.

An alternative method for testing hypotheses of integration and modularity is to construct the expected correlation matrix by predicting correlations of either zero or one between traits according to whether they belong to different or the same module (Wagner, 1988; Kingsolver & Wiernasz, 1991; Cheverud, 1995; Hallgrímsson et al., 2004; Young, 2004). We do not anticipate that correlations will typically be either zero or one, but the comparison between the observed and expected matrices is done by computing the matrix correlation between them (which is tested by a Mantel test). The matrix correlation will be high if the matrices are proportional to each other so it is the pattern of relatively high versus relatively low correlations that is tested. In the case of the Front/Back model, we would predict the correlation matrix shown in Table 12.4. When comparing the hypothesized to observed matrices by the Mantel test; we would randomly permute the rows and columns in one matrix and compute the correlation between matrices at each for each permutation.

TABLE 12.4 Correlation Matrix Among Six Partitions of the Mandible Predicted by the Front/Back Hypothesis

	IncD	IncP	Molar	RamusD	RamusP	CorD	CorP	CondD	CondP	Cond	AngD	AngP
IncD	1	1	1	1	0	0	0	0	0	0	0	0
IndP	1	1	1	1	0	0	0	0	0	0	0	0
Molar	1	1	1	1	0	0	0	0	0	0	0	0
RamusD	1	1	1	1	0	0	0	0	0	0	0	0
RamusP	0	0	0	0	1	1	1	1	1	1	1	1
CorD	0	0	0	0	1	1	1	1	1	1	1	1
CorP	0	0	0	0	1	1	1	1	1	1	1	1
CondD	0	0	0	0	1	1	1	1	1	1	1	1
CondP	0	0	0	0	1	1	1	1	1	1	1	1
Cond	0	0	0	0	1	1	1	1	1	1	1	1
AngD	0	0	0	0	1	1	1	1	1	1	1	1
AngP	0	0	0	0	1	1	1	1	1	1	1	1

IncD: distal incisor partition; IncP: proximal incisor partition; Molar: molar incisor partition; RamusD: distal horizontal ramus partition; RamusP: proximal horizontal ramus partition; CorD: istal coronoid process partition; CorP: proximal coronoid process partition; CondD: distal condyloid process partition; CondP: proximal condyloid process parition; Cond: condyle; AngD: distal angular process partition; AngP: proximal partition of the angular process. See Figure 12.20.

These two approaches to assessing the fit of a model to data can lead to conflicting results because the model deviance increases when the predicted zero correlations are actually high but not when the predicted high correlations are actually low. That is because the predicted zero correlations are fixed to zero whereas the predicted high correlations are estimated from the data. The consequence of erroneously predicting a high correlation is to use up a degree of freedom unnecessarily. A matrix correlation, however, can decrease when observed correlations are either lower or higher than predicted.

Examples: Evaluating Four Hypotheses of Mandibular Modularity

The first model that we will test is the Front/Back model (Figure 12.17A). The other three are derived from developmental biology. We select these four from the large array of hypotheses that could be derived from developmental biology and functional morphology because all four can be tested by all the methods. Additionally, three of the models are compatible with each other but the fourth is not compatible with any of the others. One developmental model is based on a proposal by Fish and colleagues (2011) that the *Satb2*-positive cell population is a developmental (and macroevolutionary) module. Fish and colleagues propose that the mandible can be divided into four modules, identifying them with skeletal units. One corresponds to the "Back" of the Front/Back model, the other three are subdivisions of the front into three modules. One of these is the small distal region that gives rise to the mandibular symphysis, the second is the incisor alveolar module and the third the molar module, which includes the molar alveolus and the portion of the ramus ventral to that (Figure 12.17B). The second model derived from developmental biology (Figure 12.17C) contains the three Front modules of the *Satb2* hypothesis, adding one to the back, corresponding to the expression domain of *goosecoid*, which has also been proposed to be a developmental module (Gaunt et al., 1993). The fourth hypothesis (Figure 12.17D) is the one proposed by Atchley and Hall (1991), who regarded mesenchymal condensations as the basic units of mandibular morphogenesis and condensations have been explicitly identified as modules (Hall and Miyake, 2000; Hall, 2003). This one

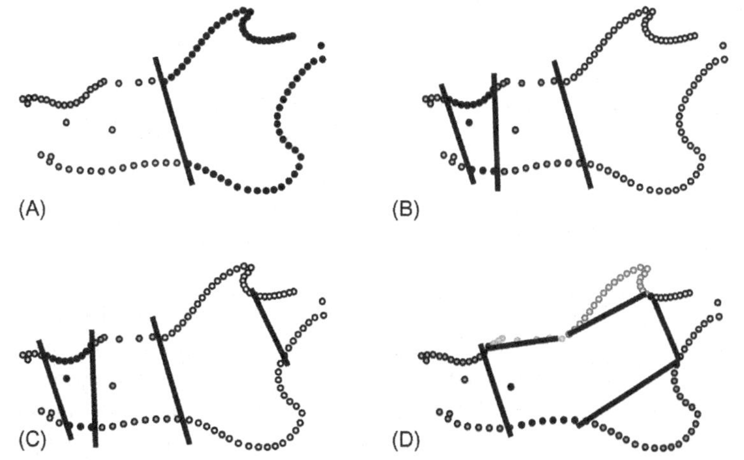

FIGURE 12.17 Four hypotheses of mandibular modularity. (A) Front/Back model; (B) *Satb2* model; (C) *Satb2* + Gsc model; (D) Condensation model. The black lines show the subdivisions between the modules.

differs from the others because there is no division of the front from the back, and there is a division between molar alveolus and ramus. There are six modules because six condensations give rise to the structures of the adult mandible (the seventh gives rise to the transitory Meckel's cartilage and the symphyseal region). The six modules are the incisor alveolus, the molar alveolus, the horizontal ramus (which crosses the boundary between front and back), and three modules corresponding to the three proximal processes (coronoid, condyloid and angular).

To test these four hypotheses of modularity, we use a simultaneous superimposition method because all the methods can analyze simultaneously superimposed data; Mint cannot analyze separately superimposed data because that would force the analysis to focus solely on the between-modular associations, i.e. on whether the covariances between modules are low enough to justify the conclusion that they are indeed modules. Second, because the mandible serves a mechanical function, the relative sizes and positions of its parts are functionally important.

"Minimum Intermodular Covariance Method"

The Front/Back model can be tested by the RV coefficient because it contains only two modules. The RV for this hypothesis is a moderately high 0.445 and it is highly statistically higher than expected by chance. Nevertheless, when tested using a large sample of random permutations (restricted to contiguous partitions with the same number of landmarks as contained in the hypothesized modules), none had a lower value. The distribution of the RV coefficient (Figure 12.18) including this one and the 10 000 random alternative partitions suggests that the alternatives vary little in RV and none of them differ much from the observed value. This is not surprising because the precise dividing line between the modules is biologically ambiguous (between which ventral semilandmarks should we divide the front from the back?). Many alternatives might be consistent with the biological hypothesis rather than conflict with it.

The remaining models are tested with the multiset RV_M coefficient. The RV_M for the *Satb2* hypothesis is moderately high 0.327, although lower than the RV for the Front/Back model. It, too, is the lowest in the distribution that includes it and RV_M of 10 000 random alternatives. The third hypothesis, *Satb2* + Gsc, yields a lower RV_M of 0.290, but it is not clear whether this lower value indicates a better-fitting model or is an artifact of the larger number of modules. The RV_M of this model is also the lowest in the distribution that includes it and 10 000 random (contiguous) alternatives. The Condensation hypothesis yields an RV_M of 0.230, which is the lowest of the four, and is also lower than all the values obtained by 10 000 random permutations.

"Minimum Deviance Method"

The rankings of the four models are shown in Table 12.5. These agree with the rankings based on the RV and RV_M. The best fitting is the Condensation model, which is ranked first with 100% jackknife support, followed by the *Satb* + Gsc model, which is consistently the second best, followed in turn by the *Satb2* model, which is consistently the third best, and then by the Front/Back model, which is consistently the worst. The difference

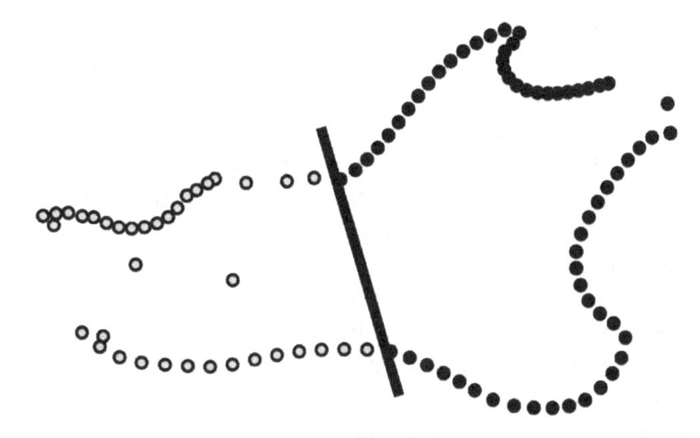

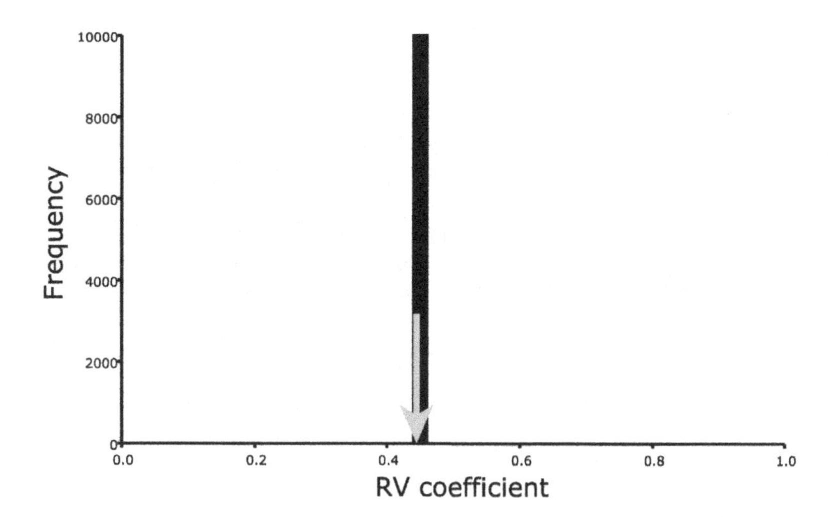

FIGURE 12.18 The distribution of the *RV* for the model and 10 000 randomly selected alternative subdivisions of the mandible into two parts having the same numbers of landmarks within each partition as contained in the model.

TABLE 12.5 Results of the Minimum Deviance Method for the Four Models of Mandibular Modularity Shown in Figure 12.16

Hypothesis	γ^*	p-Value	Rank	Jackknife Support
Null	0	0	5	100%
Front/Back	−0.2628	1	4	100%
Satb2	−0.2909	1	3	100%
Satb2/Gsc	−0.3217	0.999	2	100%
Condensation	−0.3522	0.388	1	100%

Shown are the measures of deviance scaled by the number of fixed parameters, γ^*, p-values for the null hypothesis that the difference between the modeled and observed covariance matrices is no greater than expected by chance, the ranks of the four models (with 1 being the best fitting and 5 being the worst fitting) and the jackknife support for the ranks.

between the models appears very small, so it is not clear whether the best is all that much better than the worst. However, we can evaluate these models against a much larger array of alternatives by constructing new models by mixing the modules proposed by the hypotheses into new combinations. When assessed in light of the 792 possible combinations, some of which differ only by the landmarks marking the ramal boundaries, the Condensation model is not the best of all the models. Instead, it ranks 16th. The next best fitting of the developmental models, the *Satb2* + Gsc model, ranks 203rd and the *Satb2* model ranks 363rd; the Front/Back model ranks 488th.

The best-fitting model of the 792 contains four modules, (Figure 12.19A), the small distal region of the mandibular symphysis, the molar alveolus, and the three proximal processes (coronoid, condyloid and angular). For this model, $\gamma^* = -0.3656$. It contains four modules of the Condensation hypothesis, but contrary to that hypothesis (and consistent with the *Satb2* and *Satb2* + Gsc models), it divides the incisor alveolus into two parts. The next best-fitting model ($\gamma^* = -0.3635$) differs from the best by including the horizontal ramus as a module.

Distance-Matrix Method

To examine the integration within modules as well as between, we subdivided each hypothesized module into two parts, whenever possible, producing 12 partitions (the

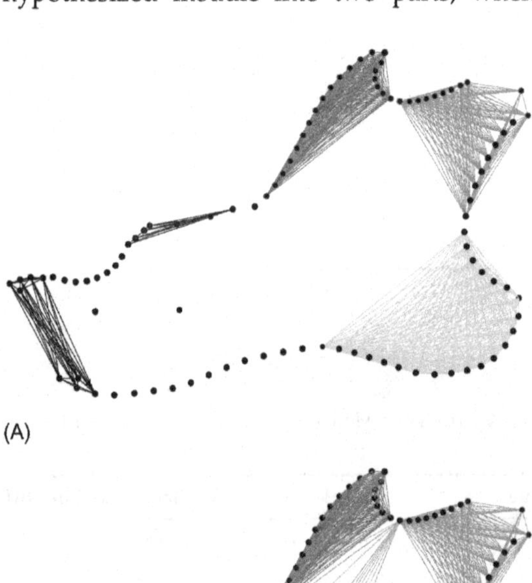

FIGURE 12.19 The two best models found by mixing the modules contained in the four *a priori* models. (A) Model ranked 1st; (B) model ranked 2nd.

molar alveolus could not be subdivided because it contains too few landmarks, and the condyloid was divided into three parts, one of which is the condyle). Three of the models fit equally well based on their nearly equal ΔAIC; the only one that we could at least tentatively reject from further consideration is the *Satb2* model (Table 12.6). We appear to get more resolution using the correlation between the observed and expected correlation matrix; based on this statistic, the Condensation and *Satb2* + Gsc models fit best, followed by the Front/Back and then by the *Satb2* model (Table 12.7). But it is not clear if this greater resolution is an artifact of the test statistic.

Covariance model selection was used to find the model that least deviates from the data; i.e. the criterion used to select the best model was minimum deviance rather than a minimum AIC. The search was conducted in a stepwise fashion by adding or deleting an edge one by one, in random order, until no further changes improved the fit of the model. This search produced the model shown in Figure 12.20, which resembles none of the *a priori* models. It also does not resemble the best-fitting model produced by mixing the modules contained in the *a priori* models. What the present result suggests is that parts of the hypothesized modules are integrated with parts of other modules. For example, the distal incisor alveolus (but not the proximal incisor alveolus) is integrated with the distal ramus and the distal coronoid process. The result does not support the hypothesis that mandibular variation is structurally modular.

TABLE 12.6 Results of the Models Fitted to the Observed Correlation Matrix, Obtained by the Matrix Correlations Between Distance Matrices

Model	χ^2	df	p	ΔAIC
Front/Back	22.58	32	0.891	−41.43
Satb2	37.57	37	0.443	−36.42
Satb2 + Gsc	62.69	51	0.126	−39.31
Condensation	82.11	61	0.037	−39.89

Models are evaluated by the model deviance, which is approximately distributed as a chi-square (χ^{2}). The relative fit of the models is assessed by the ΔAIC, which is the difference between the AIC of each model and the AIC of the fully saturated model.

TABLE 12.7 Models Evaluated by the Matrix Correlation, R_M, Between the Observed and Expected Correlation Matrices; Statistical Significance of the Matrix Correlation Tested by the Mantel Test

Model	R_M	p
Front/Back	0.291	0.031
Satb2	0.135	0.169
Satb2/Gsc	0.383	0.001
Condensation	0.392	0.003

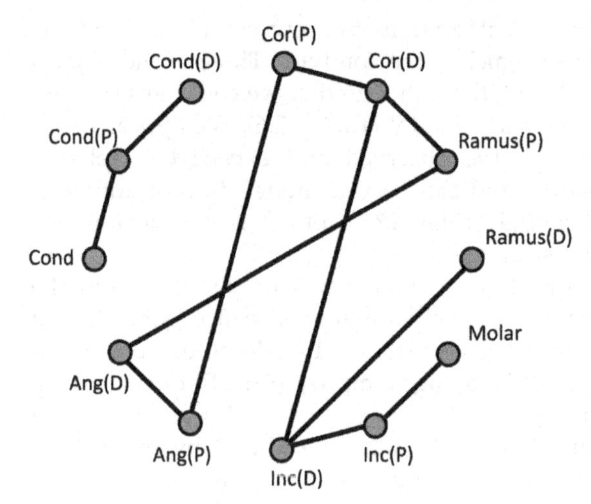

FIGURE 12.20 The model with the lowest deviance, obtained by covariance model selection.

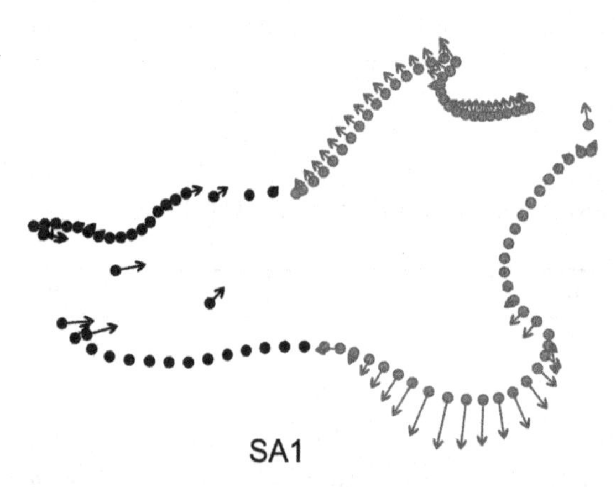

FIGURE 12.21 Two-block within-configuration Partial Least Squares Analysis. Shown is the dominant axis of covariation between the two blocks, Singular Axis 1, which accounts for 46.6% of the covariance between the front and back of the mandible.

What Do We Do Next to Interpret These Results?

Some results are consistent across methods but others suggest that we need to consider additional models. A consistent result is that the Condensation model fits better than the Front/Back model, and two of the three methods find more support for it than for *Satb2* + Gsc model, which is more highly supported than the *Satb2* model. But the exploratory results suggest that none of the developmental models predicts the variational architecture of the mandible. One obvious question is whether support for the unexpected patterns seen in the exploratory results can be seen in the covariances of landmarks. In particular, is there any support for the hypothesis of integration between distal incisor alveolus and distal coronoid process? We could begin to answer that question by returning to PLS, using the within-configuration method to look at the relationship between the front and back (Figure 12.21). In that figure, we do see support for the covariance between distal incisor alveolus, coronoid and angular processes, and perhaps also the molar alveolus. We could pursue this further by finer subdivisions of the data, isolating the landmarks of the two parts we wish to examine further and conducting a within-configuration PLS analysis. Alternatively, or additionally, we could construct a model that allows spatially disjunct regions to belong to the same module, and also allows for partial overlap between modules.

The analysis of modularity does pose real challenges, but it is not because we lack methods for analyzing modularity of shape data. Rather, it is because of the complexity of patterns of covariance. As should be evident from this example, testing only one model is not sufficient – had we tested only the Front/Back model, it would have been apparently confirmed by all the methods. Each model yielded the lowest RV or RV_M of all the models to which it was compared and each deviated less from the data than expected by chance. With the sole exception of the *Satb2* model, which was rejected by one test statistic (the correlation between the observed and expected correlation matrices), any of these models would be confirmed by all three methods. Yet, the models do not fit equally well, and the exploratory analysis raises the possibility that no modular structure fits the data as well as a non-modular one. Thus, an important part of the hypothesis-testing strategy is to consider multiple hypotheses, and to consider the possibility of integration without modularity.

References

Adams, D. C., & Collyer, M. L. (2009). A general framework for the analysis of phenotypic trajectories in evolutionary studies. *Evolution, 63*, 1143–1154.

Adams, D. C., Rohlf, F. J., & Slice, D. E. (2004). Geometric morphometrics: ten years of progress following the 'revolution'. *Italian Journal of Zoology, 71*, 5–16.

Akaike, H. (1974). A new look at the statistical model identification. *IEEE Transactions on Automatic Control, 19*, 716–723.

Ancel, L. W., & Fontana, W. (2000). Plasticity, evolvability, and modularity in RNA. *Journal of Experimental Zoology, 288*, 242–283.

Atchley, W. R., & Hall, B. K. (1991). A model for development and evolution of complex morphological structures. *Biological Reviews of the Cambridge Philosophical Society, 66*, 101–157.

Box, G. E. P. (1949). A general distribution theory for a class of likelihood criteria. *Biometrika, 36*, 317–346.

Breno, M., Leirs, H., & Van Dongen, S. (2011). No relationship between canalization and developmental stability of the skull in a natural population of *Mastomys natalensis* (Rodentia: Muridae). *Biological Journal of the Linnean Society, 104*, 207–216.

Breuker, C. J., Patterson, J. S, & Klingenberg, C. P. (2006). A single basis for developmental buffering of *Drosophila* wing shape. *Plos One, 1*.

Cheverud, J. M. (1982). Phenotypic, genetic, and environmental morphological integration in the cranium. *Evolution, 36*, 499–516.

Cheverud, J. M. (1995). Morphological integration in the saddle-back tamarin (*Saguinus fuscicollis*) cranium. *American Naturalist, 145*, 63–89.

Cheverud, J. M. (2004). Modular pleiotropic effects of quantitative trait loci on morphological traits. *Modularity in Development and Evolution*, 132–153.

Cheverud, J. M., Ehrich, T. H., Vaughn, T. T., Koreishi, S. F., Linsey, R. B., & Pletscher, L. S. (2004). Pleiotropic effects on mandibular morphology II: Differential epistasis and genetic variation in morphological integration. *Journal of Experimental Zoology Part B-Molecular and Developmental Evolution, 302B*, 424–435.

Cheverud, J. M., Routman, E. J., & Irschick, D. J. (1997). Pleiotropic effects of individual gene loci on mandibular morphology. *Evolution, 51*, 2006–2016.

Clarke, G. M. (1993). Fluctuating asymmetry of invertebrate populations as a biological indicator of environmental quality. *Environmental Pollution, 82*, 207–211.

Clarke, G. M. (1998). The genetic basis of developmental stability. V. Inter- and intra-individual character variation. *Heredity, 80*, 562–567.

Cowley, D. E., & Atchley, W. R. (1990). Development and quantitative genetics of correlation structure among body parts of *Drosophila melanogaster*. *American Naturalist, 135*, 242–268.

de Visser, J., Hermisson, J., & Wagner, G. P., et al. (2003). Perspective: Evolution and detection of genetic robustness. *Evolution, 57*, 1959–1972.

Debat, V., Alibert, P., David, P., Paradis, E., & Auffray, J. C. (2000). Independence between developmental stability and canalization in the skull of the house mouse. *Proceedings of the Royal Society of London Series B-Biological Sciences, 267*, 423–430.

Debat, V., & David, P. (2001). Mapping phenotypes: Canalization, plasticity and developmental stability. *Trends in Ecology & Evolution, 16*, 555–561.

Dworkin, I. (2005a). Evidence for canalization of Distal-less function in the leg of Drosophila melanogaster. *Evolution & Development, 7*, 89–100.

Dworkin, I. (2005b). A study of canalization and developmental stability in the sternopleural bristle system of Drosophila melanogaster. *Evolution, 59*, 1500–1509.

Ehrich, T. H., Vaughn, T. T., Koreishi, S., Linsey, R. B., Pletscher, L. S., & Cheverud, J. M. (2003). Pleiotropic effects on mandibular morphology I. Developmental morphological integration and differential dominance. *Journal of Experimental Zoology Part B-Molecular and Developmental Evolution, 296B*, 58–79.

Elena, S. F., & Lenski, R. E. (2001). Epistasis between new mutations and genetic background and a test of genetic canalization. *Evolution, 55*, 1746–1752.

Escoufier, Y. (1973). Le traitement des variables vectorielles. *Biometrics, 29*, 751–760.

Estes, E. C. J., Katholi, C. R., & Angus, R. A. (2006). Elevated fluctuating asymmetry in eastern mosquitofish (Gambusia holbrooki) from a river receiving paper mill effluent. *Environmental Toxicology and Chemistry, 25*, 1026–1033.

Fish, J. L., Villmoare, B., & Koebernick, K., et al. (2011). Satb2, modularity, and the evolvability of the vertebrate jaw. *Evolution & Development, 13*, 549–564.

Gaunt, S. J., Blum, M., & De Robertis, E. M. (1993). Expression of the mouse goosecoid gene during mid-embryogenesis may mark mesenchymal cell lineages in the developing head, limbs and ventral body wall. *Development, 117*, 769–778.

Graham, J. H., Freeman, D. C., & Emlen, J. M. (1993). Developmental stability: A sensitive indicator of populations under stress. In W. G. Landis, J. S. Hughes, & M. A. Lewis (Eds.), *Environmental toxicology and risk assessment* (pp. 136–158). Philadelphia: American Society of Testing and Materials.

Hall, B. K. (2003). Unlocking the black box between genotype and phenotype: Cell condensations as morphogenetic (modular) units. *Biology & Philosophy, 18*, 219–247.

Hall, B. K., & Miyake, T. (2000). All for one and one for all: Condensations and the initiation of skeletal development. *Bioessays, 22*, 138–147.

Hallgrimsson, B., Jamniczky, H., & Young, N. M., et al. (2009). Deciphering the palimpsest: Studying the relationship between morphological integration and phenotypic covariation. *Evolutionary Biology, 36*, 355–376.

Hallgrímsson, B., Willmore, K., Dorval, C., & Cooper, D. M. L. (2004). Craniofacial variability and modularity in macaques and mice. *Journal of Experimental Zoology Part B-Molecular and Developmental Evolution, 302B*, 207–225.

Hallgrímsson, B., Lieberman, D. E., Liu, W., Ford-Hutchinson, A. F., & Jirik, F. R. (2007a). Epigenetic interactions and the structure of phenotypic variation in the cranium. *Evolution & Development, 9*, 76–91.

Hallgrímsson, B., Lieberman, D. E., Young, N. M., Parsons, T., & Wat, S. (2007b). Evolution of covariance in the mammalian skull. In G. Bock, & J. Goode (Eds.), *Novartis foundation symposium* (pp. 164–185). New York: John Wiley & Sons, discussion, pp. 185–190.

Hallgrímsson, B., Willmore, K., & Hall, B. K. (2002). Canalization, developmental stability, and morphological integration in primate limbs. *Yearbook of Physical Anthropology, 45*, 131–158.

Hansen, T. F. (2003). Is modularity necessary for evolvability? Remarks on the relationship between pleiotropy and evolvability. *Biosystems, 69*, 83–94.

Hansen, T. F., Armbruster, W. S., Carlson, M. L., & Pelabon, C. (2003). Evolvability and genetic constraint in Dalechampia blossoms: Genetic correlations and conditional evolvability. *Journal of Experimental Zoology Part B-Molecular and Developmental Evolution, 296B*, 23–39.

Herrera, C. M., Cerda, X., Garcia, M. B., Guitian, J., Medrano, M., Rey, P. J., & Sanchez-Lafuente, A. M. (2002). Floral integration, phenotypic covariance structure and pollinator variation in bumblebee-pollinated *Helleborus foetidus*. *Journal of Evolutionary Biology, 15*, 108–121.

Hoffmann, A. A., & Woods, R. (2001). Trait variability and stress: Canalization, developmental stability and the need for a broad approach. *Ecology Letters, 4*, 97–101.

Hoffmann, A. A., Woods, R. E., Collins, E., Wallin, K., White, A., & McKenzie, J. A. (2005). Wing shape versus asymmetry as an indicator of changing environmental conditions in insects. *Australian Journal of Entomology, 44*, 233–243.

Hollander, J., Collyer, M. L., Adams, D. C., & Johannesson, K. (2006). Phenotypic plasticity in two marine snails: Constraints superseding life history. *Journal of Evolutionary Biology, 19*, 1861–1872.

Kawecki, T. J. (2000). The evolution of genetic canalization under fluctuating selection. *Evolution, 54*, 1–12.

Kingsolver, J. G., & Wiernasz, D. C. (1991). Development, function, and the quantitative genetics of wing melanin pattern in *Pieris* butterflies. *Evolution, 45*, 1480–1492.

Klingenberg, C. P. (2004). Integration, modules, and development: Molecules to morphology to evolution. In M. Pigliucci, & K. Preston (Eds.), *Phenotypic integration: Studying the ecology and evolution of complex phenotypes* (pp. 213–230). Oxford University Press.

Klingenberg, C. P. (2005). Developmental constraints, modules, and evolvability. In B. Hallgrímsson, & B. K. Hall (Eds.), *Variation: A central concept in biology* (pp. 219–247). San Diego: Elsevier Academic Press.

Klingenberg, C. P. (2008). Morphological integration and developmental modularity. *Annual Review of Ecology, Evolution and Systematics, 39*, 115–132.

Klingenberg, C. P. (2009). Morphometric integration and modularity in configurations of landmarks: Tools for evaluating a priori hypotheses. *Evolution & Development, 11*, 405–421.

Klingenberg, C. P. (2011). MorphoJ: An integrated software package for geometric morphometrics. *Molecular Ecology Resources, 11*, 353–357.

Klingenberg, C. P., Badyaev, A. V., Sowry, S. M., & Beckwith, N. J. (2001). Inferring developmental modularity from morphological integration: Analysis of individual variation and asymmetry in bumblebee wings. *American Naturalist, 157*, 11–23.

Klingenberg, C. P., Barluenga, M., & Meyer, A. (2002). Shape analysis of symmetric structures: Quantifying variation among individuals and asymmetry. *Evolution, 56*, 1909–1920.

Klingenberg, C. P., Duttke, S., Whelan, S., & Kim, M. (2012). Developmental plasticity, morphological variation and evolvability: A multilevel analysis of morphometric integration in the shape of compound leaves. *Journal of Evolutionary Biology, 25*, 115–129.

Klingenberg, C. P., Leamy, L. J., & Cheverud, J. M. (2004). Integration and modularity of quantitative trait locus effects on geometric shape in the mouse mandible. *Genetics, 166*, 1909–1921.

Klingenberg, C. P., & McIntyre, G. S. (1998). Geometric morphometrics of developmental instability: Analyzing patterns of fluctuating asymmetry with Procrustes methods. *Evolution, 52*, 1363–1375.

Klingenberg, C. P., Mebus, K., & Auffray, J. C. (2003). Developmental integration in a complex morphological structure: How distinct are the modules in the mouse mandible? *Evolution & Development, 5*, 522–531.

Klingenberg, C. P., & Monteiro, L. R. (2005). Distances and directions in multidimensional shape spaces: Implications for morphometric applications. *Systematic Biology, 54*, 678–688.

Klingenberg, C. P., & Zaklan, S. D. (2000). Morphological integration between developmental compartments in the Drosophila wing. *Evolution, 54*, 1273–1285.

Leamy, L. (1984). Morphometric studies in inbred and hybrid house mice .5. Directional and fluctuating asymmetry. *American Naturalist, 123*, 579–593.

Leamy, L. J., & Klingenberg, C. P. (2005). The genetics and evolution of fluctuating asymmetry. *Annual Review of Ecology Evolution and Systematics, 36*, 1–21.

Levene, H. (1960). Robust tests for equality of variances. In I. Olkin, S. G. Ghurye, W. Hoefding, W. G. Madow, & H. B. Mann (Eds.), *Contributions to probability and statistics* (pp. 278–292). Stanford: Stanford University Press.

Magwene, P. M. (2001). New tools for studying integration and modularity. *Evolution, 55*, 1734–1745.

Mantel, N. (1967). The detection of disease clustering and a generalized regression approach. *Cancer Research, 27*, 209–220.

Mardia, K. V., Bookstein, F. L., & Moreton, I. J. (2000). Statistical assessment of bilateral symmetry of shapes. *Biometrika, 87*, 285–300.

Marquez, E. (2012). Mint: Modularity and integration analysis tool for morphometric data. 2012.

Marquez, E. J. (2008). A statistical framework for testing modularity in multidimensional data. *Evolution, 62*, 2688–2708.

Marquez, E. J., & Knowles, L. L. (2007). Correlated evolution of multivariate traits: Detecting co-divergence across multiple dimensions. *Journal of Evolutionary Biology, 20*, 2334–2348.

McCullagh, P., & Nelder, J. (1989). *Generalized Linear Models* (2nd ed.). Boca Raton: Chapman & Hall/CRC.

Mezey, J. G., Cheverud, J. M., & Wagner, G. P. (2000). Is the genotype-phenotype map modular? A statistical approach using mouse quantitative trait loci data. *Genetics, 156*, 305–311.

Mezey, J. G., & Houle, D. (2003). Comparing G matrices: Are common principal components informative? *Genetics, 165*, 411–425.

Monteiro, L. R., Bonato, V., & dos Reis, S. F. (2005). Evolutionary integration and morphological diversification in complex morphological structures: Mandible shape divergence in spiny rats (Rodentia, Echimyidae). *Evolution & Development, 7*, 429–439.

Monteiro, L. R., & Nogueira, M. R. (2009). Adaptive radiations, ecological specialization, and the evolutionary integration of complex morphological structures. *Evolution, 64*, 724–743.

Muschick, M., Barluenga, M., Salzburger, W., & Meyer, A. (2011). Adaptive phenotypic plasticity in the Midas cichlid fish pharyngeal jaw and its relevance in adaptive radiation. *Bmc Evolutionary Biology, 11*, 116.

Palmer, A. R., & Strobeck, C. (1986). Fluctuating asymmetry – measurement, analysis, patterns. *Annual Review of Ecology and Systematics, 17*, 391–421.

Palmer, A. R., & Strobeck, C. (2003). Fluctuating asymmetry analyses revisited. In M. Polack (Ed.), *Developmental Instability (DI): Causes and consequences* (pp. 279–319). Oxford: Oxford University Press.

Parsons, K. J., Marquez, E., & Albertson, R. C. (2012). Constraint and opportunity: The genetic basis and evolution of modularity in the cichlid mandible. *American Naturalist, 179*, 64–78.

Parsons, K. J., Sheets, H. D., Skulason, S., & Ferguson, M. M. (2011). Phenotypic plasticity, heterochrony and ontogenetic repatterning during juvenile development of divergent Arctic charr (Salvelinus alpinus). *Journal of Evolutionary Biology, 24*, 1640–1652.

Reeve, E. C. R. (1960a). Some genetic tests on asymmetry of sternopleural chaeta number in Drosophila. *Genetical Research, 1*, 151–172.

Reeve, E. C. R. (1960b). Some genetic tests on asymmetry of sternopleural chaetae number in *Drosophila*. *Genetical Research, 1*, 151–172.

Rice, W. R. (1989). Analyzing tables of statistical tests. *Evolution, 43*, 223–225.

Richtsmeier, J. T., Lele, S. R., & Cole, T. M. (2005). Landmark morphometrics and the analysis of variation. In B. Hallgrímsson, & B. K. Hall (Eds.), *Variation: A central concept in biology* (pp. 49–69). New York: Elsevier.

Rohlf, F. J. (2003). Bias and error in estimates of mean shape in geometric morphometrics. *Journal of Human Evolution, 44*, 665–683.

Rohlf, F. J., & Slice, D. E. (1990). Extensions of the Procrustes method for the optimal superimposition of landmarks. *Systematic Zoology, 39*, 40–59.

Roseman, C. C., Kenney-Hunt, J. P., & Cheverud, J. M. (2009). Phenotypic integration without modularity: Testing hypotheses about the distribution of pleiotropic quantitative trait loci in a continuous space. *Evolutionary Biology, 36*, 282–291.

Santos, M., Iriarte, P. F., & Cespedes, W. (2005). Genetics and geometry of canalization and developmental stability in Drosophila subobscura. *Bmc Evolutionary Biology, 5*, 7.

Savriama, Y., & Klingenberg, C. P. (2011). Beyond bilateral symmetry: Geometric morphometric methods for any type of symmetry. *Bmc Evolutionary Biology, 11*, 280.

Scheiner, S. M., Caplan, R. L., & Lyman, R. F. (1991). The genetics of phenotypic plasticity .3. Genetic correlations and fluctuating asymmetries. *Journal of Evolutionary Biology, 4*, 51–68.

Schmalhausen, I. I. (1949). *Factors of evolution: The theory of stabilizing selection*. Philadelphia: The Blakeston Co.

Van Dongen, S., & Lens, L. (2000). The evolutionary potential of developmental instability. *Journal of Evolutionary Biology, 13*, 326–335.

Waddington, C. H. (1942). Canalization of development and the inheritance of acquired characters. *Nature, 150*, 563–565.

Wagner, G. P. (1988). The influence of variation and of developmental constraints on the rate of multivariate phenotypic evolution. *Journal of Evolutionary Biology, 1*, 45–66.

Wagner, G. P. (1996). Homologues, natural kinds and the evolution of modularity. *American Zoologist, 36*, 36–43.

Wagner, G. P., & Altenberg, L. (1996). Complex adaptations and the evolution of evolvability. *Evolution, 50*, 967–976.

Wagner, G. P., Booth, G., & Bagheri, H. C. (1997). A population genetic theory of canalization. *Evolution, 51*, 329–347.

Willmore, K. E., Klingenberg, C. P., & Hallgrímsson, B. (2005). The relationship between fluctuating asymmetry and environmental variance in rhesus macaque skulls. *Evolution, 59*, 898–909.

Willmore, K. E., Young, N. M., & Richtsmeier, J. T. (2007). Phenotypic variability: Its components, measurement and underlying developmental processes. *Evolutionary Biology, 34*, 99–120.

Wishart, J. (1928). The generalised product moment distribution in samples from a normal multivariate population. *Biometrika, 20A*, 32–52.

Wolterek, R. (1909). Weitere experimentelle Untersuchungen über Artveränderung, speziell über das Wesen quantitativer Artenunterschiede bei Daphniden. *Versuch Deutsche Zoolgische Gesellschaft, 19*, 110–172.

Wund, M. A., Baker, J. A., Clancy, B., Golub, J. L., & Fosterk, S. A. (2008). A test of the "Flexible stem" model of evolution: Ancestral plasticity, genetic accommodation, and morphological divergence in the threespine stickleback radiation. *American Naturalist, 172*, 449–462.

Young, N. (2004). Modularity and integration in the hominoid scapula. *Journal of Experimental Zoology Part B-Molecular and Developmental Evolution, 302B*, 226–240.

Young, N. M., & Hallgrímsson, B. (2005). Serial homology and the evolution of mammalian limb covariation structure. *Evolution, 59*, 2691–2704.

Zakharov, V. M. (1992). Population phenogenetics: Analysis of developmental stability in natural populations. *Acta Zoologica Fennica, 191*, 7–30.

Zelditch, M. L., Wood, A. R., Bonett, R. M., & Swiderski, D. L. (2008). Modularity of the rodent mandible: Integrating bones, muscles and teeth. *Evolution & Development, 10*, 756–768.

Zelditch, M. L., Wood, A. R., & Swiderski, D. L. (2009). Building developmental integration into functional systems: Function-induced integration of mandibular shape. *Evolutionary Biology, 36*, 71–87.

CHAPTER

13

Morphometrics and Systematics

Systematists use morphometrics to answer three types of questions. The first, which we label "taxonomic", asks whether samples are drawn from multiple taxa and, if so, by what variable(s) they are most effectively discriminated. The second, which we label "phylogenetic", seeks to identify traits distributed among taxa in patterns that may be consistent with their phylogeny and then infer that phylogeny from a consensus of those traits. The third, which we label "evolutionary", seeks to describe the historical evolutionary transformations of the features of interest. These are all interrelated issues, but there are important distinctions that bear on choosing appropriate analytic methods, and also on the suitability of different kinds of traits or trait descriptions.

One critical distinction is between discriminators and characters. It might seem obvious that taxonomic discriminators are potential characters because both are features that differ among taxa, however, taxonomic discriminators are not characters because discriminators describe the net difference between taxa; they are vectors extending between (or among) terminal taxa. The vector describes the direction in which the taxa can be distinguished from each other, regardless of whether the features distinguishing them are unique to one species, are shared by a group containing two species in the analysis, or are more broadly shared (with taxa not included in the analysis). That vector need not be aligned with a direction of evolutionary change; all that matters is that the discriminator exists (telling us that the taxa are indeed different) and is successful (allowing us to identify unknowns correctly). In contrast to a discriminator, a character is a feature shared by members of a monophyletic group. In principle, a character is a feature that is recognized as transforming at the node of a cladogram, and thus represents a hypothesis of the direction of evolutionary change. If shapes could be measured at successive nodes, characters could be found by simple pairwise comparisons between them, but samples of taxa at successive nodes usually are not available and, even if they were, the fact that those taxa represented nodes cannot be determined before reconstructing the cladogram.

Another critical difference between types of systematic questions is between identifying shape characters and reconstruction evolutionary shape changes. As noted above, changes in shape characters, traced on a cladogram, are intended to represent evolutionary

Geometric Morphometrics for Biologists
DOI: http://dx.doi.org/10.1016/B978-0-12-386903-6.00013-7

transformations in shape. However, finding (and tracking) shape characters, and reconstructing the evolution of shape, are different exercises. When looking for characters, particular features of the whole are selected as informative, making no effort to provide a complete description of the changes in shape (or of the ancestral shape). Consequently, characters usually comprise a subset of the features that evolve, and tracing characters on a cladogram does not fully reconstruct the evolution of shape. For example, all members of a particular group might have a shallow body compared to the other species, and "shallow body" is then selected as a character. But, no inference is made concerning amount of change in body depth or how shallow their ancestor was, or what the ancestor's head shape was. In contrast, reconstructing the evolution of shape requires an inference of the ancestral configuration of landmarks as well as the direction and magnitude of change in the complete configuration along each branch.

Of the three types of questions, only those relating to taxonomic discrimination are so straightforward that they require nothing more than standard morphometric tools. This does not mean taxonomic discrimination is easy; on the contrary, it can be very difficult. However, the difficulties of taxonomic discrimination pale in comparison to the problems of finding characters or reconstructing the evolution of shape. Because taxonomic discrimination is a straightforward problem, our discussion of it focuses on some of the practical issues that complicate its application to shape data.

The problems of evolutionary analysis, on the other hand, raise questions that are largely outside the scope of morphometric theory. The methods used to infer evolutionary transformations of shape either (1) minimize a distance or squared distance over the cladogram (which, in our case, would be a Procrustes distance); or (2) use an explicit model of the evolutionary process and estimate values of the model's parameters that maximize the likelihood of the data, given the model (an accessible, general overview of these approaches can be found in Felsenstein, 2002, and a discussion of them in context of geometric shape data can be found in Rohlf, 2002). Whatever the model, when these methods are used to infer shape evolution, the whole shape (the complete set of shape variables) is always used. The primary issue facing users of these methods is to choose (or develop) a realistic, justifiable model of shape evolution – a matter that involves considerations of evolutionary biology rather than morphometrics. Accordingly, we do not discuss this topic beyond a brief listing of the models that could be used.

Unlike the taxonomic and evolutionary questions, phylogenetic analysis using shape continues to raise profound methodological questions with no satisfying answers. The central problem is that there is no generally accepted method for finding characters in shape data, and it is not even clear what a method of character discovery would look like. For systematists, the lack of progress in this area since the first edition will mean that this remains a disappointing chapter. However, we chose to use the opportunity of the second edition to clarify some important points that we made in the first edition. The most important of these is that decomposing shapes into variables that will be treated as independent traits is a fundamentally flawed approach to inferring phylogenetic relationships from shape data. It does not matter whether the decomposition is into partial warps or principal components or any other means of defining vectors in a shape space or its tangent spaces. This flaw lies at the heart of multiple methods of character analysis, including the approach previously offered by us (Fink and Zelditch, 1995; Zelditch et al., 1995). The majority of this chapter focuses on this issue because understanding *why* a method fails is as important as realizing that it does fail, particularly when the aim is to avoid making the same kind of mistake again. Thus, we have kept this chapter in the second edition primarily to make a stronger and clearer argument against the use of this and related methods (see also Adams and Rosenberg, 1998; Rohlf, 1998; Adams et al., 2011).

TAXONOMIC DISCRIMINATION

The fundamental taxonomic question can be divided into two parts:

1. Are the samples different enough to warrant judging them to be different species?
2. In what do they differ?

(For the sake of simplicity, we focus on discriminating between species; however, comparable challenges may arise at any level of the taxonomic hierarchy.) To answer the first part of the question, one must decide what would be "different enough". Having stated that criterion, it is possible to ask whether the data meet it. For example, "different enough" might be that no more than 2% of the specimens are misclassified, or that the means of the samples differ statistically significantly, or even that the Procrustes distance between the means is minimally 0.03 (or any other favored value). Choosing a criterion also determines which method will tell if the data meet it, which could be MANOVA, CVA, computing Procrustes distances between means, or other valid method of evaluating the difference between species.

The more difficult decisions that need to be made concern the handling of the various potential sources of within-group variation, including geographic variation, ontogeny, and sexual dimorphism. Any of these factors could complicate distinguishing species. Obviously, you do not want to claim to have evidence for two species when the samples differ only in average developmental age or body size. If that might be the case, it would be useful to design the sampling scheme to ensure that the samples are homogeneous and comparable, or else to standardize the data to a common age or size using regression. The results can be very different. For example, Figure 13.1 shows results from three analyses: (1) samples are compared without standardizing by ontogenetic stage (Figure 13.1A); (2) samples are compared at a common juvenile stage (Figure 13.1B); and (3) samples are compared at a common adult stage (Figure 13.1C). In all three analyses, all eight CVs are significant and, with one exception (the unstandardized data), the misclassification rate is extremely low. For the unstandardized data, out of 390 specimens as many as 12 are misclassified, all of which are *Pygocentrus nattereri* that are classified either as *P. cariba* or *P. piraya*. However, for both standardized data sets no more than four individuals are misclassified (also *P. nattereri*). Not surprisingly, all species differ from all others significantly (in all pairwise comparisons, $P < 0.002$). In general, species differ by a Procrustes distance of more than 0.030, except for the three *Pygocentrus*, whose adults differ from each other by Procrustes distances as small as 0.027–0.028 (and by even less in comparisons of unstandardized specimens). Thus, the same conclusion would be drawn from all three analyses about the taxonomic status of these samples, but the results still differ because the variables discriminating among the species are different.

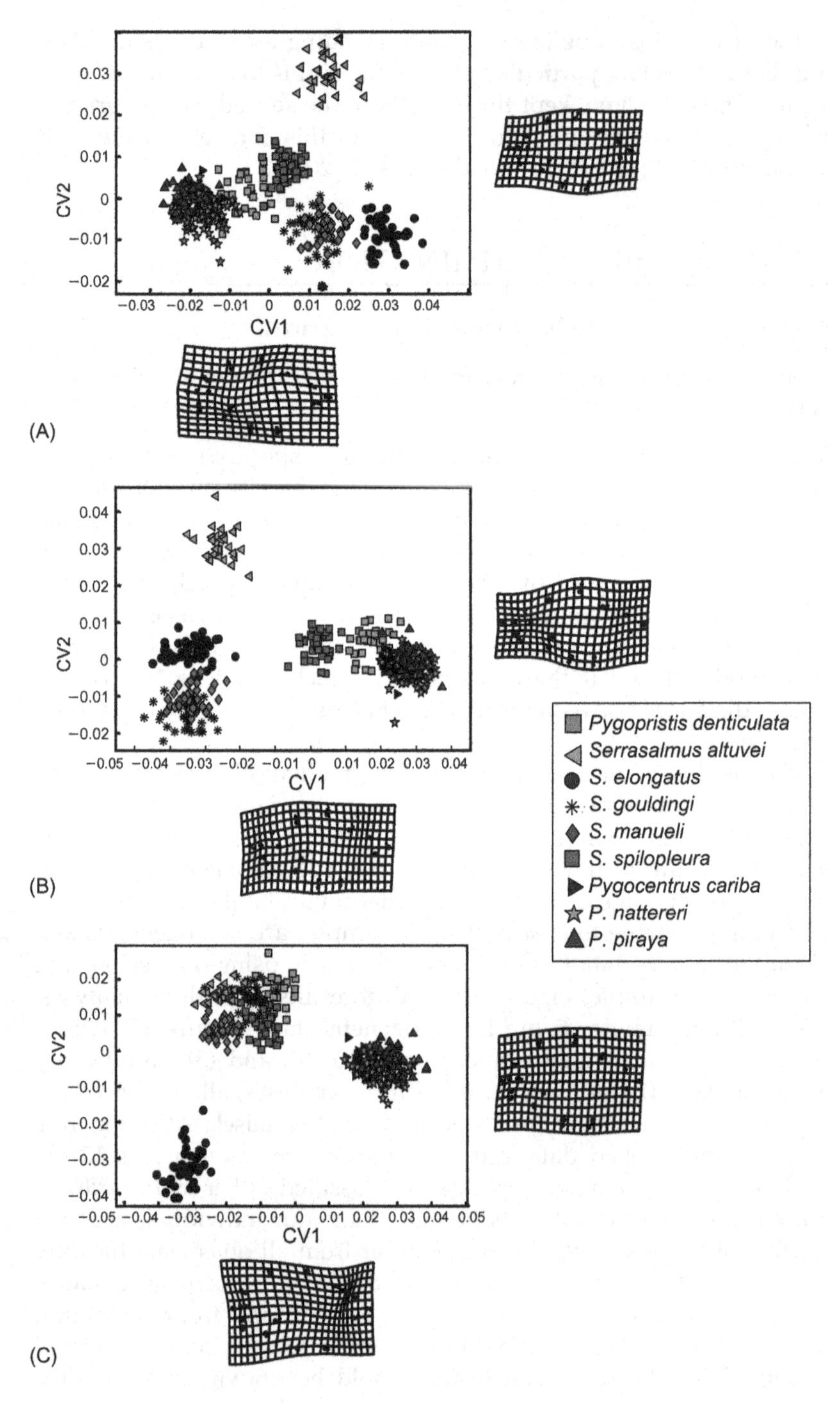

FIGURE 13.1 CVA of body shape of nine species of piranhas: (A) unstandardized data; (B) data are standardized and comparisons are made among juvenile shapes (at the transition from larval to juvenile phases); (C) data are standardized and comparisons are made among adult shapes (at the maximum body size regularly attained by each species).

After applying morphometric techniques to the data, there is still the problem of interpretation. Even if the data meet the designated criteria, the samples might not come from different species – they could come from geographically differentiated populations that were sampled only at the extremes of their range (e.g. the most northern and the most southern localities). Had they been sampled throughout the entire range, it might turn out that there is no statistically significant difference between geographically adjacent populations. Conversely, they might not meet the criterion but, nonetheless, be distinct species; it is just that the distinguishing features do not lie in shape. CVA provides a useful method for discrimination, but finding that samples can be discriminated is only part of the answer to the first taxonomic question.

As Viscosi and Cardini (2011) nicely demonstrate, structures that are repeated within an individual, such as leaves on a tree, add even more layers of complexity to this problem. Variability of repeated structures is not just a concern of botanists; many animal taxa also have repeated structures (e.g. eye facets, body segments, vertebrae), and colonial organisms like bryozoans and corals also have repeated individuals. Not only might repeated structures vary within an individual or colony, perhaps as a function of position within the individual, the pattern of intra-individual variation might have an ontogeny of its own. Intra-individual variation may also differ between individuals in response to all the same factors that might account for inter-individual variation. Even with all these potential layers, the question of "are species different?" still can be answered by well-reasoned application of convention morphometric methods.

Although CVA can be useful for answering both the first part of the taxonomic question (Are they different?), and the second part (In what are they different?), it is important to bear in mind the limitations discussed in Chapter 6. Those limitations mean that the likelihood that CVA will produce a discriminant function that is entirely an artifact of rotating and rescaling the data increases with the number of variables. But rather than reduce or eliminate semilandmarks and lose potentially informative shape data, cross-validation can be used to confirm reliability of the discrimination. Indeed, cross-validation should be used for large sets of landmarks even if no semilandmarks are used.

Whether taxonomic differences are discovered by CVA or other analyses, combinations of shape variables may not be the most useful descriptions of those differences for the biologist comparing specimens in the field or in the museum. If it is intended to be useful for field biologists, no purpose is served by writing a key that requires digitizing specimens, entering their data in a CVA, and allocating them to species according to the discriminant function. Although that could be considered a merely technological limitation, a taxonomic key serves a pragmatic purpose and therefore must be useful. The key must be applicable to the specimens in hand, under the conditions when they are in hand, while still accounting for potential sources of intraspecific variation. These requirements make writing a useful key a challenging problem, but turning a geometric analysis into a useful key adds no further difficulties. That can be done by using geometric morphometrics to determine the shape variables that best discriminate, then translating them into terms of traditional morphometric variables that can be measured with calipers or rulers. For example, if relative body depth discriminates between species, two lengths can be calculated from the landmark coordinates; one for the depth measured between two landmarks (such as anterior bases of the dorsal and anal fin), and one between the landmarks that capture standard length. That ratio does not fully describe the shape differences among species, but it suffices to identify unknown specimens.

FINDING CHARACTERS

The use of morphometric data in phylogenetic studies has long been controversial. Most often, debates among phylogenetic systematists have focused on two issues: (1) methods for coding variables that overlap, sometimes considerably; and (2) the reliability of the information obtained from the data for inferring phylogenies. Morphometric data have been viewed with suspicion partly because it is difficult to determine where to draw the line when there are no distinct gaps between the observed values. A wide variety of techniques has been proposed and debated heatedly (see, for example, Colless, 1980; Simon, 1983; Archie, 1985; Goldman, 1988; Chappill, 1989; Thiele, 1993; Swiderski et al., 1998). Only very recently has the discussion begun to focus on a more fundamental problem: what to code? What is being extracted from the data and treated as a character? Clearly, this issue must be addressed before the first one is even relevant; coding becomes a moot issue if there are no characters to code, and if there are no characters, there is nothing to test for homoplasy.

It is clear that partial warps should not be used as characters (for the reasons discussed below), but it is not clear what ought to be used instead. It is not even clear that the problem has a solution. The major objective of the first part of this section is to define the problem we had hoped to solve using partial warps, then to explain why our approach was flawed. In the next section, we discuss two alternatives; both rely on conventional multivariate methods, but neither is precisely tailored to the problem.

Defining the Problem

The general problem is to find features that differ among taxa and are shared by a subset of them. The differences indicate evolutionary novelties and the similarities indicate common ancestry, although it is not possible to determine which are novelties until the phylogenetic analysis has been completed. You would not expect that an entire shape is a character because species rarely have exactly the same shape (whether comparing the whole organism or a single part); indeed, shape analysis is used precisely because shape and its variation are complex. This suggests that if the problem is cast in terms of whole landmark configurations, it will not be possible to make any progress. Yet, that is precisely what the theory of shape demands of us; if we do not think of the problem in terms of whole landmark configurations, we will be led to theoretically invalid solutions. This, then, is the heart of the problem: to analyze entire configurations of landmarks and find features that differ among taxa and are similar among a subset of taxa. Furthermore, to say that a feature is a character, it is necessary to satisfy the criteria for recognizing phylogenetic or evolutionary homology, that is, to say *where* the feature is, and over how large a spatial expanse it extends. A primary objection to traditional morphometric variables is that they are lines, having no spatial extent as individual variables. Attempting to determine the spatial location and extent of a difference captured by these data, even using multivariate analyses, uncovers one of the most severe limitations of traditional morphometric data – their poor ability to localize morphological differences.

A further obstacle to finding characters in shape data is that neither the magnitude of the difference, nor the magnitude of its spatial extent, are relevant to its utility as a character. Small magnitude differences (so long as they are big enough to be considered a difference at all) count as much as large magnitude ones, and small-scale differences (such as the tip of an appendage) count as much as large-scale ones (perhaps spanning the whole organism). Consequently, neither the Procrustes distance between taxa nor the bending energy of the transformation has any relevance to the problem. This is one of the reasons why the problem is so difficult to solve – neither of the metrics used in geometric morphometrics is germane to the problem, and if there is a relevant metric, it has yet to be defined.

When we first approached this problem, we focused on one major limitation of conventional (qualitative) approaches: that organisms are often dissected arbitrarily, along lines of convention. Conventional anatomical subdivisions are often not biologically meaningful except in the context of a particular problem. For example, if we are interested in locomotion and foraging, we can subdivide an organism into parts that are used in locomotion and parts that are used in foraging. Alternatively, if we are interested in development, we can subdivide the organism into parts that have a common germ-layer origin, or that develop from the same type of bone, or that undergo the same kinds of epigenetic interactions, etc. These subdivisions have long been regarded as arbitrary, except to the extent that they are useful in a particular investigation. These subdivisions often are not suitable for dissecting an organism in systematic studies because what differs between taxa may cross several such divisions, and may not be wholly within any them. Our goal in using partial warps was to find a more objective basis for dissection. We did not succeed (for reasons discussed below), but the problem we defined remains a fundamental and unresolved difficulty for character analysis. Our method had fatal flaws, but so do others that require us to decompose the organism using biologically arbitrary mathematical rules.

The approach we took is similar to one that is standard in cladistic studies using morphometric data. We defined a set of variables *a priori*, and compared taxa with respect to them. A similar tactic is applied to conventional morphometric variables, when a set of lengths or ratios is defined and measured on taxa, then the values of those lengths or ratios is compared among the taxa. Most attention has focused on the problem of coding those variables, but coding is the least of the problems. Such variables do not solve the problems we had hoped to address, but share with them the flaw that we inadvertently introduced: they score taxa on arbitrarily selected components of shape, one component at a time.

Why Not to Use Partial Warps as Characters

Even though partial warps have a geometric scale, are a function of homologous landmarks, and do not emphasize differences of large magnitude at the expense of small ones, they cannot be used as characters for at least two reasons. The first is obvious (in hindsight at least): partial warps have a spatial scale only in so far as spatial scale refers to the relative proximity of landmarks with the largest contrasting displacements. In reality, every partial warp spans the entire organism and extends to infinity as the implied deformation asymptotically approaches zero. Moreover, an individual partial warp (PW) describes only part of a small-scale anatomical feature. No matter how anatomically localized the change, its description by PWs will entail scores on the largest scale PW that reflects the best fit of that PW to the data. Scores on progressively smaller scale PWs each reflect their fit to the *difference* between the data and the sum of the higher scale PWs. Consequently, partitioning a change by PWs does not correspond to partitioning it by anatomy or by characters. Even if a difference between taxa did (fortuitously) closely match a single lower scale PW, its description would be composed of scores on higher and lower scale PWs that represent that single, spatially coherent change only when taken together. Furthermore, having a high score on a localized PW does not mean that there is a localized change. Instead, it may simply mean that in this particular region a large scale anatomical change is not well described by the large scale PWs and the localized PW supplements that description. If the smaller scale PW is taken out of context of the larger-scale PWs, we cannot make anatomical sense of the one at smaller scale. Two taxa that have identical values for a small-scale PW might differ anatomically *in that same region* because the differences between the taxa cannot be seen without looking at all PWs.

All that may be obvious to readers who worked through the first several chapters, but to clarify the point (and for those who jumped straight here) we can re-examine the example that we found most promising at the time – the ontogenetic change in scores on one PW (Figure 13.2). Two of the taxa, which were used as outgroups (*Pygopristis denticulata* and *Serrasalmus gouldingi*), have statistically significant ontogenetic change on that PW (in both *X* and *Y* directions), whereas the three *Pygocentrus* do not. We would not normally be concerned about similarities among outgroups, but this example shows that similarities implied by individual PWs are not found in complete descriptions. That *P. denticulata* and *S. gouldingi* have anything in common in their development of that region is not at all obvious when looking at more complete descriptions of the five ontogenies (Figure 13.3). They are similar to each other, and differ from the three *Pygocentrus*, only in that they undergo an ontogenetic change in the caudal peduncle region that is not fully described by PWs at higher spatial scales. However, *P. denticulata* and *S. gouldingi* are not similar to each other in the changes described by the higher spatial scales (and neither are the three *Pygocentrus*). Being similar in one PW does not mean being similar in shape (or ontogeny of shape) in a particular anatomical region. When looking at one PW we lose the context supplied by all the others, and PWs are all context-dependent. Therefore, we cannot describe what happens within any one region of the body without placing every PW in context of every other. Even judged by what the method was supposed to do, it fails; it does not provide an objective, non-arbitrary method for decomposing changes (except in a purely geometric sense).

The second reason partial warps cannot be used as characters, which is related to the one above but important in a broader context, is that interpretations based on individual variables violate the fundamental principles of geometric shape analysis – that results be

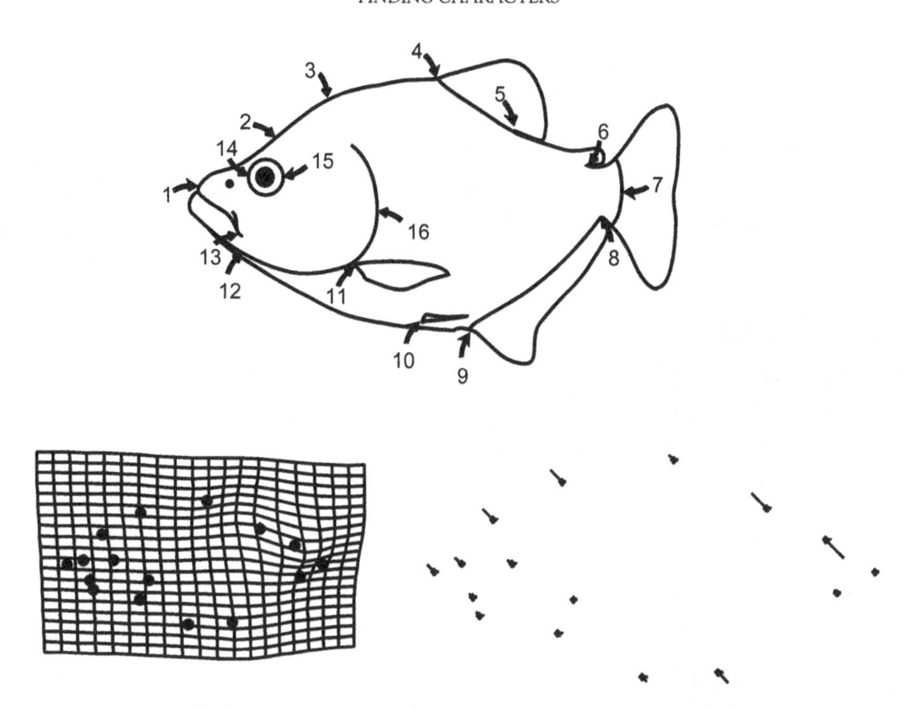

FIGURE 13.2 A single PW used to exemplify the procedure for finding systematic characters in ontogenies of shape in Fink and Zelditch, 1995.

invariant to the selection of variables. A result that depends on using partial warps is invalid (even if the phylogenetic inference based on it happens to be valid) because a partial warp score is a single variable, a one-dimensional projection onto a particular basis (defined by the reference shape), and our results cannot depend on that choice. Adhering to this basic principle does not mean that our *phylogenetic results* will be invariant to our choice of *characters* – the results of a phylogenetic analysis always depend on the characters. Rather, it means that our recognition of characters must be invariant to the selection of variables – and for that reason, a morphometric variable cannot be a character in its own right.

The obvious question is: how *can* we discover characters when we cannot look at individual morphometric variables? If variables do not provide a legitimate basis for subdividing the organism, and if conventional anatomical lines of dissection are also viewed as biologically arbitrary, then where can we look for characters? We end this section with that question because we have no satisfying answer. In the next section, we discuss two possible lines of attack. One uses a standard multivariate ordination method, principal components analysis (PCA), to explore similarities and differences, the other uses pairwise contrasts to find differences, which are then compared to find similarities among taxa in their differences from others. Neither method is tailored to the problem, but both represent feasible approaches that can be used in the interim, until we have a more satisfying method.

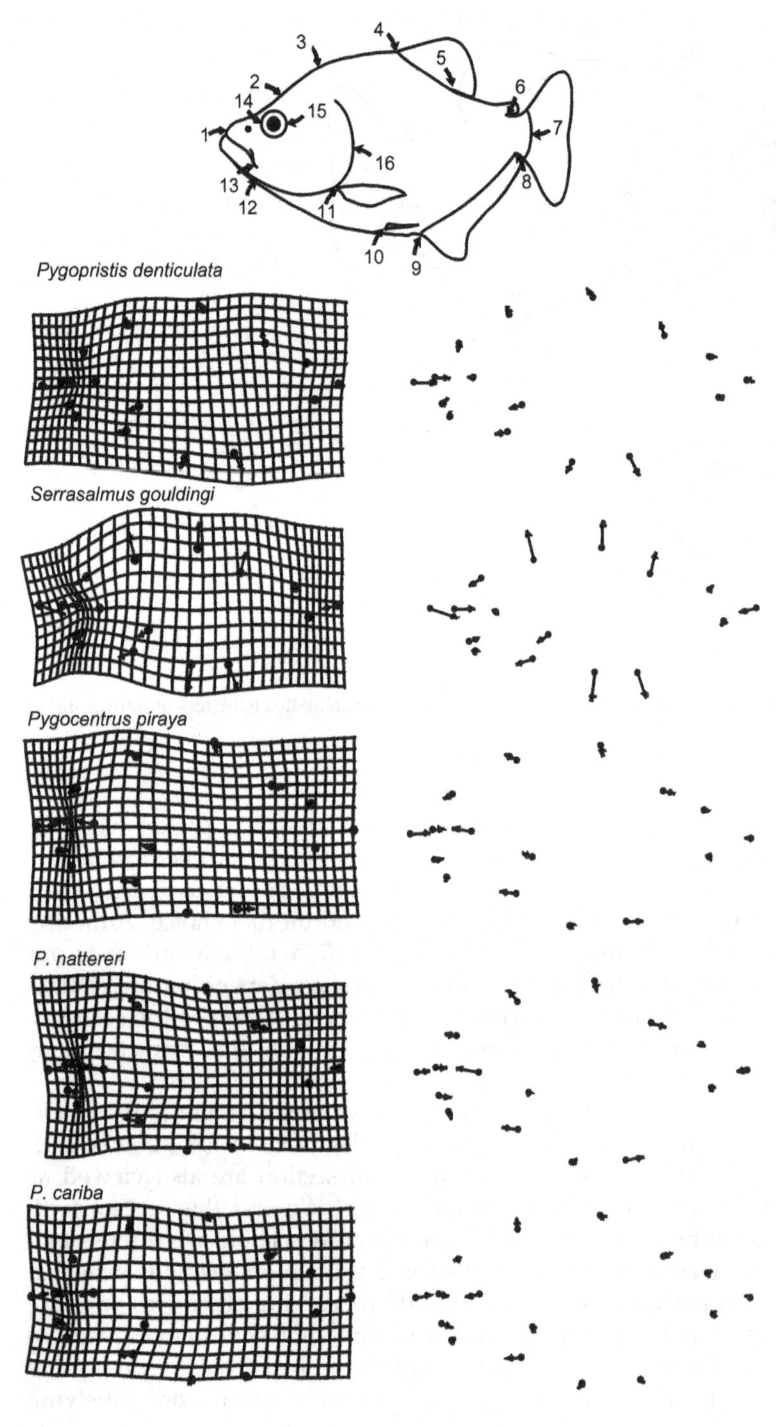

FIGURE 13.3 Ontogenies of shape for the species analyzed in Fink and Zelditch, 1995. The inference drawn from the PW shown in Figure 13.2 is that the outgroup species *P. denticulata* and *S. gouldingi* have a localized ontogenetic change in the length and depth of the caudal peduncle relative to the region between dorsal and adipose fins, whereas the three *Pygocentrus* do not.

Using PCA to Find Characters

PCA provides a coordinate system for shape analysis, and may be useful for finding characters, but first we must state an important caveat: individual PCs (like individual PWs) cannot be viewed as characters in their own right. Just as a partial warp score is a projection onto a single axis, so is a principal component score, and just as a similarity on one PW does not indicate a similarity in shape, similarity on a single PC might not demonstrate a sufficiently general (or detailed) similarity. Like PWs, PCs are context-dependent, and thus we would not expect an individual PC to be a character any more than an individual partial warp is.

Despite the similarities we just pointed out, analyses by PC and PW are not strictly comparable – there is a major difference between them. PCs are orthogonal directions of variation rather than orthogonal components of bending energy, and variation is biologically relevant to the problem at hand while bending energy is useful only in that it is used by the method for depicting the results. PCs have a biological meaning, as orthogonal dimensions of variance, even though that is not equivalent to the meaning of a character. They are not likely to be characters in their own right because they are directions of variation that are constrained to be orthogonal (by definition), not directions of evolutionary change. Directions of evolutionary change are likely to be oblique to the PCs – they are within the space spanned by the PCs, but they need not lie along an axis nor must they be orthogonal.

Although PCs are not likely to be characters, we may still find PCA useful for exploring similarities and differences. The scatter-plots allow us to see the variation among taxa, and their overlap, and both are important for finding characters. However, just as we need to interpret partial warps in combination, so we also need to interpret PCs in combination. Just because two or more species overlap in their PC1 scores does not mean that they are similar with respect to all features described by PC1. They may differ in some, so that PC1 splits the difference between them and the other PCs describe what is specific to their deviations from PC1. Taxa located in different quadrants of a scatter-plot may differ considerably in shape, depending on the proportion of the variance described by each PC and on how the PCs overlap in their descriptions of variation within the same regions. For example, we can look at a case that should be familiar by this point – the first two PCs of piranha juvenile body shape. The first, which accounts for 62% of the variance, clearly distinguishes three species (*S. manueli*, *S. elongatus* and *S. gouldingi*) from all others (Figure 13.4). Looking at the deformation that depicts the direction of greatest variance, we can see that body depth contributes heavily to it. However, PC1 is not only body depth; it also describes differences in proportions of the posterior body correlated with body depth. Species with high scores on this axis have relatively long caudal peduncles compared to the region between dorsal and adipose fins, as well as deep bodies, but we cannot necessarily say that species with high scores on PC1 have long caudal peduncles if other PCs also describe variation in posterior body proportions and scores on those PCs differ among species with similar scores on PC1. PC2, which accounts for only 8.3% of the variance, also describes variation in caudal body proportions and, on this component, species with high scores have very short caudal peduncles relative to more anterior region. Consequently, species with high scores on both components have a short caudal peduncle

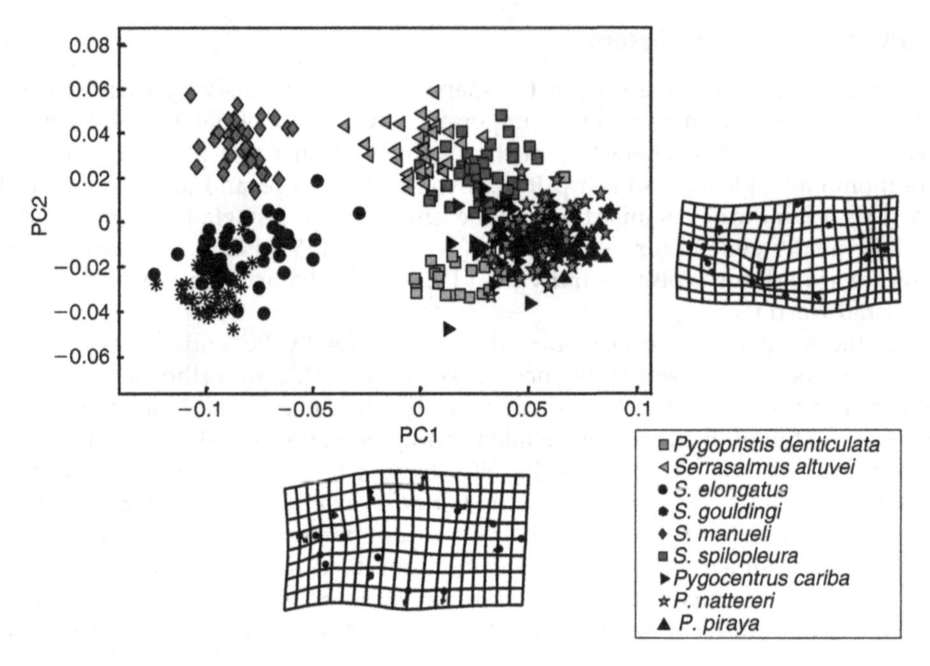

FIGURE 13.4 Principal components of body shape of nine species of piranhas; data were standardized and variation is examined among juvenile shapes (at the transition from larval to juvenile phases).

relative to other species with equally high scores on PC1. In effect, PC2 partially "compensates" for PC1.

When interpreting PCs, it is also important to consider that they describe variation around an average shape. However, the average is not a "typical" piranha; rather, it is the shape of the consensus, the point having the coordinates 0, 0 (on all PCs). Obviously, the consensus is not a typical piranha since there are no specimens at the 0, 0 point. The outgroup (*P. denticulata*) is fairly near it, but if we want to describe differences between *P. denticulata* and other species (or to make any other comparisons among species), we cannot describe changes along one PC, then along another. The direction of the difference between particular species is often oblique to several PCs.

The importance of considering scores on several PCs becomes evident when comparing the three shallow-bodied species. All three have high scores on PC1, but they differ in scores on PC2. One of the three, *S. manueli*, has high scores on PC2 (as do *S. altuvei* and *S. spilopleura*). To see how *S. manueli* differs from *S. gouldingi* and *S. elongatus* with respect to their differences from other taxa, we can draw the vector extending from those other taxa, e.g. *P. denticulata* to *S. manueli* (Figure 13.5A) and to *S. elongatus* and *S. gouldingi* (Figure 13.5B). The reason for doing this is to determine what differences from other taxa are shared by *S. elongatus*, *S. gouldingi* and *S. manueli*. We will then eliminate from the character the features peculiar to one species. Although we are making this comparison to the outgroup, we are not assuming that it has the primitive body shape. Comparisons to other species will also be necessary, and no decisions about polarity are made at this point.

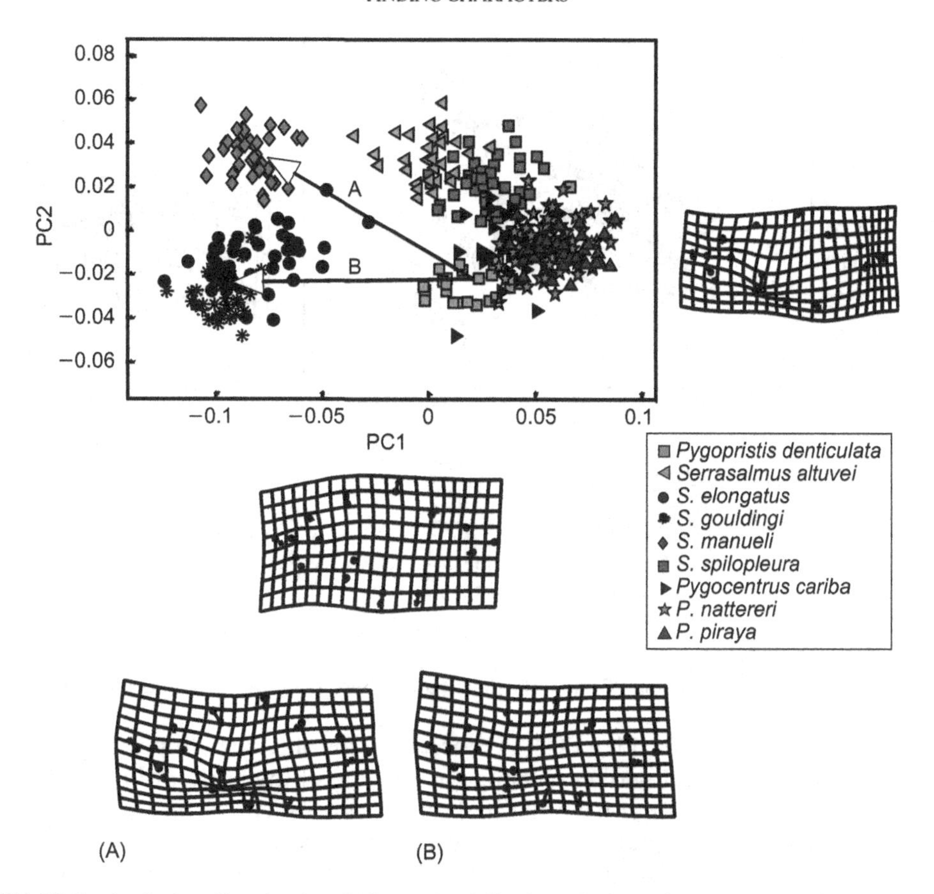

FIGURE 13.5 Analyzing direction in which species differ from *P. denticulata* in juvenile shape, to determine whether species with overlapping scores on PC1, but different scores on PC2, are similar with respect to features varying along PC1. (A) The direction of difference from *P. denticulata* to *S. manueli*; (B) the direction of difference from *P. denticulata* to *S. gouldingi* and *S. elongatus*.

Based upon the similarities between the two vectors (Figures 13.5A, 13.5B), what all three taxa share is their shallow body. There also may be a second similarity not described by either PC – a shortening of the mid-body relative to the head and posterior body. We could include that in the description of the character, but we would exclude the proportions of the caudal peduncle from that character description because *S. manueli* differs from the other two species in that clearly, the character is not equivalent to a PC.

The reason for not treating PC2 as a character in its own right is the same as the one we used to rule out treating individual PWs as characters. *S. manueli*, *S. altuvei* and also *S. spilopleura* have high scores on this one, which primarily describes a displacement of the opercle landmark towards the pectoral fin and a shortening of the caudal peduncle relative to the anal fin. However, *S. manueli* and *S. altuvei* differ along PC1 and are not similar in caudal peduncle proportions; they also differ along PC3 (Figure 13.6). Differences along PC3, as well as those along PC1, might belie the inference of morphological similarity

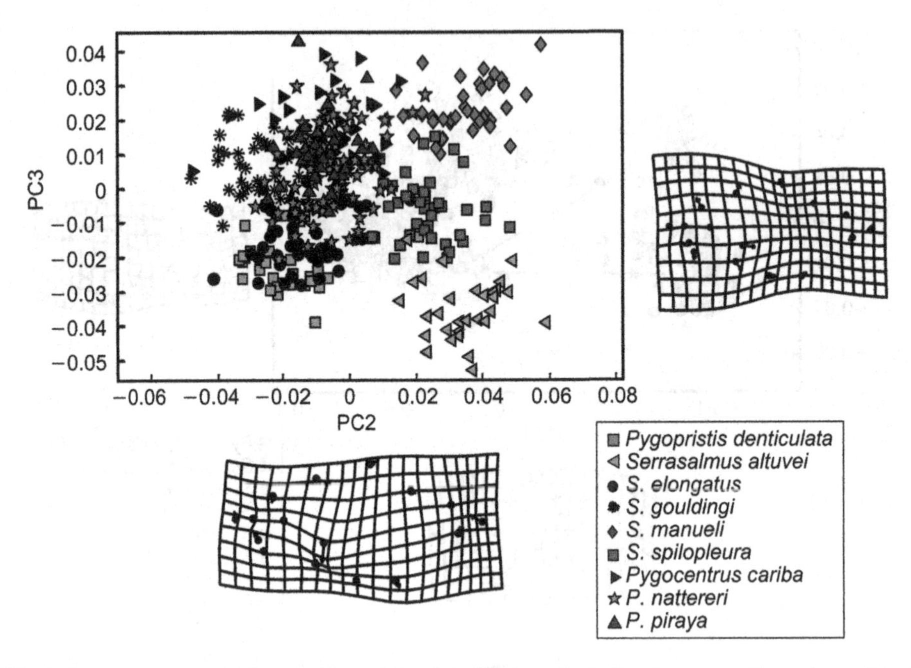

FIGURE 13.6 Scatter-plot of PC3 on PC2, and the deformations depicting these two dimensions of variation.

implied by similar PC2 scores. Like PC2, PC3 accounts for only a small portion of the variance (5.6%), but like PC2 it describes a change in location of the pectoral fin relative to the opercle. *S. manueli* and *S. altuvei* have the highest and lowest scores on PC3, respectively, which means that their pectoral fins are displaced in opposite directions relative to the opercle, which needs to be taken into account when assessing their similarity on PC2. Despite their similar scores on PC2, a feature that might have been judged a morphological similarity might not be similar by virtue of the differences along PC1 and PC3. Because of their different scores on PC1, we would also avoid construing their caudal peduncle proportions as similar, despite their similar values on PC2. Because species can be similar along one component and differ substantially along others, we cannot interpret one component at a time.

The strategy for combining PCs, outlined above, is undeniably tedious, but it might be successful at finding the features shared by two or more taxa. In cases like our example, when over 60% of the variation is along a single PC, two or more taxa have high scores and two or more have low ones, and there is virtually no overlap among the low and high scores, PC1 points to a character. When the variation is more evenly spread out across components, it will be necessary to combine many more of them because similarities on one may be outweighed by differences on the others. An obvious problem is that the comparison of vectors in a single plane, such as we used to compare the similarities among the three shallow-bodied species with respect to their difference from the outgroup, examines only the differences among them that are in that particular plane. We might prefer to look at *all* the differences between each species and the outgroup (or any other species taken as a standard), comparing *those* vectors among taxa.

Using Comparisons Between Interspecific Vectors to Find Characters

The basic idea of this approach is to compare all species to one other species (which is held constant). These pairwise contrasts can then be examined for similarities. By comparing the differences between one species and each of the others, we can then inspect the differences for similarities. The logic of the method is that we are looking for similarities in the differences – i.e. similarities among taxa in features specific to them. To exemplify this approach, we will continue the analysis of piranha juvenile body shape, comparing each species to the outgroup (Figure 13.7). Of course, it is not necessary to use the outgroup in these comparisons; any species could be used as that "other," and it may be useful to use more than one before drawing conclusions.

From these comparisons, it is obvious that *S. elongatus*, *S. gouldingi* and *S. manueli* are shallow-bodied compared to all other piranhas. They differ profoundly from *P. denticulata* in this, whereas none of the other species do. This is the feature that dominated the PCA (and it is obvious by qualitative visual inspection as well). Additionally, these three

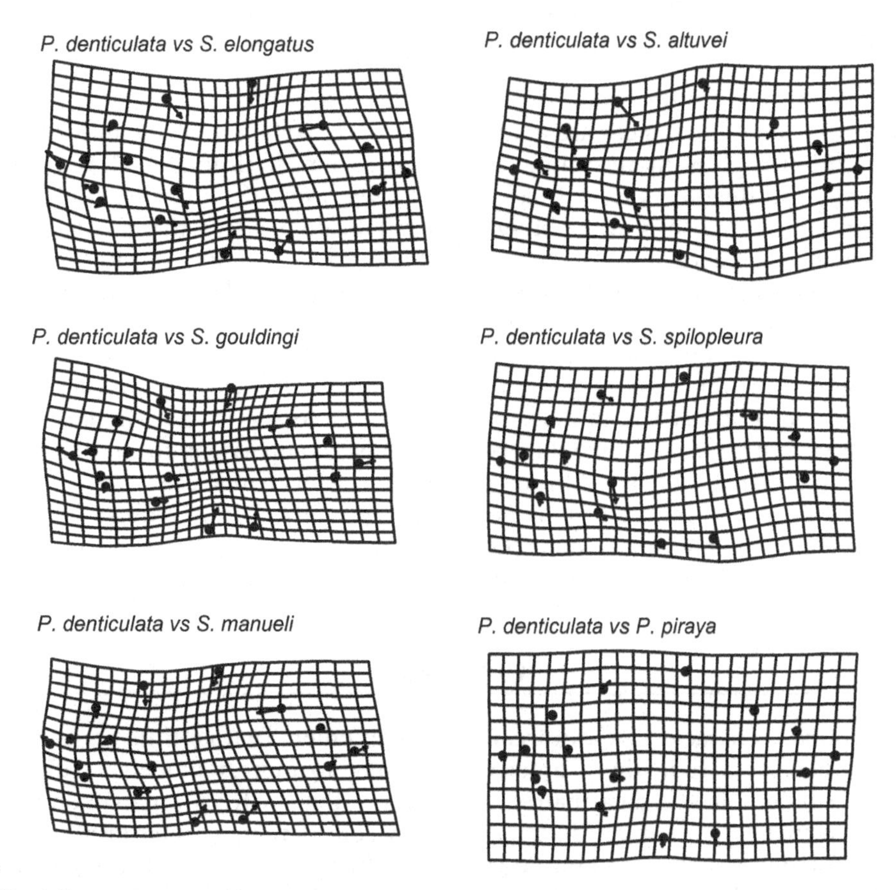

FIGURE 13.7 Pairwise comparisons between mean juvenile body shapes of *P. denticulata* and six other species. Comparisons to *P. nattereri* and *P. cariba* are not distinct from the comparison to *P. piraya*.

species have a relatively short mid-body relative to the more posterior body, a feature hinted at but not so clearly presented when the vectors were drawn from *P. denticulata* to the species in the PC1–PC2 plane (see Figure 13.5). This particular feature might reflect a decrease in the length of the dorsal fin relative to the posterior body (dorsally) and the posterior displacement of the pectoral fin and pelvic fins (rather than changes in proportions of body between them). The three shallow-bodied species appear to vary in the degree of "mid-body contraction", but they appear to be consistently more contracted than the others. The possibility that these three species are similar in having a relatively shortened mid-body is worth examining further because, unlike their shallow body, it is not obvious from a purely qualitative analysis.

To pursue that possibility further, we can compare the vectors of pairwise contrasts to each other, asking if a more contracted mid-body (compared to that of *P. denticulata*) is characteristic of the shallow-bodied species but not of the others. This is done by subtracting one of the pairwise vectors from another; where species are identical to *S. gouldingi* (in the differences from *P. denticulata*) the grid is square (Figure 13.8). Large differences indicate that the direction of change from *P. denticulata* to *S. gouldingi* is not shared by another taxon. Subtracting each contrast from the contrast between *P. denticulata* to *S. gouldingi* shows that *S. gouldingi* is not much shallower or deeper than either *S. elongatus* or *S. manueli*. All three differ from *P. denticulata* by nearly the same degree, and in that same direction. Some differences are evident in the relative length of the mid-body, however. The grid is slightly more contracted in that region, indicating that *S. gouldingi* is more extreme than the others in that feature. However, the differences are slight. In striking contrast, the comparisons to the other species indicate not only that *S. gouldingi* is far shallower than the others, but also that all differ from *S. gouldingi* in either the degree or the location of mid-body contraction. We could either take these results to mean that *S. elongatus, S. gouldingi* and *S. manueli* are all shallow-bodied and contracted in the mid-body compared to the other species, or we could continue the analysis, doing additional pairwise contrasts – this time between *P. denticulata* and *S. elongatus*, and also between *P. denticulata* and *S. manueli* – to determine that all three species are similarly different from the others. Of course, we would need additional comparisons to find features that are more widely shared, or specific to some of the deeper-bodied species.

Unlike the shallow body, which is so evident visually that it requires no detailed quantitative study, the mid-body contraction discerned in these comparisons is the kind of subtle feature that justifies the effort of a morphometric analysis.

CODING

Having found a character, we can treat it like any other. That is, if using conventional cladistic methods, we code the characters according to our preliminary judgments of homology, include it in the data matrix, and analyze that matrix by parsimony. Coding methods are a contentious subject; systematists vary considerably in their preferred criteria for coding. The debates have nothing to do with morphometrics except to the extent that the methods are applied to quantitative data, and that statistical methods are sometimes favored to decide whether species are different (and should therefore not be coded as

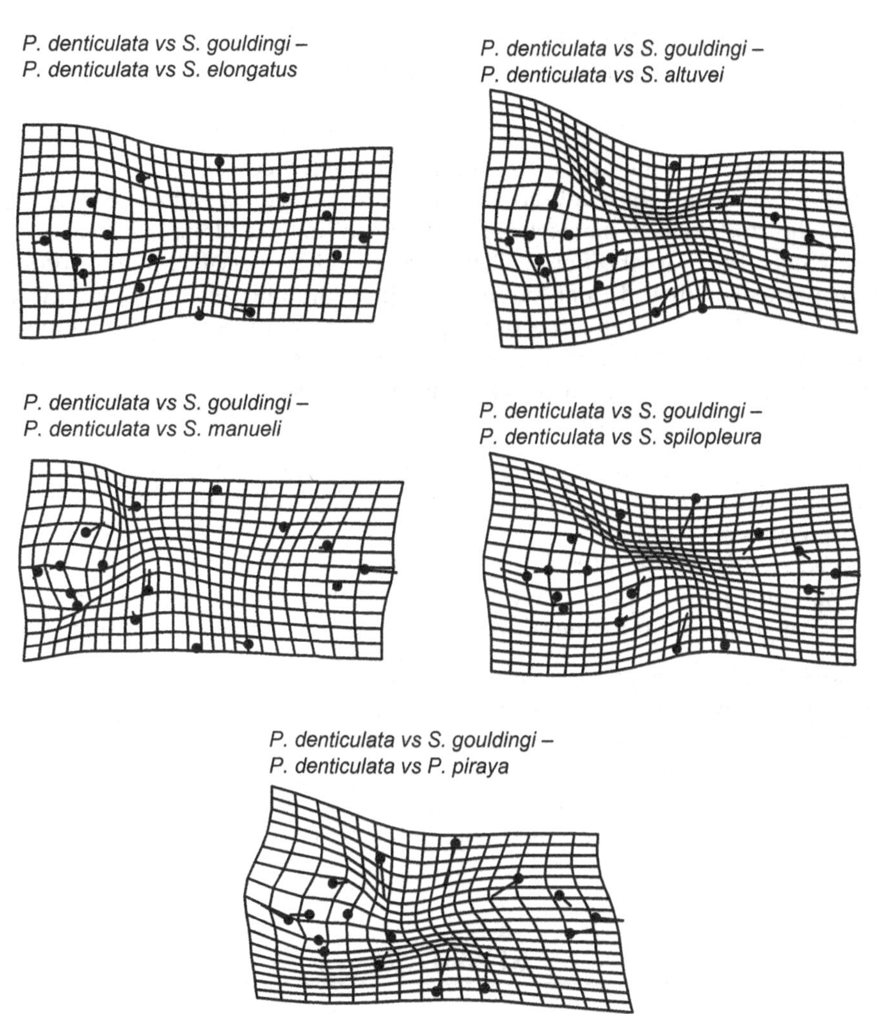

FIGURE 13.8 Comparisons among vectors describing the difference between *P. denticulata* and *S. gouldingi*, and the vector describing the difference between *P. denticulata* and each of the other six species shown in Figure 13.7. Each frame shows the contrast between the two vectors: where the squares of the grid are square, the two vectors are the same; where the grid shows large differences, the difference between that species and *P. denticulata* does not resemble the difference between *P. denticulata* and *S. gouldingi*.

having a homologous character). The literature on coding is large; interested readers can find general critiques of coding methods in several papers (e.g. Farris, 1990; Thiele, 1993; Gift and Stevens, 1997; Swiderski et al., 1998).

Any favored method can be applied to a variable that represents a character. In the special case where PC1 is a reasonable proxy of the character "body depth", the favored method can be applied to scores on PC1. It is not so easy to make decisions about characters that are combinations of several variables because it is not easy to examine the variation within species in more than two or three dimensions at a time, but doing so may be important for deciding whether species are similar enough to code their features as homologous.

The appropriateness of coding is, itself, a subject of debate. Not only are methods for coding controversial, even the idea of coding is. As mentioned earlier in this chapter, there are methods for inferring the evolution of shape that do not require coding characters. Some apply the trait values themselves, without coding, in standard tree construction algorithms (Farris, 1970; Goloboff et al., 2006). Others take a very different approach to the problem. In particular, they do not make preliminary hypotheses of homology, then formalize them by codes, then infer the phylogeny that minimizes the net number of extra steps (a step is considered "extra" if it means that a putatively homologous character is reinterpreted as arising more than once). Instead, they use explicit models of the evolutionary process, among which are:

1. Randomly varying directions of natural selection in different lineages
2. Random genetic drift of species around a single, stable optimum
3. Randomly wandering optima
4. Constrained wandering of optima
5. Wandering optima whose paths have a correlation that diminishes over time
6. Bursts of change around the time of speciation with little or no change thereafter.

(For a more detailed synopsis of the models, see Felsenstein, 2002.) Using one of these approaches avoids the whole issue of coding, but instead requires one to confront the problem of deciding which model is reasonable and justifiable. Such models have not been widely used to infer cladograms from morphological data and, like the methods which minimize a net morphometric distance (linear or squared) over a tree, model-based methods might best be considered as methods for reconstructing the evolution of shape given a cladogram.

SUMMARY

At present, no method is tailored to the problem of finding characters in morphometric data, and the available methods are cumbersome and involve an uncomfortable degree of subjectivity. Each could be improved by refining the part of the procedure that involves making linear combinations of variables, such as combining PC1, PC2 and PC3 to see whether similarities inferred from scores on one component are belied by scores on others. However, rather than improving methods that were devised to use standard morphometric techniques, it might be better to start at the beginning and develop a method tailored to our purposes. Doing so will require refining the statement of the problem. Currently, we cannot state the problem in mathematical terms, and that is necessary before we can find a mathematical solution. Our original statement of the problem focused on one particular element of it: finding characters without having to dissect organisms arbitrarily into parts prior to the phylogenetic analysis. However, that dissection need not be an integral part of a method for finding characters. We could instead use partial least squares analysis (Chapter 7) to test the hypothesis that the blocks of landmarks do not covary; if they do not, we can analyze them separately. Even though PLS does not test the hypothesis that a block constitutes an integrated unit, it may provide a more informed dissection than one based purely on anatomical conventions.

Clearly, we need additional methodological research – we should not be limited to the methods currently available when others are feasible. We also need to complement the methodological investigations by a discussion of what our concepts mean. If we do not, we may find that we have a rich array of methods that all do something interesting, but none that do what we intended. It can be bewildering to read discussions about morphometric characters, because it sometimes appears that nearly every author has a different idea of the meaning of "character" (as well of "morphometric"). Until we can define "character" precisely, in terms just as comprehensible to mathematicians as to systematists, we will not make further progress towards a mathematical solution. We also need more than a definition of the term; we need to articulate more fully the process by which we find characters, in general. Most discussions of systematic methods focus on how to analyze the data, given the data matrix. Our problem is to get that matrix in the first place. One value of morphometric data is that we find them using mathematical methods, and these are necessarily explicit. By making our methods of character analysis explicit, just like our methods of phylogenetic inference, we will enhance the rigor of morphological systematics in general.

References

Adams, D. C., & Rosenberg, M. S. (1998). Partial-warps, phylogeny, and ontogeny: A comment on Fink and Zelditch (1995). *Systematic Biology, 47*, 167–172.

Adams, D. C., Cardini, A., Monteiro, L. R., O'Higgins, P., & Rohlf, F. J. (2011). Morphometrics and phylogenetics: Principal components of shape from cranial modules are neither appropriate nor effective cladistic characters. *Journal of Human Evolution, 60*, 240–243.

Archie, J. W. (1985). Methods for coding variable morphological features for numerical taxonomic analysis. *Systematic Zoology, 34*, 236–345.

Chappill, J. A. (1989). Quantitative characters in phylogenetic analysis. *Cladistics, 5*, 217–234.

Colless, D. H. (1980). Congruence between morphometric and allozyme data for *Menidia* species: a reappraisal. *Systematic Zoology, 29*, 288–299.

Farris, J. S. (1970). Methods for computing Wagner Trees. *Systematic Zoology, 19*, 83–92.

Farris, J. S. (1990). Phenetics in camouflage. *Cladistics, 6*, 91–100.

Felsenstein, J. (2002). Quantitative characters, phylogenies, and morphometrics. In N. MacLeod, & P. L. Forey (Eds.), *Morphology, shape and phylogeny* (pp. 27–44). Taylor & Francis.

Fink, W. L., & Zelditch, M. L. (1995). Phylogenetic analysis of ontogenetic shape transformations: A reassessment of the piranha genus *Pygocentrus* (Teleostei). *Systematic Biology, 44*, 343–360.

Gift, N., & Stevens, P. F. (1997). Vagaries in the delimitation of character states in quantitative variation – an experimental study. *Systematic Biology, 46*, 112–125.

Goldman, N. (1988). Methods for discrete coding of morphological characters for numerical analysis. *Cladistics, 4*, 59–71.

Goloboff, P. A., Mattoni, C. I., & Quinteros, A. S. (2006). Continuous characters analyzed as such. *Cladistics, 22*, 589–601.

Rohlf, F. J. (1998). On applications of geometric morphometrics to studies of ontogeny and phylogeny. *Systematic Biology, 47*, 147–158.

Rohlf, F. J. (2002). Geometric morphometrics and phylogeny. In N. MacLeod, & P. L. Forey (Eds.), *Morphology, shape and phylogeny* (pp. 175–193). Taylor & Francis.

Simon, C. (1983). A new coding procedure for morphometric data with an example from periodical cicada wing veins. In J. Felsenstein (Ed.), *Numerical taxonomy* (pp. 378–382). Springer-Verlag.

Swiderski, D. L., Zelditch, M. L., & Fink, W. L. (1998). Why morphometrics isn't special: Coding quantitative data for phylogenetic analysis. *Systematic Biology, 47*, 508–519.

Thiele, K. (1993). The holy grail of the perfect character: The cladistic treatment of morphometric data. *Cladistics, 9*, 275–304.

Viscosi, V., & Cardini, A. (2011). Leaf morphology, taxonomy and geometric morphometrics: A simplified protocol for beginners. *PLoS ONE, 6*, e25630.

Zelditch, M. L., Fink, W. L., & Swiderski, D. L. (1995). Morphometrics, homology and phylogenetics: Quantified characters as synapomorphies. *Systematic Biology, 44*, 179–189.

CHAPTER 14

Forensic Applications of Geometric Morphometrics

In recent years, the forensic sciences have become enormously popular with the public due to developments in DNA technology (Lempert, 1997; Lynch, 2003) its use in high profile cases and the popularity of the CSI television series, which has raised public expectations of the performance of forensic sciences to the point that it has become a concern in jury selection (Schweitzer and Saks, 2007). At the same time, there has been increasing awareness of the tremendous variation in how forensic science is carried out in the USA (National Academy of Science, 2009; Budowle et al., 2009). Some types of forensic science, such as DNA and other chemical analyses, or methods from physical anthropology, are based on large scale laboratory science outside the forensic community as well as extensive use in the field by forensic agencies. Other areas of forensic analysis, including those of bitemarks or footwear impressions, have been developed largely by forensic or clinical practitioners based on clinical practice, case reports and anecdotal evidence. For these areas of investigation, there is little evidence or background information drawn from large-scale, systematic research projects and little involvement of scientists from national laboratories or academic positions (National Academy of Science, 2009). Apart from DNA and other chemically based analytic methods, a thorough knowledge of error rates and repeatability in many forensic areas is often lacking (National Academy of Science, 2009).

There have been a number of high profile exonerations based on DNA evidence (see The Innocence Project, www.innocenceproject.org, for examples). In addition to freeing a number of innocents, post-conviction DNA analyses have provided a one-time natural experiment that permits the analysis of factors contributing to wrongful convictions overturned by DNA. Several of the studies of these cases have indicated that errors in forensic science have been prominent contributors to these wrongful convictions, perhaps partially because of the high value placed on scientific evidence (Saks et al., 2000, 2001; Findley, 2002; Huff, 2004; Risinger, 2007). The 2009 report by the National Academy of Sciences (National Academy of Science, 2009) was a critical look at the practice of forensic science in the USA, prompted in

DOI: http://dx.doi.org/10.1016/B978-0-12-386903-6.00014-9

part by these cases. One of the outcomes of the 2009 NAS report was a call for more outside scientists to become involved in forensics, and for forensics workers to develop more connections with the general scientific community. Given these current trends, and some of the interesting questions posed in forensics, we chose to add a chapter on these topics.

In the forensic sciences, as in biology, anthropology and paleontology, there is much interest in comparing the shapes of objects. There are several specific areas of forensics in which morphometric methods, both traditional and geometric, have been applied. The first is the biometry of human form, where morphometric methods have been used to identify human remains and to estimate the age, sex and ethnic background of both pre- and antemortem humans. They can also aid the diagnosis of developmental or cognitive disorders which might be factors in sentencing. Non-invasive age estimation of living humans is currently of tremendous interest because immigrant and refugee policies in many countries, differ according to the applicant's age. Juveniles are often given a different, typically more advantageous status in consideration of refugee status or eligibility for immigration and social services (Solheim and Vonen, 2006; Crowley, 2007). Age estimation may also be a factor in criminal cases because sentencing guidelines are usually different for adults and juveniles. The second application of morphometrics in forensics is in the area of impression and pattern evidence. Impression evidence is created when two objects come in contact forcibly enough to create an image or disruption of the distribution of material on one or both objects, recording their interaction. Examples include fingerprints, palm prints, tool-marks and footprints.

In this chapter, we will first look at a morphometric approach that retains size information in the analysis (Procrustes Size Preserving, or Procrustes-SP). Next we will discuss what it means to say that shape data from different sources "match" one another; which differs from the usual concerns of biological studies with differences or variation. We will then examine three research programs that have used geometric morphometrics to address forensic problems in different areas and which also introduce new conceptual material. The first of these programs focuses on forensic bitemark analysis, primarily on the issue of variability in the anterior dentition as it is recorded in a bitemark. The question is whether this is a unique identifier of an individual. Studies directed at answering that question use the Procrustes-SP and Procrustes-based matching procedures. The second research project uses geometric morphometrics to estimate the sex of an individual post-mortem; we compare a discriminant function analysis with the *k*-means method of identifying group membership based on shape data. The third project applies three-dimensional geometric morphometric analysis to examination of human brain scans for evidence of Fetal Alcohol Spectrum Disorder (FASD) in a forensic context. The FASD studies illustrate the use of relative eigenanalysis to allow classification based on excess variation (hypervariability) within the FASD population relative to the normal population, and the use of likelihood ratios to report the strength of forensic evidence in the courtroom.

SIZE AND SHAPE

In most biological applications of morphometrics, there have been efforts to separate size and shape, and often to analyze shape independently of size. However, within the forensics community, size is viewed as an important part of the evidence. If the size of two impressions are not consistent with one another, that serves as grounds to conclude that the impressions arose from different sources (leaving aside the possibility that the recording medium has shrunk or expanded). Thus, in forensics settings, there can be an advantage to working with size and shape simultaneously. Dryden and Mardia (1992, 1998) developed distribution models using Procrustes shape variables and centroid size, treated jointly in distribution functions, but as separate quantities. This approach (sometimes referred to as Procrustes Form Space) has been used in a PCA (Mitteroecker et al., 2004), in which shape data and the log of centroid size were placed in a single data matrix and analyzed.

An alternative approach to size and shape analysis is to superimpose landmark coordinates of two or more configurations using only translation and rotation, but not using size changes (scaling). This produces landmark coordinates with $2k-3$ or $3k-6$ degrees of freedom, in superimpositions that could be called Procrustes Size Preserving (Procrustes-SP, Bush et al., 2011a; Sheets et al., 2013). Suppose we have k landmarks in two dimensions and want to superimpose the landmarks ($x_{t1}, y_{t1}, x_{t2}, y_{t2}\ldots x_{tk}, y_{tk}$) of a target ($t$) specimen on a reference (r) specimen ($x_{r1}, y_{r1}, x_{r2}, y_{r2}\ldots x_{rk}, y_{rk}$) by translations along the x and y directions (a_x and a_y) with a rotation through some angle θ. We can assume, without loss of generality, that the reference specimen is centered on the origin (0,0) so that:

$$\sum_{i=1}^{k} x_{ri} = 0 \tag{14.1}$$

$$\sum_{i=1}^{k} y_{ri} = 0 \tag{14.2}$$

but note that the reference is not scaled to a centroid size of one. Rather, it is left in the original measurement units. It is assumed that the target and reference are both measured in the same units. If we translate and rotate the target specimen, the landmark points are mapped to:

$$x'_{ti} = (x_{ti} - a_x)\cos(\theta) - (y_{ti} - a_y)\sin(\theta) \tag{14.3}$$

$$y'_{ti} = (x_{ti} - a_x)\sin(\theta) + (y_{ti} - a_y)\cos(\theta) \tag{14.4}$$

We can then calculate the squared distance between the reference and the rotated and translated target specimen:

$$D^2 = \sum_{i=1}^{k} (x_{ri} - (x_{ti} - a_x)\cos(\theta) + (y_{ti} - a_y)\sin(\theta))^2 + (y_{ri} - (x_{ti} - a_x)\sin(\theta) - (y_{ti} - a_y)\cos(\theta))^2 \tag{14.5}$$

Next we minimize this squared distance with respect to a_x, a_y and θ. This yields:

$$a_x = \frac{\sum_{i=1}^{k} x_{ti}}{k} \tag{14.6}$$

$$a_y = \frac{\sum_{i=1}^{k} y_{ti}}{k} \tag{14.7}$$

which is simply the requirement that the target specimen also has a centroid at the origin.

The rotation angle becomes:

$$\theta = \text{arctangent}\left(\frac{\sum_{i=1}^{k} (y_{ri}(x_{ti} - a_x) - x_{ri}(y_{ti} - a_y))}{\sum_{i=1}^{k} (x_{ri}(x_{ti} - a_x) + y_{ri}(y_{ti} - a_y))}\right) \tag{14.8}$$

which is identical to the expression for the angle of rotation found for a Procrustes superimposition using translation, rotation and scaling.

When one specimen is superimposed on a reference using this Procrustes-SP superimposition, the minimized squared distance between the target and reference is the squared Procrustes-SP distance between them. The translated landmark coordinates produced by this method may then be analyzed using multivariate statistical methods; bearing in mind the issues posed by the degrees of freedom and the lack of an underlying statistical model of the distribution of this type of variable. These limitations on this superimposition method and the resulting coordinates and distance measure mean that statistical studies that use the data obtained by Procrustes-SP methods should use tests based on permutation or other re-sampling methods, not analytic statistical models.

WHAT DOES IT MEAN FOR SHAPES TO “MATCH”?

One issue that arises in the analysis of pattern evidence is to determine whether one item of such evidence (e.g., a bitemark, a tool-mark or footwear pattern) “matches” another, so that the two patterns could have been produced by a common source (Bunch et al., 2009). Careful thought is needed to decide what it means for two measurements to match, whether those be traditional morphometric measurements, outlines, or landmarks superimposed by Procrustes or Procrustes-SP methods. One approach is to start from the premise that:

> If the observed difference between two measured specimens is no larger than the difference observed in repeated measurements of a single specimen, then there is no evidence that the two measured specimens are different, and thus they may be said to match.

In all continuous measurements made of physical objects, there is always some level of error or uncertainty, no matter how carefully it is measured. If we ask two people their heights, and both answer honestly that they are 5 foot 10 inches (178 cm) tall, then we would reasonably state that their heights are equal. However neither individual is likely to be exactly 5 foot 10 inches. They are probably within a span of 1 or ½ inch (2.54 or 1.27 cm) of 5 foot 10 inches at any given time. The claim that their heights are equal is due to our inability to measure a difference between them, given our measurement instruments, not to there being no difference at all in their heights. The issue of individuation in forensic science is surprisingly contentious (Saks and Koehler, 2008; Bunch et al., 2009;

Cole, 2009), that of matching perhaps slightly less so, we therefore proceed based on the simple discussion above.

In comparing two items of impression evidence based on shape, the first step is to determine the level of variation present in repeated measurements of the same specimen. A series of shape (or size and shape) measurements is made of a set of impressions produced by a single specimen. The variability within this set of repeated measures may be characterized by the variance within the collection or by looking at the distribution of all pair-wise distances between specimens in the repeated measurements group to develop an estimate of a confidence interval of distances within repeated measurements. If the distance between two shapes (measured on different impressions) lies within this repeated measures range of values, then there would appear to be little difference in shape, and hence no reason to claim that the impressions could not be have been produced by a common source. Detection of the "match" is thus based on a failure to reject a null hypothesis that the observed difference between the two arose from random variation in the measurement process. Unfortunately, this leaves us in the rather precarious position of accepting a null hypothesis, of simply stating that one cannot tell the objects apart.

Certainly we need to be cautious when taking this approach. For example, it may be that virtually all the difference between two shapes is localized to one landmark, inducing a Pinocchio effect. Because Procrustes methods tend to map variance across the whole configuration, a substantial change at one location might be mapped onto smaller changes at many landmarks. The result could be a distance between the two shapes comparable to one observed in repeated measures even though a large excursion at just one landmark is highly unlikely in repeated measurements. In such situations, Procrustes plots could be examined to check that the distribution of relative shape changes across landmarks produced by repeated measurements is consistent with the observed Pinocchio effect difference.

Using Procrustes (or Procustes SP) distances in the match criteria, albeit subject to concerns about the Pinocchio effect, allows for automated searches for matches in databases of specimens. It is straightforward to compute the Procrustes distance between a given specimen and all specimens in a database, and then to determine the number of matches found, given the match criteria.

MATCHING SHAPES IN THE HUMAN DENTITION

Use of bitemark analysis in criminal cases arose out of the highly successful practice of using dental records to identify the remains of individuals where extensive damage or decomposition has made it difficult to establish identity. Bitemark analysis is used to investigate particularly violent events, typically sexual assaults and child abuse, making this application a highly sensitive topic, not only because false identifications could imprison an innocent person, but also because a violent offender may be left at large. The analysis of bitemarks yields a range of forensic information in these disturbing crimes, but there are real concerns about its use as individuating evidence, i.e., as evidence uniquely linking a suspect to a bitemark. The National Academy of Sciences report (2009) noted that the unique association of a suspect with a bitemark requires that: (1) the human biting dentition must contain enough information to identify the suspect uniquely; and (2) the information in the dentition must be faithfully recorded in the bitemark to allow successful use of it as evidence.

In the effective post-mortem identifications, the forensic dentist often has access to detailed x-rays of the entire dentition, as well as information about the location and nature of dental work including fillings, root canals and bridges. This information is generally accepted as unique and adequate to identify a specific individual, having a long and non-controversial history of success. In contrast, in bitemark analysis, typically all that is available are impressions left by the incisal (cutting) surfaces of the twelve most anterior teeth in the human dentition (i.e., the six most forward teeth of the upper and lower jaws). Thus, a bitemark impression typically contains far less information than is available from all surfaces and internal structures of the whole dentition (normally 28–32 teeth, ± wisdom teeth).

The study most often cited to support the claim of uniqueness of the portion of the human dentition that creates bitemarks examined bitemarks in wax produced by 384 dentitions (Rawson et al., 1984). In this study, Rawson et al. superimposed or registered all the measured dentitions in a standard orientation, and then determined the x and y coordinates of the midpoint of each tooth (to ±1mm) and the relative angular orientation of each tooth within ±2.5 degrees (see Rawson et al., 1984; or Bush et al., 2011b, for complete details). Rawson et al. used a form of baseline registration that was fairly sophisticated for 1984, but not consistent with methods developed since then. From this information, they calculated the number of possible states (or positions) each tooth position and angle could occupy by the range of observed values divided by the measurement resolution. From the very large calculated number of possible states, Rawson and colleagues concluded that the human anterior dentition was effectively unique to individuals.

In reconsidering this work, Bush et al. (2011b) noted that the Rawson et al. model did not include correlation among teeth and assumed uniform distribution models. Bush and colleagues also noted that no attempt was made to search through the data to see if any dentitions matched. When Bush et al. (2011b) searched their collected data, they found matches between lower dentitions in relatively small data sets. Later, using two- and three-dimensional geometric morphometric data, they looked for matches in collections of measured human dentitions (Bush et al., 2011a,b,c; Sheets et al., 2011, 2013). In a later study (Sheets et al., 2013), a collection of 1106 paired sets of three-dimensional scanned maxillary and mandibular dentitions (upper and lower, respectively) were obtained from a commercial dental laboratory, which had used these scans to produce occlusal guards (night guards). The dental models were three-dimensional scans from private practice patients across the USA. (All necessary Human Subject Institutional Review Board protocols were completed for this project and exemption was granted and all patient identifying information was stripped from the data). This was a sample of convenience that contained a wide range of alignment patterns, from relatively straight to fairly mal-aligned. After initial matching studies, seven of these specimens were identified as being repeated scans of the same individuals and removed from the study, leaving 1099 distinct individuals.

Landmarks were recorded on the three-dimensional scanned dentitions by placing 10 data points along the incisal edge of each of the six anterior teeth, using the digitizing program Landmark (Institute for Data Analysis and Visualization, UC Davis, 2011, Figure 14.1). The dentitions were rotated in three-dimensional space within the software

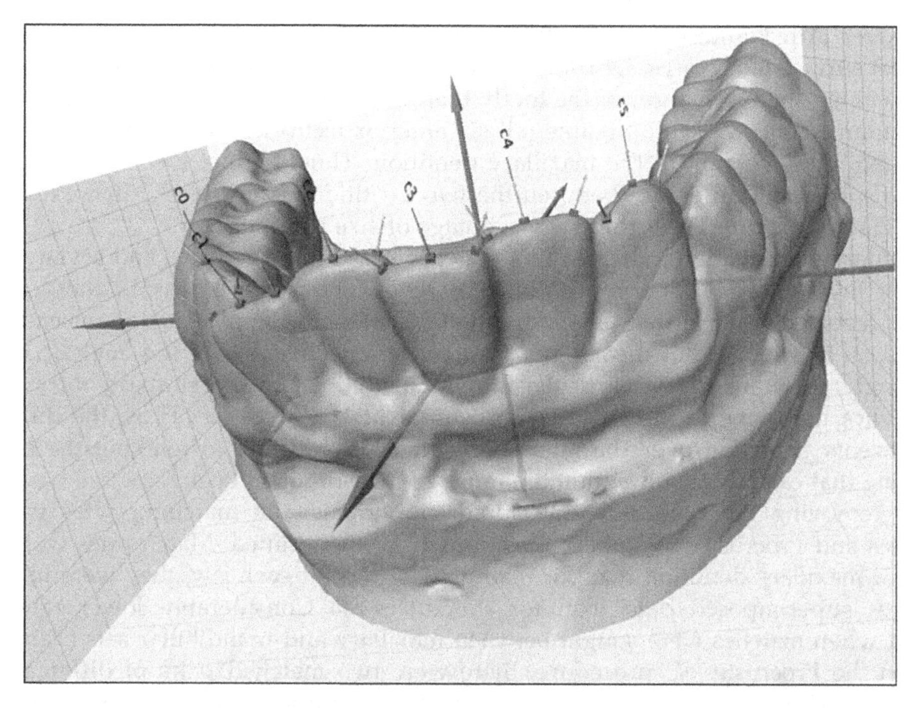

FIGURE 14.1 Landmark and semilandmark placement on three-dimensional scans of a cast made of a human dentition. Semilandmarks were placed along the occlusal surfaces of the six anterior teeth in the maxilla and mandible.

while landmarks were being placed so changes in three-dimensional perspective could be used to verify accurate placement. This resulted in a total of 60 points along the incisal edges of the anterior teeth of both the maxillary and mandibular arches.

To determine the measurement resolution, the scans of three maxillary and three mandibular specimens were digitized ten times by the same operator, and the scatter of Procrustes and Procrustes-SP distances for each specimen about the mean for that specimen was determined. This is an underestimate of the actual measurement error because it does not include estimates of the contributions from the creation of the dental casts and three-dimensional scanning. From the data of these distances, the average root mean square (RMS) scatter of repeated measurement specimens was calculated. The RMS scatter is analogous to a standard deviation, although it is not measured in the same way, nor does it have the same statistical properties. However, experiments have found that 93–96% of repeatedly digitized specimens lie within twice the RMS scatter level.

In Procrustes units the RMS scatter in this data was 0.02 for both the maxillary and mandibular dentition, so the matching criteria were twice this, 0.04. Because landmarks are scaled to the centroid size, this measurement indicates that our typical error in measuring all 60 landmarks was about 2% of the total size of the dentition. When using Procrustes-SP methods, the RMS scatter per landmark point was roughly 0.2 mm, still about 2% of centroid size, which is consistent with an approximate 1 mm width of the incisal edges. A maximum error of ±0.4 mm per point would still leave each landmark on the incisal surface in all cases. Of course there might also be a lateral component, a sliding of points along the top margin of the tooth, that contributes to this RMS scatter. The RMS scatter summed up over all 60 points (all six anterior teeth) in the mandibular dentition, was 3.1 mm and 4.28 mm in the maxillary dentition. Thus, the net error was 2% of total size. The substantial difference between the two dentitions is due to the increased size of the maxillary structure. The error as a percentage of size was similar.

An initial determination indicated five matches based on Procrustes and seven matches based on Procrustes-SP superimposition, but seven of the specimens in these matches were the repeated scans of the same individuals. That inclusion of repeated measurements provided an accidental, but meaningful, test of the effectiveness of the analytic method. Interestingly, the scans of one particular individual were taken a year apart, and were still within twice the range of the repeated measurements RMS scatter. Thus, the differences due to events occurring over the intervening time, in addition to retaking the cast and rescanning that cast, were still within the range of measurement error.

After removing the repeatedly measured individuals, the matching rates based on Procrustes and Procrustes-SP superimpositions were determined. Match rates were much higher for maxillary dentition than for mandibular, and for each jaw, they were higher for Procrustes superimposed data than for Procrustes-SP. Considerably lower rates were obtained when matches were sought between maxillary and mandibular sets (Table 14.1). Based on the Procrustes-SP procedure, there were two matched pairs of different specimens for a total of four individuals that were not unique (in only 1099). It is also notable that the two pairs that matched using the Procrustes-SP were not the same individuals who matched using the Procrustes method.

The dependence of match rate on measurement error is summarized in Table 14.2. In actual forensic cases, the information about the incisal edges will probably not be recorded with a resolution as high as that obtained from the three-dimensional laboratory data, so an alteration of the match rate is expected as measurement error increases. The level of error in repeated measurements was known to be underestimated because error in the casting and scanning operations was not included. Clearly, the rates of matches in this large population increases rapidly as measurement resolution or repeatability decreases, as represented by the increased RMS errors used to calculate this table.

TABLE 14.1 Match Rates in the Maxillary and Mandibular Dentitions, and in Both Combined, Based on Procrustes and Procrustes-SP Superimpositions

	Procrustes		Procrustes-SP	
	Number of Matches	**Number of Individuals**	**Number of Matches**	**Number of Individuals**
Maxillary	1691	487	763	396
Mandibular	129	131	75	83
Both	1	2	2	4

The matching criterion used was twice the RMS scatter of repeatedly measured specimens.

TABLE 14.2 The Dependence of the Number of Matches, and the Number of Individuals Matching in this Data Set, Under Both Criteria, for the Maxilla and Mandible, and as a Function of Increasing RMS Scatter Level, or Decreased Measurement Resolution

	Fraction of RMS Scatter					
	100%		125%		150%	
	Number of Matches	Number of Individuals	Number of Matches	Number of Individuals	Number of Matches	Number of Individuals
Procrustes						
Maxillary	1691	487	21358	873	85282	1007
Mandibular	129	131	3119	500	18543	769
Both	1	2	526	246	6826	579
Procrustes						
Maxillary	763	396	9660	826	39854	1001
Mandibular	75	83	1658	451	9510	759
Both	2	4	166	144	2056	502

The percentages are of the actual repeated measure scatter, so 125% represents a 25% increase in measurement error (603 351 total comparisons, 1099 individuals).

The authors of this study interpreted the results to mean that the human anterior dentition is not unique in an open population—the characteristics of the dentition are not generally suitable for uniquely identifying an individual. There is a tremendous amount of work left to be done to understand the effectiveness of forensic bitemark comparisons, but these results contradict the earlier claims of Rawson et al. (1984) and Kieser et al. (2007) about the uniqueness in this portion of the human dentition.

SEX ESTIMATION IN A FORENSIC CONTEXT

Sex estimation of humans is of interest in many forensic contexts (Krogman and Iscan, 1986) as well as in anthropological studies. One classic approach to sex estimation in postmortem settings has been to examine the pelvic structure, particularly the sciatic notch, but also the ischiopubic complex (MacLaughlin and Bruce, 1990; Bruzek, 2002). Approaches have been based on qualitative statements about these structures, on traditional morphometric methods and on geometric morphometric methods (Gonzalez et al., 2009; Steyn and Iscan, 2008). In the study carried out by Gonzalez and colleagues, a total of 121 human pelvic specimens were obtained from museum anthropological collections. The individuals were of European ancestry and had been buried in the 19th and 20th centuries. Photographs were taken with the camera oriented perpendicular to the largely two-dimensional plane of the structures of interest: the sciatic notch and the margin of the ischiopubic complex. Two landmarks and 14 semilandmarks were placed along the sciatic notch, with the landmarks placed at the base of the ischial spine and the tip of the piriform tubercle (Figure 14.2). The semilandmarks were placed along this arch using a "fan" guideline produced using the MakeFan program. Similarly, two landmarks and 35 semilandmarks were used to capture information about the ischiopubic region. One was placed at the intersection of the upper edge of the pubis with the perpendicular line that reaches the uppermost point of the obdurator groove. The second was placed at the intersection between the external margin of the ischium and the inferior border of the acetabulum. Thirty-five semilandmarks were then placed along the curve between these points. The placement of the first landmark on the ischiopubic region was difficult in this study as it required an estimated location based on a perpendicular line due to a lack of a clearly defined landmark in this region.

After landmark and semilandmark placement was completed, a semilandmark alignment procedure was done, using distance minimization on a mean form estimated using Generalized Procrustes Analysis methods (which also iteratively estimates the mean). PCA analysis of each data set showed substantial evidence of segregration of males and females along the first axis (58.2% of variance in the sciatic notch data, and 56.9% of the variance in the ischiopubic complex). These results were compared to those obtained from discriminant analysis and *k*-means clustering (MacQueen, 1967). Each of these analyses was performed on the same shape data as the PCA and on a combined size + shape data set (Procrustes form space). For the discriminant analysis (similar to CVA, but with only a single axis), correct classification rates were obtained using a resampling-based cross-validation analysis. In the *k*-means clustering method, the specimens are divided into

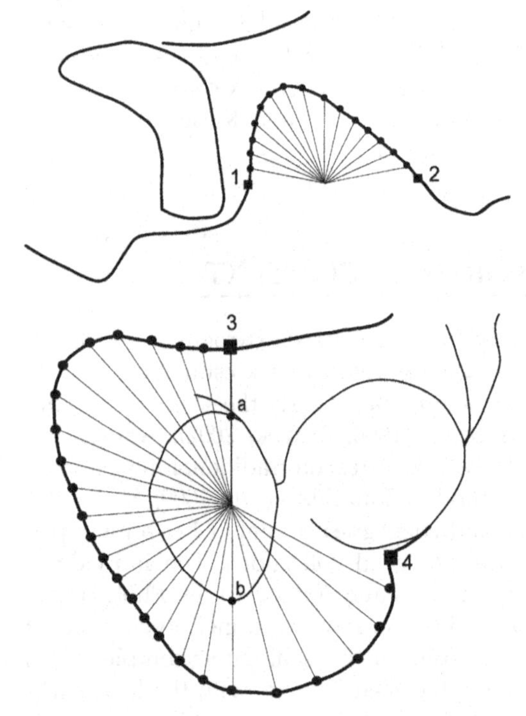

FIGURE 14.2 Landmark and semilandmark placement on the sciatic notch (a) and the ischiopubic complex (b) (From Gonzalez et al., 2009).

k groups *a priori* ($k = 2$ for sex estimation), the centroid of each group is determined and each specimen is reassigned to the nearest centroid. Then an iterative process is initiated in which a new group centroid is computed and specimens are re-assigned to the nearest centroid. The process continues until group memberships and the estimates of the centroids are stable. The resulting rate of assignments is assessed in the same manner as a CVA or discriminant analysis.

The assignment rates for the Gonzalez et al. study are shown in Tables 14.3 and 14.4. The overall cross-validation rates (frequencies of correct assignments) based on discriminant analysis of shape data only were 90.9% for the sciatic arch and 93.4% for the ischiopubic arch. The k-means rates were 90.9% and 90.1%. Females were misclassified slightly more often than males using ischiopubic data in both methods of analysis; combining ischiopubic and sciatic notch data yielded a slight improvement in classification rates for the discriminant function but not for k-means clustering. In this last result, the misclassified females were the three youngest but the misclassified males spanned a wide range of ages. Interestingly, including size in the analysis usually yielded worse cross-validation rates (Table 14.4) indicating that size was not helpful in forming these classifications.

Classification rates based on visual scoring rubrics have been reported as ranging from 96% correct (Phenice, 1969) to 60–90% correct (Bruzek, 2002; MacLaughlin and Bruce, 1990). Methods based on traditional morphometrics measurements have produced rates from 73% to 79.1% (Patriquin et al., 2003; Steyn and Iscan, 2008). Because these rates are probably

TABLE 14.3 Rates of Correct Assignment of Specimens to Gender Based on Sciatic Notch and Ischiopubic Complex Shape Using Discriminant Analysis and k-Means Clustering

	Discriminant Analysis			**k-Means Clustering**		
	Correct Assignments	**Incorrect Assignments**	**Percent Correct**	**Correct Assignments**	**Incorrect Assignments**	**Percent Correct**
Sciatic Notch						
F	47	5	90.4	46	6	88.5
M	63	6	91.4	64	5	92.3
Total	110	11	90.9	110	11	90.9
Ischiopubic Complex						
F	46	6	88.5	45	7	86.5
M	67	2	97.1	67	2	97.1
Total	113	8	93.4	112	9	90.1
Sciatic Notch & Ischiopubic Complex						
F	49	3	94.2	46	6	88.5
M	65	4	94.2	65	5	94.2
Total	114	7	94.2	111	11	91.7

Data taken from Gonzalez et al. (2009), Table 2.

TABLE 14.4 Rates of Correct Assignment of Specimens to Gender Based on Sciatic Notch and Ischiopubic Complex Size and Shape Using Discriminant Analysis and *k*-Means Clustering

	Discriminant Analysis			*k*-Means Clustering		
	Correct Assignments	Incorrect Assignments	Percent Correct	Correct Assignments	Incorrect Assignments	Percent Correct
Sciatic Notch						
F	46	6	88.4	45	7	86.5
M	63	6	91.3	64	5	92.8
Total	109	12	90.1	109	12	90.9
Ischiopubic Complex						
F	39	13	75	49	3	94.2
M	53	16	76.8	67	2	97.1
Total	92	29	76	116	5	95.9
Sciatic Notch & Ischiopubic Complex						
F	47	5	90.4	45	7	86.5
M	63	6	91.3	61	8	88.4
Total	110	11	90.1	106	15	87.6

Data taken from Gonzalez et al. (2009), Table 3.

somewhat population dependent, it is difficult to say with certainty that the geometric morphometric approaches are superior; they are certainly on par with the best results reported using the other methods, being perhaps less subjective (Bruzek, 2002) and more reliable, given the difficulties with traditional measures noted by Steyn and Iscan (2008).

LIKELIHOOD RATIOS AND FETAL ALCOHOL SYNDROME

Fetal Alcohol Spectrum Disorders (FASD) are a complex set of developmental, behavioral and anatomical disorders resulting from fetal exposure to alcohol. They encompass Fetal Alcohol Syndrome (FAS), Fetal Alcohol Effect (FAE), and Alcohol Related Neurodevelopmental Disorder (ARND). As with many multiple organ abnormality syndromes, diagnosis is often based on subjective assessment of several features, with terminologies that reflect the number of affected systems and the relative severity of dysfunction. Medicolegal concern with FASD centers on the tendency of affected individuals to exhibit a variety of cognitive deficits, including poor memory, attention deficits and impulsive behavior. Individuals with FAS have diminished executive functioning, a deficit in the skills and capabilities needed to self-regulate in an appropriate, context-dependent manner. These capabilities include planning, learning, processing sensory input, decision making, and response inhibition.

It has been argued in court that FASD is similar in many ways to other neural deficits (such as epilepsy) over which the sufferer has no control, and that the presence of FASD is a factor both in the treatment of offenders with FASD, and where the victim of a crime had FASD. Like other disabilities, individuals with FASD may have cognitive limitations which make it difficult for them to participate effectively in their own criminal defense, or which should be considered in the sentencing phase of a trial. Individuals with FASD may also be less capable of dealing with social situations, making them more vulnerable to some criminal acts. The vulnerability of persons with cognitive disorders is often taken into account in sentencing when they are victims of crimes. For these reasons, it may be important in a criminal case to determine whether or not an individual should be considered to have FASD.

Diagnosis of FASD in adults can be difficult, particularly if neither the mother nor others who can testify about her alcohol consumption during pregnancy are available. In such cases, diagnosis of FASD in an adult relies on physical measurements of facial features and neuroanatomy, as well as behavioral and cognitive testing (Streissguth et al., 1998). Information about the course of the individual's life, such as participation in special educational settings in school, may also be considered. In such cases, the diagnosis is often based on a preponderance of evidence.

To assess neuroanatomical consequences of fetal alcohol exposure, Streissguth, Bookstein and colleagues analyzed shape changes in the corpus callosum and other portions of the brain (Bookstein et al., 2001, 2002a,b). MRI images of three-dimensional brain scans were obtained from 180 subjects aged 14 to 37, with 60 individuals in each of three categories: (1) those diagnosed with fetal alcohol syndrome (FAS), (2) those diagnosed with fetal alcohol effects (FAE, i.e., having signs of neurological deficits but lacking obvious facial features required for diagnosis of FAS) (iii) and a control group of unexposed normal individuals without other diagnosed cognitive or neurological abnormalities or known history of fetal alcohol exposure.

Two data sets were used, one consisting of 12 landmark points along the mid-sagittal plane and 8 bilaterally paired points in the human brain measured in three-dimensions (see Bookstein et al., 2001 for details) and a set of mostly semilandmark measurements of the corpus callosum, (the large white matter structure in the brain that coordinates communication between the left and right cerebral hemispheres). One landmark was defined on the corpus callosum: a sharp corner near its anterior end called the rostrum. Most of the other landmarks were extremal points on recognized anatomical features – that is the anterior-most or dorsal-most point on the structure. The semilandmarks on the corpus callosum were digitized in three dimensions along a curve approximating its midline. This curve was not defined by an external reference plane; instead each point was placed on the surface in a position that appeared to be above (or below) the local plane of symmetry of the structure (Bookstein et al., 2002a). The semilandmarks were then slid using bending-energy alignment. In the initial study (Bookstein et al., 2001) differences in amount of asymmetry were a focus of analysis, so sides were not reflected and averaged, but this was done for some analyses in later studies.

For both the brain landmarks and the corpus callosum semilandmarks, means of exposed (FAS and FAE) individuals were not significantly different from unexposed (normal) individuals. However, there was a clear increase in variability (scatter) of exposed individuals about the mean. To characterize the elevated variability of the exposed populations and identify the most affected features, a *relative eigenanalysis* was performed. This technique compares within-group covariance matrices, expressing dimensions of greatest variability in one data set (here, the alcohol exposed population) as a multiple of the variance in the other data set (unexposed). When the data are shapes, the results can be used to produce a picture that shows which regions are more variable and the direction of that elevated variation. Because this study had more dimensions in either set of shape data than individuals in any subject group, the analysis was restricted to a subset of dimensions, the first 11 PCA axes. Relative eigenanalysis is similar to canonical variates analysis (CVA) in some respects, but instead of computing axes of between group difference as CVA does, relative eigenanalysis computes axes of within-group variance that are amplified in one group relative to the other.

The relative eigenalysis of the landmark data set indicated that the largest axis showing increased variance in exposed individuals could be expressed in terms of length ratios using only four anterior landmarks: the interior genu, the caudate, the rostrum and the genu. This was using measurements taken from a baseline registration with the genu and rostrum as endpoints. Scatter-plots of the relative height of the interior genu vs the height of the caudate show a tight cluster of most unexposed individuals, within a much larger scatter of exposed individuals (and a few unusual unexposed individuals). Relative eigenanalysis of the corpus callosum semilandmarks revealed a similar pattern of excess variance that was not as narrowly circumscribed anatomically, but scatterplots of scores on these eigenvectors again showed that many exposed individuals were outside the range of variation of most unexposed individuals.

Although the results is somewhat disappointing with regards to using brain scans as diagnostic tool for fetal alcohol exposure (the broad overlaps result in poor classification efficiency), they are interesting for their biological significance. Based on the evidence of elevated variation in the affected populations, the authors concluded that fetal alcohol exposure caused disruptions in the developmental process in the maturing fetal brain. This manifested as increased variation in structure, rather than specific defects or alterations, as there was no change in the mean shapes of the examined structures. Thus these data appear to be an example of disruption of a developmental process resulting in increased variation, rather than a specific alteration in shape.

In a later discussion, Bookstein and Kowell (2010) explain how a likelihood ratio can be constructed to express the strength of evidence that a given individual has FASD based on geometric measurements of brain structure. In a likelihood ratio, one calculates the probability of the data X given two different hypotheses H_1 and H_2, such as the two claims made by opposing sides in a courtroom (Lindley, 1977):

$$\text{Likelihood} = P(X|H_1)/P(X|H_2) \tag{14.9}$$

The likelihood ratio approach to describing the strength of forensic evidence has seen a rapid growth in other forensic settings as well (Lindley, 1977; Neumann et al., 2007; Su and Srihari, 2009, 2010).

In the case of the brain scan data, X is the set of processed measurements as described above, the classifying data. H_1 is the hypothesis that the individual suffered some level of fetal alcohol exposure and H_2 is the probability that the individual did not. In the case report discussed by Bookstein and Kowell (2010), the transect vector across the isthmus of the outline of the corpus callosum was used as a discriminating variable, based on a comparison of the individual defendant to normal and exposed individuals in the studies by Bookstein and collaborators (Bookstein et al., 2001, 2002a,b). The authors calculated a likelihood ratio of 800 to 1 for this particular set of measurements in this individual. They discuss how this type of evidence may then be combined with other evidence to diagnose FASD, such as behavioral tests, which can likewise yield likelihood ratios.

References

Bookstein, F. L., Sampson, P. D., Streissguth, A. P., & Connor, P. D. (2001). Geometric morphometrics of Corpus Callosum and subcortical structures in the fetal-alcohol-affected brain. *Teratology, 64*, 4–32.

Bookstein, F. L., Sampson, P. D., Connor, P. D., & Streissguth, A. P. (2002a). Midline Corpus Callosum is a neuro-anatomical focus of fetal alcohol damage. *The Anatomical Record - The New Anatomist, 269*, 169–174.

Bookstein, F. L., Streissguth, A. P., Sampson, P. D., Connor, P. D., & Barr, H. M. (2002b). Corpus Callosum shape and neurophyschological deficits in adult males with heavy fetal alcohol exposure. *NeuroImage, 15*, 233–251.

Bookstein, F. L., & Kowell, A. P. (2010). Bringing morphometrics into the fetal alcohol report: Statistical language for the forensic neurologist or psychiatrist. *Journal of Psychiatry and Law, 38*, 449–472.

Bruzek, J. (2002). A method for visual determination of sex, using the human hip bone. *American Journal of Physical Anthropology, 117*, 157–168.

Budowle, B., Bottrell, M. C., & Bunch, S. G., et al. (2009). A perspective on errors, bias and interpretation in the forensic sciences and direction for continuing advancement. *Journal of Forensic Science, 54*, 798–809.

Bunch, S. G., Smith, E. D., Giroux, B. N., & Murphy, P. D. (2009). Is a match really a match? A primer on the procedures and validity of firearm and toolmark identification. *Forensic Science Communications, 11* (3) <http://www.fbi.gov/about-us/lab/forensic-science-communications/fsc/july2009> Accessed 3.12.2012.

Bush, M. A., Bush, P. J., & Sheets, H. D. (2011a). A study of multiple bitemarks inflicted in human skin by a single dentition using geometric morphometrics analysis. *Forensic Science International, 211*, 1–8.

Bush, M. A., Bush, P. J., & Sheets, H. D. (2011b). Statistical evidence for the similarity of the human dentition. *International Journal of Forensic Science, 56*, 118–123.

Bush, M. A., Bush, P. J., & Sheets, H. D. (2011c). Similarity and match rates of the human dentition in three dimensions: relevance to bitemark analysis. *International Journal of Legal Medicine, 125*, 779–885.

Cole, S. A. (2009). Forensics without uniqueness, conclusions without individualization: the new epistemology of forensic identification. *Law, Probability and Risk*, 1–23.

Crowley, H. (2007). When is a child not a child? Asylum, age disputes and the process of age assessment. Immigration Law Pracitioners' Association, Research Report, London. <http://www.ilpa.org.uk/data/resources/13266/ILPA-Age-Dispute-Report.pdf> Accessed 3.14.2012.

Dryden, I. L., & Mardia, K. V. (1992). Size and shape analysis of landmark data. *Biometrika, 79*, 57–68.

Dryden, I. L., & Mardia, K. V. (1998). *Statistical shape analysis*. Chichester: Wiley.

Findley, K. A. (2002). Learning from our mistakes: A criminal justice commission to study wrongful convictions. *California Western Law Review, 38*, 333–353.

Gonzalez, P. N., Bernal, V., & Perez, S. I. (2009). Geometric morphometrics approach to sex estimation of human pelvis. *Forensic Science International, 189*, 68–74.

Huff, C. R. (2004). Wrongful conviction: Causes and public policy issues. *Criminal Justice, 15*, 15–19.

Institute for Data Analysis and Visualization. (2011). Landmark Software System. UC Davis. <http://graphics.idav.ucdavis.edu/research/projects/EvoMorph> Accessed 4.9.11.

Kieser, J. A., Bernal, V., Waddell, J. N., & Raju, S. (2007). The uniqueness of the human anterior dentition: A geometric morphometric analysis. *Journal of Forensic Science, 52*, 671–677.

Krogman, W. M., & Iscan, M. Y. (1986). *The human skeleton in forensic medicine.* Springfield, Illinois: Charles C. Thomas (publisher).

Lempert, R. (1997). After the DNA wars: Skirmishing with NRC II. *Jurimetrics Journal, 37*, 439–468.

Lindley, D. V. (1977). A problem in forensic science. *Biometrika, 64*, 207–213.

Lynch, M. (2003). God's signature: DNA profiling, the new gold standard in forensic science. *Endeavor, 27*, 93–97.

MacLaughlin, S. M., & Bruce, M. F. (1990). The accuracy of sex identification in European skeletal remains using the Phenice characters. *Journal of Forensic Science, 35*, 1384–1392.

MacQueen, J. (1967). Some methods for classification and analysis of multivariate observations. In *Proceedings of the 5th Berkeley symposium on mathematics, statistics and probability* (Vol. 1, pp. 281–297). Berkeley: University of California Press.

Mitteroecker, P., Gunz, P., Bernhard, M., Schaefer, K., & Bookstein, F. L. (2004). Comparison of cranial ontogenetic trajectories among great apes and humans. *Journal of Human Evolution, 46*, 679–698.

National Academy of Science. (2009). Committee on identifying the needs of the forensic sciences Community, National Research Council. *Strengthening forensic science in the United States: A path forward*. National Academies Press, February 2009.

Neumann, C., Champod, C., Puch-Solis, R., Egli, N., Anthonioz, A., & Bromage-Griffiths, A. (2007). Computation of likelihood ratios in fingerprint identification for configurations of any number of minutiae. *Journal of Forensic Sciences, 51*, 1255–1266.

Patriquin, M., Loth, S. R., & Steyn, M. (2003). Sexually dimorphic pelvic morphology in South African whites and blacks. *Homo, 53*, 255–262.

Phenice, T. W. (1969). A newly developed visual method of sexing os pubis. *American Journal of Physical Anthropology, 30*, 297–301.

Rawson, R. D., Ommen, R. K., Kinard, G., Johnson, J., & Yfantis, A. (1984). Statistical evidence for the individuality of the human dentition. *Journal of Forensic Science, 29*, 245–253.

Risinger, D. M. (2007). Convicting the innocent: An empirically justified wrongful conviction rate. *Journal of Criminal Law and Criminology, 97*, 761–817.

Saks, M. J., & Koehler, J. J. (2008). The individualization fallacy in forensic science evidence. *Vanderbilt Law Review, 61*, 199–219.

Saks, M. J., Constantine, L., & Dolezal, M., et al. (2000). Toward a model act for the prevention and remedy of erroneous convictions. *New England Legal Review, 35*, 669–683.

Saks, M. J., Constantine, L., Dolezal, M., & Garcia, J. (2001). Model prevention and remedy of erroneous convictions. *Arizona State Law Journal, 33*, 669–719.

Schweitzer, N. J., & Saks, M. J. (2007). The CSI effect: Popular fiction about forensic science affects the public's expectations about real forensic science. *Jurimetrics, 47*, 357–364.

Sheets, H. D., Bush, P. J., & Bush, M. A. (2011). Mathematical matching of a dentition to bitemarks: Use and evaluation of affine methods. *Forensic Science International, 207*, 111–118.

Sheets, H. D., Bush, P. J., & Bush, M. A. (2013). Patterns of variation and match rates of the anterior biting dentition: Characteristics of a database of 3D scanned dentitions. *Journal of Forensic Science*, in press

Solheim, T., & Vonen, A. (2006). Dental age estimation, quality assurance and age estimation of asylum seekers in Norway. *Forensic Science International, 159*(Suppl), S56–S60.

Steyn, M., & Iscan, M. Y. (2008). Metric sex determination from the pelvis in modern Greeks. *Forensic Science International, 179*, 86.

Streissguth, A. P., Bookstein, F. L., Barr, H., Press, S., & Sampson, P. (1998). A fetal alcohol behavior scale. *Alcoholism: Clinical and Experimental Research, 22*, 325–333.

Su, C., & Srihari, S. N. (2009). *Probability of random correspondence for fingerprints*. In *IWCF 2009 Proceedings* (pp. 55–66). Berlin: Springer-Verlag.

Su, C., & Srihari, S. N. (2010). Evaluation of rarity of fingerprints in forensics. *Proceedings of neural information processing systems*. Vancouver, Canada, December 6–9, 2010.

Bibliography

Abdala, F., Flores, D. A., & Giannini, N. P. (2001). Postweaning ontogeny of the skull of *Didelphis albiventris*. *Journal of Mammalogy, 82*, 190–200.

Ackermann, R. R., & Krovitz, G. E. (2002). Common patterns of facial ontogeny in the hominid lineage. *Anatomical Record, 269*, 142–147.

Adams, D. C., & Collyer, M. L. (2009). A general framework for the analysis of phenotypic trajectories in evolutionary studies. *Evolution, 63*, 1143–1154.

Adams, D. C., & Nistri, A. (2010). Ontogenetic convergence and evolution of foot morphology in European cave salamanders (Family: Plethodontidae). *Bmc Evolutionary Biology, 10*.

Adams, D. C., & Rosenberg, M. S. (1998). Partial-warps, phylogeny, and ontogeny: A comment on Fink and Zelditch (1995). *Systematic Biology, 47*, 167–172.

Adams, D. C., Rohlf, F. J., & Slice, D. E. (2004). Geometric morphometrics: ten years of progress following the 'revolution'. *Italian Journal of Zoology, 71*, 5–16.

Adams, D. C., Cardini, A., Monteiro, L. R., O'Higgins, P., & Rohlf, F. J. (2011). Morphometrics and phylogenetics: Principal components of shape from cranial modules are neither appropriate nor effective cladistic characters. *Journal of Human Evolution, 60*, 240–243.

Akaike, H. (1974). A new look at the statistical model identification. *IEEE Transactions on Automatic Control, 19*, 716–723.

Alberch, P., Gould, S. J., Oster, G. F., & Wake, D. B. (1979). Size and shape in ontogeny and phylogeny. *Paleobiology, 5*, 296–317.

Albrecht, G. (1978). Some comments on the use of ratios. *Systematic Zoology, 27*, 67–71.

Ancel, L. W., & Fontana, W. (2000). Plasticity, evolvability, and modularity in RNA. *Journal of Experimental Zoology, 288*, 242–283.

Anderson, M. J. (2001a). A new method for non-parametric multivariate analysis of variance. *Austral Ecology, 26*, 32–46.

Anderson, M. J. (2001b). Permutation tests for univariate or multivariate analysis of variance and regression. *Canadian Journal of Fisheries and Aquatic Sciences, 58*, 626–639.

Anderson, M. J. (2003). *XMATRIX: A FORTRAN computer program for calculating design matrices for terms in ANOVA designs in a linear model.* Department of Statistics, University of Auckland.

Anderson, M. J. (2004). *DISTLM: Distance-based multivariate analysis for a linear model.* Department of Statistics, University of Auckland.

Anderson, M. J. (2006). Distance-based tests for homogeneity of multivariate dispersions. *Biometrics, 62*, 245–253.

Anderson, M. J., & Robinson, J. (2001). Permutation tests for linear models. *Australian & New Zealand Journal of Statistics, 43*, 75–88.

Anderson, M. J., & ter Braak, C. J. F. (2003). Permutation tests for multi-factorial analysis of variance. *Journal of Statistical Computation and Simulation, 73*, 85–113.

Anderson, T. W. (1958). *An introduction to multivariate analysis.* New York: Wiley.

Angielczyk, K. D., & Sheets, H. D. (2007). Investigation of simulated tectonic deformation in fossils using geometric morphometrics. *Paleobiology, 33*, 125–148.

Anstey, R. L., & Pachut, J. F. (1995). Phylogeny, diversity history, and speciation in Paleozoic Bryozoans. In D. H. Erwin, & R. L. Anstey (Eds.), *New approaches to speciation in the fossil record* (pp. 239–284). Columbia University Press.

Archie, J. W. (1985). Methods for coding variable morphological features for numerical taxonomic analysis. *Systematic Zoology, 34*, 236–345.

Arendt, J. (2010). Morphological correlates of sprint swimming speed in five species of spadefoot toad tadpoles: Comparison of morphometric methods. *Journal of Morphology, 271*, 1044–1052.

Arif, S., Adams, D. C., & Wicknick, J. A. (2007). Bioclimatic modelling, morphology, and behaviour reveal alternative mechanisms regulating the distributions of two parapatric salamander species. *Evolutionary Ecology Research, 9*, 843–854.

Arnqvist, G., & Martensson, T. (1998). Measurement error in geometric morphometrics: Empirical strategies to assess and reduce its impact on measures of shape. *Acta Zoologica Academiae Scientiarum Hungaricae, 44*, 73–96.

Atchley, W. R., & Anderson, D. (1978). Ratios and the statistical analysis of biological data. *Systematic Zoology, 27*, 71–78.

Atchley, W. R., & Hall, B. K. (1991). A model for development and evolution of complex morphological structures. *Biological Reviews of the Cambridge Philosophical Society, 66*, 101–157.

Atchley, W. R., & Hall, B. K. (1991). A model for development and evolution of complex morphological structures. *Biological Reviews of the Cambridge Philosophical Society, 66*(2), 101–157.

Atchley, W. R., Gaskins, C. T., & Anderson, D. (1976). Statistical properties of ratios. I. Empirical results. *Systematic Zoology, 25*, 137–148.

Axler, S. (1996). *Linear algebra done right*. New York: Springer-Verlag.

Barker, M., & Rayens, W. (2003). Partial least squares for discrimination. *Journal of Chemometrics, 17*, 166–173.

Barrow, E., & Macleod, N. (2008). Shape variation in the mole dentary (Talpidae : Mammalia). *Zoological Journal of the Linnean Society, 153*, 187–211.

Bartlett, M. S. (1947). Multivariate analysis. *Journal of the Royal Statistical Society, Series B, 9*, 176–197.

Bastir, M., & Rosas, A. (2004). Comparative ontogeny in humans and chimpanzees: Similarities, differences and paradoxes in postnatal growth and development of the skull. *Annals of Anatomy-Anatomischer Anzeiger, 186*, 503–509.

Bastir, M., & Rosas, A. (2005). Hierarchical nature of morphological integration and modularity in the human posterior face. *American Journal of Physical Anthropology, 128*, 26–34.

Bastir, M., & Rosas, A. (2006). Correlated variation between the lateral basicranium and the face: A geometric morphometric study in different human groups. *Archives of Oral Biology, 51*, 814–824.

Bastir, M., Rosas, A., & Sheets, H. D. (2005). The morphological integration of the hominoid skull: A partial least squares and PC analysis with morphogenetic implications for European mid-pleistocene mandibles. In D. Slice (Ed.), *Modern morphometrics in physical anthropology* (pp. 265–284). New York: Kluwever Academic/Plenum Publishers.

Bastir, M., O'Higgins, P., & Rosas, A. (2007). Facial ontogeny in Neanderthals and modern humans. *Proceedings of the Royal Society B-Biological Sciences, 274*, 1125–1132.

Bastir, M., Sobral, P. G., Kuroe, K., & Rosas, A. (2008). Human craniofacial sphericity: A simultaneous analysis of frontal and lateral cephalograms of a Japanese population using geometric morphometrics and partial least squares analysis. *Archives of Oral Biology, 53*, 295–303.

Benitez-Vieyra, S., Medina, A. M., & Cocucci, A. A. (2009). Variable selection patterns on the labellum shape of *Geoblasta pennicillata*, a sexually deceptive orchid. *Journal of Evolutionary Biology, 22*, 2354–2362.

Berg, R. L. (1960). The ecological significance of correlation pleiades. *Evolution, 14*, 171–180.

Bininda-Emonds, O. R. P., Jeffrey, J. E., & Richardson, M. K. (2003). Inverting the hourglass: quantitative evidence against the phylotypic stage in vertebrate development. *Proceedings of the Royal Society of London, Series B, 270*, 341–346.

Birch, J. M. (1999). Skull allometry in the marine toad, *Bufo marinus*. *Journal of Morphology, 241*, 115–126.

Blackstone, N. W. (1987). Allometry and relative growth: Pattern and process in evolutionary studies. *Systematic Zoology, 36*, 76–78.

Blomberg, S. P., Garland, T., Jr., & Ives, A. R. (2003). Testing for phylogenetic signal in comparative data: Behavioral traits are more labile. *Evolution, 57*, 717–745.

Bookstein, F. L. (1982). The geometric meaning of soft modeling, with some generalizations. In K. G. Jöreskog, & H. Wold (Eds.), *Systems under indirect observation: causality, structure, prediction* (pp. 55–74). New York: North Holland Publishing Co.

Bookstein, F. L. (1989). "Size and shape": A comment on semantics. *Systematic Zoology, 38*, 173–190.

Bookstein, F. L. (1989). Principal warps: Thin-plate splines and the decomposition of deformations. *IEEE Transactions on Pattern Analysis and Machine Intelligence, 11*, 567–585.

Bookstein, F. L. (1991). *Morphometric tools for landmark data: Geometry and biology*. Cambridge: Cambridge University Press.

Bookstein, F. L. (1996). Combining the tools of geometric morphometrics. In L. F. Marcus, M. Corti, A. Loy, G. J. P. Naylor, & D. E. Slice (Eds.), *Advances in morphometrics* (pp. 131–151). New York: Plenum.

Bookstein, F. L. (1996). Standard formula for the uniform shape component in landmark data. In L. F. Marcus, M. Corti, A. Loy, G. J. P. Naylor, & D. E. Slice (Eds.), *Advances in morphometrics* (pp. 53–168). New York: Plenum.

Bookstein, F. L. (1997). Landmark methods for forms without landmarks: Morphometrics of group differences in outline shape. *Medical Image Analysis, 1*, 97–118.

Bookstein, F. L., & Kowell, A. P. (2010). Bringing morphometrics into the fetal alcohol report: Statistical language for the forensic neurologist or psychiatrist. *Journal of Psychiatry and Law, 38*, 449–472.

Bookstein, F. L., Chernoff, B. L., Elder, R. L., Humphries, J. M., Jr., Smith, G. R., & Strauss, R. E. (1985). *Morphometrics in evolutionary biology*. Philadelphia: The Academy of Natural Sciences Philadelphia.

Bookstein, F. L., Sampson, P. D., Streissguth, A. P., & Connor, P. D. (2001). Geometric morphometrics of Corpus Callosum and subcortical structures in the fetal-alcohol-affected brain. *Teratology, 64*, 4–32.

Bookstein, F. L., Streissguth, A. P., Sampson, P. D., Connor, P. D., & Barr, H. M. (2002). Corpus callosum shape and neuropsychological deficits in adult males with heavy fetal alcohol exposure. *Neuroimage, 15*, 233–251.

Bookstein, F. L., Sampson, P. D., Connor, P. D., & Streissguth, A. P. (2002a). Midline Corpus Callosum is a neuroanatomical focus of fetal alcohol damage. *The Anatomical Record - The New Anatomist, 269*, 169–174.

Bookstein, F. L., Streissguth, A. P., Sampson, P. D., Connor, P. D., & Barr, H. M. (2002b). Corpus Callosum shape and neurophyschological deficits in adult males with heavy fetal alcohol exposure. *NeuroImage, 15*, 233–251.

Bookstein, F. L., Gunz, P., & Ingeborg, H., et al. (2003). Cranial integration in Homo: Singular warps analysis of the midsagittal plane in ontogeny and evolution. *Journal of Human Evolution, 44*, 167–187.

Both, C., & Visser, M. E. (2005). The effect of climate change on the correlation between avian life-history traits. *Global Change Biology, 11*, 1606–1613.

Both, C., van Asch, M., Bijlsma, R. G., van den Burg, A. B., & Visser, M. E. (2009). Climate change and unequal phenological changes across four trophic levels: Constraints or adaptations? *Journal of Animal Ecology, 78*, 73–83.

Boughner, J. C., & Dean, M. C. (2008). Mandibular shape, ontogeny and dental development in bonobos *(Pan paniscus)* and chimpanzees (*Pan troglodytes*). *Evolutionary Biology, 35*, 296–308.

Bowers, C. M. (2006). Problem-based analysis of bitemark misidentifications: The role of DNA. *Forensic Science International, 159S*, S104–109.

ter Braak, C. J. F. (1992). Permutation versus bootstrap significance test in multiple regression and ANOVA. In K. H. Jöckel, G. Rothe, & W. Sendler (Eds.), *Bootstrapping and related techniques* (pp. 79–86). Berlin: Springer-Verlag.

Breno, M., Leirs, H., & Van Dongen, S. (2011). No relationship between canalization and developmental stability of the skull in a natural population of *Mastomys natalensis* (Rodentia: Muridae). *Biological Journal of the Linnean Society, 104*, 207–216.

Breuker, C. J., Patterson, J. S, & Klingenberg, C. P. (2006). A single basis for developmental buffering of *Drosophila* wing shape. *Plos One, 1*.

Bruzek, J. (2002). A method for visual determination of sex, using the human hip bone. *American Journal of Physical Anthropology, 117*, 157–168.

Budowle, B., Bottrell, M. C., & Bunch, S. G., et al. (2009). A perspective on errors, bias and interpretation in the forensic sciences and direction for continuing advancement. *Journal of Forensic Science, 54*, 798–809.

Bulygina, E., Mitteroecker, P., & Aiello, L. (2006). Ontogeny of facial dimorphism and patterns of individual development within one human population. *American Journal of Physical Anthropology, 131*, 432–443.

Bunch, S. G., Smith, E. D., Giroux, B. N., & Murphy, P. D. (2009). Is a match really a match? A primer on the procedures and validity of firearm and toolmark identification. *Forensic Science Communications, 11*. (3)<http://www.fbi.gov/about-us/lab/forensic-science-communications/fsc/july2009> Accessed 3.12.2012

Bush, M. A. (2011). Forensic dentistry and bitemark analysis: Sound science or junk science? *Journal of the American Dental Association, 142*, 997–999.

Bush, M. A., Bush, P. J., & Sheets, H. D. (2011a). A study of multiple bitemarks inflicted in human skin by a single dentition using geometric morphometrics analysis. *Forensic Science International, 211*, 1–8.

Bush, M. A., Bush, P. J., & Sheets, H. D. (2011b). Statistical evidence for the similarity of the human dentition. *International Journal of Forensic Science, 56*, 118–123.

Bush, M. A., Bush, P. J., & Sheets, H. D. (2011c). Similarity and match rates of the human dentition in three dimensions: relevance to bitemark analysis. *International Journal of Legal Medicine, 125*, 779–885.

Campbell, N. A., & Atchley, W. R. (1981). The geometry of canonical variates analysis. *Systematic Zoology, 30*, 268–280.

Cardini, A., & Thorington, R. W. (2006). Postnatal ontogeny of marmot (Rodentia, Sciuridae) crania: Allometric trajectories and species divergence. *Journal of Mammalogy, 87*, 201–215.

Cardini, A., & Tongiorgi, P. (2003). Yellow-bellied marmots (*Marmota flaviventris*) "in the shape space" (Rodentia, Sciuridae): Sexual dimorphism, growth and allometry of the mandible. *Zoomorphology, 122*, 11–23.

Carlson, R. L., & Wainwright, P. C. (2010). The ecological morphology of darter fishes (Percidae: Etheostomatinae). *Biological Journal of the Linnean Society, 100*, 30–45.

Caumul, R., & Polly, P. D. (2005). Phylogenetic and environmental components of morphological variation: Skull, mandible, and molar shape in marmots (Marmota, Rodentia). *Evolution, 59*, 2460–2472.

Chapman, R. E. (1990). In F. J. Rohlf, & F. Bookstein (Eds.), *Conventional procrustes approaches* (pp. 251–267). Ann Arbor: University of Michigan Museum of Zoology.

Chappill, J. A. (1989). Quantitative characters in phylogenetic analysis. *Cladistics, 5*, 217–234.

Chatfield, C., & Collins, A. J. (1980). *Introduction to multivariate analysis*. London: Chapman & Hall.

Cheverud, J. M. (1982). Phenotypic, genetic and environmental integration in the cranium. *Evolution, 36*, 499–512.

Cheverud, J. M. (1984). Quantitative genetics and developmental constraints on evolution by selection. *Journal of Theoretical Biology, 110*, 155–172.

Cheverud, J. M. (1995). Morphological integration in the saddle-back tamarin (*Saguinus fuscicollis*) cranium. *American Naturalist, 145*, 63–89.

Cheverud, J. M. (2004). Modular pleiotropic effects of quantitative trait loci on morphological traits. *Modularity in Development and Evolution*, 132–153.

Cheverud, J. M., Dow, M. M., & Leutenegger, W. (1985). The quantitative assessment of phylogenetic constraints in comparative analyses: Sexual dimorphism in body weight among primates. *Evolution, 39*, 1335–1351.

Cheverud, J. M., Hartman, S. E., Richtsmeier, J. T., & Atchley, W. R. (1991). A quantitative genetic analysis of localized morphology in mandibles of inbred mice using finite-element scaling analysis. *Journal of Craniofacial Genetics and Developmental Biology, 11*, 122–137.

Cheverud, J. M., Routman, E. J., & Irschick, D. J. (1997). Pleiotropic effects of individual gene loci on mandibular morphology. *Evolution, 51*, 2006–2016.

Cheverud, J. M., Ehrich, T. H., Vaughn, T. T., Koreishi, S. F., Linsey, R. B., & Pletscher, L. S. (2004). Pleiotropic effects on mandibular morphology II: Differential epistasis and genetic variation in morphological integration. *Journal of Experimental Zoology Part B-Molecular and Developmental Evolution, 302B*, 424–435.

Ciampaglio, C. N. (2002). Determining the role that ecological and developmental constraints play in controlling disparity: Examples from the crinoid and blastozoan fossil record. *Evolution & Development, 4*, 170–188.

Ciampaglio, C. N., Kemp, M., & McShea, D. W. (2001). Detecting changes in morphospace occupation patterns in the fossil record: Characterization and analysis of measures of disparity. *Paleobiology, 27*, 695–715.

Clarke, G. M. (1993). Fluctuating asymmetry of invertebrate populations as a biological indicator of environmental quality. *Environmental Pollution, 82*, 207–211.

Clarke, G. M. (1998). The genetic basis of developmental stability. V. Inter- and intra-individual character variation. *Heredity, 80*, 562–567.

Claude, J. (2008). *Morphometrics. R.* New York: Springer.

Claverie, T., Chan, E., & Patek, S. N. (2011). Modularity and scaling in fast movements: Power amplication in mantis shrimp. *Evolution, 65*, 443–461.

Cobb, S., & O'Higgins, P. (2004). Hominins do not share a common postnatal facial ontogenetic shape trajectory. *Journal of Experimental Zoology (Mol Dev Evol), 302B*, 302–321.

Cole, S. A. (2009). Forensics without uniqueness, conclusions without individualization: the new epistemology of forensic identification. *Law, Probability and Risk*, 1–23.

Collar, D. C., Schulte, J. A., O'Meara, B. C., & Losos, J. B. (2010). Habitat use affects morphological diversification in dragon lizards. *Journal of Evolutionary Biology, 23*, 1033–1049.

Colless, D. H. (1980). Congruence between morphometric and allozyme data for *Menidia* species: a reappraisal. *Systematic Zoology, 29*, 288–299.

Cope, E. D. (1887). *The origin of the fittest*. New York: McMillan.

Corner, B. D., Lele, S., & Richtsmeier, J. T. (1992). Measuring precision of three-dimensional landmark data. *Journal of Quantitative Anthropology, 3*, 347–359.

Corruccini, R. S. (1977). Correlation properties of morphometric ratios. *Systematic Zoology*, *26*, 211–214.

Costa, C., Aguzzi, J., Menesatti, P., Antonucci, F., Rimatori, V., & Mattoccia, M. (2008). Shape analysis of different populations of clams in relation to their geographical structure. *Journal of Zoology*, *276*, 71–80.

Cronier, C., & Courville, P. (2003). Developmental variations in neodevonian phacopide trilobites. *Comptes Rendus Palevol*, *2*, 577–585.

Crowley, H. (2007). When is a child not a child? Asylum, age disputes and the process of age assessment. Immigration Law Pracitioners' Association, Research Report, London. <http://www.ilpa.org.uk/data/resources/13266/ILPA-Age-Dispute-Report.pdf> Accessed 3.14.2012.

Debat, V., & David, P. (2001). Mapping phenotypes: Canalization, plasticity and developmental stability. *Trends in Ecology & Evolution*, *16*, 555–561.

Debat, V., Alibert, P., David, P., Paradis, E., & Auffray, J. C. (2000). Independence between developmental stability and canalization in the skull of the house mouse. *Proceedings of the Royal Society of London Series B-Biological Sciences*, *267*, 423–430.

Debat, V., Debelle, A., & Dworkin, I. (2009). Plasticity, canalization, and developmental stability of the *Drosophila* wing: Joint effects of mutations and developmental temperature. *Evolution*, *63*, 2864–2876.

Diaz-Uriarte, R., & Garland, T., Jr. (1998). Effects of branch length errors on the performance of phylogenetically independent contrasts. *Systematic Biology*, *47*, 654–672.

Dodson, P. (1978). On the use of ratios in growth studies. *Systematic Zoology*, *27*, 62–67.

Drake, A. G. (2011). Dispelling dog dogma: An investigation of heterochrony in dogs using 3D geometric morphometric analysis of skull shape. *Evolution & Development*, *13*, 204–213.

Drake, A. G., & Klingenberg, C. P. (2008). The pace of morphological change: Historical transformation of skull shape in St Bernard dogs. *Proceedings of the Royal Society B-Biological Sciences*, *275*, 71–76.

Dryden, I. L., & Mardia, K. V. (1992). Size and shape analysis of landmark data. *Biometrika*, *79*, 57–68.

Dryden, I. L., & Mardia, K. V. (1998). *Statistical shape analysis*. New York: John Wiley & Sons.

Dryden, I. L., & Mardia, K. V. (1998). *Statistical shape analysis*. Chichester: Wiley.

Dworkin, I. (2005a). Evidence for canalization of Distal-less function in the leg of Drosophila melanogaster. *Evolution & Development*, *7*, 89–100.

Dworkin, I. (2005b). A study of canalization and developmental stability in the sternopleural bristle system of Drosophila melanogaster. *Evolution*, *59*, 1500–1509.

Eastman, J. M., Alfaro, M. E., Joyce, P., Hipp, A. L., & Harmon, L. J. (2011). A novel comparative method for identifying shifts in the rate of character evolution on trees. *Evolution*, *65*, 3578–3589.

Eble, G. J. (2003). Developmental morphospaces and evolution. In J. P. Crutchfield, & P. Schuster (Eds.), *Evolutionary dynamics* (pp. 35–65). Oxford: Oxford University Press.

Edgington, E. S. (1995). *Randomization tests*. New York: Marcel Dekker.

Edginton, E. S. (1995). *Randomization tests*. New York: Marcel Dekker.

Efron, B. (1979). Computers and the theory of statistics, thinking the unthinkable. *Society for Industrial and Applied Mathematics Review*, *21*, 460–480.

Efron, B. (1983). Estimating the error rate of a prediction rule: Improvement on cross-validation. *American Statistician*, *37*, 36–48.

Efron, B. (1987). Better bootstrap confidence intervals. *Journal of the American Statistical Association*, *82*, 171–185.

Efron, B. (1992). Jackknife-after-bootstrap standard errors and influence functions. *Journal of the Royal Statistical Society Series B, Methodological*, *54*, 83–127.

Efron, B., & Tibshirani, R. (1995). *Cross-validation and the bootstrap: Estimating the error rate of a prediction rule*. Canada: Department of Statistics, University of Toronto.

Efron, B., & Tibshirani, R. J. (1993). *An introduction to the bootstrap*. London: Chapman & Hall.

Ehrich, T. H., Vaughn, T. T., Koreishi, S., Linsey, R. B., Pletscher, L. S., & Cheverud, J. M. (2003). Pleiotropic effects on mandibular morphology I. Developmental morphological integration and differential dominance. *Journal Of Experimental Zoology Part B – Molecular And Developmental Evolution*, *296B*, 58–79.

Elena, S. F., & Lenski, R. E. (2001). Epistasis between new mutations and genetic background and a test of genetic canalization. *Evolution*, *55*, 1746–1752.

Emerson, S. B., Travis, J., & Blouin, M. (1988). Evaluating a hypothesis about heterochrony: Larval life-history traits and juvenile hindlimb morphology in *Hyla crucifer*. *Evolution*, *42*, 68–78.

Escoufier, Y. (1973). Le traitement des variables vectorielles. *Biometrics*, *29*, 751–760.

Estes, E. C. J., Katholi, C. R., & Angus, R. A. (2006). Elevated fluctuating asymmetry in eastern mosquitofish (Gambusia holbrooki) from a river receiving paper mill effluent. *Environmental Toxicology and Chemistry, 25*, 1026–1033.

Fadda, C., & Corti, M. (1998). Geographic variation of Arvicanthis (Rodentia, Muridae) in the Nile Valley. *Zeitschrift für Saugetierkunde-International Journal of Mammalian Biology, 63*, 104–113.

Falsetti, A. B., & Cole, T. M. (1992). Relative growth of the postcranial skeleton in callitrichines. *Journal of Human Evolution, 23*, 79–92.

Farris, J. S. (1970). Methods for computing Wagner Trees. *Systematic Zoology, 19*, 83–92.

Farris, J. S. (1990). Phenetics in camouflage. *Cladistics, 6*, 91–100.

Feller, W. (1968). *An introduction to probability theory and its applications*. John Wiley and Sons.

Felsenstein, J. (1985). Phylogenies and the comparative method. *American Naturalist, 125*, 1–15.

Felsenstein, J. (2002). Quantitative characters, phylogenies, and morphometrics. In N. MacLeod, & P. L. Forey (Eds.), *Morphology, shape and phylogeny* (pp. 27–44). Taylor & Francis.

Findley, K. A. (2002). Learning from our mistakes: A criminal justice commission to study wrongful convictions. *California Western Law Review, 38*, 333–353.

Fink, W. L. (1993). Revision of the piranha genus *Pygocentrus* (Teleostei, Characiformes). *Copeia*, 665–687.

Fink, W. L., & Zelditch, M. L. (1995). Phylogenetic analysis of ontogenic shape transformations – a reassessment of the piranha genus *Pygocentrus* (Teleostei). *Systematic Biology, 44*(3), 343–360.

Fink, W. L., & Zelditch, M. L. (1995). Phylogenetic analysis of ontogenetic shape transformations: A reassessment of the piranha genus *Pygocentrus* (Teleostei). *Systematic Biology, 44*, 343–360.

Fink, W. L., & Zelditch, M. L. (1996). Historical patterns of developmental integration in piranhas. *American Zoologist, 36*, 61–69.

Fiorello, C. V., & German, R. Z. (1997). Heterochrony within species: Craniofacial growth in giant, standard and dwarf rabbits. *Evolution, 51*, 250–261.

Fish, J. L., Villmoare, B., & Koebernick, K., et al. (2011). Satb2, modularity, and the evolvability of the vertebrate jaw. *Evolution & Development, 13*, 549–564.

Fisher, R. A. (1935). *The design of experiments*. Edinburgh: Oliver & Boyd.

Foote, M. (1986). Developmental buffering as a mechanism for stasis. *Evolution, 42*, 396–399.

Foote, M. (1990). Nearest-neighbor analysis of trilobite morphospace. *Systematic Zoology, 39*, 371–382.

Foote, M. (1992). Paleozoic record of morphological diversity in blastozoan echinoderms. *Proceedings of the National Academy of Sciences of the United States of America, 89*, 7325–7329.

Foote, M. (1993a). Contributions of individual taxa to overall morphological disparity. *Paleobiology, 19*, 403–419.

Foote, M. (1993b). Discordance and concordance between morphological and taxonomic diversity. *Paleobiology, 19*, 185–204.

Foote, M. (1994). Morphological disparity in Ordovician-Devonian crinoids and the early saturation of morphological space. *Paleobiology, 20*, 320–344.

Foote, M. (1997). The evolution of morphological diversity. *Annual Review of Ecology and Systematics, 28*, 129–152.

Foote, M., & Gould, S. J. (1992). Cambrian and recent morphological disparity. *Science, 258*, 1816.

Fornel, C., & Bookstein, F. (1982). *Two structural equation models: LISREL and PLS applied to consumer exit-voice theory, Journal of Marketing Research, 19*, 440–452.

Fornel, R., Cordeiro-Estrela, P., & De Freitas, T. R. O. (2010). Skull shape and size variation in Ctenomys minutus (Rodentia: Ctenomyidae) in geographical, chromosomal polymorphism, and environmental contexts. *Biological Journal of the Linnean Society, 101*, 705–720.

Frederich, B., & Sheets, H. D. (2010). Evolution of ontogenetic allometry shaping giant species: A case study from the damselfish genus Dascyllus (Pomacentridae). *Biological Journal of the Linnean Society, 99*, 99–117.

Frederich, B., & Vandewalle, P. (2011). Bipartite life cycle of coral reef fishes promotes increasing shape disparity of the head skeleton during ontogeny: An example from damselfishes (Pomacentridae). *Bmc Evolutionary Biology, 11*.

Frederich, B., Adriaens, D., & Vandewalle, P. (2008). Ontogenetic shape changes in Pomacentridae (Teleostei, Perciformes) and their relationships with feeding strategies: A geometric morphometric approach. *Biological Journal of the Linnean Society, 95*, 92–105.

Freund, J. E., & Walpole, R. E. (1980). *Mathematical statistics* (3rd ed.). Englewood Cliffs, NJ: Prentice-Hall.

Fuiman, L. A. (1983). Growth gradients in fish larvae. *Journal of Fish Biology, 23*, 117–123.

Gaillard, J. -M., Pontier, D., Allaine, D., Loison, A., Herve, J. -C., & Heizmann, A. (1997). Variation in growth form and precocity at birth in eutherian mammals. *Proceedings of the Royal Society B-Biological Sciences, 264*, 859–868.

Galatius, A., Berta, A., Frandsen, M. S., & Goodall, R. N. P. (2011). Interspecific variation of ontogeny and skull shape among porpoises (Phocoenidae). *Journal of Morphology, 272*, 136–148.

Garland, T., Jr., Dickerman, A. W., Janis, C. M., & Jones, J. A. (1993). Phylogenetic analysis of covariance by computer simulation. *Systematic Biology, 42*, 265–292.

Garland, T., Jr., Bennett, A. F., & Rezende, E. L. (2005). Phylogenetic approaches in comparative physiology. *Journal of Experimental Biology, 208*, 3015–3035.

Gaunt, S. J., Blum, M., & De Robertis, E. M. (1993). Expression of the mouse goosecoid gene during mid-embryogenesis may mark mesenchymal cell lineages in the developing head, limbs and ventral body wall. *Development, 117*, 769–778.

Gavrilets, S. (1999). Dynamics of morphological diversification on the morphological hypercube. *Proceedings of the Royal Society of London, Series B, 266*, 817–824.

Gerber, S. (2011). Comparing the differential filling of morphospace and allometric space through time: The morphological and developmental dynamics of early Jurassic ammonoids. *Paleobiology, 37*, 369–382.

Gerber, S., Eble, G. J., & Neige, P. (2008). Allometric space and allometric disparity: A developmental perspective in the macroevolutionary analysis of morphological disparity. *Evolution, 62*, 1450–1457.

Gerber, S., Eble, G. J., & Neige, P. (2011). Developmental aspects of morphological disparity dynamics: A simple analytical exploration. *Paleobiology, 37*, 237–251.

German, R. Z., Hertweck, D. W., Sirianni, J. E., & Swindler, D. R. (1994). Heterochrony and sexual dimorphism in the pigtailed macaque (*Macaca nemestrina*). *American Journal of Physical Anthropology, 93*, 373–380.

Gift, N., & Stevens, P. F. (1997). Vagaries in the delimitation of character states in quantitative variation – an experimental study. *Systematic Biology, 46*, 112–125.

Gkantidis, N., & Halazonetis, D. J. (2011). Morphological integration between the cranial base and the face in children and adults. *Journal of Anatomy, 218*, 426–438.

Goldman, N. (1988). Methods for discrete coding of morphological characters for numerical analysis. *Cladistics, 4*, 59–71.

Goloboff, P. A., Mattoni, C. I., & Quinteros, A. S. (2006). Continuous characters analyzed as such. *Cladistics, 22*, 589–601.

Gomez, J. M., Perfectti, F., & Camacho, J. P. M. (2006). Natural selection on *Erysimum mediohispanicum* flower shape: Insights into the evolution of zygomorphy. *American Naturalist, 168*, 531–545.

Gomez, J. M., Abdelaziz, M., Munoz-Pajares, J., & Perfectti, F. (2009). Heritability and genetic correlation of corolla shape and size in *Erysimum mediohispanicum*. *Evolution, 63*(7), 1820–1831.

Gonzalez, P. N., Bernal, V., & Perez, S. I. (2009). Geometric morphometrics approach to sex estimation of human pelvis. *Forensic Science International, 189*, 68–74.

Good, P. (1994). *Permutation tests: A practical guide to resampling methods for testing hypotheses.* New York: Springer-Verlag.

Goodall, C. (1991). Procrustes methods in the statistical analysis of shape. *Journal of the Royal Statistical Society, Series B: Methodological, 53*, 285–339.

Gottfries, J., Blennow, K., Wallin, A., & Gottfries, C. G. (1995). Diagnosis of dementias using partial least squares discriminant analysis. *Dementia, 6*, 83–88.

Gould, S. J. (1966). Allometry and size in ontogeny and phylogeny. *Biological Reviews of the Cambridge Philosophical Society, 41*, 587–640.

Gould, S. J. (1966). Allometry and size in ontogeny and phylogeny. *Biological Reviews, 41*, 587–640.

Gould, S. J. (1977). *Ontogeny and phylogeny*. Cambridge, MA: Harvard University Press.

Gould, S. J. (1982). Change in developmental timing as a mechanism of macroevolution. In J. T. Bonner (Ed.), *Evolution and development* (pp. 333–346). Berlin: Springer-Verlag.

Gould, S. J. (1984). Morphological channeling by structural constraint: Convergence in styles of dwarfing and gigantism in *Cerion*, with a description of two new fossil species and a report on the discovery of the largest *Cerion*. *Paleobiology, 10*, 172–194.

Gould, S. J. (1988). The uses of heterochrony. In M. L. McKinney (Ed.), *Heterochrony in evolution: A multidisciplinary approach* (pp. 1–13). New York: Plenum Press.

Gould, S. J., & Garwood, R. A. (1969). Levels of integration in mammalian dentitions: An analysis of correlations in *Nesophantes micrus* (Insectivora) and *Oryzomys couesi* (Rodentia). *Evolution, 23*, 276–300.

Graham, J. H., Freeman, D. C., & Emlen, J. M. (1993). Developmental stability: A sensitive indicator of populations under stress. In W. G. Landis, J. S. Hughes, & M. A. Lewis (Eds.), *Environmental toxicology and risk assessment* (pp. 136–158). Philadelphia: American Society of Testing and Materials.

Green, W. D. K. (1996). The thin-plate spline and images with curving features. In K. V. Mardia, C. A. Gill, & I. L. Dryden (Eds.), *Image fusion and shape variability* (pp. 79–87). Leeds: University of Leeds Press.

Hall, B. K. (2003). Unlocking the black box between genotype and phenotype: Cell condensations as morphogenetic (modular) units. *Biology & Philosophy, 18*, 219–247.

Hall, B. K., & Miyake, T. (2000). All for one and one for all: Condensations and the initiation of skeletal development. *Bioessays, 22*, 138–147.

Hallgrímsson, B., Willmore, K., & Hall, B. K. (2002). Canalization, developmental stability, and morphological integration in primate limbs. *Yearbook of Physical Anthropology, 45*, 131–158.

Hallgrímsson, B., Lieberman, D. E., Liu, W., Ford-Hutchinson, A. F., & Jirik, F. R. (2007a). Epigenetic interactions and the structure of phenotypic variation in the cranium. *Evolution & Development, 9*, 76–91.

Hallgrímsson, B., Lieberman, D. E., Young, N. M., Parsons, T., & Wat, S. (2007b). Evolution of covariance in the mammalian skull. In G. Bock, & J. Goode (Eds.), *Novartis foundation symposium* (pp. 164–185). New York: John Wiley & Sonsdiscussion pp. 185–90.

Hallgrimsson, B., Jamniczky, H., & Young, N. M., et al. (2009). Deciphering the palimpsest: Studying the relationship between morphological integration and phenotypic covariation. *Evolutionary Biology, 36*, 355–376.

Hansen, T. F. (2003). Is modularity necessary for evolvability? Remarks on the relationship between pleiotropy and evolvability. *Biosystems, 69*, 83–94.

Hansen, T. F., Armbruster, W. S., Carlson, M. L., & Pelabon, C. (2003). Evolvability and genetic constraint in Dalechampia blossoms: Genetic correlations and conditional evolvability. *Journal of Experimental Zoology Part B-Molecular and Developmental Evolution, 296B*, 23–39.

Harmon, L. J., & Glor, R. E. (2010). Poor statistical performance of the Mantel test in phylogenetic comparative analyses. *Evolution, 64*, 2173–2178.

Hills, M. (1978). On ratios – a response to Atchley, Gaskins and Anderson. *Systematic Zoology, 27*, 61–62.

Hingst-Zaher, E., Marcus, L. F., & Cerqueira, R. (2000). Application of geometric morphometrics to the study of postnatal size and shape changes in the skull of *Calomys expulsus. Hystrix, The Italian Journal of Mammalogy, 11*, 99–113.

Hoeffding, W. (1952). The large-sample power of tests based on permutation of observations. *Annals of Mathematical Statistics, 23*, 169–192.

Hoffmann, A. A., & Woods, R. (2001). Trait variability and stress: Canalization, developmental stability and the need for a broad approach. *Ecology Letters, 4*, 97–101.

Hoffmann, A. A., Woods, R. E., Collins, E., Wallin, K., White, A., & McKenzie, J. A. (2005). Wing shape versus asymmetry as an indicator of changing environmental conditions in insects. *Australian Journal of Entomology, 44*, 233–243.

Hollander, J., Collyer, M. L., Adams, D. C., & Johannesson, K. (2006). Phenotypic plasticity in two marine snails: Constraints superseding life history. *Journal of Evolutionary Biology, 19*, 1861–1872.

Hopkins, M. J., & Thurman, C. L. (2010). The geographic structure of morphological variation in eight species of fiddler crabs (Ocypodidae: genus Uca) from the eastern United States and Mexico. *Biological Journal of the Linnean Society, 100*, 248–270.

Houle, D., Mezey, J., & Galpern, P. (2002). Interpretation of the results of common principal components analysis. *Evolution, 56*, 433–440.

Huff, C. R. (2004). Wrongful conviction: Causes and public policy issues. *Criminal Justice, 15*, 15–19.

Hulsey, C. D., & Wainwright, P. C. (2002). Projecting mechanics into morphospace: Disparity in the feeding mechanics of labrid fishes. *Proceedings of the Royal Society of London, Series B, 269*, 317–326.

Huxley, J. S. (1932). *Problems of relative growth*. New York: MacVeagh. The Dial Press.

Hylander, W. L. (1985). Mandibular function and biomechanical stress and scaling. *American Zoologist, 25*, 315–330.

Institute for Data Analysis and Visualization. (2011). Landmark Software System. UC Davis. <http://graphics.idav.ucdavis.edu/research/projects/EvoMorph> Accessed 4.9.11.

Ivanovic, A., Vukov, T. D., Dzukic, G., Tomasevic, N., & Kalezic, M. L. (2007). Ontogeny of skull size and shape changes within a framework of biphasic lifestyle: A case study in six Triturus species (Amphibia, Salamandridae). *Zoomorphology, 126*, 173–183.

Ivanovic, A., Cvijanovic, M., & Kalezic, M. L. (2011). Ontogeny of body form and metamorphosis: Insights from the crested newts. *Journal of Zoology, 283*, 153–161.

Jackson, D. A., & Somers, K. M. (1989). Are probability estimates from the permutation models of Mantel's test stable? *Canadian Journal of Zoology, 67*, 766–779.

Johansson, F., Soderquist, M., & Bokma, F. (2009). Insect wing shape evolution: independent effects of migratory and mate guarding flight on dragonfly wings. *Biological Journal of the Linnean Society, 97*, 362–372.

Johnson, C. R., & Field, C. A. (1993). Using fixed-effects model multivariate analysis of variance in marine biology and ecology. *Oceanography and Marine Biology Annual Review, 31*, 177–221.

Jolicoeur, P. (1963). The multivariate generalization of the allometry equation. *Biometrics, 19*, 497–499.

Jöreskog, K. G., & Wold, H. (1982). *Systems under direct observation: Causality–structure–prediction.* North Holland Publishing Co.

Jungers, W. L., & Cole, M. S. (1992). Relative growth and shape of the locomotor skeleton in lesser apes. *Journal of Human Evolution, 23*, 93–105.

Katz, J. M. (1980). Allometry formula: A cellular model. *Growth, 44*, 89–96.

Kawecki, T. J. (2000). The evolution of genetic canalization under fluctuating selection. *Evolution, 54*, 1–12.

Kemsley, E. K. (1996). Discriminant analysis of high-dimensional data: A comparison of principal components analysis and partial least squares data reduction methods. *Chemometrics and Intelligent Laboratory Systems, 33*, 47–61.

Kendall, D. G. (1977). The diffusion of shape. *Advances in Applied Probability, 9*, 428–430.

Kendall, D. G., & Kendall, W. S. (1980). Alignments in two-dimensional random sets of points. *Advances in Applied Probability, 12*, 380–424.

Kieser, J. A., Bernal, V., Waddell, J. N., & Raju, S. (2007). The uniqueness of the human anterior dentition: A geometric morphometric analysis. *Journal of Forensic Science, 52*, 671–677.

Kim, K., Sheets, H. D., Haney, R. A., & Mitchell, C. E. (2002). Morphometric analysis of ontogeny and allometry of the Middle Ordovician trilobite *Triarthrus becki*. *Paleobiology, 28*(3), 364–377.

Kim, K., Sheets, H. D., Haney, R. A., & Mitchell, C. E. (2002). Morphometric analysis of ontogeny and allometry of the middle ordovician trilobite *Triarthrus becki*. *Paleobiology, 28*, 364–377.

Kimmel, C. B., Ballard, W. W., Kimmel, S. R., Ullmann, B., & Schilling, T. F. (1995). Stages of embryonic development of the zebrafish. *Developmental Dynamics, 203*, 253–310.

Kingsolver, J. G., & Wiernasz, D. C. (1991). Development, function, and the quantitative genetics of wing melanin pattern in *Pieris* butterflies. *Evolution, 45*, 1480–1492.

Klingenberg, C. P. (1998). Heterochrony and allometry: The analysis of evolutionary change in ontogeny. *Biological Reviews, 73*, 99–123.

Klingenberg, C. P. (2004). Integration, modules, and development: Molecules to morphology to evolution. In M Pigliucci, & K. Preston (Eds.), *Phenotypic integration: Studying the ecology and evolution of complex phenotypes* (pp. 213–230). Oxford University Press.

Klingenberg, C. P. (2005). Developmental constraints, modules, and evolvability. In B. Hallgrímsson, & B. K. Hall (Eds.), *Variation: A central concept in biology* (pp. 219–247). San Diego: Elsevier Academic Press.

Klingenberg, C. P. (2008). Morphological integration and developmental modularity. *Annual Review of Ecology, Evolution and Systematics, 39*, 115–132.

Klingenberg, C. P. (2008). Novelty and "homology-free" morphometrics: What's in a name? *Evolutionary Biology, 35*(3), 186–190.

Klingenberg, C. P. (2009). Morphometric integration and modularity in configurations of landmarks: Tools for evaluating a priori hypotheses. *Evolution & Development, 11*, 405–421.

Klingenberg, C. P. (2011). MorphoJ: an integrated software package for geometric morphometrics. *Molecular Ecology Resources, 11*, 353–357.

Klingenberg, C. P., & Froese, R. (1991). A multivariate comparison of allometric growth patterns. *Systematic Zoology, 40*, 410–419.

Klingenberg, C. P., & Gidaszewski, N. A. (2010). Testing and quantifying phylogenetic signals and homoplasy in morphometric data. *Systematic Biology, 59*, 245–261.

Klingenberg, C. P., & McIntyre, G. S. (1998). Geometric morphometrics of developmental instability: Analyzing patterns of fluctuating asymmetry with Procrustes methods. *Evolution, 52,* 1363–1375.

Klingenberg, C. P., & Monteiro, L. R. (2005). Distances and directions in multidimensional shape spaces: Implications for morphometric applications. *Systematic Biology, 54,* 678–688.

Klingenberg, C. P., & Zaklan, S. D. (2000). Morphological integration between developmental compartments in the Drosophila wing. *Evolution, 54,* 1273–1285.

Klingenberg, C. P., Badyaev, A. V., Sowry, S. M., & Beckwith, N. J. (2001). Inferring developmental modularity from morphological integration: Analysis of individual variation and asymmetry in bumblebee wings. *American Naturalist, 157,* 11–23.

Klingenberg, C. P., Barluenga, M., & Meyer, A. (2002). Shape analysis of symmetric structures: Quantifying variation among individuals and asymmetry. *Evolution, 56,* 1909–1920.

Klingenberg, C. P., Mebus, K., & Auffray, J. C. (2003). Developmental integration in a complex morphological structure: How distinct are the modules in the mouse mandible? *Evolution & Development, 5,* 522–531.

Klingenberg, C. P., Leamy, L. J., & Cheverud, J. M. (2004). Integration and modularity of quantitative trait locus effects on geometric shape in the mouse mandible. *Genetics, 166,* 1909–1921.

Klingenberg, C. P., Duttke, S., Whelan, S., & Kim, M. (2012). Developmental plasticity, morphological variation and evolvability: a multilevel analysis of morphometric integration in the shape of compound leaves. *Journal of Evolutionary Biology, 25,* 115–129.

Kluge, A. G., & Kerfoot, C. (1973). The predictability and regularity of character divergence. *American Naturalist, 107,* 426–464.

Knoke, J. D. (1986). The robust estimation of classification error rates. *Computers and Mathematics with Applications, 2,* 253–260.

Krogman, W. M., & Iscan, M. Y. (1986). *The human skeleton in forensic medicine.* Springfield, Illinois: Charles C. Thomas (publisher).

La Croix, S., Holekamp, K. E., Shivik, J. A., Lundrigan, B. L., & Zelditch, M. L. (2011a). Ontogenetic relationships between cranium and mandible in coyotes and hyenas. *Journal of Morphology, 272,* 662–674.

La Croix, S., Zelditch, M. L., Shivik, J. A., Lundrigan, B. L., & Holekamp, K. E. (2011b). Ontogeny of feeding performance and biomechanics in coyotes. *Journal of Zoology, 285,* 301–315.

Laffont, R., Renvoise, E., Navarro, N., Alibert, P., & Montuire, S. (2009). Morphological modularity and assessment of developmental processes within the vole dental row (Microtus arvalis, Arvicolinae, Rodentia). *Evolution & Development, 11,* 302–311.

Lagler, K. F., Bardach, J. E., & Miller, R. R. (1962). *Ichthyology.* New York: John Wiley & Sons.

Laird, A. K. (1965). Dynamics of relative growth. *Growth, 29,* 249–263.

Laird, A. K., Barton, A. D., & Tyler, S. A. (1968). Growth and time: An interpretation of allometry. *Growth, 32,* 347–354.

Lande, R. (1979). Quantitative genetic analysis of multivariate evolution, applied to brain: Body size allometry. *Evolution, 33,* 402–416.

Lande, R. (1980). The genetic covariance between characters maintained by pleiotropic mutations. *Genetics, 94,* 314–334.

Lande, R., & Arnold, S. J. (1983). The measurement of selection on correlated characters. *Evolution, 37,* 1210–1226.

Larson, P. M. (2005). Ontogeny, phylogeny, and morphology in anuran larvae: Morphometric analysis of cranial development and evolution in Rana tadpoles (Anura : Ranidae). *Journal of Morphology, 264,* 34–52.

Lauder, G. V. (1981). Form and function – structural analysis in evolutionary morphology. *Paleobiology, 7,* 430–442.

Leamy, L. (1984). Morphometric studies in inbred and hybrid house mice .5. Directional and fluctuating asymmetry. *American Naturalist, 123,* 579–593.

Leamy, L. J., & Klingenberg, C. P. (2005). The genetics and evolution of fluctuating asymmetry. *Annual Review of Ecology Evolution and Systematics, 36,* 1–21.

Lele, S., & Richtsmeier, J. T. (1991). Euclidean distance matrix analysis – A coordinate-free approach for comparing biological shapes using landmark data. *American Journal of Physical Anthropology, 86,* 415–427.

Lele, S., & Richtsmeier, J. T. (2001). *An invariant approach to statistical analysis of shapes.* London: Chapman & Hall-CRC Press.

Lempert, R. (1997). After the DNA wars: Skirmishing with NRC II. *Jurimetrics Journal, 37,* 439–468.

Levene, H. (1960). Robust tests for equality of variances. In I. Olkin, S. G. Ghurye, W. Hoefding, W. G. Madow, & H. B. Mann (Eds.), *Contributions to probability and statistics* (pp. 278–292). Stanford: Stanford University Press.

Lieberman, D. E., Carlo, J., de Leon, M. P., & Zollikofer, C. P. E. (2007). A geometric morphometric analysis of heterochrony in the cranium of chimpanzees and bonobos. *Journal of Human Evolution, 52*, 647–662.

Liem, K. F. (1973). Evolutionary strategies and morphological innovations – cichlid pharyngeal jaws. *Systematic Zoology, 22*, 425–441.

Lindley, D. V. (1977). A problem in forensic science. *Biometrika, 64*, 207–213.

Lorenzen, T. J., & Anderson, M. J. (1993). *Design of experiments: A no-name approach.* New York: Marcel Dekker.

Lowe, A. A., Özbeck, M. M., Miyamoto, K., & Fleetham, J. A. (1997). Cephalometric and demographic characteristics of obstructive sleep apnea: An evaluation with partial least squares analysis. *The Angle Orthodontist, 67*, 143–154.

Lundrigan, B. (1996). Morphology of horns and fighting behavior in the family Bovidae. *Journal of Mammalogy, 77*, 462–475.

Lynch, M. (2003). God's signature: DNA profiling, the new gold standard in forensic science. *Endeavor, 27*, 93–97.

MacLaughlin, S. M., & Bruce, M. F. (1990). The accuracy of sex identification in European skeletal remains using the Phenice characters. *Journal of Forensic Science, 35*, 1384–1392.

MacQueen, J. (1967). Some methods for classification and analysis of multivariate observations. In *Proceedings of the 5th Berkeley symposium on mathematics, statistics and probability* (Vol. 1, pp. 281–297). Berkeley: University of California Press.

Maddison, W. P. (1991). Squared-change parsimony reconstructions of ancestral states for continuous-valued characters on a phylogenetic tree. *Systematic Zoology, 40*, 304–314.

Maderson, P. F. A., Alberch, P., & Goodwin, B. C., et al. (1982). The role of development in macroevolutionary change. In J. T. Bonner (Ed.), *Evolution and development* (pp. 279–312). Berlin: Springer-Verlag.

Magwene, P. M. (2001). New tools for studying integration and modularity. *Evolution, 55*, 1734–1745.

Manly, B. F. (1997). *Randomization, bootstrap and Monte Carlo methods in biology.* London: Chapman & Hall.

Mantel, N. (1967). The detection of disease clustering and a generalized regression approach. *Cancer Research, 27*, 209–220.

Mardia, K. V., Kent, J. T., & Bibby, J. M. (1979). *Multivariate analysis.* San Diego: Academic Press.

Mardia, K. V., Bookstein, F. L., & Moreton, I. J. (2000). Statistical assessment of bilateral symmetry of shapes. *Biometrika, 87*, 285–300.

Marquez, E. (2012). Mint: Modularity and integration analysis tool for morphometric data. 2012.

Marquez, E. J. (2008). A statistical framework for testing modularity in multidimensional data. *Evolution, 62*, 2688–2708.

Marquez, E. J., & Knowles, L. L. (2007). Correlated evolution of multivariate traits: Detecting co-divergence across multiple dimensions. *Journal of Evolutionary Biology, 20*, 2334–2348.

Marroig, G., & Cheverud, J. M. (2001). A comparison of phenotypic variation and covariation patterns and the role of phylogeny, ecology, and ontogeny during cranial evolution of New World monkeys. *Evolution, 55*, 2576–2600.

Marroig, G., & Cheverud, J. M. (2005). Size as a line of least evolutionary resistance: Diet and adaptive morphological radiation in New World monkeys. *Evolution, 59*, 1128–1142.

Martins, E. P., & Garland, T., Jr. (1991). Phylogenetic analyses of correlated evolution of continuous characters: A simulation study. *Evolution, 45*, 534–557.

Martins, E. P., & Hansen, T. F. (1997). Phylogenies and the comparative method: A general approach to incorporating phylogenetic information into the analysis of interspecific data. *American Naturalist, 149*, 646–667.

Maynard Smith, J., Burian, R., & Kauffman, S., et al. (1985). Developmental constraints and evolution: A perspective from the Mountain Lake conference on development and evolution. *Quarterly Review of Biology, 60*, 265–287.

McArdle, B. H., & Anderson, M. J. (2001). Fitting multivariate models to community data: A comment on distance-based redundancy analysis. *Ecology, 82*, 290–297.

McGuire, J. L. (2010). Geometric morphometrics of vole (Microtus californicus) dentition as a new paleoclimate proxy: Shape change along geographic and climatic clines. *Quaternary International, 212*, 198–205.

McKinney, M. L. (1986). Ecological causation of heterochrony: A test and implications for evolutionary theory. *Paleobiology, 12*, 282–289.

McKinney, M. L. (1988). Classifying heterochrony: Allometry, size and time. In M. L. McKinney (Ed.), *Heterochrony in evolution: A multidisciplinary approach* (pp. 17–34). New York: Plenum Press.

McKinney, M. L. (1999). Heterochrony: Beyond words. *Paleobiology, 25*, 149–153.

McKinney, M. L., & McNamara, K. J. (1991). *Heterochrony: The evolution of ontogeny*. New York: Plenum Press.

Menesatti, P., Costa, C., & Paglia, G., et al. (2008). Shape-based methodology for multivariate discrimination among Italian hazelnut cultivars. *Biosystems Engineering, 4*, 417–424.

Mercer, J. M., & Roth, V. L. (2003). The effects of Cenozoic global change on squirrel phylogeny. *Science, 299*, 1568–1572.

Mevic, B. -H., & Wehrens, R. (2007). The pls package: Principal component and partial least squares regression in R. *Journal of Statistical Software, 18*, 1–23.

Mezey, J. G., & Houle, D. (2003). Comparing G matrices: Are common principal components informative? *Genetics, 165*, 411–425.

Mezey, J. G., Cheverud, J. M., & Wagner, G. P. (2000). Is the genotype-phenotype map modular? A statistical approach using mouse quantitative trait loci data. *Genetics, 156*, 305–311.

Michaux, J., Hautier, L., Simonin, T., & Vianey-Liaud, M. (2008). Phylogeny, adaptation and mandible shape in Sciuridae (Rodentia, Mammalia). *Mammalia, 72*, 286–296.

Miller, A. I., & Foote, M. (1996). Calibrating the Ordovician radiation of marine life: Implications for Phanerozoic diversity trends. *Paleobiology, 22*, 304–309.

Miller-Rushing, A. J., Hoye, T. T., Inouye, D. W., & Post, E. (2010). The effects of phenological mismatches on demography. *Philosophical Transactions of the Royal Society B-Biological Sciences, 365*, 3177–3186.

Mitteroecker, P., & Bookstein, F. (2007). The conceptual and statistical relationship between modularity and morphological integration. *Systematic Biology, 56*, 818–836.

Mitteroecker, P., & Bookstein, F. (2008). The evolutionary role of modularity and integration in the hominoid cranium. *Evolution, 62*, 943–958.

Mitteroecker, P., & Bookstein, F. (2011). Linear discrimination, ordination, and the visualization of selection gradients in modern morphometrics. *Evolutionary Biology, 38*, 100–114.

Mitteroecker, P., Gunz, P., Bernhard, M., Schaefer, K., & Bookstein, F. L. (2004). Comparison of cranial ontogenetic trajectories among great apes and humans. *Journal of Human Evolution, 46*, 679–698.

Mitteroecker, P., Gunz, P., Bernhard, M., Schaefer, K., & Bookstein, F. L. (2004a). Comparison of cranial ontogenetic trajectories among great apes and humans. *Journal of Human Evolution, 46*, 679–697.

Mitteroecker, P., Gunz, P., Weber, G. W., & Bookstein, F. L. (2004b). Regional dissociated heterochrony in multivariate analysis. *Annals of Anatomy-Anatomischer Anzeiger, 186*, 463–470.

Mitteroecker, P., Gunz, P., & Bookstein, F. L. (2005). Heterochrony and geometric morphometrics: A comparison of cranial growth in *Pan paniscus* versus *Pan troglodytes*. *Evolution & Development, 7*, 244–258.

Monteiro, L. R., & Nogueira, M. R. (2009). Adaptive radiations, ecological specialization, and the evolutionary integration of complex morphological structures. *Evolution, 64*, 724–743.

Monteiro, L. R., & Nogueira, M. R. (2009). Adaptive radiations, ecological specialization, and the evolutionary integration of complex morphological structures. *Evolution, 64*, 724–744.

Monteiro, L. R., Lessa, L. G., & Abe, A. S. (1999). Ontogenetic variation in skull shape of *Thrichomys apereoides* (Rodentia : Echimyidae). *Journal of Mammalogy, 80*, 102–111.

Monteiro, L. R., Duarte, L. C., & dos Reis, S. F. (2003). Environmental correlates of geographical variation in skull and mandible shape of the punare rat *Thrichomys apereoides* (Rodentia : Echimyidae). *Journal of Zoology, 261*, 47–57.

Monteiro, L. R., Bonato, V., & dos Reis, S. F. (2005). Evolutionary integration and morphological diversification in complex morphological structures: Mandible shape divergence in spiny rats (Rodentia, Echimyidae). *Evolution & Development, 7*, 429–439.

Morrison, D. F. (1990). *Multivariate statistical methods*. New York: McGraw Hill.

Mullin, S. K., & Taylor, P. J. (2002). The effects of parallax on geometric morphometrics data. *Computers in Biology and Medicine, 32*, 455–464.

Muschick, M., Barluenga, M., Salzburger, W., & Meyer, A. (2011). Adaptive phenotypic plasticity in the Midas cichlid fish pharyngeal jaw and its relevance in adaptive radiation. *Bmc Evolutionary Biology, 11*, 116.

Myers, P., Lundrigan, B. L., Gillespie, B. W., & Zelditch, M. L. (1996). Phenotypic plasticity in skull and dental morphology in the prairie deer mouse (*Peromyscus maniculatus bairdii*). *Journal of Morphology, 229*, 229–237.

Myers, P., Lundrigan, B. L., Gillespie, B. W., & Zelditch, M. L. (1996). Phenotypic plasticity in skull and dental morphology in the prairie deer mouse (*Peromyscus maniculatus bairdii*). *Journal of Morphology, 229*, 229–237.

National Academy of Science, Committee on Identifying the Needs of the Forensic Sciences Community, National Research Council (February 2009). *Strengthening forensic science in the United States: A path forward*. National Academies Press.

Neige, P., Marchand, D., & Laurin, B. (1997). Heterochronic differentiation of sexual dimorphs among Jurassic ammonite species. *Lethaia, 30*, 145–155.

Neumann, C., Champod, C., Puch-Solis, R., Egli, N., Anthonioz, A., & Bromage-Griffiths, A. (2007). Computation of likelihood ratios in fingerprint identification for configurations of any number of minutiae. *Journal of Forensic Sciences, 51*, 1255–1266.

Noback, M. L., Harvati, K., & Spoor, F. (2011). Climate-related variation of the human nasal cavity. *American Journal of Physical Anthropology, 145*, 599–614.

Nolte, A. W., & Sheets, H. D. (2005). Shape based assignment tests suggest transgressive phenotypes in natural sculpin hybrids (Teleostei, Scorpaeniformes, Cottidae). *Frontiers in Zoology, 2*, 11 <http://www.frontiersinzoology.com/content/2/1/11>

Oksanen, J., Blanchet, F. G., & Kindt, R., et al. (2001). vegan: Community Ecology Package.

Olson, E. C., & Miller, R. L. (1958). *Morphological integration*. University of Chicago Press.

Ortiz, M., Sarabia, L., Symington, C., Santamaria, F., & Iniguez, M. (1996). Analysis of ageing and typification of vintage ports by partial least squares and soft independent modeling class analogy. *Analyst*1009–1013.

Oxnard, C. E. (1968). The architecture of the shoulder in some mammals. *Journal of Morphology, 126*, 249–290.

O'Higgins, P., Chadfield, P., & Jones, N. (2001). Facial growth and the ontogeny of morphological variation within and between the primates *Cebus apella* and *Cercocebus torquatus*. *Journal of Zoology, 254*, 337–357.

Palmer, A. R., & Strobeck, C. (1986). Fluctuating asymmetry – measurement, analysis, patterns. *Annual Review of Ecology and Systematics, 17*, 391–421.

Palmer, A. R., & Strobeck, C. (2003). Fluctuating asymmetry analyses revisited. In M. Polack (Ed.), *Developmental Instability (DI): Causes and consequences* (pp. 279–319). Oxford: Oxford University Press.

Parsons, K. J., Sheets, H. D., Skulason, S., & Ferguson, M. M. (2011). Phenotypic plasticity, heterochrony and ontogenetic repatterning during juvenile development of divergent Arctic charr (Salvelinus alpinus). *Journal of Evolutionary Biology, 24*, 1640–1652.

Parsons, K. J., Marquez, E., & Albertson, R. C. (2012). Constraint and opportunity: The genetic basis and evolution of modularity in the cichlid mandible. *American Naturalist, 179*, 64–78.

Patriquin, M., Loth, S. R., & Steyn, M. (2003). Sexually dimorphic pelvic morphology in South African whites and blacks. *Homo, 53*, 255–262.

Perez, S. I., Diniz, J. A. F., Rohlf, F. J., & Dos Reis, S. F. (2009). Ecological and evolutionary factors in the morphological diversification of South American spiny rats. *Biological Journal of the Linnean Society, 98*, 646–660.

Pfaller, J. B., Herrera, N. D., Gignac, P. M., & Erickson, G. M. (2010). Ontogenetic scaling of cranial morphology and bite-force generation in the loggerhead musk turtle. *Journal of Zoology, 280*, 280–289.

Phenice, T. W. (1969). A newly developed visual method of sexing os pubis. *American Journal of Physical Anthropology, 30*, 297–301.

Piras, P., Colangelo, P., & Adams, D. C., et al. (2010). The Gavialis–Tomistoma debate: The contribution of skull ontogenetic allometry and growth trajectories to the study of crocodylian relationships. *Evolution & Development, 12*, 568–579.

Piras, P., Maiorino, L., & Raia, P., et al. (2010). Functional and phylogenetic constraints in Rhinocerotinae craniodental morphology. *Evolutionary Ecology Research, 12*, 897–928.

Piras, P., Salvi, D., & Ferrar, S., et al. (2011). The role of post-natal ontogeny in the evolution of phenotypic diversity in Podarcis lizards. *Journal of Evolutionary Biology, 24*, 2705–2720.

Ponce de Leon, M. S., & Zollikofer, C. P. E. (2001). Neanderthal cranial ontogeny and its implications for late hominid diversity. *Nature, 412*, 534–538.

Post, E., & Forchhammer, M. C. (2008). Climate change reduces reproductive success of an Arctic herbivore through trophic mismatch. *Philosophical Transactions of the Royal Society B-Biological Sciences, 363*, 2369–2375.

Post, E., Pedersen, C., Wilmers, C. C., & Forchhammer, M. C. (2008). Warming, plant phenology and the spatial dimension of trophic mismatch for large herbivores. *Proceedings of the Royal Society B-Biological Sciences, 275*, 2005–2013.

Pretty, I. A. (2006). The barriers to achieving an evidence base for bitemark analysis. *Forensic Science International, Suppl. 1,* S110–S120.

Pretty, I. A., & Sweet, D. (2001). The scientific basis for human bitemark analyses: A critical review. *Science and Justice, 41,* 85–92.

Pulcini, D., Costa, C., Aguzzi, J., & Cataudella, S. (2008). Light and shape: A contribution to demonstrate morphological differences in diurnal and nocturnal teleosts. *Journal of Morphology, 269,* 375–385.

Quenouille, M. (1949). Approximate tests of correlation in time series. *Journal of the Royal Statistical Society B, 11,* 18–44.

Quinn, G. P., & Keogh, M. J. (2002). *Experimental design and data analysis for biologists.* Cambridge: Cambridge University Press.

R_Development_Core _Team (2011). *R: A language and environment for statistical computing.* Vienna, Austria: R Foundation for Statistical Computing.

Rao, C. R. (1973). *Linear statistical inference and its applications.* New York: John Wiley & Sons.

Raspé, R. E. (1785). Baron Münchhausen's narrative of his Marvelous Travels and Campaigns in Russia.

Rawson, R. D., Ommen, R. K., Kinard, G., Johnson, J., & Yfantis, A. (1984). Statistical evidence for the individuality of the human dentition. *Journal of Forensic Science, 29,* 245–253.

Reeve, E. C. R. (1960a). Some genetic tests on asymmetry of sternopleural chaeta number in Drosophila. *Genetical Research, 1,* 151–172.

Reeve, E. C. R. (1960b). Some genetic tests on asymmetry of sternopleural chaetae number in *Drosophila. Genetical Research, 1,* 151–172.

Rencher, A. C. (1995). *Methods of multivariate analysis.* New York: John Wiley & Sons.

Rencher, A. C., & Schaalje, G. B. (2008). *Linear models in statistics.* New York: John Wiley & Sons.

Rencher, A. C., & Schaalje, G. B. (2008). *Linear models in statistics.* Hoboken, NJ: Wiley.

Revell, L. J. (2009). Size-correction and principal components for interspecific comparative studies. *Evolution, 63,* 3258–3268.

Rice, W. R. (1989). Analyzing tables of statistical tests. *Evolution, 43,* 223–225.

Richtsmeier, J. T., & Lele, S. (1993). A coordinate-free approach to the analysis of growth-patterns: Models and theoretical considerations. *Biological Reviews of the Cambridge Philosophical Society, 68,* 381–411.

Richtsmeier, J. T., Lele, S. R., & Cole, T. M. (2005). Landmark morphometrics and the analysis of variation. In B. Hallgrímsson, & B. K. Hall (Eds.), *Variation: A central concept in biology* (pp. 49–69). New York: Elsevier.

Ricklefs, R. E., & Cox, G. W. (1977). Morphological similarity and ecological overlap among passerine birds on St Kitts, British West Indies. *Oikos, 28,* 60–66.

Ricklefs, R. E., & Travis, J. (1980). A morphological approach to the study of avian community organization. *The Auk, 97,* 321–338.

Risinger, D. M. (2007). Convicting the innocent: An empirically justified wrongful conviction rate. *Journal of Criminal Law and Criminology, 97,* 761–817.

Robinson, J. (1973). Large-sample power of permutation tests for randomization models. *Annals of Statistics, 1,* 291–296.

Rohlf, F. J. (1998). On applications of geometric morphometrics to studies of ontogeny and phylogeny. *Systematic Biology, 47,* 147–158.

Rohlf, F. J. (1999). Shape statistics: Procrustes superimpositions and shape spaces. *Journal of Classifiation, 16,* 197–225.

Rohlf, F. J. (2000). Statistical power comparisons among alternative morphometric methods. *American Journal of Physical Anthropology, 111,* 463–478.

Rohlf, F. J. (2001). Comparative methods for the analysis of continuous variables: Geometric interpretations. *Evolution, 55,* 2143–2160.

Rohlf, F. J. (2002). Geometric morphometrics and phylogeny. In N. MacLeod, & P. L. Forey (Eds.), *Morphology, shape and phylogeny* (pp. 175–193). Taylor & Francis.

Rohlf, F. J. (2003). Bias and error in estimates of mean shape in geometric morphometrics. *Journal of Human Evolution, 44,* 665–683.

Rohlf, F. J. (2006). A comment on phylogenetic correction. *Evolution, 60,* 1509–1515.

Rohlf, F. J. (2009). tpsRegr 1.37. *Ecology and evolution.* State University of New York at Stony Brook.

Rohlf, F. J., & Bookstein, F. L. (2003). Computing the uniform component of shape variation. *Systematic Biology, 52*, 66–69.

Rohlf, F. J., & Corti, M. (2000). Use of two-block partial least-squares to study covariation in shape. *Systematic Biology, 49*, 740–753.

Rohlf, F. J., & Slice, D. E. (1990). Extensions of the Procrustes method for the optimal superimposition of landmarks. *Systematic Zoology, 39*, 40–59.

Rohlf, F. J., Gilmartin, A. J., & Hart, G. (1983). The Kluge–Kerfoot phenomenon: A statistical artifact? *Evolution, 37*, 180–202.

Romano, J. P. (1989). Bootstrap and randomization tests of some non-parametric hypotheses. *Annals of Statistics, 17*, 141–159.

Rosell, J. A., & Olson, M. E. (2007). Testing implicit assumptions regarding the age vs. size dependence of stem biomechanics using Pittocaulon (similar to Senecio) praecox (Asteraceae). *American Journal of Botany, 94*, 161–172.

Roseman, C. C., Kenney-Hunt, J. P., & Cheverud, J. M. (2009). Phenotypic integration without modularity: Testing hypotheses about the distribution of pleiotropic quantitative trait loci in a continuous space. *Evolutionary Biology, 36*, 282–291.

Roth, V. L. (1993). On three-dimensional morphometrics, and on the identification of landmark points. In L. F. Marcus, M. Corti, A. Loy, G. J. P. Naylor, & D. E. Slice (Eds.), *Advances in morphometrics* (pp. 41–62). New York: Plenum.

Rothwell, B. R. (1995). Bitemarks in forensic dentistry: A review of legal, scientific issues. *Journal of the American Dental Association, 126*, 223–232.

Ruber, L., & Adams, D. C. (2001). Evolutionary convergence of body shape and trophic morphology in cichlids from Lake Tanganyika. *Journal of Evolutionary Biology, 14*, 325–332.

Saks, M. J., & Koehler, J. J. (2008). The individualization fallacy in forensic science evidence. *Vanderbilt Law Review, 61*, 199–219.

Saks, M. J., Constantine, L., & Dolezal, M., et al. (2000). Toward a model act for the prevention and remedy of erroneous convictions. *New England Legal Review, 35*, 669–683.

Saks, M. J., Constantine, L., Dolezal, M., & Garcia, J. (2001). Model prevention and remedy of erroneous convictions. *Arizona State Law Journal, 33*, 669–719.

Sampson, P. D., Streissguth, A. P., Barr, H. M., & Bookstein, F. L. (1989). Neurobehavioral effects of prenatal alcohol: Part II. Partial least squares analysis. *Neurotoxicology and Teratology, 11*, 477–491.

Sampson, P. D., Bookstein, F. L., Sheehan, H., & Bolson, E. L. (1996). Eigenshape analysis of left ventricular outlines from contrast ventriculograms. In L. F. Marcus, M. Corti, A. Loy, G.J.P. Naylor, & D.E. Slice (Eds.), *Advances in Morphometrics* (pp. 131–152). Nato ASI Series, Series A: Life Science. New York.

Sander, K. (1983). The evolution of patterning mechanisms: Gleanings from insect embryogenesis and spermatogenesis. In B. C. Goodwin, N. Holder, & C. C. Wylie (Eds.), *Development and evolution* (pp. 137–159). Cambridge: Cambridge University Press.

Sanfelice, D., & De Freitas, T. R. O. (2008). A comparative description of dimorphism in skull ontogeny of *Arctocephalus australis, Callorhinus ursinus*, and *Otaria byronia* (Carnivora : Otariidae). *Journal of Mammalogy, 89*, 336–346.

Santos, M., Iriarte, P. F., & Cespedes, W. (2005). Genetics and geometry of canalization and developmental stability in Drosophila subobscura. *Bmc Evolutionary Biology, 5*, 7.

Savriama, Y., & Klingenberg, C. P. (2011). Beyond bilateral symmetry: Geometric morphometric methods for any type of symmetry. *Bmc Evolutionary Biology, 11*, 280.

Schaefer, S. A., & Lauder, G. V. (1996). Testing historical hypotheses of morphological change: Biomechanical decoupling in loricariod catfishes. *Evolution, 50*, 1661–1675.

Scheiner, S. M., Caplan, R. L., & Lyman, R. F. (1991). The genetics of phenotypic plasticity .3. Genetic correlations and fluctuating asymmetries. *Journal of Evolutionary Biology, 4*, 51–68.

Schiavo, R. A., & Hand, D. J. (2000). Ten more years of error rate research. *International Statistical Review, 68*, 295–310.

Schluter, D. (1996). Adaptive radiation along genetic lines of least resistance. *Evolution, 50*, 1766–1774.

Schmalhausen, I. I. (1949). *Factors of evolution: The theory of stabilizing selection.* Philadelphia: The Blakeston Co.

Schweitzer, N. J., & Saks, M. J. (2007). The CSI effect: Popular fiction about forensic science affects the public's expectations about real forensic science. *Jurimetrics*, *47*, 357–364.

Schweitzer, P. N., & Lohmann, G. P. (1990). Life-history and the evolution of ontogeny in the ostracode genus. Cyprideis. Paleobiology, *16*, 107–125.

Searle, S. R. (1997). *Linear models*. New York: John Wiley & Sons.

Searle, S. R. (2006). *Linear models for unbalanced data*. New York: Wiley.

Seidl, F. (1960). Körpergrundgestalt und Keimstruktur. Eine Erörterung über die Gundlagen der vergleichenden und experimentellen Embryologie un deren Gültigkeit bei phylogeneticschen Überlegungen. *Zoologische Anzeiger*, *164*, 245–305.

Seilacher, A. (1979). Constructional morphology of sand dollars. *Paleobiology*, *5*, 191–221.

Shea, B. T. (1983a). Allometry and heterochrony in the African apes. *American Journal of Physical Anthropology*, *62*, 275–289.

Shea, B. T. (1983b). Paedomorphosis and neoteny in the pygmy chimpanzee. *Science*, *222*, 521–522.

Shea, B. T. (1985). Bivariate and multivariate growth allometry: Statistical and biological considerations. *Journal of Zoological Society of London (A)*, *206*, 367–390.

Shea, B. T. (1992). Ontogenetic scaling of skeletal proportions in the talapoin monkey. *Journal of Human Evolution*, *23*, 283–307.

Sheets, H. D., & Mitchell, C. E. (2001). Why the null matters: Statistical tests, random walks and evolution. *Genetica*, *112*, 105–125.

Sheets, H. D., Bush, P. J., & Bush, M. A. Patterns of variation and match rates of the anterior biting dentition: Characteristics of a database of 3D scanned dentition. *Journal of Forensic Science* (in press).

Sheets, H. D., Kim, K., & Mitchell, C. E. (2004). A combined landmark and outline-based approach to ontogenetic shape change in the Ordovician trilobite *Triarthrus becki*. In A. M. T. Elewa (Ed.), *Morphometrics: Applications in biology and paleontology* (pp. 67–82). New York: Springer.

Sheets, H. D., Covino, K. M., Panasiewicz, J. M., & Morris, S. R. (2006). Comparison of geometric morphometric outline methods in the discrimination of age-related differences in feather shape. *Frontiers in Zoology*, *3*, 15.

Sheets, H. D., Corvino, K. M., Panasiewicz, J. M., & Morris, S. R. (2006). Comparison of geometric morphometric outline methods in the discrimination of age-related differences in feather shape. *Frontiers in Zoology*, *3*, 15 <http://www.frontiersinzoology.com/content/3/1/115>

Sheets, H. D., Bush, P. J., & Bush, M. A. (2011). Mathematical matching of a dentition to bitemarks: Use and evaluation of affine methods. *Forensic Science International*, *207*, 111–118.

Sheets, H. D., Bush, P. J., & Bush, M. A. (2013). Patterns of variation and match rates of the anterior biting dentition: Characteristics of a database of 3D scanned dentitions. *Journal of Forensic Science*in press

Simon, C. (1983). A new coding procedure for morphometric data with an example from periodical cicada wing veins. In J. Felsenstein (Ed.), *Numerical taxonomy* (pp. 378–382). Springer-Verlag.

Simpson, G. G. (1953). *Major features of evolution*. New York: Columbia University Press.

Singleton, M., Rosenberger, A. L., Robinson, C., & O'Neill, R. (2011). Allometric and metameric shape variation in *Pan* mandibular molars: A digital morphometric analysis. *Anatomical Record – Advances in Integrative Anatomy and Evolutionary Biology*, *294*, 322–334.

Skinner, M. M., Gunz, P., Wood, B. A., & Hublin, J. J. (2008). Enamel–dentine junction (EDJ) morphology distinguishes the lower molars of *Australopithecus africanus* and *Paranthropus robustus*. *Journal of Human Evolution*, *55*, 979–988.

Slack, J. M., Holland, P. W., & Graham, C. F. (1993). The zootype and the phylotypic stage. *Nature*, *361*, 490–492.

Slice, D. E. (2001). Landmark coordinates aligned by Procrustes analysis do not lie in Kendall's shape space. *Systematic Biology*, *50*, 141–149.

Slice, D. E., Bookstein, F. L., Marcus, L. F., & Rohlf, F. J. (1996). A glossary for geometric morphometrics. In L. F. Marcus, M. Corti, A. Loy, G. J. P. Naylor, & D. E. Slice (Eds.), *Advances in morphometrics* (pp. 531–551). New York: Plenum.

Small, C. G. (1996). *The statistical theory of shape*. New York: Springer-Verlag.

Smith, L. H., & Lieberman, B. S. (1999). Disparity and constraint in olenelloid trilobites and the Cambrian radiation. *Paleobiology*, *25*, 248–272.

Sneath, P. H. A., & Sokal, R. R. (1973). *Numerical taxonomy*. San Francisco: Freeman.

Snedecor, G. W., & Cochran, W. G. (1980). *Statistical methods*. Iowa State University Press.

van Snik, G. M. J., van den Boogaart, J. G. M., & Osse, W. M. (1997). Larval growth patterns in *Cyprinus carpio* and *Clarias gariepinus* with attention to the finfold. *Journal of Fish Biology, 50*, 1339–1352.

Sokal, R. R. (1976). The Kluge–Kerfoot phenomenon reexamined. *American Naturalist, 110*, 1077–1091.

Sokal, R. R., & Rohlf, F. J. (1995). *Biometry: The principals and practice of statistics in biological research* (3rd ed.). New York: W.H. Freeman and Co.

Solheim, T., & Vonen, A. (2006). Dental age estimation, quality assurance and age estimation of asylum seekers in Norway. *Forensic Science International, 159*(Suppl. 1), S56–S60.

Stein, B. R. (1981). Comparative limb myology of two opossums, *Didelphis* and *Chironectes. Journal of Morphology, 169*, 113–140.

Steyn, M., & Iscan, M. Y. (2008). Metric sex determination from the pelvis in modern Greeks. *Forensic Science International, 179*, 86.

Stone, E. A. (2011). Why the phylogenetic regression appears robust to tree misspecification. *Systematic Biology, 60*, 245–260.

Strand Vioarsdottir, U., O'Higgins, P., & Stringer, C. (2002). A geometric morphometric study of regional differences in the ontogeny of the modern human facial skeleton. *Journal of Anatomy, 201*, 211–229.

Strauss, R. E. (1987). On allometry and relative growth in evolutionary studies. *Systematic Zoology, 36*, 72–75.

Strauss, R. E., & Altig, R. (1992). Ontogenetic body form changes in 3 ecological morphotypes of anuran tadpoles. *Growth Development and Aging, 56*, 3–16.

Strauss, R. E., & Bookstein, F. L. (1982). The truss–body form reconstructions in morphometrics. *Systematic Zoology, 31*, 113–135.

Strauss, R. E., & Fuiman, L. A. (1985). Quantitative comparisons of body form and allometry in larval and adult Pacific sculpins (Teleostei: Cottidae). *Canadian Journal of Zoology, 63*, 1582–1589.

Streissguth, A. P., Bookstein, F. L., Sampson, P. D., & Barr, H. M. (1993). *The enduring effects of prenatal alcohol exposure on child development: Birth through seven years, a partial least squares solution.* Ann Arbor, MI: University of Michigan Press.

Streissguth, A. P., Bookstein, F. L., Barr, H., Press, S., & Sampson, P. (1998). A fetal alcohol behavior scale. *Alcoholism: Clinical and Experimental Research, 22*, 325–333.

Su, C., & Srihari, S. N. (2010). Evaluation of rarity of fingerprints in forensics. *Proceedings of neural information processing systems.* Vancouver, Canada, December 6–9, 2010.

Su, C., & Srihari, S. N. (2009). *Probability of random correspondence for fingerprints* (pp. 55–66). *IWCF 2009 Proceedings.* Berlin: Springer-Verlag.

Swiderski, D. L. (1993). Morphological evolution of the scapula in tree squirrels, chipmunks, and ground squirrels (Sciuridae): An analysis using thin-plate splines. *Evolution, 47*, 1854–1873.

Swiderski, D. L., & Zelditch, M. L. (2010). Morphological diversity despite isometric scaling of lever arms. *Evolutionary Biology, 37*, 1–18.

Swiderski, D. L., Zelditch, M. L., & Fink, W. L. (1998). Why morphometrics isn't special: Coding quantitative data for phylogenetic analysis. *Systematic Biology, 47*, 508–519.

Symonds, M. R. E. (2002). The effects of topological inaccuracy in evolutionary trees on the phylogenetic comparative method of independent contrasts. *Systematic Biology, 51*, 541–553.

Tanner, J. B., Zelditch, M. L., Lundrigan, B. L., & Holekamp, K. E. (2010). Ontogenetic change in skull morphology and mechanical advantage in the spotted hyena (*Crocuta crocuta*). *Journal of Morphology, 271*, 353–365.

Taylor, M. E. (1974). The functional anatomy of the forelimb of some African Viverridae (Carnivora). *Journal of Morphology, 143*, 307–336.

Thiele, K. (1993). The holy grail of the perfect character: The cladistic treatment of morphometric data. *Cladistics, 9*, 275–304.

Thompson, D. A. W. (1992). *On growth and form: The complete revised edition* (2nd ed.). Dover reprint of 1942. New York.

Tukey, J. W. (1958). Bias and confidence in not quite large samples. (Abstract). *Annals of Mathematical Statistics, 29*, 614.

Underwood, A. J. (1997). *Experiments in ecology: Their logical design and interpretation using analysis of variance.* Cambridge: Cambridge University Press.

Van Bocxlaer, B., & Schultheiß, R. (2010). Comparison of morphometrics techniques for shapes with few homologous landmarks based on machine-learning approaches to biological discrimination. *Paleobiology, 36*, 497–515.

Van Buskirk, J. (2009). Natural variation in morphology of larval amphibians: Phenotypic plasticity in nature? *Ecological Monographs, 79*, 681–705.

Van Dongen, S., & Lens, L. (2000). The evolutionary potential of developmental instability. *Journal of Evolutionary Biology, 13*, 326–335.

Van Valen, L. (1962). Developmental gradients in the dentition of *Peromyscus*. *Evolution, 16*, 272–277.

Van Valen, L. (1970). An analysis of developmental fields. *Developmental Biology, 23*, 456–477.

Velhagen, W. A., & Roth, V. L. (1997). Scaling of the mandible in squirrels. *Journal of Morphology, 232*, 107–132.

Viguier, B. (2002). Is the morphological disparity of lemur skulls (Primates) controlled by phylogeny and/or environmental constraints? *Biological Journal of the Linnean Society, 76*, 577–590.

Vincent, S. E., Brandley, M. C., Herrel, A., & Alfaro, M. E. (2009). Convergence in trophic morphology and feeding performance among piscivorous natricine snakes. *Journal of Evolutionary Biology, 22*, 1203–1211.

Vinyard, C. J., & Ravosa, M. (1998). Ontogeny, function, and scaling of the mandibular symphysis in papionin primates. *Journal of Morphology, 235*, 157–175.

Viscosi, V., & Cardini, A. (2011). Leaf morphology, taxonomy and geometric morphometrics: A simplified protocol for beginners. *PLoS ONE, 6*, e25630.

de Visser, J., Hermisson, J., & Wagner, G. P., et al. (2003). Perspective: Evolution and detection of genetic robustness. *Evolution, 57*, 1959–1972.

Voss, R. S., & Marcus, L. F. (1992). Morphological evolution in muroid rodents II. Craniometric factor divergence in seven neotropical general, with experimental results from *Zygodontomys*. *Evolution, 46*, 1918–1934.

Waddington, C. H. (1942). Canalization of development and the inheritance of acquired characters. *Nature, 150*, 563–565.

Wagner, G. P. (1988). The influence of variation and of developmental constraints on the rate of multivariate phenotypic evolution. *Journal of Evolutionary Biology, 1*, 45–66.

Wagner, G. P. (1996). Homologues, natural kinds and the evolution of modularity. *American Zoologist, 36*, 36–43.

Wagner, G. P., & Altenberg, L. (1996). Complex adaptations and the evolution of evolvability. *Evolution, 50*, 967–976.

Wagner, G. P., Booth, G., & Bagheri, H. C. (1997). A population genetic theory of canalization. *Evolution, 51*, 329–347.

Wagner, P. J. (1995). Testing evolutionary constraint hypotheses with early Paleozoic gastropods. *Paleobiology, 21*, 459–470.

Wagner, P. J. (1997). Patterns of morphologic diversification among the Rostroconchia. *Paleobiology, 23*, 115–150.

Ward, J. H. (1963). Hierarchical grouping to optimize an objective function. *Journal of the American Statistical Association, 58*, 236–244.

Wayne, R. K. (1986). Cranial morphology of domestic and wild canids: The influence of development on morphological change. *Evolution, 40*, 243–261.

Webber, A. J., & Hunda, B. R. (2007). Quantitatively comparing morphological trends to environment in the fossil record (cincinnatian series; upper ordovician). *Evolution, 61*, 1455–1465.

Webster, M. (2007). Ontogeny and evolution of the early Cambrian trilobite genus *Nephrolenellus* (Olenelloidea). *Journal of Paleontology, 81*, 1168–1193.

Webster, M. (2009). Ontogeny, systematics, and evolution of the effaced early Cambrian trilobites peachella Walcott, 1910 and Eopeachella new genus (Olenelloidea). *Journal of Paleontology, 83*, 197–218.

Webster, M., & Hughes, N. C. (1999). Compaction-related deformation in Cambrian olenelloid trilobites and its implications for fossil morphometry. *Journal of Paleontology, 73*(2), 355–371.

Webster, M., & Zelditch, M. L. (2005). Evolutionary modifications of ontogeny: heterochrony and beyond. *Paleobiology, 31*, 354–372.

Webster, M., Sheets, H. D., & Hughes, N. C. (2001). Allometric patterning in trilobite ontogeny: testing for heterochrony in *Nephrolenellus*. In M. L. Zelditch (Ed.), *Beyond heterochrony: The evolution of development* (pp. 105–144). New York: John Wiley & Sons.

Williams, S. T., Hall, A., & Kuklinski, P. (2012). Unraveling cryptic diversity in the Indo-West Pacific gastropod genus *Lunella* (Turbinidae) using elliptic Fourier analysis. *American Malacological Bulletin, 30*, 189–206.

Willmore, K. E., Klingenberg, C. P., & Hallgrímsson, B. (2005). The relationship between fluctuating asymmetry and environmental variance in rhesus macaque skulls. *Evolution, 59*, 898–909.

Willmore, K. E., Young, N. M., & Richtsmeier, J. T. (2007). Phenotypic variability: Its components, measurement and underlying developmental processes. *Evolutionary Biology, 34*, 99–120.

Willmore, K. E., Roseman, C. C., Rogers, J., Cheverud, J. M., & Richtsmeier, J. T. (2009). Comparison of mandibular phenotypic and genetic integration between baboon and mouse. *Evolutionary Biology, 36*, 19–36.

Wills, M. A. (2001). Morphological disparity: A primer. In J. M. Adrain, G. D. Edgecombe, & B. S. Lieberman (Eds.), *Fossils, phylogeny, and form: An analytical approach* (pp. 55–144). Kluwer Academic/Plenum Publishers.

Wills, M. A., Briggs, D. E. G., & Fortey, R. A. (1994). Disparity as an evolutionary index – a comparison of Cambrian and recent arthropods. *Paleobiology, 20*, 93–130.

Wilson, L. A. B., & Sanchez-Villagra, M. R. (2010). Diversity trends and their ontogenetic basis: an exploration of allometric disparity in rodents. *Proceedings of the Royal Society B-Biological Sciences, 277*, 1227–1234.

Wishart, J. (1928). The generalised product moment distribution in samples from a normal multivariate population. *Biometrika, 20A*, 32–52.

Witten, I. H., & Frank, E. (2005). *Data mining: Practical machine learning tools and techniques.* San Francisco: Morgan Kaufmann.

Wold, H. (1966). Estimation of principal components and related models by iterative least squares. In P. R. Krishnaiaah (Ed.), *Multivariate analysis* (pp. 391–420). New York: Academic Press.

Wold, S., Sjöström, M., & Erickson, L. (2001). PLS-regression: A basic tool of chemometrics. *Chemometrics and Intelligent Laboratory Systems, 58*, 109–130.

Wolterek, R. (1909). Weitere experimentelle Untersuchungen über Artveränderung, speziell über das Wesen quantitativer Artenunterschiede bei Daphniden. *Versuch Deutsche Zoolgische Gesellschaft, 19*, 110–172.

Wood, A. R., Zelditch, M. L., Rountrey, A. N., Eiting, T. P., Sheets, H. D., & Gingerich, P. D. (2007). Multivariate stasis in the dental morphology of the Paleocene-Eocene condylarth. Ectocion. Paleobiology, 33, 248–260.

Wund, M. A., Baker, J. A., Clancy, B., Golub, J. L., & Fosterk, S. A. (2008). A test of the "Flexible stem" model of evolution: Ancestral plasticity, genetic accommodation, and morphological divergence in the threespine stickleback radiation. *American Naturalist, 172*, 449–462.

Yee, W. L., Chapman, P. S., Sheets, H. D., & Unruh, T. R. (2009). Analysis of body measurements and wing shape to discriminate *Rhagoletis pomonella* and *Rhagoletis zephyria* (Diptera: Tephritidae) in Washington State. *Annals of the Entomological Society of America, 102*, 1013–1028.

Young, N. M. (2008). A comparison of the ontogeny of shape variation in the anthropoid scapula: Functional and phylogenetic signal. *American Journal of Physical Anthropology, 136*, 247–264.

Zakharov, V. M. (1992). Population phenogenetics: Analysis of developmental stability in natural populations. *Acta Zoologica Fennica, 191*, 7–30.

Zelditch, M. L. (1988). Ontogenetic variation in patterns of phenotypic integration in the laboratory rat. *Evolution, 42*, 28–41.

Zelditch, M. L., & Carmichael, A. C. (1989). Ontogenetic variation in patterns of developmental and functional integration in skulls of *Sigmodon fulviventer. Evolution, 43*, 814–824.

Zelditch, M. L., Bookstein, F. L., & Lundrigan, B. L. (1992). Ontogeny of integrated skull growth in the cotton rat *Sigmodon fulviventer. Evolution, 46*, 1164–1180.

Zelditch, M. L., Bookstein, F. L., & Lundrigan, B. L. (1993). The ontogenetic complexity of developmental constraints. *Journal of Evolutionary Biology, 6*, 121–141.

Zelditch, M. L., Fink, W. L., & Swiderski, D. L. (1995). Morphometrics, homology and phylogenetics: Quantified characters as synapomorphies. *Systematic Biology, 44*, 179–189.

Zelditch, M. L., Sheets, H. D., & Fink, W. L. (2000). Spatiotemporal reorganization of growth rates in the evolution of ontogeny. *Evolution, 54*, 1363–1371.

Zelditch, M. L., Sheets, H. D., & Fink, W. L. (2003). The ontogenetic dynamics of shape disparity. *Paleobiology, 29*, 139–156.

Zelditch, M. L., Lundrigan, B. L., Sheets, H. D., & Garland, T. (2003). Do precocial mammals develop at a faster rate? A comparison of rates of skull development in *Sigmodon fulviventer* and *Mus musculus domesticus. Journal of Evolutionary Biology, 16*, 708–720.

Zelditch, M. L., Lundrigan, B. L., Sheets, H. D., & Garland, T. (2003a). Do precocial mammals develop at a faster rate? A comparison of rates of skull development in *Sigmodon fulviventer* and *Mus musculus domesticus*. *Journal of Evolutionary Biology*, *16*, 708–720.

Zelditch, M. L., Sheets, H. D., & Fink, W. L. (2003a). The ontogenetic dynamics of shape disparity. *Paleobiology*, *29*, 139–156.

Zelditch, M. L., Sheets, H. D., & Fink, W. L. (2003b). The ontogenetic dynamics of shape disparity. *Paleobiology*, *29*, 139–156.

Zelditch, M. L., Lundrigan, B. L., Sheets, H. D., & Garland, J. T. (2003b). Do precocial mammals have a fast developmental rate? A comparison between *Sigmodon fulviventer* and *Mus musculus domesticus*. *Journal of Evolutionary Biology*, *16*, 708–720.

Zelditch, M. L., Wood, A. R., Bonett, R. M., & Swiderski, D. L. (2008). Modularity of the rodent mandible: Integrating bones, muscles and teeth. *Evolution & Development*, *10*, 756–768.

Zelditch, M. L., Wood, A. R., & Swiderski, D. L. (2009). Building developmental integration into functional systems: Function-induced integration of mandibular shape. *Evolutionary Biology*, *36*, 71–87.

Zollikofer, C. P. E., & Ponce de Leon, M. S. (2010). The evolution of hominin ontogenies. *Seminars in Cell & Developmental Biology*, *21*, 441–452.

Zullinger, E. M., Ricklefs, R. R., Redford, K. H., & Mace, G. M. (1884). Fitting sigmoidal equations to mammalian growth curves. *Journal of Mammalogy*, *65*, 607–636.

Glossary

Affine transformation (Also called "uniform"). Transformation (or mapping) that leaves parallel lines parallel. The possible affine transformations include those that do not alter shape (scaling, translation, rotation) and those that do (shear and contraction/dilation). See also **Explicit uniform terms, Implicit uniform terms** (Chapter 5).

Allometry Shape change correlated with size change, sometimes more narrowly defined as a change in the size of a part according to the power law $Y = bX^k$, where Y is the size of the part, X is either the size of another part or overall body size, and k and b are constants. There are three distinct types of allometry: (1) ontogenetic, an ontogenetic change in shape correlated with an ontogenetic increase in size; (2) static, variation in shape correlated with variation size among individuals at a common developmental stage; and (3) evolutionary, an evolutionary change in shape correlated with evolutionary changes in size (Chapters 9, 11).

***Alpha* (α)** (1) The acceptable Type I error rate, typically 5%; (2) a factor multiplying partial warps before computing principal components of them; if $\alpha = 0$, principal components of partial warps are conventional principal components; when $\alpha \neq 0$, the partial warps are differentially weighted. Either those with lower bending energy are weighted more highly ($\alpha > 0$) or those with greater bending energy are weighted more highly ($\alpha < 0$). Typically, values of $+1$ or -1 are used. See also **Relative warps**.

ANCOVA Analysis of covariance. A method for testing the hypothesis that samples do not differ in their means when the effects of a covariate are taken into account. See also **ANOVA, MANOVA** and **MANCOVA** (Chapters 8, 9).

Anisotropic Not isotropic, having a preferred direction. In general, anisotropy is a measure of the degree to which variation in some parameter is a function of its direction relative to some axis. In geometric morphometrics, anisotropy usually refers to a measure of an affine transformation–either the ratio between principal strains, or a ratio of variances along principal axes. See also **Isotropic** (Chapter 3).

ANOVA Analysis of variance. A method for testing the hypothesis that samples do not differ in their means. ANOVA differs from MANOVA in that the means are unidimensional scalars. See also **ANCOVA, MANOVA** and **MANCOVA** (Chapter 9).

Balanced Design An experimental design in which the sample size for each combination of factors is equal. This makes it relatively straightforward to partition the variance (Chapter 9).

Baseline A line joining two landmarks, used in some superimposition methods to register shapes by assigning fixed values to one or more coordinates of those landmarks. See also **Baseline registration, Bookstein coordinates, Sliding baseline registration** (Chapters 3, 4).

Baseline registration A method of superimposing landmark configurations by assigning two landmarks fixed values (the two landmarks are the endpoints of the baseline). The most common method of baseline registration is the two-point registration developed by Bookstein, in which the ends of the baseline are fixed at (0, 0) and (1, 0), yielding Bookstein coordinates. Other methods of baseline registration fix the endpoints at different values (see Dryden and Mardia, 1998) or only fix one coordinate of each baseline point (see **Sliding baseline registration**) (Chapters 3, 4).

Basis A set of linearly independent vectors that span the entire vector space, also the smallest necessary set of vectors that span the space. The basis can serve as a coordinate system for the space because every vector in that space is a unique linear combination of the basis vectors. However, the basis itself is not unique; any vector space has infinitely many bases that differ by a rotation. An orthonormal basis is a set of mutually orthogonal axes, all of unit length. Partial warps and principal components are two common orthonormal bases used in shape analysis. See also **Eigenvectors** (Chapters 5, 6).

Bending energy (1) A measure of the amount of non-uniform shape difference based on the thin-plate spline metaphor. In this metaphor, bending energy is the amount of energy required to bend an ideal, infinite and infinitely thin steel plate by a given amplitude between chosen points. Applying this concept to the deformation of a two-dimensional configuration of landmarks involves modeling the displacements of landmarks in the X, Y plane as if they were displacements above or below the plane ($\pm Z$). (2) Eigenvalues of the bending-energy matrix, representing the amount of bending energy per unit deformation along a single principal warp (eigenvector of the bending-energy matrix). This concept of bending energy is useful because it provides a measure of spatial scale; it takes more energy to bend the plate by a given amount between closely spaced landmarks than between more distantly spaced landmarks. Thus, principal warps with large eigenvalues represent more localized components of deformation than principal warps with smaller eigenvalues. The total bending energy (definition 1) of an observed deformation is a sum of multiples of the eigenvalues, and accounts for the non-uniform deformation of the reference shape into the target shape. See also **Thin-plate spline, Principal warps, Partial warps** (Chapter 5).

Bending-energy matrix The matrix used to compute principal warps and their bending energies (eigenvectors and eigenvalues, respectively). This matrix is a function of the distances between landmarks in the reference shape. See also **Principal warps, Partial warps** (Chapter 5).

Between Groups Principal Components Analysis A method for reducing the dimensionality of multivariate data, performed by extracting the eigenvectors of the variance–covariance matrix of the group means. All individuals in the samples are then scored on these axes; as in ordinary PCA, the scores of individuals are calculated by taking the dot product between that principal component and the data for that specimen (Chapter 6).

Biorthogonal directions Principal axes of a deformation; the term was used in Bookstein et al., 1985; more recently, workers refer to principal axes (Chapter 3).

Black Book Marcus, L.F., Bello, E. and Garcia-Valdcasas, A. (eds) (1993). *Contributions to Morphometrics.* Madrid, Monografias del Museo Nacional de Ciencias Naturales 8. See also **Blue Book, Orange Book, Red Book** and **White Book.**

Blue Book Rohlf, F.J. and Bookstein, F.L. (eds) (1990). *Proceedings of the Michigan Morphometrics Workshop.* University of Michigan Museum of Zoology, Special Publication No. 2. See also **Black Book, Orange Book, Red Book** and **White Book.**

Bonferroni correction, Bonferroni adjustment An adjustment of the α-value to protect against inflating Type I error rate when testing multiple *a posteriori* hypotheses. The adjustment is done by dividing the acceptable Type I error rate (α) by the number of tests. That quotient is the adjusted α-value for each of the *a posteriori* hypotheses. For example, if the desired Type I error rate is 5%, and there are 10 *a posteriori* hypotheses to test, $0.05/10 = 0.005$ is the α-value for each of those 10 tests. A less conservative approach is a sequential Bonferroni adjustment in which the desired α-value is divided by the number of remaining tests. Thus, the adjusted α for the first test would be 0.05/10; for the second it would be 0.05/9; for the third it would be 0.05/8, etc. To apply this sequential adjustment, hypotheses are ordered from lowest to highest p-value; the null hypothesis is rejected for each in turn until reaching one that cannot be rejected (the analysis stops at that point).

Bookstein coordinates (BC) The shape variables produced by the two-point registration, in which the configuration is translated to fix one end of the baseline at (0, 0), and then rescaled and rigidly rotated to fix the other end of the baseline at (1, 0). See also **Baseline registration** (Chapter 3).

Bookstein two-point registration (BTR) See **Two-point registration, Bookstein coordinates.**

Bootstrap test A statistical test based on random resampling (with replacement) of the data. Usually, the method is used to simulate the null model that one wishes to test. For example, if using a bootstrap test of the difference between means, the null hypothesis of no difference is simulated. Bootstrap tests are used when the data are expected to violate distributional assumptions of conventional analytic statistical tests. Rather than assuming that the data meet the distributional assumptions, bootstrapping produces an empirical distribution that can be used either for hypothesis testing or for generating confidence intervals. See also **Jackknife test, Permutation test** (Chapters 8, 9).

Canonical variates analysis (CVA) A method for finding the axes along which groups are best discriminated. These axes (canonical variates) maximize the between-group variance relative to the within-group variance. Scores for individuals along these axes can be used to assign specimens (including unknowns) to the groups, and can be plotted to depict the distribution of specimens along the axes. CVA is an ordination rather than statistical method. See also **Ordination methods, Principal components analysis** (Chapter 6).

Cartesian coordinates Coordinates that specify the location of a point as displacements along fixed, mutually perpendicular axes. The axes intersect at the origin, or zero point, of all axes. Two Cartesian coordinates are needed to specify positions in a plane (flat surface); three are required to specify positions in a three-dimensional space. These coordinates are called "Cartesian" after the philosopher Descartes, a pioneer in the field of analytic geometry.

Centered A matrix is centered when its centroid is at the origin of a Cartesian coordinate system; i.e. at (0, 0) of a two-dimensional system or at (0, 0, 0) of a three-dimensional system (Chapter 4).

Centroid See **Centroid position**.

Centroid position The position of the averaged coordinates of a configuration of landmarks. The centroid position has the same number of coordinates as each of the landmarks. The X-component of the centroid position is the average of the X-coordinates of all landmarks of an individual configuration. Similarly, the Y-component is the average of the Y-coordinates of all landmarks of an individual configuration. It is common to place the centroid position at (0, 0), because this often simplifies other computations (Chapter 4).

Centroid size (CS) A measure of geometric scale, calculated as the square root of the summed squared distances of each landmark from the centroid of the landmark configuration. This is the size measure used in geometric morphometrics. It is favored because centroid size is uncorrelated with shape in the absence of allometry, and also because centroid size is used in the definition of the Procrustes distance (Chapter 4).

Coefficient A number multiplying a function. For example, in the equation $Y = mX$, m is the coefficient for the slope, which is the function that relates X and Y (Chapters 8, 9).

Column vector A vector whose entries are arranged in a column. Contrast to a **Row vector**.

Complex numbers A number consisting of both a real and an imaginary part. An imaginary number is a real number multiplied by i, where i is $\sqrt{-1}$. A complex number is written as $Z = X + iY$, where X and Y are real numbers. In that notation, X is said to be the real part of Z and Y is the imaginary part. A complex number is often used to represent a vector in two dimensions. The mathematics of two-dimensional vectors and complex numbers are similar, so it is sometimes useful to perform calculations or derivations in complex number form.

Configuration see **Landmark configuration**.

Configuration matrix A matrix representing the configuration of K landmarks, each of which has M dimensions. A configuration matrix is a $K \times M$ matrix in which each row represents a landmark and each column represents one Cartesian coordinate of that landmark; $M = 2$ for landmarks of two-dimensional configurations (planar shapes), and $M = 3$ for landmarks of three-dimensional configurations. Two configuration matrices can differ in location, size and orientation, as well as shape (Chapter 4).

Configuration space The set of all possible configuration matrices describing all possible configurations of K landmarks with M coordinates (all with the same values of K and M). Because there are $K \times M$ elements in the configuration matrices, there are $K \times M$ dimensions in the configuration space. In statistical analyses, the configuration space accounts for $K \times M$ degrees of freedom because that is the number of independent pieces of information (e.g. landmark coordinates) needed to specify a particular configuration (Chapter 4).

Consensus configuration The mean (average) configuration of landmarks in a sample of configurations. Usually, this is calculated after superimposing coordinates. See also **Generalized Procrustes superimposition**, **Reference form** (Chapter 4).

Contraction A mathematical mapping that "shrinks" a configuration along one axis. A contraction along the X-axis would map the point (X, Y) to the point (AX, Y), where A is less than one. A contraction along the Y-axis would map (X, Y) to (X, AY). **Expansion** or **dilation** is the opposite of contraction ($A > 1$).

Coordinates The set of values that specify the location of a point along a set of axes (see **Cartesian coordinates**).

Correlation A measure of the association between two or more variables. In morphometrics, correlation is most often measured using Pearson's product-moment correlation, which is the covariance divided by the product of the variances:

$$R_{XY} = \frac{\sum(X - X_{\text{mean}})(Y - Y_{\text{mean}})}{\sqrt{\sum(X - X_{\text{mean}})^2 \sum(Y - Y_{\text{mean}})^2}}$$

where the sums are taken over all specimens. When variables are highly correlated we can predict one from the other (e.g. Y from X), and the more highly correlated they are, the better our predictions will be. Uncorrelated variables are considered independent. See also **Covariance**.

Covariance Like correlation, a measure of the association between variables. The sample estimate of the covariance between X and Y is:

$$S_{XY} = \left(\frac{1}{N-1}\right) \sum (X - X_{mean})(Y - Y_{mean})$$

where the summation is over all N specimens.

Curved space A metric space in which the distance measure is not linear. The ordinary rules of Euclidean geometry do not apply in such spaces. The consequences of the curvature depend upon the distance between points; we can treat the surface of the earth as flat as long as the maps cover only small areas, but in long-distance navigation, the curvature must be taken into account. Shape space is curved, so the rules of Euclidean geometry do not apply, which is why shapes are mapped onto a Euclidean space tangent to shape space. The reference form is the point of tangency of the linear tangent space to the underlying curved space, so using the mean of all specimens as the reference forms acts to minimize the differences between the two spaces.

D A generalized statistical distance between means of two groups (**X1** and **X2**) relative to the variance within the groups:

$$D = \sqrt{(\mathbf{X1}-\mathbf{X2})^T \mathbf{S}_p^{-1} (\mathbf{X1}-\mathbf{X2})}$$

where $(\,)^T$ refers to the transpose of the enclosed matrix, and $\mathbf{S}_p^{-1}$ is the inverse of the pooled variance–covariance matrix. This distance takes into account the correlations among variables when computing the distance between means. The generalized distance is used in Hotelling's T^2-test. Also known as the **Mahalanobis' distance**.

D² The squared generalized distance, **D**. See **D**.

Deformation A smooth, continuous mapping or transformation; in morphometrics, it is usually the transformation of one shape into another. The deformation refers not only to the change in positions of landmarks, but also to the interpolated changes in locations of unanalyzed points between landmarks (Chapter 5).

Degrees of freedom In general, the number of independent pieces of information. In statistical analyses, the total degrees of freedom are approximately the product of the number of variables and the number of individuals (the total may be partitioned into separate components for some tests). If every measurement on every individual were completely independent, the degrees of freedom would be the product of the number of variables and the number of individuals, but if one statistic is known (or estimated), the number of degrees of freedom that remain to estimate a second statistic will be reduced. For example, the estimate of the mean height of N individuals in a sample will have $N \times 1 = N$ degrees of freedom, because all N measurements are needed and there is only one measured variable. In contrast, the estimate of the variance in height will have $N-1$ degrees of freedom because only $N-1$ deviations from mean height are independent (the deviation of the Nth individual can be calculated from the mean and the other $N-1$ observed heights). In geometric morphometrics, when configurations of landmarks are superimposed, degrees of freedom are lost for a different reason; namely, information that is not relevant to comparison of shapes (location, scale and rotation) is removed from the coordinates.

Dilation Opposite of **Contraction**.

Dimensionality reduction There is a common need to reduce the dimensionality of a data set, both for display and to reduce the number of variables used to less than the degrees of freedom in the data, thus allowing inversion of a variance–covariance matrix. If a PCA is performed on the data, and the scores corresponding to all PCs with non-zero eigenvalues are retained, and the rest discarded, the degrees of freedom in the remaining scores will equal the degrees of freedom in the data.

Discriminant function The linear combination of variables optimally discriminating between two groups. It is produced by discriminant function analysis. Scores on the discriminant function can be used to identify members of the groups (Chapter 14).

Discriminant function analysis A two-group **canonical variates analysis**. See **Canonical variates analysis** (Chapter 14).

Disparity, morphological disparity (*MD*) Phenotypic variety, usually morphological. Several metrics can be used to measure disparity, but the one most commonly used in studies of continuous variables is:

$$MD = \frac{\sum_{j=1}^{N} D_j^2}{(N-1)}$$

where D_j is the distance of species j from the overall centroid (i.e. the grand mean calculated over N groups, e.g. species) (Chapters 10, 11).

Distance A function measuring the separation between points. Within any space there are multiple possible distances. For this reason, it is necessary to specify the type of distance used. See also **D, $\mathbf{D^2}$, Euclidean distance, Generalized distance, Geodesic distance, Great circle distance, Partial Procrustes distance, Full Procrustes distance, Mahalanobis' distance** (Chapter 4).

Dot product (Also called inner product.) Given two vectors $A = \{A_1, A_2, A_3 ... A_N\}$, $\mathbf{B} = \{B_1, B_2, B_3 ... B_N\}$, the dot product of **A** and **B** is:

$$\mathbf{A} \cdot \mathbf{B} = A_1 B_1 + A_2 B_2 + A_3 B_3 + \cdots + A_N B_N$$

and

$$\mathbf{A} \cdot \mathbf{B} = |A||\mathrm{B}| \cos(\theta)$$

where $|A|$ is the magnitude of **A**, $|\mathrm{B}|$ is the magnitude of **B**, and θ is the angle between **A** and **B**. If the magnitude of **A** is 1, then $\mathbf{A} \cdot \mathbf{B} = |\mathrm{B}| \cos(\theta)$, which is the component of **B** along the direction specified by **A**. The dot product is used to calculate scores on coordinate axes, by projecting the data onto those axes (this is how partial warp scores and scores on principal components are calculated). It is also used to find the vector correlation, R_V, between two vectors (that correlation is the cosine of the angle between vectors).

Edge registration See **Baseline registration**.

Eigenvalues See **Eigenvectors**.

Eigenvectors Eigenvectors are the non-zero vectors, **A**, satisfying the eigenvector equation:

$$(\mathbf{X} - \lambda \mathbf{I})\mathbf{A} = 0$$

The values of λ that satisfy this equation are eigenvalues of **X**. Eigenvectors are orthogonal to one another, and provide the smallest necessary set of axes for a vector space (i.e. they provide a basis for that space). The eigenvectors of a variance–covariance matrix are called principal components; the eigenvalue corresponding to each axis gives the variance associated with it. The eigenvectors of the bending-energy matrix are the principal warps; the eigenvalue corresponding to each axis gives the bending energy associated with it. See also **Basis** (Chapters 5, 6).

Element of a matrix A number in a matrix, typically referenced by the symbol designating the matrix with subscripts indicating its row and column; for example, $X_{4,5}$ refers to the element on the fourth row and fifth column of the matrix **X**.

Euclidean distance The square root of the summed squared distances along all orthogonal axes. A Euclidean distance does not change when the axes of the space are rotated (in contrast to a Manhattan distance, which is simply the sum of the distances). See also **D, $\mathbf{D^2}$, Distance, Generalized distance, Geodesic distance, Great circle distance, Procrustes distance, Full Procrustes distance, Partial Procrustes distance** (Chapter 4).

Euclidean space A coordinate space in which the metric is a Euclidean distance.

Exchangeable A concept used in designing permutation tests. If the null hypothesis used in the test states that a property, such as group membership or size, is not a statistically significant factor, then that property is said to be exchangeable under that hypothesis. If age is exchangeable in a given test, then the age of the specimens would be permuted as part of the test. If age is significant, then this would be evident from the permutation test, as the observed statistic would be unlikely (i.e. have a low probability, p) based on the distribution of the statistic over the permuted data, and we would reject the null hypothesis as having a low probability of being true.

Explicit uniform term, explicit uniform component A uniform component describes affine or uniform deformations. Some of these do not alter shape (i.e. rotation, translation and rescaling) whereas others do (i.e. shear and dilation). Accordingly, we divide affine deformations into two sets: (1) implicit uniform terms, which do not alter shape and are used in superimposing forms but are not explicitly recorded; and (2) explicit uniform terms, which do alter shape and therefore are typically reported as components of the deformation. All uniform terms must be known to model a deformation correctly (Chapter 5).

Fiber In geometric morphometrics, the set of all points in pre-shape space representing all possible rigid rotations of a landmark configuration that has been centered and scaled to unit centroid size; in other words, the set of pre-shapes that have the same shape. Fibers are collapsed to a point in shape space (Chapter 4).

Form Size-plus-shape of an object; form includes all the geometric information not removed by rotation and translation. Form is also called **Size-and-shape**.

Full Procrustes distance (D_F) The distance between two landmark configurations when, after partial Procrustes superimposition, the shape superimposed on the reference is rescaled to cs = cos(ρ) further to minimize the sum of squared distances between their corresponding landmarks. See also **Partial Procrustes distance, Procrustes distance** (Chapter 4).

Full Procrustes superimposition The superimposition that yields the Full Procrustes distance of a shape from the reference, achieved by reducing centroid size to cos(ρ). See also **Full Procrustes distance** (Chapter 4).

F-test One of a variety of test statistics, formed as a ratio of the variance explained by a model or a factor to the unexplained or error variance estimate. In some cases, the observed *F*-value is compared to an applicable analytic model, otherwise permutation tests are commonly used to determine the associated p-value.

Generalized distance See **D**.

Generalized least squares superimposition A generalized superimposition method that uses a least squares fitting criterion, meaning that the parameters are estimated to minimize the sum of squared distances over all landmarks over all specimens. Usually, in geometric morphometrics, GLS refers specifically to a generalized least squares Procrustes superimposition – a different approach is used in generalized resistant-fit methods (Chapter 4).

Generalized least squares Procrustes superimposition (GLS) A generalized superimposition minimizing the partial Procrustes distance over all shapes in the sample, using a least squares fitting function. This is the method usually used in geometric morphometrics; it is now usually termed Generalized Procrustes Analysis (Chapter 4).

General Linear Models (GLM) A generalization of the ideas behind MANOVA, MANCOVA and regression models, to encompass any model that is linear in its fitted parameters. The statistical significance of the covariates and factors in the model are typically assessed using *F*-tests (Chapter 9).

Generalized Procrustes Analysis (GPA) A Procrustes-based analysis using generalized superimposition to estimate iteratively the mean and then superimpose all specimens on it. This is now the standard term for a Generalized least squares Procrustes superimposition. See **Generalized superimposition**.

Generalized superimposition The superimposition of a set of specimens onto their mean. This involves an iterative approach because the mean cannot be calculated without superimposing specimens, which cannot be superimposed on the mean before the mean is calculated (an alternative approach is used in ordinary Procrustes analysis). See also **Consensus configuration** (Chapter 4).

Geodesic distance The shortest distance between points in a space. On a flat planar surface, this is the length of the straight line joining the points – i.e. the Euclidean distance. On curved surfaces, this distance is the length of an arc.

Great circle The intersection of the surface of a sphere and a plane passing through its center. A great circle divides the surface of the sphere in half. On the surface of the sphere, the shortest distance between two points lies along the great circle that passes through those points. If the Earth were perfectly spherical, the equator and all lines of latitude would be great circles.

Great circle distance The arc length of the segment of the great circle connecting two points on the surface of a sphere; this is the geodesic distance between those points, the shortest distance between the points in the space of the surface of the sphere.

Homology (1) In biology, similarity due to common evolutionary origin. (2) In morphometrics, the correspondences between landmarks, sometimes imputed by a mathematical function, called a "homology function" (e.g. see Bookstein et al., 1985). Homology is the primary criterion for selecting landmarks (Chapter 2).

Hypersphere The generalization of a three-dimensional sphere to more than three dimensions. In three dimensions, points on the surface of a sphere of radius R that is centered at the origin satisfy the equation $X^2 + Y^2 + Z^2 = R^2$.

Implicit uniform terms See **Explicit uniform terms**.

Induced correlation A correlation induced by dividing two values by a third which is common to both. The induced correlation between the (rescaled) variables is not present in the original variables.

Inner product See **Dot product**.

Invariant A quantity is invariant under a mathematical operation or transformation when it is not changed by that operation. For example, centroid size is invariant under translation, centroid position is not.

Isometric In general, a transformation that leaves distances between points unaltered. In morphometrics, isometry usually means that shape is uncorrelated with size. In statistical tests of allometry, isometry is the null hypothesis (Chapters 8, 11).

Isotropic A property is said to be isotropic if it is uniform in all directions, i.e. if it does not differ as a function of direction. When an error is isotropic, it is equal in all directions, and there is no correlation among errors. Isotropic is the opposite of anisotropic.

Jackknife test An approach to statistical testing that involves resampling the original observations to generate an empirical distribution. Jackknifing is carried out by omitting one specimen at a time. See also **Bootstrap test**, **Permutation test** (Chapter 8).

Kendall's shape space The space in which the distance between landmark configurations is the Procrustes distance. This space is constructed by using operations that do not alter shape to minimize differences between all configurations of landmarks that have the same values of K (number of landmarks) and M (number of coordinates of a landmark). Kendall's shape space is the curved surface of a hypersphere, so conventional statistical analyses are conducted in a Euclidean tangent space (Chapter 4).

Landmark Biologically, landmarks are discrete, homologous anatomical loci; mathematically, landmarks are points of correspondence, matching within and between populations (Chapter 2).

Landmark configuration The positions (coordinates) of a set of landmarks representing a single object, containing information about size, shape, location and orientation. The number of landmarks is typically represented by K, and the dimensionality of the landmarks (number of coordinates) is typically represented by M. Therefore, if there are 16 landmarks, each with an X- and Y-coordinate, then $K = 16$ and $M = 2$ (Chapter 4).

Least squares A method of choosing parameters that minimizes the summed square differences over all individuals (and variables) (Chapters 4, 7, 8).

Linear A function $f(X)$ is linear if it depends only on the first power of X; e.g. $f(X) = 2(X)$ is linear, but $f(X) = (X)^2$ is not.

Linear combination A vector produced by multiplying and summing coefficients of one or more vectors. For example, given the vector $\mathbf{X}^T = \{X_1, X_2 \ldots X_N\}$ and $\mathbf{A}^T = \{A_1, A_2 \ldots A_N\}$, then $\mathbf{Y} = A_1X_1 + A_2X_2 + \ldots A_NX_N$ is a linear combination of the vectors. We can write this as $\mathbf{Y} = \mathbf{A}^T\mathbf{X}$.

Linear transformation A transformation producing a set of new vectors that are linear combinations of the original variables. See **Linear combination**.

Linear vector space The set of all linear combinations of a set of vectors. The space spans all possible linear combinations of the basis vectors, as well as all sums or differences of any linear combination of those basis vectors. The two-dimensional Cartesian plane is the linear vector space formed by the linear combinations of two vectors of unit length, one along the X-axis, the other along the Y-axis.

Mahalanobis' distance (D^2) The squared distance between two means divided by the pooled sample variance–covariance matrices. This is a generalized statistical distance, adjusting for correlations among variables. See also **D**, **Generalized distance**.

MANCOVA Multivariate analysis of covariance. A method for testing the hypothesis that samples do not differ in their means when the effects of a covariate are taken into account. See also **General Linear Models (GLM)**, **ANOVA**, **ANCOVA** and **MANOVA** (Chapters 8, 9).

MANOVA Multivariate analysis of variance. A method for testing the hypothesis that samples do not differ in their means; MANOVA differs from ANOVA in that the means are multidimensional vectors. See also **General Linear Models (GLM), ANOVA, ANCOVA** and **MANCOVA** (Chapters 8, 9).

Map A mathematical function relating **X** to **Y** by stating the correspondence between elements in **X** and **Y**. Each element in **X** is placed in correspondence with one element in **Y**. Multiple elements in **X** may map to the same element in **Y** (landmark configurations differing only in rotation for example would all map to the same shape). A map is written as: $f : \mathbf{X} \rightarrow \mathbf{Y}$ where f is the map from the set **X** to the set **Y**.

Matrix A rectangular array of numbers (real or complex). The numbers in a matrix are referred to as elements of the matrix. The size of a matrix is always given as the number of rows followed by the number of columns; e.g. a 4×2 matrix has four rows and two columns.

Mean Also known as the average; an estimate of the center of the distribution calculated by summing all observations and dividing by the sample size.

Median An estimate of the center of a distribution calculated such that half the observed values are above and the other half are below.

Metric A non-negative real-valued function, $D(X, Y)$, of the points X and Y in a space such that:

1. The only time that the function is zero is when X and Y are the same point, i.e. $D(X, Y) = 0$, if and only if $X=Y$
2. If we measure from X to Y, we get the same distance as when we measure from Y to X, so $D(X, Y) = D(Y, X)$ for all X and Y
3. The triangle inequality holds true. The triangle inequality states the distance between any two points, X and Y, is less than or equal to the sum of distances from each to a third point, Z, so $D(X, Y) \leq D(X, Z) + D(Y, Z)$, for all X, Y and Z.

Multiple regression Regression of a single (univariate) dependent variable on more than one independent variable. See also **Multivariate regression, Multivariate multiple regression, Regression.**

Multivariate analysis of variance See **MANOVA**.

Multivariate multiple regression Regression of several dependent variables on more than one independent variable. In morphometrics, this method is used to regress shape (the dependent variables) onto multiple independent variables. See also **General Linear Models, Multiple regression, Multivariate regression, Regression.**

Multivariate regression Regression of several dependent variables onto one independent variable. In morphometrics, this method is used to regress shape onto a single independent variable, such as size. The coefficients obtained by multivariate regression are the same as those estimated by simple bivariate regression of each dependent variable on the independent variable. However, the statistical test of the null hypothesis differs. See also **General Linear Models, Multiple regression, Multivariate multiple regression, Regression** (Chapters 8, 9).

Non-uniform Non-isotropic, or localized, not Uniform; Non-affine. See **Non-uniform deformation.**

Non-uniform deformation The component of a deformation that is not uniform. In contrast to a uniform deformation, which leaves parallel lines parallel and has the same effect everywhere across a form, a non-uniform deformation turns squares into trapezoids or diamonds (shapes that do not have parallel sides) and has different effects over different regions of the form. Most deformations comprise both uniform and non-uniform parts. The non-uniform component can be further subdivided, see **Partial warps** (Chapter 5).

Normalize To set the magnitude to one. Normalizing a vector sets the length of the vector to one; this is done by dividing each component of the vector by the length of the vector, calculated by taking the square root of the summed squared coefficients.

Null hypothesis, or **null model** Usually, the hypothesis that the factor of interest has no effect beyond that expected by chance. For example, in an analysis of allometry, the null hypothesis being tested by regression of shape on size is that shape does not depend on size (i.e. isometry). Similarly, in a comparison of two means using Hotelling's T^2-test, the null hypothesis is that the two groups do not differ beyond what is expected by chance.

Orange Book Bookstein, F.L. (1991). *Morphometric Tools for Landmark Data. Geometry and Biology.* Cambridge University Press. See also **Black Book, Blue Book, Red Book** and **White Book.**

Ordinary Procrustes analysis (OPA) An approach to superimposition in which one landmark configuration is fitted to another, which differs from a **Generalized superimposition** in that it involves only two forms. This approach has rarely been used once iterative methods became available for generalized superimpositions. See also **Generalized Procrustes Analysis (GPA), Generalized superimposition, Consensus form** (Chapter 4).

Ordination Ordering specimens along one or more axes based on some criterion (e.g. from youngest to oldest, or shortest to tallest). Ordination methods include principal components analysis and canonical variates analysis; the scores on the axes provide a basis for ordering specimens (Chapter 6).

Orthogonal Perpendicular (at right angles to each other). Two vectors are orthogonal if the angle between them is 90°; when they are, their dot product is zero.

Orthonormal Perpendicular and of unit length; vectors are orthonormal if they are mutually orthogonal and of unit length.

Orthonormal basis See **Basis**. An orthonormal basis is comprised of a set of mutually perpendicular basis vectors normalized so that their magnitudes are all set to one.

Outline A curve around the perimeter of an object (or around a distinct part of it).

Partial least squares analysis A method of exploring patterns of covariance or correlation between two blocks of variables measured on the same set of specimens. A singular value decomposition is used to determine the pair of vectors (each a linear combination of variables within one of the blocks) that expresses the greatest proportion of the covariance between blocks. See also **Singular value decomposition, Singular warps** (Chapter 7).

Partial Procrustes distance (D_p) The distance between two landmark configurations when both shapes are centered, fixed to unit centroid size, and rotated to minimize the sum of squared distances between their corresponding landmarks. See also **Full Procrustes distance, Procrustes distance** (Chapter 4).

Partial Procrustes superimposition The superimposition that yields the Partial Procrustes distance of a shape from the reference, achieved by fixing centroid size at 1. See also **Partial Procrustes distance** (Chapter 4).

Partial warps The term partial warps sometimes refers solely to the components of the non-uniform deformation, which are computed as eigenvectors of the bending-energy matrix projected onto the X, Y-plane of the data (they are projections of principal warps), ordered from least to most bending energy. These eigenvectors provide an orthonormal basis for the non-uniform part of a deformation. Sometimes "partial warps" also includes the components of the uniform deformation, as the zeroth partial warp – in which case the scores on this component are included among the partial warp scores (Chapter 5).

Partial warp scores Coefficients indicating the position of an individual, relative to the reference, along partial warps. They are calculated by taking the dot product between the partial warps and the data for a specimen. When appropriate scores on the uniform component are also included among the partial warps scores, the sum of the squared scores equals the squared partial Procrustes distance of that specimen from the reference. This full set of scores can be used as shape variables in any conventional statistical analysis because they are based on the appropriate distance measure and have the same number of coordinates as degrees of freedom. See also **Non-uniform deformation, Partial warps, Principal warps, Uniform deformation** (Chapter 5).

Permutation test An approach to statistical testing that involves permuting (rather than randomly sampling) observed values. Unlike many bootstrapping methods, the resampling done is without replacement, each specimen appears only once in the permutation set. See also **Bootstrap test, Jackknife test, Monte Carlo simulations** (Chapters 8, 9).

Phylogenetic generalized least squares A method of accounting for the expected non-independence of observations on taxa by incorporating the phylogenetic variance–covariance matrix (a function of branch lengths between taxa) in statistical analyses (Chapter 10).

Phylogenetic independent contrasts A method of computing independent evolutionary changes from observations of taxa that are related by varying degrees of common ancestry. Sometimes called simply independent contrasts, these are net differences between daughter nodes (including tips) of the phylogenetic tree, weighted by branch lengths between the nodes (Chapter 10).

Pinocchio effect A large change concentrated at one landmark, with little or none at others; a highly localized change. In the presence of the Pinocchio effect, Procrustes superimpositions imply that the shape difference is distributed over all landmarks. Resistant-fit methods, such as RFTRA, were devised to avoid that implication (Chapter 4).

Population The set of all possible individuals of a specific type, such as all members of a species, or all leaves on a particular kind of tree. See also **Sample** (Chapter 8).

Position See **Centroid position.**

Pre-shape A centered landmark configuration, scaled to unit centroid size (Chapter 4).

Pre-shape space The set of all possible pre-shapes for a given number of landmarks with a given number of dimensions. This is the surface of a sphere of $KM - M - 1$ dimensions, where K is the number of landmarks and M is the number of dimensions of each landmark (Chapter 4).

Principal axes The set of orthogonal axes used in modeling the change of one shape into another as an affine transformation. This transformation can be parameterized by its effect on a circle or sphere (for two or three dimensional shapes, respectively). In two dimensions, an affine transformation takes a circle into an ellipse and the principal axes are the directions of the circle that undergo the greatest relative elongation or shortening mapped onto the major and minor axes of the ellipse. The ratio of the lengths of these axes is the anisotropy, a measure of the amount of affine shape change. Principal axes are invariant under a change in the coordinate system. See also **Principal strains** (Chapters 3, 4).

Principal components analysis (PCA) A method for reducing the dimensionality of multivariate data, performed by extracting the eigenvectors of the variance–covariance matrix. These eigenvectors are called principal components. Their associated eigenvalues are the variance explained by each axis. Principal components provide an orthonormal basis. The position of a specimen along a principal component is represented as its principal component score, calculated by taking the dot product between that principal component and the data for that specimen (Chapter 6).

Principal strain In an affine deformation, the ratio of the length of a principal axis in the ellipse to the original diameter of the circle. See also **Principal axes** (Chapter 3).

Principal warp An eigenvector of the bending-energy matrix interpreted as a warped surface over the surface of the X, Y-plane of the landmark coordinates. Principal warps are ordered from least to most bending energy (smallest to largest eigenvalue), which corresponds to the least to most spatially localized deformation. Principal warps differ from partial warps in that partial warps are projections of principal warps onto the X, Y-plane of the data. See also **Bending energy, Bending-energy matrix, Orthonormal basis, Partial warp, Thin-plate spline** (Chapter 5).

Probability distribution A mathematical function that describes the probability of a measurement taking on either a particular value or a range of values, depending on whether the variable is discrete or continuous, respectively (Chapter 8).

Procrustes distance This term has been used to refer to the sum of squared distances between corresponding points of two superimposed shapes after one shape has been centered on the other and rotated to minimize that sum of squares. When the shape being superimposed is reduced in centroid size to minimize further the difference between it and the target, the distance may be called a **Full Procrustes distance** (D_F). When both sizes are held at centroid size = 1, the distance may be called a **Partial Procrustes distance** (D_p). Both D_F and D_p are related to the arc distance (ρ) between configurations in the space of aligned preshapes with centroid sizes fixed at 1. This arc length has also been called a Procrustes distance, with the others called full (or partial) Procrustes *chord* distances to distinguish them from the arc length. See also **Full Procrustes distance, Partial Procrustes distance** (Chapter 4).

Procrustes Form Space An approach to analyzing both size and shape. In this method, a data matrix is formed of the coordinates in Procrustes superimposition and a column of the log centroid size values is added. This data matrix is then analyzed using a PCA or in hypothesis testing procedures (Chapter 14).

Procrustes methods A general term referring to the superimposition of matrices based on a least squares criterion. The term comes from the Greek mythological figure, Procrustes, who fitted visitors to a bed by stretching them or amputating overhanging parts (Chapter 4). See **Full Procrustes, Partial Procrustes**.

Procrustes residuals Coordinates of a landmark configuration obtained by a Procrustes superimposition. They are residuals in the sense that they indicate the deviation of each specimen from the mean (i.e. the consensus configuration) or other reference. See also **Consensus configuration, Procrustes superimposition, Reference** (Chapter 5).

Procrustes Size Preserving or Procrustes SP, is a Procrustes superimposition variant in which the superimposition is done using only translations and rotation, not scaling. Dryden and Mardia (1998) refer to this as the "size and shape" of a configuration. It is possible to define a Procrustes SP distance as well. See also **Procrustes Form Space** (Chapter 14).

Procrustes superimposition A superimposition of shapes that minimizes the Procrustes distances over the sample. The term is used whether the distance being minimized is the full or the partial Procrustes distance (Chapter 4).

Red Book Bookstein, F.L., Chernoff, B., Elder, R.L. et al. (eds) (1985). *Morphometrics in Evolutionary Biology: The Geometry of Size and Shape Change, with Examples from Fishes.* Academy of Natural Sciences of Philadelphia, Special Publication No. 15. See also **Black Book, Blue Book, Orange Book** and **White Book**)

Reference, Reference form The shape to which all others are compared. It is the point of tangency between Kendall's shape space and the tangent space. Because the linear approximation to Kendall's shape space may be inaccurate when the point of tangency is far from the center of the distribution of specimens, the reference is usually chosen to minimize the distances between it and the other specimens–i.e. it is chosen to be the consensus shape (Chapter 4).

Regression An analytic procedure for fitting a predictive model to data and assessing the validity of that model. One variable is expressed as a function of the other, e.g. $Y = mX + b$ expresses Y as a linear function of X. The predictor variable(s) are the independent variable(s), and those variables predicted by the model are the dependent variable(s). In the linear model above, X is the independent variable that predicts the dependent variable, Y. The term "regression" comes from Francis Galton (1889), who concluded that offspring tend towards (regress towards) the mean of the population. As stated by Galton in his law of universal regression, "each peculiarity in a man is shared by his kinsman, but *on the average*, in a less degree". Thus, the offspring of unusually tall fathers regress towards the mean height of the population (Chapters 8, 9).

Relative warps Principal components of partial warp scores, sometimes weighted to emphasize components of low or high bending energy (that weighting is done by setting the parameter α to a value other than 0). Originally, the term referred to an eigenanalysis of the variance–covariance matrix relative to the bending-energy matrix, hence a new term was coined for these components (Bookstein, 1991). Currently, the term usually refers to a conventional principal components analysis of partial warp scores. See also ***Alpha*** **(α), Bending energy, Partial warp scores, Principal components analysis** (Chapter 5).

Repeated measurement error The level of variation that appears when a single specimen is repeatedly measured, this quantity indicates the achievable level of precision in a particular approach to measurement. In landmark-based methods, it is measured as the summed squared Procrustes distances of repeated measurements of a single specimen about the mean of those measurements, or as the square root of this value, the RMS scatter.

Repeated median The median of medians, used in estimating the scaling factor and rotation angle by resistant-fit superimposition methods such as RFTRA. The repeated median is more robust to large deviations than the median or a least squares estimator. See also **Resistant-fit superimposition, RFTRA** (Chapter 4).

Resampling A method whereby a new data set is constructed by randomly selecting from the original data (either values recorded on specimens or residuals from a model). Construction of a large series of resampled data sets can be used to simulate either the distribution of measured values or the distribution of a test statistic under the null model. Under some conditions, resampling can also be used to produce confidence intervals around the statistic. This approach permits hypothesis tests when the data are expected to deviate from the distributional assumptions of conventional analytic tests. Resampling may be done with replacement, meaning that each observation can appear more than once in a resampled data set; resampling without replacement means each observation appears only once in a set. See also **Bootstrap test, Jackknife test, Permutation test** (Chapters 8, 9).

Rescale Multiply or divide by a scalar value; used in geometric morphometrics to change the centroid size of a configuration (Chapters 3, 4).

Residual Deviation of an observation from the expected value under a model. For example, a residual from a regression is the deviation between the observed and expected values of the dependent variable at a given value of the independent variable. The term is also used for the coordinates obtained by a Procrustes superimposition, the Procrustes residuals, which are deviations between individual specimens and the reference.

Resistant-fit superimposition A superimposition method that uses medians or repeated medians (rather than a least squares error criterion) to superimpose forms. The method is intended to be resistant to large localized shape differences, such as those produced by the Pinocchio effect. RFTRA is an example of this type of method. See also **Repeated medians, RFTRA** (Chapter 4).

RFTRA (Resistant-fit theta-rho analysis) A resistant-fit superimposition method using the method of repeated medians to determine the scaling factor and rotational angle. See also **Resistant fit** and **Repeated median** (Chapter 4).

Rigid rotation A rotation of an entire vector or matrix by a single angle. Rigid rotations do not alter the size, shape or location of the object. Rotations are often represented by square matrices. The rotation matrix:

$$R = \begin{bmatrix} \cos\theta & -\sin\theta \\ \sin\theta & \cos\theta \end{bmatrix}$$

rotates a $2 \times N$ matrix through an angle θ. When different vectors are multiplied by different angles, the rotation is oblique, not rigid.

RMS scatter The square root of the mean of the summed squared distances of specimens about their mean (root mean square, RMS). It is thus the square root of the variance as measured using Procrustes distances, and a linear measure of the variability of a group (Chapter 14).

Row vector A vector with coefficients in a row. Contrast to a **Column vector**.

Sample The collection of observed individuals representing members of a population. An individual observation is the smallest sampling unit in the study, which might be an individual organism or one of its parts, or a collection of organisms such as a species or a bacterial colony (Chapter 8).

Scalar A real or complex number.

Scale (1) Noun – size of an object (given some definition of size); (2) verb – to change the size of an object (equivalent to rescale).

Scaling factor A constant which is used to change the scale or size of a matrix or vector. This is done by multiplying or dividing the matrix or vector by the constant.

Score In morphometrics, a coefficient locating a specimen along a vector, calculated by projecting the specimen onto an axis. Usually, scores locate the position of a specimen relative to the axes of a coordinate system. They are calculated by taking the dot product between an axis of the coordinate system and the data of a specimen. The scores are linear combinations of the original variables. Partial warp scores locate the position of an individual specimen relative to the coordinate system provided by the partial warps. Similarly, principal component scores locate the position of an individual specimen relative to the coordinate system provided by the principal components. Scores can be calculated relative to any basis of a vector space because each basis provides a coordinate system for that space. See **Dot product**.

Semilandmark A point on a geometric feature (curve, edge or surface) defined in terms of its position on that feature (e.g. at 10% of the length of the curve from one end). Semilandmarks are used to incorporate information about curvature in a geometric shape analysis. Because semilandmarks are defined in terms of other features, they contain less information (fewer degrees of freedom) than landmarks (Chapter 2).

Shape In geometric morphometrics, following Kendall, shape is all the geometric information remaining in an landmark configuration after differences in location, scale and rotational effects are removed (Chapters 1, 2, 3).

Shape coordinates Within geometric morphometrics, coordinates of landmarks after superimposition (Chapters 3, 4)

Shape space Within geometric morphometrics, Kendall's shape space. The term is more general, however, as it can apply to any space defined by a particular mathematical definition of shape. There are shape spaces for outline measurements, for example. There are also shape spaces based on different definitions of size. The characteristics of these various shape spaces are not necessarily the same as those of Kendall's (Chapter 4).

Shape variable A general term for any variable expressing the shape of an object, including ratios, angles, shape coordinates obtained by a superimposition method, or vectors of coefficients obtained from partial warp analysis, principal components analysis, regression, etc. Shape variables are invariant under translation, scaling and rotation.

Shear An affine (or uniform) deformation that leaves the *Y*-coordinate fixed while the *X*-coordinate is displaced along the *X*-axis by a multiple of *Y*. Under a shear, the point (X, Y) maps to $(X + AY, Y)$, where *A* is the magnitude of the shear. Visually, this looks like altering a square by sliding the top side to the left or right, without altering its height or the lengths of the top and bottom (Chapter 5).

Singular axes Orthonormal vectors produced by singular value decomposition. See **Singular value decomposition** (Chapter 7).

Singular value In a singular value decomposition, a quantity expressing a relationship between two singular axes; an element λ_i of the diagonal matrix **S**. In partial least squares analysis, each singular value represents the covariance explained by the corresponding pair of singular axes. See **Singular value decomposition** (Chapter 7).

Singular value decomposition (SVD) A mathematical technique for taking an $M \times N$ matrix **A** (where *N* is greater than or equal to *M*) and decomposing it into three matrices:

$$A = USV^T$$

where U is an $M \times N$ matrix whose columns are orthonormal vectors, S is an $N \times N$ diagonal matrix with on-diagonal elements λ_i, and V is an $N \times N$ matrix whose columns are orthonormal vectors. The values λ_i are called the *singular values* of the decomposition, and the columns of U and V are called the *singular vectors* or *singular axes* corresponding to a given singular value. In partial least squares analysis, A is the matrix of covariances between the two blocks, the columns of U are linear combinations of the variables in one of the two data sets, the columns of V are linear combinations of the variables in the other data set, and each λ_i is the portion of the total covariance explained by the corresponding pair of singular axes (Chapter 7).

Singular warps Sometimes used in geometric morphometrics for singular axes computed from shape data (partial warp scores or residuals of a Procrustes superimposition), so that the singular axes describe patterns of differences in shape. See **Singular value decomposition** (Chapter 7).

Size Any positive real valued function g(**X**), where **X** is a configuration or set of points, such that $g(A\mathbf{X}) = Ag(\mathbf{X})$, where *A* is any positive, real scalar value. In other words, multiplying every element in **X** by *A* multiplies g(**X**) by *A*. There are a wide variety of measures of size, including lengths measured between landmarks, sums or differences of interlandmark distances, square roots of area, etc. The size measure used in geometric morphometrics is centroid size. See also **Centroid size** (Chapters 3, 4).

Size-and-shape All the geometric information remaining in an object (such as a landmark configuration) after differences in location and rotational effects are removed. See **Procrustes SP, Form**.

Space A set of objects (or measurements thereof) that satisfies some definition. For example, a space might be defined as the set of all four-landmark configurations measured in two dimensions.

Statistic Any mathematical function based on an analysis of all measured individuals, e.g. the mean, standard deviation, variance, maximum, minimum, and range. The true value of the statistic in the population is called the parameter, which we are trying to estimate from our sample (Chapter 8).

Superimposition A method for matching two landmark configurations (or matrices) prior to further analysis, sometimes also called a registration. A number of different optimality criteria may be used. See also **Bookstein coordinates, Procrustes superimposition** (also **Full Procrustes superimposition** and **Partial Procrustes superimposition, RFTRA, Sliding baseline registration** (Chapters 3, 4).

Strain See **Principal strain.**

Tangent space The linear vector space tangent to a curved space. In geometric morphometrics, the Euclidean space tangent to Kendall's shape space. In the tangent space, distances between shapes are linear functions, which allows for analysis of shape variation by ordinary multivariate statistical methods. When the linear approximation to the curved surface is accurate (when all shapes in a study are close to the point of tangency), distances in the tangent space approximate distances in the curved space. The point of tangency between Kendall's shape space and the tangent space is the reference form. See also **Kendall's shape space, Reference form** (Chapter 4).

Target shape A shape being compared to the reference shape. See **Reference**.

Thin-plate spline An interpolation function used to predict the difference in shape between a reference and another shape over all points on the form, not just at landmarks. This interpolation function minimizes the bending energy of the deformation, which is equivalent to modeling that deformation as smoothly as possible given the observed landmarks (thus taking a parsimonious approach to interpolation). Thin-plate spline analysis produces scores for the non-uniform component of the deformation–scores for the uniform component are produced by a different analysis (Chapter 5).

Transformation See **Map**.

Two-point shape coordinates See **Bookstein coordinates**.

Type I, Type II error Type I error is rejecting a true null hypothesis. Type II error is failing to reject a false null hypothesis. Most approaches to statistics are careful to state and analyze Type I error, much less attention is paid to Type II error, and it is typically harder to estimate the rate of Type II errors.

Type 1 landmark A landmark that can be defined in terms of local information, such as a landmark located at the junction of three bones or two bones and a muscle (i.e. anatomical features that meet at a point). There is no need to refer to any distant structures or maxima/minima of curvature. The typology of landmarks is based on Bookstein, 1991. See also **Type 2** and **Type 3 landmarks** (Chapter 2).

Type 2 landmark A landmark defined by a relatively local property, such as the maximum or minimum of curvature of a small bulge or at the endpoint of a structure. These are considered less useful than Type 1 landmarks because their evidence of homology is at least partly geometric rather than purely histological or osteological. See also **Type 1** and **Type 3 landmarks** (Chapter 2).

Type 3 landmark A landmark defined in terms of extremal points, such as the landmark on the rostrum *furthest away from* the foramen magnum. Such landmarks are regarded as deficient because they have one less degree of freedom than they have coordinates (the other degree of freedom is lost when specifying how to locate the landmark). Such landmarks can be used in geometric morphometric studies, but the loss of a degree of freedom must be taken into account when conducting statistical tests. See also **Type 1** and **Type 2 landmarks** (Chapter 2).

Unbalanced Design An experimental design in which the sample size within each combination of the factors is not equal. This makes it difficult to partition the variance, and gives rise to a variety of different approaches to calculating the sums of squares in a MANOVA or MANCOVA (Chapter 9).

Uniform components The components describing the uniform deformation. For two-dimensional configurations, the uniform deformation is described by two components: compression/dilation and shear. The uniform deformation is sometimes considered the zeroth partial warp (Chapter 5).

Uniform deformation A deformation that is purely uniform (or affine), or the purely uniform component of a deformation. The uniform deformations include only the uniform transformations that alter shape (compression/dilation and shear). They do not include transformations that do not alter shape (translation, scaling and rotation). See also **Uniform shape component** (Chapter 5).

Uniform component scores Scores locating a specimen, relative to the reference, along the uniform components. The summed squared scores on the uniform components and partial warps equal the Procrustes distance between each specimen and the reference. Taken together, the uniform and non-uniform scores fully describe the shape difference between the reference and that specimen (Chapter 5).

Variation, morphological variation A term used to refer to the general idea of variety, usually within a single population in contrast to disparity which refers to variation among species. Variation is typically measured as the sample variance:

$$S^2 = \frac{\sum_{i=1}^{n}(x_i - \overline{x})^2}{(n-1)}$$

that is, the mean of the summed squared deviations from the mean. See also **Disparity** (Chapter 10).

Vector A set of P coordinates that specify the location of a point in P dimensions.

Vector space A set of vectors, together with rules for adding and multiplying them (thereby obtaining all permissible linear combinations of them). Addition and scalar multiplication are required to meet eight rules:

1. $\mathbf{X} + \mathbf{Y} = \mathbf{Y} + \mathbf{X}$
2. $\mathbf{X} + (\mathbf{Y} + \mathbf{Z}) = (\mathbf{X} + \mathbf{Y}) + \mathbf{Z}$
3. A unique zero vector exists such that $\mathbf{X} + \mathbf{0} = \mathbf{X}$, for all $\mathbf{X}$
4. For each $\mathbf{X}$ there exists a unique vector $-\mathbf{X}$ such that $\mathbf{X} + (-\mathbf{X}) = \mathbf{0}$
5. $1\mathbf{X} = \mathbf{X}$
6. $(C_1C_2)\mathbf{X} = C_1(C_2\mathbf{X})$
7. $C(\mathbf{X} + \mathbf{Y}) = C\mathbf{X} + C\mathbf{Y}$
8. $(C_1 + C_2)\mathbf{X} = C_1\mathbf{X} + C_2\mathbf{X}$.

White Book Marcus, L.F., Corti, M., Loy, A. et al. (1996). *Advances in Morphometrics*. Plenum Press. See also **Black Book**, **Blue Book**, **Orange Book** and **Red Book**.

Index

Note: Page numbers followed by "f" and "t" refer to Figures and Tables, respectively.

G

H

I

J

K

L

N

O

P

R

S

T

U

W

Printed in the United States
By Bookmasters